Climatic Change *and the* Mediterranean

Climatic Change and the Mediterranean

Environmental and Societal
Impacts of Climatic Change and sea-level
Rise in the Mediterranean Region

Edited by

L Jeftic
*Senior Marine Scientist, Co-ordinating Unit for
the Mediterranean Action Plan, United Nations
Environment Programme, Athens, Greece*

J D Milliman
Woods Hole Oceanographic Institution, MA, USA

and

G Sestini
*Applied Earth Science Consultant
London, UK*

Edward Arnold
A division of Hodder & Stoughton
LONDON   NEW YORK   MELBOURNE   AUCKLAND

© 1992 United Nations Environment Programme
First published in Great Britain 1992

Distributed in the USA by Routledge, Chapman and Hall, Inc.
29 West 35th Street, New York, NY 10001

Disclaimer
The designations employed and the presentation of the material in
this book do not imply the expression of any opinion whatsoever on
the part of UNEP, concerning the legal status of any state, territory,
city or area, or of its authorities, or concerning the delimitations of
their frontiers or boundaries. The book contains the views
expressed by the authors and may not necessarily reflect the views
of UNEP.

British Library Cataloguing in Publication Data

Climatic change and the level of seas: Vol 1. The Mediterranean.
 I. Milliman, John D.
 551.460163

 ISBN 0-340-55329-4

Library of Congress Cataloging in Publication Data

Climatic change and the Mediterranean: environmental and societal
 impacts of climate change and sea-level rise in the Mediterranean
 region/edited by L. Jeftic, J.D. Milliman, and G. Sestini; United
 Nations Environment Programme.
   p.   cm.
 Includes bibliographical references and index.
 1. Climatic changes–Environmental aspects–Mediterranean
Region. 2. Climatic changes–Social aspects–Mediterranean Region.
3. Sea level– Mediterranean Region. I. Jeftic, L. II. Milliman, John D.
III. Sestini, G. (Giuliano), 1921–   . IV. United Nations
Environment Programme.
 QC981.8.C5C555   1992
 551.69182'2–dc20   92-20006
                                                                  CIP

All rights reserved. No part of this publication may be reproduced
or transmitted in any form or by any means, electronically,
including photocopying, recording or any information storage and
retrieval system, without either prior permission in writing from
the publisher or a licence permitting restricted copying. In the
United Kingdom such licences are issued by the Copyright
Licensing Agency: 90 Tottenham Court Road, London W1P 9HE.

Typeset in Palatino 10/11 pt. by Anneset, Weston-super-Mare,
Avon. Printed and bound in Great Britain for Edward Arnold, a
division of Hodder and Stoughton Limited, Mill Road, Dunton
Green, Sevenoaks, Kent TN13 2YA by Butler and Tanner Limited,
Frome, Somerset.

Preface

The greenhouse effect is among Man's potentially most pressing environmental problems, one that represents major scientific, environmental, economic, social and political challenges. Changes in global climate between now and the middle of the twenty-first Century are likely to be dominated by the influence of global warming due to the increasing concentrations of carbon dioxide and other greenhouse gases in the atmosphere. These greenhouse gases individually and collectively change the radiation balance in the atmosphere, trapping more heat near the earth's surface and causing a rise in global mean surface air temperature: as a consequence, substantial global warming is virtually certain.

The matter of climate change was the main topic of the Second World Climate Conference (Geneva, 1990). The Conference agreed that the international consensus of scientific understanding of climate change points out that without actions to reduce emissions, global warming is predicted to reach 2 to 5 °C over the next century, a rate of change unprecedented in the past 10 000 years. The warming is expected to be accompanied by a sea-level rise of 65 ± 35 cm by the end of the next century. There remain uncertainties in the predictions, particularly in regard to the timing, magnitude and regional patterns of climate change.

Many important economic and social decisions being made today, such as water resources management, coastal engineering projects, urban and energy planning, nature conservation, etc., are based on the assumption that past climate data provide a reliable guide to the future. This is no longer a safe assumption; climatic change must be considered, particularly in view of the current population explosion, increasing the utilization of coastal areas for industry, agriculture, fisheries and tourism.

To assess the environmental problems associated with the potential impact of expected climatic changes on the marine environment and on adjacent coastal areas and to identify suitable policy options and response measures which may mitigate the negative consequences of the expected impacts, the Oceans and Coastal Area Programme Activity Centre (OCA/PAC) of the United Nations Environment Programme (UNEP), in cooperation with the Intergovernmental Oceanographic Commission (IOC) and several other intergovernmental and non-governmental organisations, launched, co-ordinated and financially supported number of activities, including establishment of regional task teams. By 1990 task teams were established for nine regions covered by UNEP Regional Seas Programmes

(Mediterranean, Caribbean, South Pacific, East Asian Seas, South Asian Seas, Southeast Pacific, West and Central Africa, Eastern Africa and Kuwait Action Plan Regions). The two initial objectives of the task teams were to prepare regional overviews and site specific case studies on the possible impact of predicted climate change on the ecologic systems as well as on the socio-economic structures and activities of their respective regions, and to assist governments in the identification and implementation of suitable policy options and response measures which may mitigate the negative consequences of the impact. The regional studies were intended to cover the marine environment and adjacent coastal areas influenced by or influencing the marine environment.

The initial results of the Mediterranean, Caribbean and South Pacific task teams were reviewed at the meeting of the representatives of task teams in Split, Croatia in 1988. An overview of the implications of expected climatic changes in the Mediterranean was published subsequently (G. Sestini, L. Jeftic and J.D. Milliman: Implications of expected climatic changes in the Mediterranean region: An overview. UNEP Regional Seas Reports and Studies No. 103, Nairobi, 1989). The main findings of the task teams (general assessment of potential impacts of climate change, specific regional features, future strategies), as well as the experience of the task teams in bringing their findings to the attention of relevant national authorities and international bodies and programmes were reviewed at UNEP's Joint meeting of Co-ordinators of the regional task teams on implications of climatic changes, Singapore, 12-16 November 1990.

The present book represents the first completed reginal study. It will be followed in close order by studies on the Caribbean and South West Pacific. Other task teams and task team reports are being organized and completed.

Contents

Preface   v

The Mediterranean Sea and Climate Change – An Overview   1
J D Milliman, L Jeftic and G Sestini

1	A geographic and climatic perspective	1
2	The marine and coastal environments	1
3	Population and development	5
4	Resources	5
5	The greenhouse effect and climate changes	9
6	Future changes and strategies	12
7	References	14

1. Future Climate of the Mediterranean Basin With Particular Emphasis on Changes in Precipitation   15
T M L Wigley

1	Introduction	15
2	Global-mean changes	16
3	Regional changes	23
4	Present-day Mediterranean climate	25
5	GCM projections for the Mediterranean basin	28
6	Summary and conclusions	40
7	References	42

2. Sea-level Response to Climatic Change and Tectonics in the Mediterranean Sea   45
J D Milliman

1	Introduction	45
2	Tectonic and climatic changes in sea-level	46
3	Late quaternary sea-level	46
4	Sea-level during the past 100 years	48
5	Global sea-level rise by 2025	49
6	Land movement in the Mediterranean	51
7	Summary and future studies	54
8	References	56

3. Hydrological and Water Resources Impact of Climate Change 58
G Lindh

1	Introduction	58
2	Societal consequences of climate changes	58
3	Implications for the Mediterranean area	71
4	Suggested research activities	90
5	References	91

4. Implications of Climatic Change on Land Degradation in the Mediterranean 95
A C Imeson and I M Emmer

1	Introduction	95
2	Sensitivity and vulnerability of Mediterranean soils to changes in climate	100
3	Land degradation, soil erosion and river channel changes	113
4	Conclusions	124
5	References	126

5. Implications of Climatic Change on the Socio-Economic Activities in the Mediterranean Coastal Zones 129
A Barić and F Gašparović

1	Introduction	130
2	Background	132
3	Integrated planning	133
4	The coastline and the immediate coastal zones	142
5	Distribution and dynamics of population in littoral zones	144
6	Activities	146
7	Components	160
8	Conclusions	172
9	References	174

6. Vegetation and Land-Use in the Mediterranean Basin by the Year 2050: A Prospective Study 175
H N Le Houérou

1	Introduction: the study area and its climate	175
2	Present vegetation, land use and their evolutionary trends	195
3	Consequences of a hypothetical warming of the atmosphere by 3.0 +/− 1.5°C	203
4	Demographic and socio-economic situations and their consequences on natural vegetation and land use	221
5	Conclusions	224
6	References	227

7. Aspects of the Response of the Mediterranean Sea to Long-Term Trends in Atmospheric Forcing   233
M Gačić, T S Hopkins and A Lascaratos

1	Introduction 233
2	Some remarks on the Mediterranean Sea 234
3	Aspects of wind forcing and response 235
4	Dense-water formation 240
5	A preliminary assessment 242
6	Conclusions 244
7	References 244

8. Predictions of Relative Coastal Sea-Level Change in the Mediterranean Based on Archaeological, Historical and Tide-Gauge Data   247
N C Flemming

1	Introduction 247
2	Method 248
3	Interpretation of archaeological results 263
4	Analysis of tide-gauge data 271
5	Predictions of future coastal earth movements 273
6	Conclusions and recommendations 276
7	General note 277
8	References 277

9. Implications of a Future Rise in Sea-Level on the Coastal Lowlands of the Mediterranean   282
S Jelgersma and G Sestini

1	Introduction 282
2	Environment, land use and shoreline behaviour 286
3	The impact of a future sea-level rise 299
4	Conclusions and proposed research 299
5	Recommendations 300
6	References 302

10. Implications of Climatic Change on the Ebro Delta   304
M G Mariño

1	Introduction 304
2	Mediterranean setting 305
3	The Ebro Delta 306
4	Evaluation of climatic change impacts 318
5	Conclusions and recommendations 325
6	Acknowledgements 326
7	References 326

x   Contents

11. Implications des Changements Climatiques Etude de cas: Le Golfe du Lion (France)   328
J–J Corre

1	Introduction 339	
2	Les Enjeux socio-économiques 341	
3	Tendances climatiques au Nord de la Méditerranée de 1944 à 1987 354	
4	Le domaine marin 368	
5	Le milieu terrestre littoral 372	
6	Procédés de lutte contre les effets de recul de la côte dus aux éléments naturels 409	
7	Recomméndations 413	
8	Références bibliographiques 417	
	Annexes 421	

12. Implications of Climatic Changes for the Po Delta and Venice Lagoon   428
G Sestini

1   Introduction   429
2   Geographic setting   431
3   The socio-economic setting   438
4   The physical regime   448
5   Geology   459
6   Water resources   470
7   Ecosystems   476
8   An evaluation of the impact of climatic changes   477
9   Conclusions and recommendations   488
10 References   490

13. Implication of Future Climatic Changes on the Inner Thermaikos Gulf   495
D Georgas and C Perissoratis

1   Introduction   495
2   The Coastal environment   497
3   Water cycle   505
4   Ecosystems   516
5   Coastal activities   521
6   Impacts   523
7   Conclusions – suggested actions   531
8   References   533

14. Implications of Climatic Changes for the Nile Delta   535
G Sestini
1   Introduction   536
2   Geographic setting   538
3   The socio-economic setting   548

Contents xi

4	The physical regime	556
5	Geology	566
6	Hydrology and water resources	576
7	Ecosystems	586
8	An evaluation of the impact of climatic change	588
9	Conclusions	595
10	Recommendations	595
11	References	597

15. Implications of Climatic Changes in the Mediterranean Basin 602
Garaet el Ichkeul and Lac de Bizerte, Tunisia
G E Hollis

1	Introduction	603
2	Coastal Tunisia	605
3	Manifestations of rises in temperature and sea-level by 2025 and beyond	622
4	Conclusions	659
5	References	664

Index 677

Contributors

A Barić	Institute of Oceanography and Fisheries, Split, Croatia
J-J Corre	Consultant, Institut de Botanique, Montpelliers, France
I M Emmer	Scientific Assistant, Laboratory of Physical Geography and Soil Science, University of Amsterdam, Netherlands
N C Flemming	Senior Principle Scientific officer, Institute of Oceanographic Sciences, Deacon Laboratory, Wormley, Godalming, UK
M Gačić	Instituto per lo Studio della Dimanica delle Grandi Masse - CNR, Venice, Italy. On leave from Institute of Oceanography and Fisheries, Split, Croatia
F Gašparović	Consultant on Land-Use and Environment, Zagreb, Croatia
D Georgas	Coastal Environment UNEP Consultant, Athens, Greece
G E Hollis	Department of Geography, University College, London, UK and Station Biologique de la Tour du Valet
T S Hopkins	Department of Marine, Earth and Atmospheric Sciences, North Carolina State University, Raleigh, USA
H N Le Houérou	Centre d'Ecologie Fonctionelle et Erdutive, Montpellier, France
A C Imeson	Laboratory of Physical Geography and Soil Sciences, University of Amsterdam, Netherlands
L Jeftić	Senior Marine Scientist, Co-ordinating Unit for the Mediterranean Action Plan, United Nations Environment Programme, Athens, Greece
S Jelgersma	Applied Geo-Science Advisor, Geological Survey of the Netherlands, Haarlem, Netherlands
A Lascaratos	Department of Applied Physics, University of Athens, Greece
G Lindh	Professor, Department of Water Resources Engineering, Lund University, Sweden
M G Mariño	Pollution Control specialist, Technical Department for Latin America and the Caribbean, The World Bank, Washington DC, USA. Formerly, Advisor on Sanitary and Environmental Engineering, Institute de Salud Carlos III, Majadahonda, Madrid
J D Milliman	Woods Hole Oceanographic Institution, MA, USA
C Perissoratis	Institute of Geology and Mineral Exploration, Athens, Greece
T M L Wigley	Professor, Climate Research Unit, University of East Anglia, Norwich, UK

The Mediterranean Sea and Climate Change – An Overview

J. D. Milliman
(*Woods Hole Oceanographic Institution, MA, USA*)

L. Jeftić
(*United Nations Environment Programme, Athens, Greece*)

G. Sestini
(*Applied Earth Science Consultant, London, England*)

1 A Geographic and Climatic Perspective

The Mediterranean Sea covers about 2.5 million km^2 in area, with an average water depth of about 1.5 km. The length of the Mediterranean coastline totals about 46,000 km, of which 19,000 km represents island coastlines (Fig. 1). The entire coastal region covers an area of nearly 1.5 million km^2, 17% of the total area of the 18 bordering countries: Spain, France, Monaco, Italy, Yugoslavia, Albania, Greece, Turkey, Cyprus, Syria, Lebanon, Israel, Egypt, Libya, Malta, Tunisia, Algeria and Morocco (Fig. 1). Coastlines of individual countries range from 15,000 km (Greece) to 5 km (Monaco), with three countries (Greece, Italy and Yugoslavia) accounting for two-thirds of the entire Mediterranean coastline.

Climatically, the Mediterranean is characterized by generally warm temperatures, winter-dominated rainfall, dry summers and a profusion of microclimates due to local terrain. The north tends to be relatively temperate and damp, whereas the south is hot and arid (Fig. 2). The strong summer-winter rainfall contrast that characterizes the Mediterranean climate is associated with pronounced seasonal cycles in most climatic variables. July, August and September are warm and dry, whereas the winter is characterized by cyclonic disturbances and resulting rain. Precipitation, although mainly associated with these cyclonic disturbances, is also influenced by local orographic effects.

One result of the seasonal rainfall and high evaporation is that water shortages are endemic. The problem is particularly striking in the southern parts of the Mediterranean: in contrast to seasonal shortages in the north (corresponding to the dry months), the dry season in some southern countries exceeds six months, meaning that water shortage is a permanent handicap for sociological and economic development.

2 The Marine and Coastal Environments

Because of the seasonal rains, high rates of evaporation and the correspondingly low runoff of the relatively few short rivers (Fig. 3), the Mediterranean has a deficient hydrological balance, one that has increased with the damming of the Nile and other rivers. As a result, Atlantic water that enters the Mediterranean through the Strait of Gibraltar exits as more saline (by nearly 10%) subsurface water. This basin-wide circulation, combined with the Mediterranean climate and low land runoff, explains the low biologic productivity throughout the sea.

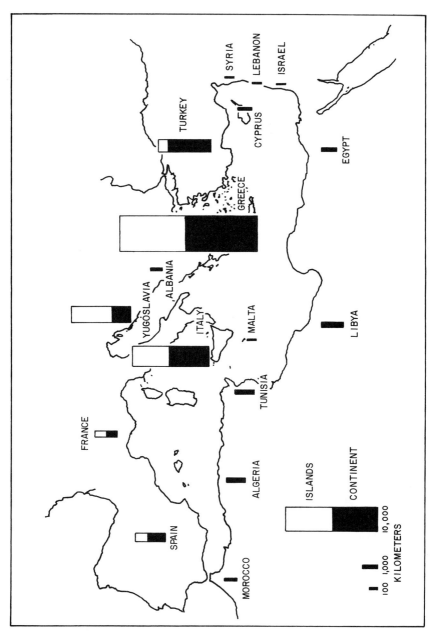

Fig. 1 The Mediterranean coastline, showing lengths of continental and island coastlines for the 18 Mediterranean countries (after UNEP, 1987c).

Fig. 2 Average annual rainfall around the Mediterranean basin (after *Times Atlas of the World*).

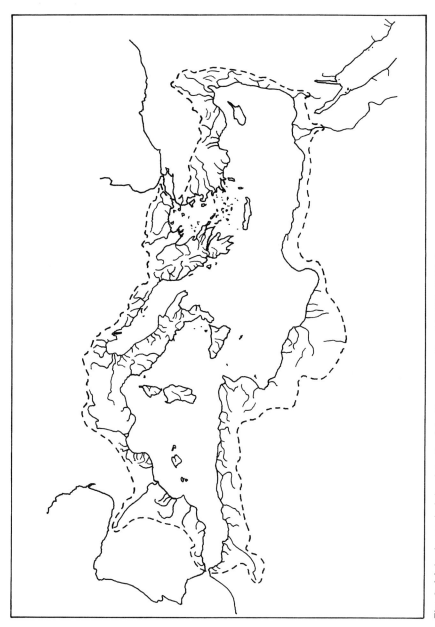

Fig. 3 Main rivers draining into the Mediterranean Sea (UNEP, 1987c).

Although alluvial and coastal plains are few and not extensive (the Nile Delta being by far the largest), most coastal plains have demographic and economic importance ranging from agriculture to industry/commerce to recreation to historical/archeological significance. Most areas still contain partly to little-modified natural ecosystems of irreplaceable value (Fig. 4). Because of their ecological fragility, related to the land-use transition, and their economic importance, these coastal lowlands are particularly vulnerable to climatic changes that can affect hydrology, ecosystems and sea-level rise.

3 Population and Development

In 1985 the 18 countries bordering the Mediterranean Sea had a combined population of 352 million people, of which 37% lived directly in the coastal zone, meaning that population densities were generally three times greater in coastal than non-coastal areas. Coastal population densities range from greater than 1000/km^2 in the Nile Delta to less than 20/km^2 along coastal Libya. According to some projections, the population in the Mediterranean is expected to reach 430 million in the year 2000, and 545 million in 2025 (UNEP, 1987a). Increasingly, the population will urbanize: in 1980, 57% of the population was urban, whereas by 2025 it is expected to be 75% urban. The economic and environmental burden on cities, therefore, will increase substantially.

Population growth, however, shows major differences between north and south. The European countries have nearly stable populations, annual growth rates often less than 1%. In contrast, population growth in southern countries ranges from 2 to 3% per year. As a result, the population in the coming years will increase and become younger in the south (Fig. 5). With this shift will come increasing problems in education and job-creation in southern countries.

The wide variation in political and economic systems as well as historic differences have led to great discrepancies in the level of development between Mediterranean countries. The highly developed industrial countries in the north (France, Italy and Spain) and countries on the way to becoming industrialized (Greece, Yugoslavia and Turkey) stand in stark contrast to the countries in the south. For instance, Gross National Product per Capita (GNPC) ranges from US$9760 in France to US$670 in Morocco (Fig. 6).

4 Resources

Mediterranean countries mostly lack natural resources. Libya, Algeria and Egypt are considered moderate-sized petroleum producers, Morocco is the world's third-largest producer of phosphates, Albania the third-largest producer of chrome, and Spain the second-largest producer of mercury. Water resources are relatively plentiful in the north but scarce in the east and south. Forests have limited economic significance, but are important for the preservation of soil as for recreation and landscape.

Mediterranean agriculture is characterized by multi-faceted crops, particularly olives, citrus fruits, grapes and hard grain; the main livestock

6 *Climatic Change in the Mediterranean*

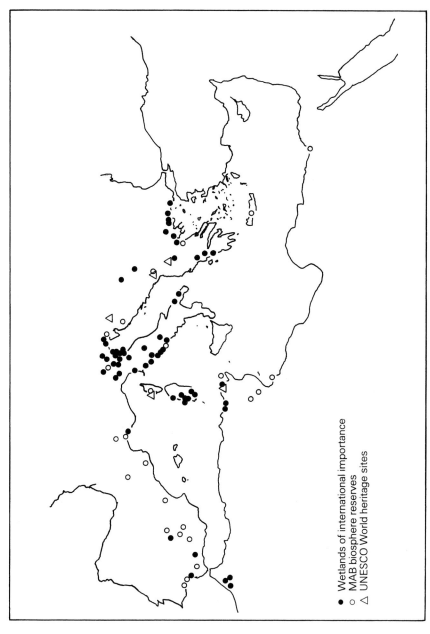

Fig. 4 Wetlands of international importance, MAB biosphere reserves, and UNESCO World heritage sites (UNEP, 1987c).

An overview 7

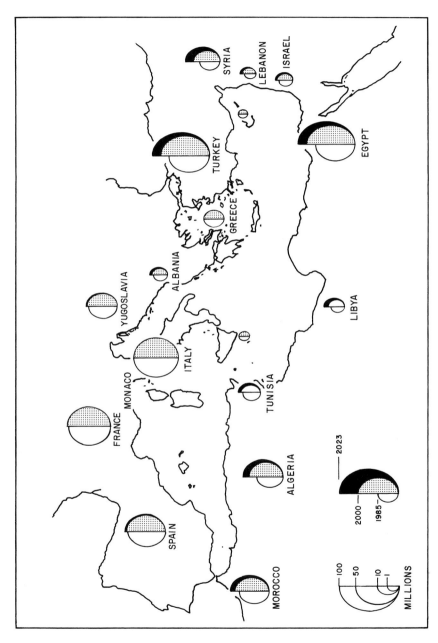

Fig. 5 Population and population trends in the Mediterranean (UNEP, 1987a).

8 *Climatic Change in the Mediterranean*

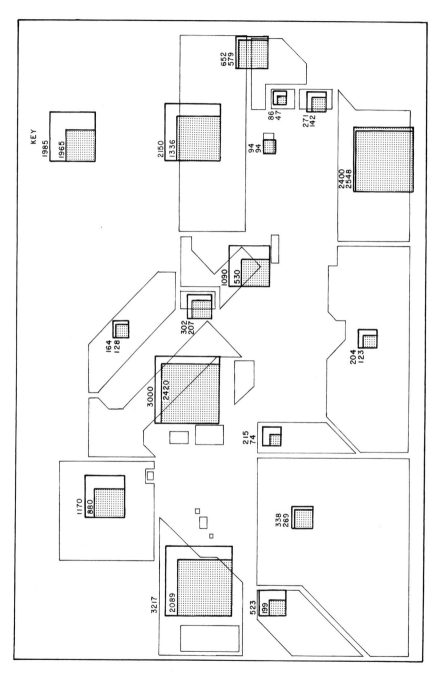

Fig. 6 Trends in Gross Domestic Product (per capita) (UNEP, 1988b).

is sheep. Increasingly irrigation is needed in the south to maintain or increase crop production. While coastal regions tend to have little agricultural land, it often is of high quality, particularly around delta areas. However, Mediterranean agriculture is also characterized by long-term misuse and overexploitation. In part this is due to generally poor soils, lack of rain and the increasing population pressures (particularly in the south).

In recent years, efficient farming and growing urbanization in the north has led to increased abandonment of farmland and rangeland and the corresponding advance of forests. This contrasts strongly in the south and east where marginal areas, such as arid steppes and rangelands, are being cleared for grain production; unfortunately, the lack of rain and continually flowing rivers has restricted the use of irrigation (Fig. 7). One result has been the increased desertification in North Africa and the Near East. Given the trends of the past 40 years, in the near future virtually all tillable land in the southern and eastern part of the Mediterranean basin will be cultivated for cereal production, even though the risk will be high and the yields low.

The Mediterranean is not capable of supplying enough fish for its inhabitants; more than 50% must come from outside areas. Most demersal stocks along the northern coasts of the Mediterranean are heavily fished, and the introduction of regulatory measures is necessary to maintain high levels of yield and catch. Coastal pelagic stocks, together with aquaculture, appear to be the best prospects for future development. Competition with coastal zones for other types of development, together with increased coastal pollution, however, may limit the extent to which aquaculture can develop.

Finally, it should be pointed out that because of both climate and historical/archeological significance, the Mediterranean continues to be the greatest tourist destination in the entire world. Conversely, tourism is the greatest consumer/user of the Mediterranean coast (Fig. 8). In 1984, for example, there were about 100 million tourists in coastal areas. Total tourist facilities take up more than 2 million m^2 of space on the coast, and water consumption in 1984 amounted to 569 million m^3 (UNEP, 1989). The number of tourists by the year 2000 could rise to 120–180 million, and 170–340 by 2025 (UNEP, 1987b). Such a growth will mean an increasing demand for coastal space as well as such necessities as electric power and water. Furthermore, the impact on certain habitats (particularly sandy beaches and dunes) will increase.

5 THE GREENHOUSE EFFECT AND CLIMATIC CHANGES

Given this background for the Mediterranean coastal region, this book deals with the problem of the greenhouse effect and projected changes in both climate and sea-level. Using existing global climate models, we have assumed a temperature change of 1.5 to 3°C and also changes in precipitation and evapotranspiration patterns. The rise of sea-level by 2025 is more problematic, but it may reach 12–18 cm, about three to five times the rate during the past 100 years; locally the rise in sea-level could be much greater, depending on land subsidence.

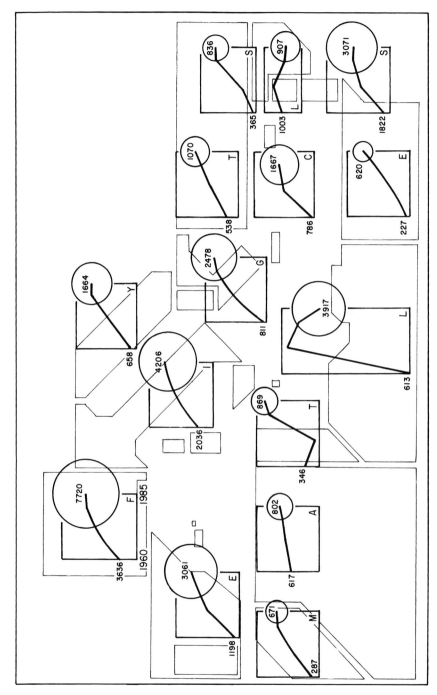

Fig. 7 Irrigated land (in thousands of ha) between 1965 and 1985 (UNEP, 1988b).

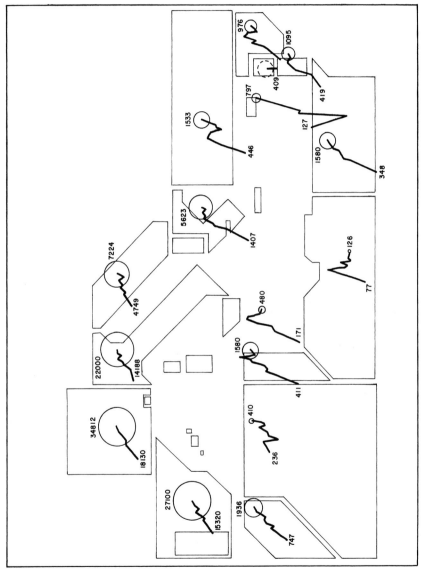

Fig. 8 Trends in international tourism in the Mediterranean in thousands, 1970–1984 (UNEP, 1988b).

The 15 chapters in this book discuss climatic change and its probable impact on the Mediterranean. The presentation is divided into three parts. The first part (Chapters 1 and 2) discusses the greenhouse effect and probable scenarios for climate change and sea-level rise. The second part (Chapters 3–9) reviews the Mediterranean area in terms of climate, land and water use, impacts from human activity and climate, sea-level and coastal lowlands. The last six chapters of the book deal in greater detail with specific coastal lowland areas in terms of both present-day environment and predicted environmental/economic impacts from climate change and sea-level rise.

While each chapter is written to "stand alone", the picture gained from all the chapters shows the complex interaction between climate change, environmental impacts and the resulting economic, sociological and political changes (e.g. Fig. 9). Some of these impacts and possible changes are discussed in the following paragraphs.

6 FUTURE CHANGES AND STRATEGIES

Most of the chapters in this book touch on the same reality, that in the Mediterranean climatic change and sea-level change most likely will only exacerbate problems that already exist and that are increasing in some coastal countries. Some coastal lowlands will be lost, ground-water balance will change as a result of altered precipitation and evaporation patterns and increased sea-water intrusion, agricultural patterns and yields will shift, and at least locally economic and sociological impacts will occur.

The physical impact of climate change and sea-level rise can be predicted, even modelled, quantitatively on the basis of the present-day parameters of morphology, hydrodynamics, sediment budgets, land subsidence and the effects of artificial structures. Similarly, assuming changes in air temperature, precipitation/evaporation and resultant soil-water parameters, impacts on the biosystems, such as agriculture, can be estimated. What is more difficult to estimate is the impact of these physical and biological changes on the future socio-economic framework of the threatened lowlands.

In most instances, however, other human-related activities will have equal or greater impact. Because of rapid population growth and migration to urban areas, southern countries face major economic, environmental, social and political problems in the coming years regardless of climate change. Warmer climate and changing precipitation patterns may decrease viable cropland and increase water problems, but probably not to the degree that demands from an expanded population will. In contrast, climate change on northern countries along the Mediterranean coasts may have much less impact simply because the population, economics or environment probably will experience less deleterious change. Perhaps the one of the greatest problems for the north will be the increased immigration from the south.

This does not mean, however, that strategies should not be developed to minimize climatic changes. Clearly, such strategies require a better knowledge of both the degree of change and the ecologic/eco-

An overview

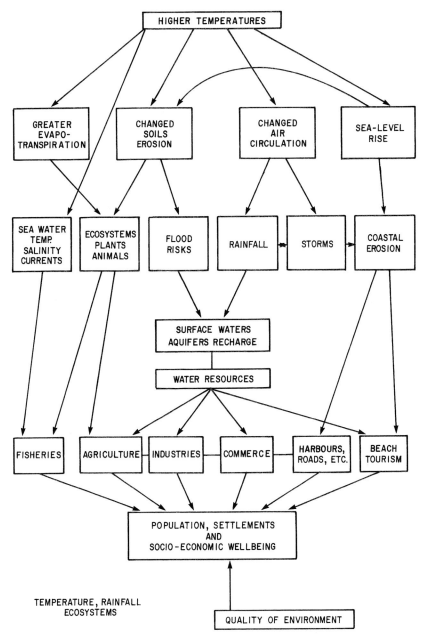

Fig. 9 Impact of climatic change on environment and society.

nomic areas upon which these changes will impact. The first need is to document more completely climatic parameters and local sea-level in order to ascertain change. Attention should be given to identifying and accessing data that can be used for climatic impact assessment. The value of long-term data series must be stressed, particularly since some of the presumed changes (for example, sea-level) require 20–30 years of continuous data to define a change in trend. Finally, these data need to be used to develop more realistic predictive models regarding both climate change and corresponding rise of sea-level (as well as its impact on coastal areas).

Close attention must be paid to the conservation of soil, ground water and wetland resources in the Mediterranean, as they are critical to environmental stability. Water management is the key area for concern in light of expanding populations, changing climate and shifting precipitation patterns. Lack of potable water clearly can hinder or prevent proper land usage. On the other hand, over withdrawal of ground water can lead to substantial subsidence which can accentuate the local rise of sea-level; the sinking of Venice is an obvious example.

One immediate task is to identify "high risk" areas with respect to sea-level rise. Reactions to sea-level change then can be preventive or reactive. However, preventive/engineering solutions, such as dikes and walls, are not likely to represent a realistic long-term solution for rising sea-level except in special/local cases. Rather, adaptation, evolution and land-use change in most instances probably will represent the most appropriate responses to sea-level rise. Such reactive actions, for example, would include shifting land use, decreasing urbanization along coasts, increasing natural reserves, etc. In the final analysis, coastal-zone management must be based on cost-effectiveness, wherein the value of the threatened land use is taken into account within the context of local needs, both at present and in the future.

Organizational and legal instruments should be developed to control coastal development, land reclamation and ground-water exploitation. In addition, data and ideas must be shared on a regional basis. Some of these instruments will be local, some national and some regional or even global. Clearly, they must be initiated as soon as possible.

7 REFERENCES

UNEP (United Nations Environment Program), 1987a. *Data base on the Mediterranean activities.* Provisional document, Blue Plan, Sophia Antipolis.

UNEP, 1987b. Preliminary report on Blue Plan scenarios (UNEP/WG. 171/3), Blue Plan, Sophia Antipolis.

UNEP, 1987c. Environmental data on Mediterranean Basin (natural environment and resources). Provisional version, Blue Plan, Sophia Antipolis.

UNEP, 1988a. Report on the Joint Meeting of the Task Team on Implications of Climatic Changes in the Mediterranean and the Coordinators of Task Teams for the Caribbean, Southeast Pacific, South Pacific, East Asian Seas and South Asian Seas Regions. Split, 3–8 October, UNEP, Nairobi, UNEP (OCA)/WG.2/25.

UNEP, 1988b. The Blue Plan, futures of the Mediterranean Basin: Executive summary and suggestions for action, Blue Plan, Sophia Antipolis.

UNEP, 1989. State of the Mediterranean marine environment, L. Jeftic, (ed.), MAP Technical Report Series No. 28, Athens.

1

Future Climate of the Mediterranean Basin with Particular Emphasis on Changes in Precipitation

T.M.L. Wigley
(*University of East Anglia, Climate Research Unit, Norwich, UK*)

ABSTRACT

Changes in global climate over the past 100 years are reviewed as background to an assessment of future climatic change under greenhouse-gas forcing. In the context of estimating future climate, the difference between equilibrium and transient temperature change is explained. The former is important in defining the potential warming commitment, while the latter defines the actual amount of warming likely to be experienced at any given future time. Transient global-mean warming between 1990 and 2030 is expected to lie in the range 0.5–1.4°C. This warming rate is between two and seven times faster than the warming that has occurred over the past 100 years. Regional-scale changes in climate can only be estimated using General Circulation Models (GCMs). The current state of the GCM art is reviewed. For the Mediterranean basin, the only GCM result that one can place confidence in is a general warming. The regional transient-response warming rate is likely to be similar to the global-mean rate, with no evidence of any marked differences between seasons. However, these greenhouse-related changes may well be masked by natural climatic variability for a number of decades into the future. For precipitation changes, there is some evidence of increased autumn precipitation occurring in the northern part of the basin and of decreases in precipitation occurring in the southern part, but it is impossible to estimate the timing or magnitude of these changes.

1 INTRODUCTION

Changes in global climate between now and the middle of the 21st century are likely to be dominated by the influence of the greenhouse effect caused by increasing concentrations of carbon dioxide (CO_2), methane (CH_4), nitrous oxide (N_2O), ozone (O_3) and halocarbons (CFCs, etc.). These greenhouse gases individually and collectively change the radiative balance of the atmosphere, trapping more heat near the Earth's surface and causing a rise in global-mean surface air temperature. Substantial global warming is virtually certain, but the attendant changes in climate at the regional level are highly uncertain. Except for a few regions, about all we can be sure of is that large changes will occur, but we cannot yet quantify these changes. Indeed, for precipitation we are as yet unable to specify even the sign of regional changes with any reliability. Nevertheless, some useful statements can be made with regard to the Mediterranean basin, and it is of some value to review our present state of knowledge. We begin by considering the global scale, highlighting the important distinction between equilibrium and transient changes in global-mean temperature. We then describe the present-day climate of the Mediterranean basin and review published projections of the future climate of the region based on General Circulation Models (GCMs). After critically reviewing these projections, we give a consensus view of future climate.

2 GLOBAL-MEAN CHANGES

2.1 Observed Climate Changes Over the Past 100 Years

Future changes in climate can be set in context by a brief description of changes that have occurred over the past 100 years. Both regionally and in terms of global-mean values, climatic conditions have fluctuated noticeably from year-to-year, decade-to-decade and on longer time-scales. Consider the largest spatial scales first.

The near-surface air temperature averaged over the globe has increased by about 0.5°C since the late 19th century (Jones *et al.*, 1986a,b,c) (Fig. 1.1). Parallel changes in the temperature of the lower troposphere have also

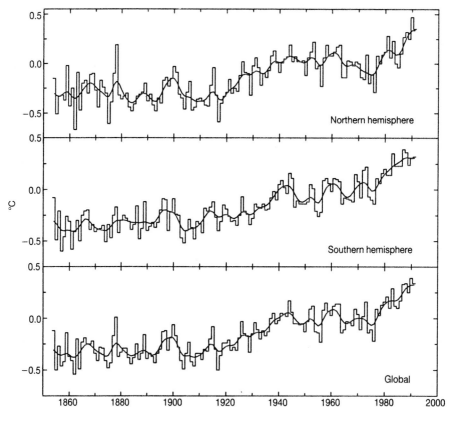

Fig. 1.1 Hemispheric and global annual mean temperature changes based on land and marine data (from Jones *et al.*, 1986c, updated).

occurred; see Folland *et al.* (1992) for a recent review. As is evident in Fig. 1.1, however, this warming has been neither continuous nor spatially homogeneous; trends have varied substantially from region to region. Over the past 40 years, for example, the whole of the Mediterranean basin, along with much of the North Atlantic and western Europe, has undergone a noticeable cooling (Fig. 1.2). That such a cooling has occurred in spite of the expected global-mean greenhouse-gas-induced warming demonstrates the degree to which external forcing influences can be masked by natural climatic variability. This is especially so at the regional scale. Similar decadal and longer time-scale "anomalies" (i.e. departures from greenhouse-effect expectations) can be expected to occur in the future. It will not be until the early decades of next century, or later, that the increasing strength of the expected greenhouse-effect signal (which will be quantified in a later section) will begin to dominate over this natural noise at the regional level.

While the temperature trend in the Mediterranean over the past 40 years has been one of cooling, the trend pattern over the past 20 years is quite different. Over the period 1967 to 1986, the Mediterranean basin has warmed in the west and cooled in the east. Clearly, with regard to future changes over coming decades, one cannot simply assume that changes in this region (or any other region) will follow the global-mean trend.

Noticeable changes in precipitation also have occurred this century on all spatial scales. Large-scale area-average precipitation changes are more difficult to quantify than temperature changes because of the higher spatial variability of precipitation, because of data homogeneity problems, and because we have no information on trends over the oceans. For land-based data in the Northern Hemisphere, Bradley *et al.* (1987) show an upward trend from 1920 to the present in mid-to-high latitudes (35–70°N) and a marked downward trend in tropical-to-subtropical latitudes (5–35°N) (Fig. 1.3). On smaller spatial scales, some regions of the world have experienced marked changes on decadal time-scales. The Sahel region of Africa is a particularly striking example (Fig. 1.4).

Almost all other climate variables show evidence of a continually changing climate. For example, there have been noticeable changes in the atmospheric circulation of the North Atlantic as measured by the pressure gradient between the Azores High and Iceland Low – a trend towards a decreasing gradient. The extent of mountain glaciers has decreased over the 20th century (Meier, 1984), albeit with marked regional differences and interdecadal variations. This has contributed noticeably to a general rise in sea-level (Gornitz *et al.*, 1982; Barnett, 1985). Of the 10–15 cm global-mean rise (the current best estimate), roughly one-third can be attributed to each of mountain glacier melting, oceanic thermal expansion and melting from large ice sheets (Greenland and Antarctica) (Wigley and Raper, 1987). The evidence, however, is scanty, and the overall rise in sea-level and the different contributions to this rise (especially the ice-sheet contribution) are subject to much larger uncertainties than those for global-mean temperature changes.

In parallel with these changes in mean conditions, most measures of climate show changes in variability (as measured by variance and/or the

18 *Climatic Change in the Mediterranean*

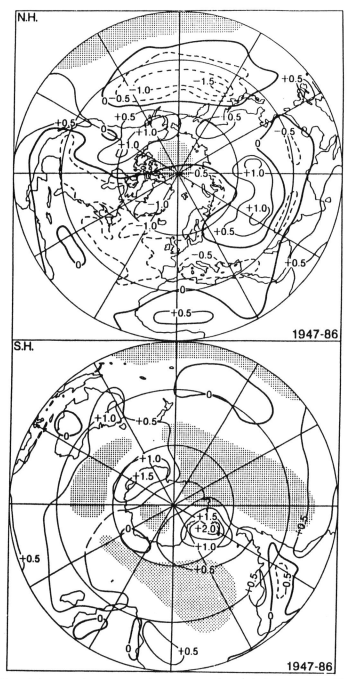

Fig. 1.2 Trends in annual mean temperature over the period 1947–1986. The isolines show the change in temperature (°C) over 1947–1986 accounted for by a linear trend fitted to each annual gridpoint time series. For the Antarctic region, the linear trend has been fitted to annual data for the 1957–1986 period. Shaded areas are regions with no data (from Jones *et al.*, 1988).

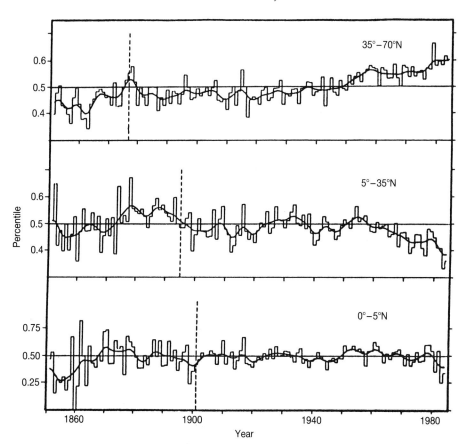

Fig. 1.3 Precipitation changes for different parts of the Northern Hemisphere (land data only) (from Bradley *et al.*, 1987).

frequencies of extreme events). Very few of these changes in variability (in contrast to changes in means) have been statistically significant. A notable exception is England and Wales area-average precipitation, where there has been a significant trend towards more wet extremes in spring and more dry extremes in summer (Wigley and Jones, 1987).

A leading question is: What is/are the cause/causes of these changes? In the absence of changes in external forcing factors, the climate system would be expected to show considerable natural variability on annual, decadal and longer time-scales simply because of the complexity of interactions between the oceans, cryosphere, atmosphere and land surface and the range of time-scales associated with these components. Indeed, at the regional level, and perhaps even at the global-mean scale, much of the observed variability of climate over the past 100 years may be attributable to internally caused natural variability. In addition, forcing undoubtedly occurs due to: changes in the reflectivity (or albedo) of the Earth-atmosphere system associated with explosive volcanic eruptions (which produce dust and aerosol layers in the stratosphere), with

20 Climatic Change in the Mediterranean

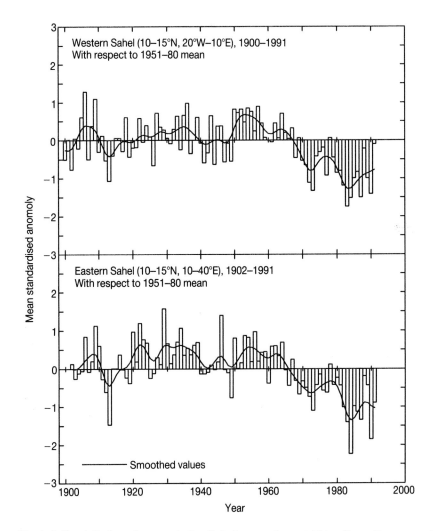

Fig. 1.4 Precipitation changes in the Sahel zone of west Africa (from Farmer and Wigley, 1985, updated by Dr M Hulme).

changes in cloudiness and with changes in surface characteristics (vegetation, snow and ice cover, etc.); changes in solar irradiance; and changes in the concentrations of greenhouse gases (CO_2, etc.) and sulphate aerosols which perturb the radiation balance of the atmosphere.

For the greenhouse gases, the global-mean radiative forcing change over the past 100 years has been large (around 2Wm^{-2}) and this would be expected to cause a global-mean warming. Although a warming has occurred and its magnitude (viz. 0.3–0.6°C) is compatible with that expected to have resulted from the greenhouse effect (viz. 0.3–0.8°C, see below), we are not yet able to positively relate cause and effect. Nevertheless, the qualitative agreement and the realistic physical basis of current climate models demand that the possibility of substantial future global-mean warming, and the multitude of regional changes that would accompany such a warming, must be taken seriously.

2.2 Future Global-Mean Changes

The magnitude of future global-mean warming will depend on future atmospheric trace gas concentrations, on the sensitivity of the climate system to external forcing and on delays due to the thermal inertia of the oceans. The greenhouse effect will cause changes in global-mean temperature on time-scales of decades to centuries, with associated trends in all climate variables at smaller spatial scales. These changes also will have superimposed on them, inter-annual to inter-decadal variability characteristic of the unperturbed climate system.

In considering future climatic change, it is important to distinguish between the equilibrium and transient response of the climate system. Consider the equilibrium response first. For any given future greenhouse-gas forcing (ΔQ) there will be a corresponding equilibrium global-mean temperature change (ΔT_e) which is the temperature change that would be achieved when the climate system reached a new steady state. The relationship between ΔQ and ΔT_e is determined by the climate sensitivity.

The climate sensitivity is usually expressed in terms of the equilibrium change in global-mean temperature (ΔT_e) that would occur for a forcing change of 1Wm^{-2} at the top of the troposphere. Alternatively, it can be expressed as the ΔT_e value that would occur if the CO_2 level doubled (a doubling of the CO_2 concentration corresponds to a forcing change of about 4.4Wm^{-2}), viz ΔT_{2x}. The higher the value of ΔT_{2x}, the higher the climate sensitivity. The value of ΔT_{2x} is uncertain, but it is thought to lie in the range 1.5–4.5°C (Houghton et al., 1990, 1992). The CO_2-doubling case is a standard case used in most modelling studies. The equilibrium change for an arbitrary concentration change from C_o to C can be related to this standard case using

$$\Delta T_e = \Delta T_{2x}(\ln(C/C_o)/\ln 2) \qquad (1)$$

Equation (1) arises because ΔQ for CO_2 is logarithmic in concentration,

$$\Delta Q = 6.3 \ln(C/C_o) \qquad (2)$$

Since CO_2 is not the only greenhouse gas that must be considered in the forcing, it is often convenient to express the total radiative forcing (ΔQ_T) due to changes in all anthropogenic greenhouse gases in terms of an equivalent CO_2 concentration (C^*). This is defined by inverting equation(2) to give

$$C^* = C_o \exp(\Delta Q_T/6.3) \qquad (3)$$

where, by convention, C_o is taken to be the preindustrial level of CO_2 alone (i.e. the other greenhouse gases are not considered in defining this reference level).

Between pre-industrial (~1750–1800) and present times (1990), the concentrations of all greenhouse gases have increased substantially: CO_2, 279–354 ppmv; CH_4, 790–1720 ppbv; N_2O, 285–310 ppbv; CFC–11, 0–280 pptv; CFC–12, 0–440 pptv. Pre-industrial levels of CO_2 and CH_4 are not known precisely, and they have certainly varied by small amounts naturally. However, at least over the past 10,000 years, these natural variations are minor compared with the more recent anthropogenic changes. Thus, relative to conditions in the mid to late 18th century, the combined radiative forcing (ΔQ_T) is about 2.4Wm^{-2} today, with approximately 1.5Wm^{-2} of this being due to CO_2 changes and the rest due to changes in the other trace gases, mainly methane and the CFCs. The 1990 value of the equivalent CO_2 concentration is just over 410 ppmv. This forcing change has been offset, however, by sulphate aerosol and stratospheric ozone depletion effects (Wigley and Raper, 1992) leaving a net forcing of about 1.5Wm^{-2}. For the future, a further forcing increase of about 2Wm^{-2} is expected between 1990 and 2030 (Wigley and Raper, 1992, Fig. 2). On the basis of these latest results, the equivalent CO_2 concentration is not expected to reach double the pre-industrial CO_2 level until around 2050.

Using ΔT_{2x} = 1.5–4.5°C, the radiative forcing changes to 1990 imply an *equilibrium* warming of 0.5–1.5°C over the period since 1765. Over the shorter period 1880–1990, for which ΔQ_T = 1.2Wm^{-2}, ΔT_e lies in the range 0.4–1.2°C. Over the 40 years 1990–2030, based on the 2.0Wm^{-2} projected forcing change, the implied equilibrium warming is 0.7–2.2°C. This projection is based on a "best estimate" of future greenhouse-gas forcing only. It incorporates the uncertainty that arises from indeterminate climate sensitivity, but it does not allow for uncertainties in future greenhouse-gas concentrations (i.e. future forcing uncertainties). Allowing for these widens the range of possible values for ΔT_e over 1990–2030 to 0.5–3.0°C. Although this is a very large range of uncertainty, even values at the lower end of the range correspond to very significant changes in climate.

The range 0.5–3.0°C represents the extreme range of possible equilibrium warming over the interval 1990–2030. This is not the change that would actually be observed over this interval (viz. the transient-response warming). This is because, due to the large thermal inertia of the oceans, observed changes lag behind the equilibrium changes. The amount of lag is determined by the rate of ocean mixing, by the climate sensitivity, and by the rate of change of forcing. Because of this thermal inertia effect, only some 50–90% of the equilibrium warming may be observed at any given time.

Let us consider the consequences of this lag effect. Over 1880–1990, the expected equilibrium warming lies in the range 0.4–1.2°C. However, the expected transient warming (i.e. the value which should be compared with observations) is much smaller, lying in the range 0.3–0.8°C. This latter range can be compared with the observed warming of about 0.5°C. Since 0.5°C lies near the middle, one might conclude that the climate sensitivity must lie near the middle of the range 1.5–4.5°C. There are, however, a number of other ways to explain the observations, so we cannot yet dismiss higher or lower sensitivity values.

An important consequence of the equilibrium/transient difference is that, no matter what happens to future greenhouse-gas concentrations, we will always be committed to an additional warming as the climate system tends towards equilibrium. Thus, the difference between the observed 1880–1990 global-mean warming of 0.5°C and the ΔT_e range of 0.4–1.2°C represents a current warming *commitment* (a phrase introduced by Mintzer, 1987), in the sense that it would result even if the concentrations of all greenhouse gases could be held constant at their 1990 values.

By the year 2030, the model-based transient-response changes relative to 1990 lie in the range 0.5–1.4°C. The range is somewhat larger (0.4–1.5°C) if one accounts for uncertainties in future greenhouse-gas concentrations. Note that, just as there is a warming commitment now, there will be a correspondingly larger commitment in 2030. (The above results are based on the model results of Wigley and Raper, 1992.)

3 REGIONAL CHANGES

3.1 Introduction

In the above, we have concentrated on global-mean temperature as a key measure of the state of the climate system. Changes in all climate variables, precipitation, evaporation, wind patterns and strengths, cloudiness, etc., will necessarily accompany changes in global-mean temperature. In the future, just as they have in the past, these changes will differ noticeably from region to region. Thus, for assessing impacts, global-mean quantities are mainly of academic interest, although they do show that future climatic changes probably will be large and probably will occur more rapidly than any previous changes. For direct assessments of impacts, regional-scale (or smaller) details of future changes are needed – not only for temperature, but for a variety of climate variables. At present, our capability for predicting these details is limited and we must resort to the use of scenarios.

3.2 Climate Scenarios

The most important method for obtaining information on possible future climates is to use an atmospheric General Circulation Model (GCM). However, because of deficiencies in current GCMs (described in more detail below), their outputs should be considered only as possible scenarios for future climatic change rather than predictions. A climate scenario, as defined in the literature, is intended to be an internally consistent picture of possible future climatic conditions. In addition to the use of GCMs, scenarios can be constructed by a number of other means (see, e.g. Pittock and Salinger, 1982; Palutikof et al., 1984). However, since these methods have not been applied to the Mediterranean basin, we will concentrate here on GCM results. But first, it is important to keep in mind the limitations of GCMs. While these limitations are well-recognized problems within the modelling community, they are not always realized by analysts who make use of GCM output.

3.3 GCM Limitations
(For further details see Mitchell et al., 1990; Gates et al., 1992.)
1) Most published GCM studies of the greenhouse effect consider only the equilibrium response to a doubled or quadrupled concentration of carbon dioxide. It is not yet known whether the regional patterns of climatic change for the transient response will correspond to those given by equilibrium studies (Bretherton et al., 1990).
2) Atmospheric GCMs must be coupled to ocean models in order to account for some of the ocean's thermal inertia effects. The ocean models currently in use for this purpose are still under development.
3) In addition to their treatment of the oceans, GCMs still have a number of recognized deficiencies in the way they model clouds, sea-ice effects and land-surface processes.
4) Different GCM studies that attempt to describe the climate of a high-CO_2 world show regions of agreement and disagreement. Since regions of agreement can occur by chance, they cannot necessarily be accepted as regions where the predictions are more reliable. For example, on the global scale, many recent models give a similar value for ΔT_{2x} (of around 4°C), but different models give these similar answers for quite different reasons.
5) The atmospheric models currently used in greenhouse-gas studies are unable to simulate the current ($1xCO_2$) atmospheric circulation at the regional level with any realism (Santer and Wigley, 1990), although they do perform reasonably well in describing some of the large-scale features of the general circulation. A reliable "control-run" simulation is thought to be a necessary condition for reliability in any perturbation experiment.
6) While GCMs can produce output with very short time steps (~10 minutes), the realism of these short-time-step results is in doubt. For current models, the limit of temporal resolution on which results might have some meaning is of order one day. However, some models currently in use do not have a diurnal insolation cycle, which necessarily precludes realism even at the daily time-scale, and the strange results produced by some models on the seasonal time-scale (see below) point to serious deficiencies on time-scales of months and longer.
7) The current spatial resolution of GCMs is too coarse for most impact studies. Coarse resolution means that the orography in the models is highly smoothed and small-scale weather systems are non-existent. This means that one would not expect a GCM to produce reliable results even at the grid-point scale. At best, one can only hope for reliability on the scale of the area covered by tens of grid points – but even this is an optimistic hope at present.

Despite the deficiencies, there are some broad-scale GCM results which can be accepted with some confidence as predictions of future climate. These are summarized in Table 1.1.

In spite of the problems that plague current GCMs, they are the best tool we have for projecting future changes in climate at the regional level. A number of such projections have been used in impact studies,

Table 1.1 A selection of model results from equilibrium GCM experiments for a doubling or quadrupling of atmospheric CO_2 concentration, together with an estimate of the confidence that can be placed in these results. "Unknown" indicates that knowledge of possible future changes is zero. This applies to all items not mentioned

Model results	Confidence
Global scale (i.e. global-mean values)	
Warming of lower troposphere	High
Increased precipitation	High
Cooling of stratosphere	High
Warming of upper troposphere (especially the tropics)	Moderate
Zonal-mean to regional scale	
Reduced sea ice	High
Enhanced polar warming in Northern Hemisphere (especially winter half year)	High
Increased P−E* in high latitudes	High
More absolute high temperature extremes	High
Increased continental summer dryness	Moderate
Stronger monsoon	Moderate
More tropical storms	Unknown
More/less blocking	Unknown
Greater/less interannual variability	Unknown
Spatial detail in general	Unknown
Rainfall extremes	Unknown

Note: * Precipitation minus evapotranspiration

sometimes with implied faith in the reliability of the projections and other times more sceptically, with a primary aim being to examine climate sensitivities and to develop tools and methods for impact assessment (e.g. Parry et al., 1988). Surprisingly, such studies have invariably relied on a single GCM, and very few detailed (i.e. regional-scale) comparisons have been made of the projections of different GCMs. Just such a comparison is made later in this chapter. Before doing so, however, the present climate of the Mediterranean basin will be described.

4 PRESENT-DAY MEDITERRANEAN CLIMATE

(The following description is based on the review by Wigley and Farmer, 1982.)

The Köppen definition of a Mediterranean climate (one of many possible definitions) is, in simple terms, one in which winter rainfall is more than three times summer rainfall. Most of the region satisfies this criterion: indeed, for much of the region summer rainfall is practically zero. This strong summer-winter rainfall contrast is associated with a well pronounced seasonal cycle in almost all climate variables. The seasonal cycle proceeds as follows: July, August and September are characterized by warm, dry conditions associated with a strong high-pressure ridge which pushes eastwards from the Azores subtropical high over the Mediterranean. The axis of this ridge is displaced southward over Egypt by a trough which extends from the Persian Gulf area north-westwards towards Greece and which is associated with the Indian summer monsoon depression. In mid-October the rainy season begins, associated with a change in the mean-wave pattern of the upper westerlies from a four- to a three- wave pattern on the five-day time-scale (Chang, 1972) and an upper air flow which is characterized by a trough over Europe (whose position is highly variable). Winter is characterized by cyclonic disturbances and low mean pressure in the Mediterranean, with higher pressure to the east associated with the Siberian High. In March, April and May, as the main features of the upper flow (jet streams, air mass discontinuities) begin to move northward from their southernmost winter positions, the rainy season continues. By May, the polar front and associated strong upper-air westerly flow is sufficiently far north that its influence is removed, and the subtropical highs and associated ridges once more exert their influence.

Precipitation, although mainly associated with cyclonic disturbances that originate in the Mediterranean basin, is strongly influenced by local orographic effects. Very few surface cyclones can be traced back to Atlantic origins. There is a common misconception that depressions moving south of the British Isles often move into the Mediterranean as recognizable surface features. In fact, most depressions originate in four specific regions within the basin (Fig. 1.5). The main region lies in the western Mediterranean, producing "Gulf of Genoa" depressions which only occasionally move eastward far enough to affect the eastern basin. Atlas Mountains lee depressions, which form in spring, do not bring rainfall: to the contrary, and especially those that follow a north African trajectory across into Egypt, they are associated with hot, dry, windy conditions. For the eastern Mediterranean, central basin depressions and eastern basin depressions ("Cyprus Lows") – both are winter and spring phenomena – are most important. Occasionally, central basin depressions may be traced back to either Gulf of Genoa or Atlas Mountains lee depressions, but this is not the general rule. Central basin depressions tend to move to the north-east (around 10 occasions per year) or eastwards (eight per year) where further intensification may occur in the eastern basin area. Eastern basin depressions also have preferred tracks either to the north-east or to the east. The latter, especially when undergoing subsequent interaction with polar continental air from the Iranian plateau, may bring rain to a wide area from Egypt to Iraq. The winter mean surface pressure pattern shows features which result from these cyclogenetic aspects.

On rare occasions, monsoon air masses may bring summer rain to the eastern Mediterranean. Gat and Magaritz (1980, p. 86) note the following example: "the 'once in a hundred year' flood which took place in the Sinai in 1974 and whose origin was traced by both synoptic and isotopic evidence (Carmi, private communication) to a freak northward intrusion of a tropical (monsoonal) air mass".

The formation of Mediterranean depressions is partly determined by transitory excursions of the polar front jet and the European trough, modified by the land-sea temperature contrast which favours cyclogenesis

Future climate of the Mediterranean basin 27

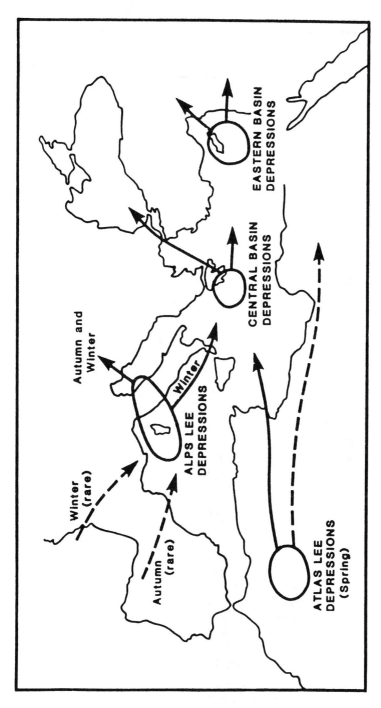

Fig. 1.5 Regions of cyclogenesis and principal cyclone tracks (from Wigley and Farmer, 1982).

over the warm ocean. Eastern basin depressions are often associated with cold northerly airflow and lee cyclogenesis. These relationships provide a link between the local rain-producing pressure systems and larger-scale aspects of the general circulation over Europe.

The movement of depressions is not well understood. In the western Mediterranean, depressions are frequently "steered" along the "Mediterranean front", the temperature contrast which results from colder continental air moving over the warmer sea (this front is particularly strong in spring). Roughly half of the central basin depressions are steered over the Black Sea, and there is some evidence of steering by the upper flow along the axis of the subtropical jet. Eastern basin depressions have two preferred tracks to the north-east and to the east (Fig. 1.5). Occasionally they may move further south. Eastward penetration may be determined by the zonality of the upper flow and/or by the strength of the Siberian High.

5 GCM Projections for the Mediterranean Basin

It is clear from the preceding section that the climate of the Mediterranean region is determined partly by larger-scale characteristics of the atmospheric circulation (the Azores subtropical high, the monsoon, the waves in the upper westerly flow and the associated polar front and subtropical jet streams, etc.), partly by interactions between the large-scale flow and orography and land-sea contrasts, and partly by more local effects. Greenhouse-gas changes will change the large-scale characteristics of climate and will affect temperatures and temperature patterns in the sea and surrounding land areas. The precise patterns of future climatic change, however, will be controlled to a considerable extent by the way these effects are modulated by geography.

In an earlier section, it was noted that GCMs currently used in greenhouse-gas studies are unable to reproduce regional-scale features of present-day climate with any reliability. These GCMs, furthermore, have quite coarse resolution and they use highly smoothed orography (which is insufficiently detailed to be able to show the orography surrounding the Mediterranean with any semblance of realism). Since they cannot properly resolve the most important regional-scale features of the general circulation (such as the areas of cyclogenesis shown in Fig. 1.5) and they cannot possibly simulate the way orography interacts with the circulation in controlling precipitation patterns, they are unlikely to produce realistic simulations of the present and/or the future climate of the Mediterranean basin.

How, then, can we gain any insights into future changes in climate over the Mediterranean basin, given the inadequacy of our primary tool the General Circulation Model? The answer is that we cannot obtain any detailed predictive insights. We can examine GCM results (as will be done below), but we must be extremely cautious in interpreting and applying these results – i.e. they should be treated strictly as scenarios of a possible future climate and not as predictions.

Figures 1.6–1.13 summarize the results from four independent GCM studies in terms of seasonal-mean values of the equilibrium changes in temperature and precipitation due to a doubling of the atmospheric CO_2 level. (Note that, for "CO_2" we can read "equivalent CO_2" in order to account for the other greenhouse gases.) The GCMs are "state-of-the-art" models in that each has a mixed-layer ocean which is essential in order to obtain information at the seasonal level. The models are: the widely used GISS model (Goddard Institute for Space Studies; Hansen et al., 1984), the GFDL model (Geophysical Fluid Dynamics Laboratory; Manabe and Stouffer, 1980), the Community Climate Model of the National Center for Atmospheric Research (CCM; Washington and Meehl, 1984) and the OSU model (Oregon State University, Schlesinger and Zhao, 1989). Further details of these models are given by Schlesinger and Mitchell (1987).

These four models are similar in their basic physics, but they differ in the way they handle sea ice, clouds and surface processes, in the way they solve the various partial differential equations (GISS and OSU solve these in grid-point form, while GFDL and CCM use a spectral method), and in their resolution (OSU has four times the horizontal resolution of GISS, with GFDL and CCM in between; but OSU has much coarser vertical resolution than any of the other models). The GISS and OSU models are known to perform poorly in simulating present-day regional-scale climatic conditions (Santer and Wigley, 1990), with GISS giving somewhat more unrealistic results than OSU. The "control run" performances of the GFDL and CCM models have not been assessed at the regional level.

Before describing the results produced by these models, we need to assign some time in the future to which they might apply. The key point here is that they are equilibrium results for $2 \times CO_2$. That is, they do not simulate the transient response to continually changing CO_2 levels. As noted earlier, the spatial details of the transient response need not look like those for the equilibrium response. If we ignore this complication, then the main equilibrium/transient difference will be one of timing. Thus, if an equivalent $2 \times CO_2$ concentration level were reached in, say, 2050, the climate would not achieve the equilibrium configuration corresponding to this level for a number of years. The magnitude of the delay is uncertain, since, amongst other factors, it depends on properties of ocean heat transport that are presently rather poorly understood. Transient response calculations by Wigley (1989), using the model of Wigley and Raper (1987), indicate that the climate changes shown in Figs 1.6–1.13 should be representative of the change between now and around 2090 or later.

But. . . can we put any faith in these results? Certainly, we cannot believe any of the spatial differences that the models show occurring between different parts of the study area. Some of these differences are so large and they occur over such small distances, that they are clearly impossible – they serve only to expose serious defects in the models concerned. In all cases, the spatial differences vary from model to model: some models show greater temperature changes in the south, some in the north; some models show marked east–west contrasts in precipitation changes, while others do not; and so on. These model-to-model differences hardly instill any confidence in the performance of any individual model.

30 *Climatic Change in the Mediterranean*

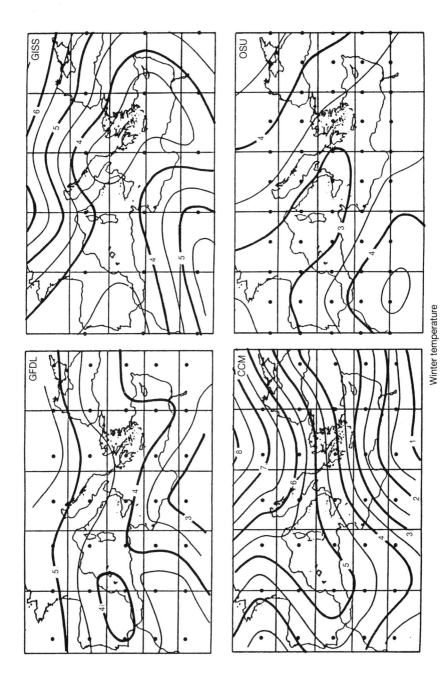

Fig. 1.6 Winter (DJF) temperature changes (in °C) due to a doubling of the CO_2 concentration; results from four independent GCMs.

Future climate of the Mediterranean basin   31

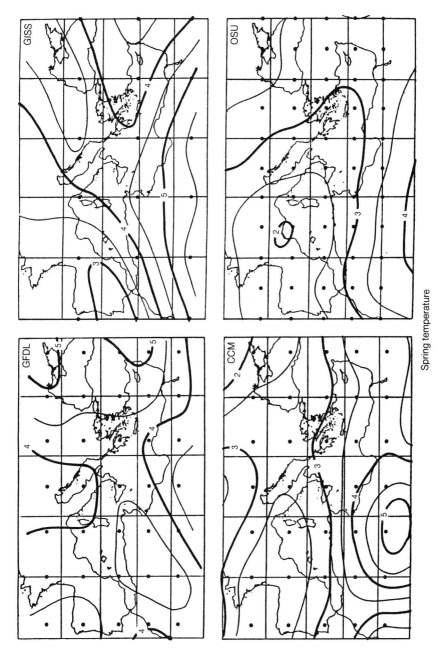

Fig. 1.7 Spring (MAM) temperature changes (in °C) due to a doubling of the CO_2 concentration; results from four independent GCMs.

32   *Climatic Change in the Mediterranean*

Fig. 1.8 Summer (JJA) temperature changes (in °C) due to a doubling of the CO_2 concentration; results from four independent GCMs.

Future climate of the Mediterranean basin 33

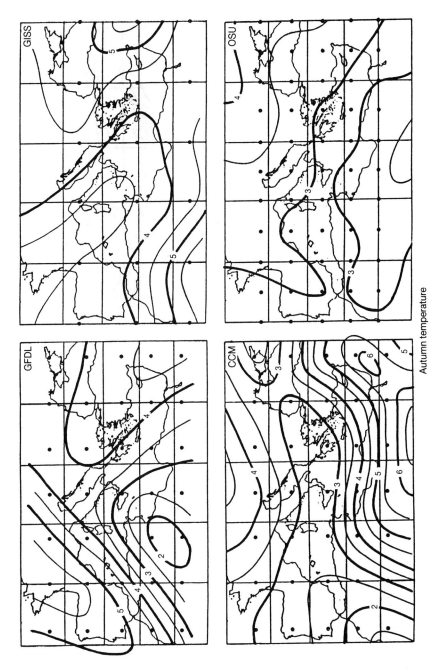

Fig. 1.9 Autumn (SON) temperature changes (in °C) due to a doubling of the CO_2 concentration; results from four independent GCMs.

34 *Climatic Change in the Mediterranean*

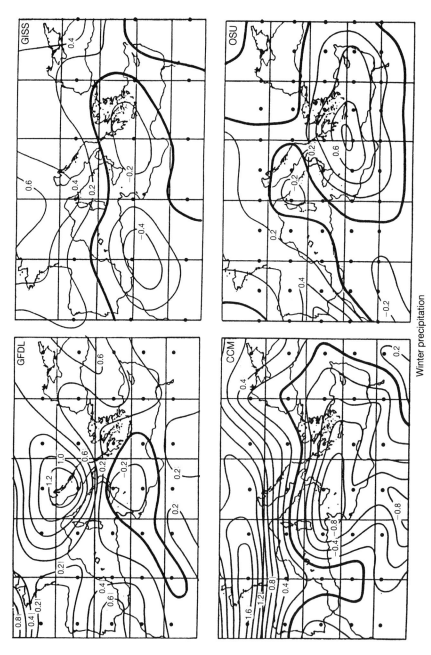

Fig. 1.10 Winter (DJF) precipitation changes (in mm/day) due to a doubling of the CO_2 concentration; results from four independent GCMs.

Future climate of the Mediterranean basin 35

Fig. 1.11 Spring (MAM) precipitation changes (in mm/day) due to a doubling of the CO_2 concentration; results from four independent GCMs.

36 *Climatic Change in the Mediterranean*

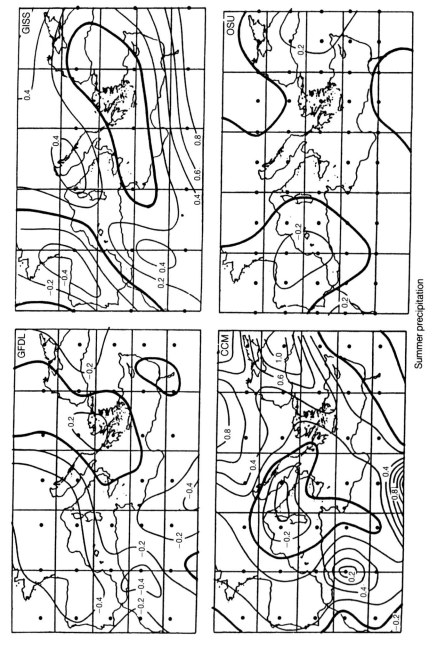

Fig. 1.12 Summer (JJA) precipitation changes (in mm/day) due to a doubling of the CO_2 concentration; results from four independent GCMs.

Future climate of the Mediterranean basin 37

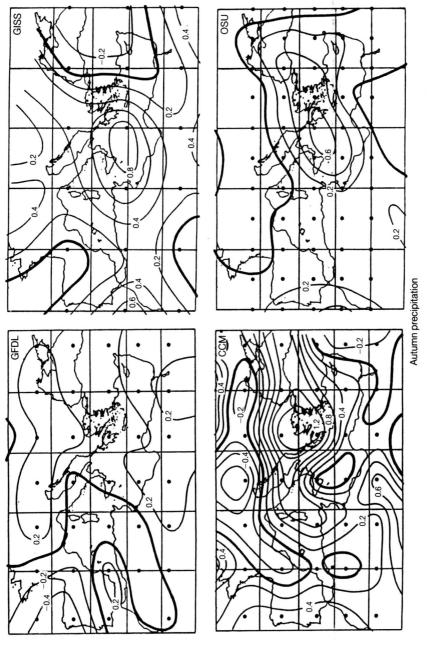

Fig. 1.13 Autumn (SON) precipitation changes (in mm/day) due to a doubling of the CO_2 concentration; results from four independent GCMs.

The only common feature in which we can have any faith, is a large-scale warming in all seasons. There are only small differences between the seasons, especially if one averages the results of all models together: a warming of about 3.5°C spread uniformly over the seasons would be the best guess. This is slightly less than the average global mean change produced by these particular models.

It is important to note that, on top of the large regional-scale uncertainty, there is a further uncertainty which is hidden here because all four models have similar climate sensitivities, viz. about 4–5°C for a doubling of the CO_2 concentration. This range is at the top end of the accepted range of climate sensitivity, viz. ΔT_{2x}=1.5–4.5°C. Thus, the results shown in Figs 1.6–1.13 probably represent upper limits to the changes likely to occur between now and the late 21st century. The estimated warming of 3.5°C by this time over the basin as a whole should therefore be translated to a warming range of about 1.2–3.5°C. At the low end of this range, the change is only a little over double the change actually observed over 1947–86 (see Fig. 1.2). If this observed change is an indication of the level of natural variability and if the greenhouse-gas warming signal were a steady warming, then it would be many decades before the signal could be distinguished convincingly from the noise of natural variability. On the other hand, if the signal were at the top end of the likely range, it would probably be clearly discernible by the early decades of next century.

What can we say about future precipitation changes on the basis of the model results? *A priori*, the answer to this question is "very little", since the models simply do not adequately resolve the factors which control precipitation. Nevertheless, it is of some interest to try to summarize the common features of the four projections, and to objectively quantify their similarities and differences. Simply averaging the four projections, season by season, is clearly inadequate, since the average could be dominated by the results of a single model and since this procedure would mask information about the degree of model-to-model variability. A better procedure is to use the four sets of results to estimate the probability of a positive or negative change in precipitation at each grid point.

The procedure is as follows. First, all results must be transformed to a common grid. The most convenient grid is that used by the NCAR CCM and the GFDL model. At each grid point, suppose the four precipitation change values to be ΔR_i (i = 1,4). From these, a mean ($\Delta \bar{R}$) and an unbiased standard deviation estimate ($\hat{s}$) can be calculated. If one then assumes that the ΔR_i values are random samples from a Normal Distribution, standard tables can be used to find the probability that ΔR is less than (or greater than) zero (viz. p or 1-p). For example, at the top left grid-point in winter (46.7°N, 7.5°W), the values of $\Delta \bar{R}$ and $\hat{s}$ are 0.50 mm/day and 0.72 mm/day. All four models show an increase in precipitation, but the mean is dominated by the NCAR CCM increase of 1.57 mm/day. The large value of $\hat{s}$ reflects the considerable model-to-model variability. For these values of $\Delta \bar{R}$ and $\hat{s}$, $\Delta R = 0$ lies 0.70 standard deviation units below the mean, so the probability that $\Delta R < 0$ is given by p = 0.24. Given the results, the most likely outcome is clearly an increase in precipitation. However, even though all four models agree qualitatively in this respect, the large

Future climate of the Mediterranean basin 39

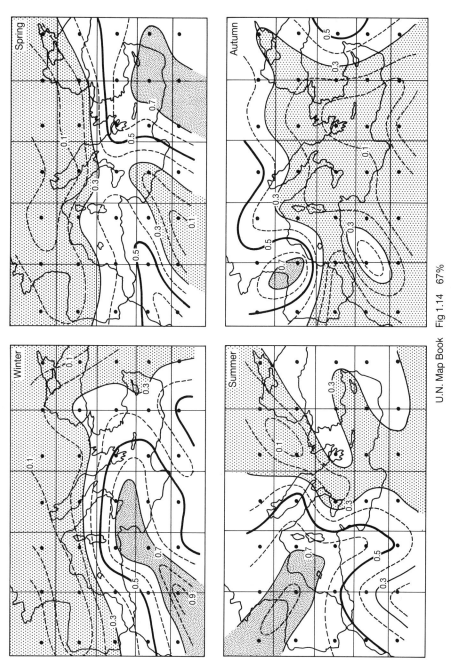

Fig. 1.14 Probabilities of a precipitation decrease (p) based on the four sets of GCM results. The shaded areas are those where p < 0.3 or p > 0.7 (i.e. where the probability of a change exceeds 70%).

inter-model differences mean that there is still a significant chance (about one in four) of a precipitation decrease.

This procedure has been applied to all grid points in all four seasons, and the results are shown in Fig. 1.14. It must be stressed that these results are presented here only as a means of summarizing the diversity of model projections. Since the individual model results are of extremely dubious quality, the summary must be equally suspect: the average of four "wrongs" cannot be a "right". Fig. 1.14 shows the following features. In winter, most of the Mediterranean basin shows an increase in precipitation. The exception is the south-central region and the adjacent North African coast, especially around Tunisia. In spring, precipitation increases everywhere except in the eastern basin. In Egypt and Israel, the probability of a precipitation decrease in spring exceeds 0.7. In summer, the western basin becomes drier (especially around the Pyrenees), while the eastern basin becomes wetter. Finally, in autumn the whole region shows an increase in precipitation except for northern Spain and western France.

It should be noted that there are few regions where the probability of an increase or decrease in precipitation exceeds 90% (i.e. where $p < 0.1$ or $p > 0.9$). Thus, nowhere are there strong indications of an overall change in either direction. Regions where the probability of a precipitation increase exceeds 90% account for only about 10% of the area. This is the sort of result one might expect to occur purely by chance, although these areas do tend to occur preferentially in the northern part of the basin. There is only one grid point (see the winter map in Fig. 1.14) where the probability of a precipitation decrease exceeds 90%.

To summarize these results, the best guess would be: no major precipitation changes, with a hint of a slight increase in autumn precipitation. Precipitation increases appear to be marginally more likely in the northern part of the basin, while the evidence for the southern part of the basin points marginally towards precipitation decreases. Whether or not there will be changes in the finer-scale details of precipitation (precipitation intensity, extreme values, dry- and wet-day probabilities, etc.) is impossible to say on the basis of these model results.

6 SUMMARY AND CONCLUSIONS

The greenhouse problem can be viewed as one involving large uncertainties but with risks that are potentially high. Even the most well-defined projections, those for global-mean temperature, are subject to considerable uncertainty. Global-mean warming projections for the interval 1990–2030 lie in the range 0.5–1.4°C, a range which widens further if one takes account of uncertain future greenhouse-gas concentrations. At the low end of the range, the change is comparable to the global-mean warming that has occurred already this century (although the future warming would be more rapid). However, the likely observed change by 2030 is not the only important parameter; the warming *commitment* also needs consideration. In 2030, there is likely to be a substantial commitment to future warming, even if greenhouse-gas concentrations could be held constant at their 2030 levels. Because of this warming commitment and because large increases in greenhouse-gas concentrations are inevitable, we are already committed to significant changes in regional climate. When coupled with changes in extreme-event frequencies that also must occur, the impacts could be considerable. At the high end of the commitment range (i.e. the low-probability/high-risk area), the projected global-mean temperature would far exceed anything previously experienced by human beings. There is clearly considerable cause for concern and a pressing need to reduce the uncertainties.

The impacts of any global-mean climatic change will depend on the regional details of changes in a wide variety of climate variables and in changes in the interannual variability of these variables. At present, we are unable to predict these changes. However, GCM results do give us data that can be used to develop scenarios of future changes. For the Mediterranean basin, GCM results point to a warming similar in magnitude to the global-mean value, with no evidence for any marked seasonal differences in the warming. Although the magnitude of this warming is uncertain, we can be fairly confident that, as a prediction, it is qualitatively correct. While it may be many decades before the change can be statistically detected above the noise of natural, regional-scale climatic variability, the existence of a background warming trend will still be of considerable importance. As time progresses, the probability of periods of extreme warmth will increase, with attendant effects on human wellbeing and on "natural" events like forest fires. Furthermore, increased warmth will probably lead to greater evapotranspiration and (in the absence of increased winter/spring precipitation) reduced growing-season soil moisture levels, adding to direct thermal stresses on vegetation.

Projected precipitation changes vary so much from model to model that one cannot say, on the basis of model results alone, whether precipitation will increase or decrease. Depending on location, model used and season considered, projected changes between now and around 2050 range between ±1 mm/day. The mean precipitation rate for the Mediterranean basin as a whole is roughly 1 mm/day. Such large changes are undoubtedly unrealistic, and must reflect model deficiencies. However, the *possibility* of substantial changes (say, up to ±30% over the next 50–100 years) must be considered as realistic.

Because the models are so uninformative with regard to precipitation, it is worthwhile speculating on possible changes using physical arguments. First, as noted earlier, cyclogenesis and rainfall are often promoted by land-sea temperature contrasts. Because land and sea have different effective thermal inertias, a large-scale warming could affect this contrast, possibly reducing it in winter months. This could in turn lead to reductions in rainfall and in storminess, particularly in the eastern basin. On the other hand, warmer sea-surface temperatures both in the Mediterranean and in the North Atlantic could lead to increases in atmospheric moisture availability, and this would tend to increase precipitation. In addition, since the intensity of the monsoon circulation is expected to increase, the rare extreme events that occur in the far east of the region (as a result of incursions of monsoon air masses) may increase in frequency. Moreover, a more intense monsoon also may lead to increased precipitation in the headwaters of the Nile, with important consequences for Egypt. The situation here, however, is unclear because the tropical easterly jet, which is an integral part of the monsoon system and which extends over the Nile headwaters area in summer, also affects precipitation amounts and patterns. A northward movement of the easterly jet could reduce rainfall in Ethiopia and the Sudan. Much of the Mediterranean region's precipitation is influenced by interactions between the large-scale flow and orography. Changes in the former are virtually certain, and a northward shift of the main upper westerly flow could reduce the length of the rainy season, particularly in the western and central parts of the basin.

All of the above suggestions must be taken as possibilities, rather than probabilities. They are, however, amenable to more detailed investigation using existing data and model results. Indeed, there is much that can be done to reduce uncertainties and to develop better methods for interpreting GCM results. At the same time, GCMs themselves are constantly being improved and much better results can be expected to appear within the next 5–10 years. The greenhouse effect is humanity's most pressing environmental problem, one which presents major scientific challenges across a wide range of disciplines. Developing better regional-scale projections of future climate is at the forefront of these challenges.

7 REFERENCES

Barnett, T.P., 1985. On long-term climate change in observed physical properties of the oceans. In: *Detecting the Climatic Effects of Increasing Carbon Dioxide*, MacCracken, M.C. and Luther, F.M. (eds), US Department of Energy, Carbon Dioxide Research Division, Washington, DC, 91–107.

Bradley, R.S., Diaz, H.F., Eischeid, J.K., Jones, P.D., Kelly, P.M. and Goodess, C.M., 1987. Precipitation fluctuations over Northern Hemisphere land areas since the mid–19th century. *Science* **237**, 171–175.

Bretherton, F.P., Bryan, K. and Woods, J.D., 1990. Time-dependent greenhouse-gas-induced climate change. In: *Climate Change: The IPCC Scientific Assessment*, Houghton J.T., Jenkins, G.J. and Ephraums, J.J. (eds), Cambridge University Press, Cambridge, 173–193.

Chang, JenHu, 1972. *Atmospheric Circulation Systems and Climate*. Oriental Publishing, Honolulu, 326pp

Farmer, G. and Wigley, T.M.L., 1985. *Climatic Trends for Tropical Africa*. Report to the UK Overseas Development Administration, 136pp.

Folland, C.K., Karl, T.R., Nicholls, N., Nyenzi, B.S., Parker, D.E. and Vinnikov, K.Ya., 1992. Observed climate variability and change. In: *Climate Change 1992: The Supplementary Report to the IPCC Scientific Assessment*, Houghton, J.T., Callander, B.A. and Varney, S.K. (eds), Cambridge University Press, Cambridge, 135–170.

Gat, J.R. and Magaritz, M., 1980. Climatic variations in the eastern Mediterranean Sea area. *Naturwissenschaften*, **67**, 80–87.

Gates, W.L., Mitchell, J.F.B., Boer, G.J., Cubasch, U. and Meleshko, V.P., 1992. Climate modelling, climate prediction and model validation. In: *Climate Change 1992: The Supplementary Report to the IPCC Scientific Assessment*, Houghton, J.T., Callander, B.A. and Varney, S.K. (eds), Cambridge University Press, Cambridge, 97–134.

Gornitz, V.L., Lebedeff, S. and Hansen, J., 1982. Global sea-level trend in the past century. *Science*, **215**, 1611–1614.
Hansen, J., Lacis, A., Rind, D., Russell, G., Stone, P., Fung, I., Ruedy, R. and Lerner, J., 1984. Climate sensitivity. Analysis of feedback mechanisms. In: *Climate Processes and Climate Sensitivity*, Maurice Ewing Series No. 5, Hansen, J. and Takahashi, T. (eds), American Geophysical Union, Washington, DC, 130–163.
Houghton, J.T., Jenkins, G.J. and Ephraums, J.J. (eds), 1990. *Climate Change: The IPCC Scientific Assessment*, Cambridge University Press, Cambridge, 365pp.
Houghton, J.T., Callander, B.A. and Varney, S.K. (eds), 1992. *Climate Change 1992: The Supplementary Report to the IPCC Scientific Assessment*, Cambridge University, Cambridge, p. 200.
Jones, P.D., Raper, S.C.B., Bradley, R.S., Diaz, H.F., Kelly, P.M. and Wigley, T.M.L., 1986a. Northern Hemisphere surface air temperature variations. 1851–1984. *J. Clim. and Applied Meteorol.*, **25**, 161–179.
Jones, P.D., Raper, S.C.B. and Wigley, T.M.L., 1986b. Southern Hemisphere surface air temperature variations. 1851–1984. *J. Clim. and Applied Meteorol.*, **25**, 1213–1230.
Jones, P.D., Wigley, T.M.L. and Wright, P.B., 1986c. Global temperature variations between 1861 and 1984. *Nature*, **322**, 430–434.
Jones, P.D., Wigley, T.M.L., Folland, C.K. and Parker, D.E., 1988. Spatial patterns in recent worldwide temperature trends. *Clim. Mon.*, **16**, 175–185.
Manabe, S. and Stouffer, R.J., 1980. Sensitivity of a global climate model to an increase of CO_2 concentration in the atmosphere. *J. Geophys. Res.*, **85**, 5529–5554.
Meier, M.F., 1984. Contribution of small glaciers to global sea-level. *Science*, **226**, 1418–1421.
Mintzer, I.M., 1987. *A Matter of Degrees. The Potential for Controlling the Greenhouse Effect*. Research Report No. 5, World Resources Institute, Washington, DC, 60pp.
Mitchell, J.F.B., Manabe, S., Meleshko, V. and Tokioka, T., 1990. Equilibrium climate change – and its implications for the future. In: *Climate Change: The IPCC Scientific Assessment*, Houghton, J.T., Jenkins, G.J. and Ephraums, J.J. (eds), Cambridge University Press, Cambridge, 131–174.
Palutikof, J.P., Wigley, T.M.L. and Lough, J.M., 1984. *Seasonal Scenarios for Europe and North America in a High CO_2, Warmer World*. Technical Report TR012, US Department of Energy, Carbon Dioxide Research Division, Washington, DC, 70pp.
Parry, M.L., Carter, T.R. and Konijn, N.T. (eds), 1988. *The Impact of Climatic Variations on Agriculture. Vol. 1., Assessments in Cool Temperate and Cold Regions.* Kluwer Academic Publishers, Dordrecht, 876 pp.
Pittock, A.B. and Salinger, J.M., 1982. Towards regional scenarios for a CO_2-warmed Earth. *Climatic Change*, **4**, 23–40.
Santer, B.D. and Wigley, T.M.L., 1990. Regional validation of means, variances, and spatial patterns in General Circulation Model control runs. *J. Geophys. Res.*, **95**, 829–850.
Schlesinger, M.E. and Mitchell, J.F.B., 1987. Climate model simulations of the equilibrium climatic response to increased carbon dioxide. *Rev. Geophys.*, **25**, 760–798.
Schlesinger, M.E. and Zhao Zongci, 1989. Seasonal climate changes induced by doubled CO_2 as simulated by the OSU atmospheric GCM/mixed layer ocean model. *J. Clim*, **2**, 459–495.
Washington, W.M. and Meehl, G.A., 1984. Seasonal cycle experiment on the climate sensitivity due to a doubling of CO_2 with an atmospheric general

circulation model coupled to a simple mixed layer ocean. *J. Geophys. Res.*, **89**, 9475–9503.

Wigley, T.M.L., 1987. The effect of model structure on projections of greenhouse-gas-induced climatic change. *Geophys. Res. Let.*, **14**, 1135–1138.

Wigley, T.M.L., 1989. When will equilibrium 2xCO$_2$ results be relevant. *Clim. Mon.*, **17**, 99–106.

Wigley, T.M.L. and Farmer, G., 1982. Climate of the eastern Mediterranean and the Near East. In: *Palaeoclimates, Palaeoenvironments and Human Communities in the Eastern Mediterranean Region in Later Prehistory*, B.A.R. International Series 133, Bintliff, J.L. and Van Zeist, W. (eds), British Archaeological Reports, Oxford, 3–37.

Wigley, T.M.L. and Jones, P.D., 1987. England and Wales precipitation. a discussion of recent changes in variability and an update to 1985. *J. Climatol.*, **7**, 231–246.

Wigley, T.M.L. and Raper, S.C.B., 1987. Thermal expansion of sea water associated with global warming. *Nature*, **330**, 127–131.

Wigley, T.M.L. and Raper, S.C.B., 1992. Implications for climate and sea level of revised IPCC emissions scenarios. *Nature*, **357**, 293–300.

2

Sea-level Response to Climate Change and Tectonics in the Mediterranean Sea

J. D. Milliman
(Woods Hole Oceanographic Institution, MA, USA)

Abstract

Over the past 100 years, global sea-level has risen between 10 and 20 cm. There appears to have been no dramatic acceleration in sea-level rise in recent years even though CO_2 content in the atmosphere has increased more than 30% and the average global temperature has risen 0.5° C.

By the years 2025–2030, in response to a doubling of atmospheric CO_2 levels, the rate of global sea-level rise is projected to accelerate by a factor of 2 to 3, mostly because of thermal expansion of sea water and the melting of alpine glaciers. The unknown response of the Antarctic ice shelves to climate change prevents a more accurate forecast. Nevertheless, eustatic sea-level is expected to rise 12 to 18 cm in the next 35–40 years, although in the Mediterranean local subsidence (e.g. Po and Nile deltas) could result in local sea-level rise by as much as 25–40 cm. Increased withdrawal of water or petroleum from coastal areas could accelerate local sea-level rise even more.

1 Introduction

Few global problems have been studied in as much detail with as few reliable data as sea-level. With the expected warming of the Earth's climate in response to man-induced build-up of greenhouse gases in the atmosphere, the possible impact on sea-level is clearly of concern, particularly with an increasing portion of the Earth's population living in coastal areas. The concern about rising sea-level is accentuated by the possibility of increased storm surges if weather patterns shift considerably.

The problem with sea-level prediction, however, is that we have a rather poor idea of how sea-level has changed over the past few hundred years (i.e. before and after the increase in atmospheric CO_2 resulting from the industrial and agricultural revolutions) or how it is changing at present. In fact, because defining a meaningful trend involves delineating short-term fluctuations in sea-surface configuration and land elevation, as well as local sea-level change, we have no clear reference for global sea-level; rather, as will be seen in this chapter, at present our terms of reference refer to local sea-level. Moreover, the major causes of short-term (in a geological sense) sea-level fluctuation (e.g. thermal expansion of sea water, ice melting) themselves depend on physical parameters that are not sufficiently understood (e.g. oceanic circulation) or climatic shifts that are not well predicted (e.g. changing precipitation patterns over Antarctica). Thus, while various numerical models forecast future sea-levels, the validity of these predictions is in question.

46 Climatic Change in the Mediterranean

In this chapter I address the types of sea-level change and discuss briefly sea-level change during the past 30,000 years, that is during the last glacial period and subsequent climatic warming. Of particular interest is the rate of sea-level change during the past 100 years. I then examine briefly the variables considered in predicting future sea-level change. Finally, these numbers are projected for the Mediterranean, with special emphasis on the role of local tectonics in defining sea-level change.

2 Tectonic and Climatic Changes in Sea-level

Over geologic time-scales, changes in sea-level occur in two distinct ways: tectonic changes related to the rate of sea-floor spreading and climatic changes related to the glacial events. Tectonic events occur over time spans on hundreds of thousands to millions of years and can involve sea-level changes of hundreds of metres, whereas glacial events occur on the time-scale of tens of thousands of years and normally involve a sea-level change on the order of 100 m or less.

Tectonic variations in sea-level occur when the rate of mid-ocean ridge formation and related sea-floor spreading change. As spreading increases and more warm (and therefore less dense) ridge is formed, a corresponding amount of sea water is displaced and global sea-level rises. As it cools and becomes more dense, the ridge subsides, losing topographic relief. Thus if the rate of ridge formation decreases (i.e. spreading lessens) sea-level will fall. One product of the the worldwide exploration petroleum and the Ocean Drilling Project (ODP) has been the understanding of global sedimentary events to identify periods of high and low stands of sea-level (e.g. Haq *et al.*, 1988 and references therein).

Superimposed on these longer term sea-level fluctuations are glacially induced changes, basically caused by the net removal of water from the ocean (via evaporation) to produce (via precipitation) advancing ice sheets on land. Sea-level falls during glacial times (as glaciers form) and rises (as the glaciers melt) during interglacial times. Formation of the late Würm ice sheets about 30 thousand years ago (ka), for example, lowered sea-level more than 100 metres; melting of the ice, beginning about 15–17 ka caused a corresponding rise in sea-level (see below).

3 Late Quaternary Sea-level

The extent of the last sea-level lowering and its subsequent transgression across the continental shelf has been delineated by carbon-14 dating of sea-level indicators (such as oyster shells or peat deposits) that lie beneath present-day sea-level. Assuming that a datable sample formed at or near sea-level, its radiometric date gives a reasonable estimate of the placement of sea-level at that time. Extreme care, of course, must be used to verify that the object has not been diagenetically altered nor transported substantially from its site of origin (e.g. Van de Plassche, 1986).

With the melting of the late Würm glaciers 15–17 ka, sea-level rose rapidly, reaching present-day sea-level about 5–6 ka (Fig. 2.1). Most workers agree that in the last 5000 years sea-level has fluctuated no more

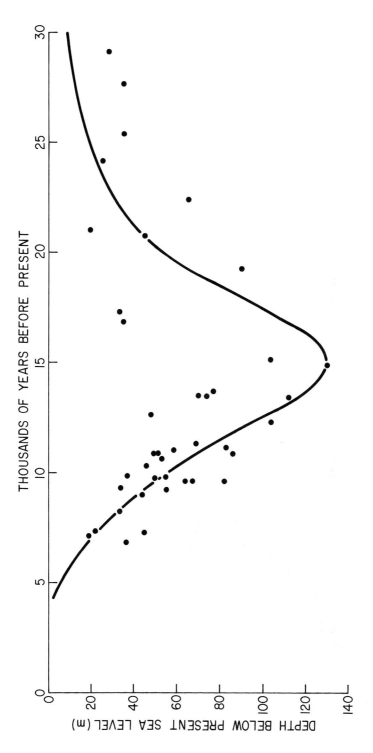

Fig. 2.1 Proposed Late Quaternary sea-level curve for the eastern United States;: after Milliman and Emery (1968). Subsequent research has shown that sea-level off the south-eastern US fell by only about half as much as off the north-eastern US, but for many other areas in the world, it appears that sea-level did fall by about 100 m relative to present-day values.

than several metres, some of these fluctuations related to slight climatic changes, but many caused by local or regional tectonic changes (i.e. uplift or subsidence). In the Mediterranean area, these sea-level changes are dramatically indicated by the submergence or emergence of archeological and historic sites (see Flemming, 1990).

4 SEA-LEVEL DURING THE PAST 100 YEARS

On the human scale, one is interested in changes over much shorter periods of time (tens of years) and smaller vertical intervals (10s of cm). To document such variations in sea-level, most workers have used variations in local mean sea-level as registered by tide gauges. Annual mean sea-level averages out such variables as tides, waves and short-term atmospheric fluctuations, but it does not take into account annual variations in river flow (many tide gauges being located on or near river mouths), longer term atmospheric or oceanic variations (for example, an El Niño event).

These disturbances in local sea-level are considered "noise", from which mean sea-level must be filtered. The problems, of course, is that the "noise" is generally orders of magnitude greater than any shift in sea-level. Tidal range, for instance, can be several metres or more, and seasonal shifts in river discharge can cause up to a 1 m shift in local sea-level. In contrast, the mean annual change in sea-level recorded by most tidal gauges is less than 1 cm. Thus one needs long-term tide-gauge records to obtain the annual means from which meaningful trends in sea-level change can be delineated. In earlier years such trends were computed graphically; now Eigenanalysis is used (e.g. Aubrey and Emery, 1983). Records at least 20 years long are necessary to recognize meaningful trends, and a 30- to 50-year period is preferable. Clearly, short-term changes, such as those associated with an accelerated rise in sea-level, are difficult if not impossible to delineate using only tide-gauge records.

The problem with identifying even a statistically valid tide-gauge change in sea-level is that it only indicates local change. In fact, rather than being used to define the absolute rise of sea-level, tide-gauge data often are used to indicate the relative motion of land (e.g. Emery and Aubrey, in preparation). For example, tectonic uplift and isostatic rebound (from glacial retreat in northern latitudes) have resulted in a drop of local sea-level, whereas basin subsidence and sediment compaction (particularly near deltas) can accentuate the local rise of sea-level.

Such problems make it difficult to define a world mean rise of sea-level or to delineate a trend in the change of sea-level. Taking into account local tectonic variations, Raper et al. (1990) conclude that sea-level over the past 100 years has changed 12 ± 12 cm, or an annual rise of 1.2 mm/yr; the scatter is such, however, that it is statistically possible to conclude that these data indicate no sea-level rise. Aubrey and Emery (oral communication) come up with a similar number to that of Raper and Wigley, about 1.5 + 1.0 mm/yr.

Realizing that the last deglaciation meant a net shift of water mass to the southern oceans as the northern ice sheets melted, Peltier and Tushingham (1989) have modelled the mantle response to this shift in water mass. Adjusting for the isostatic rebound experienced in northern latitudes, they conclude that global rise over the past 50 years has been considerably greater (2.4 + 0.90 mm/yr) than cited by other workers. They conclude that this represents the first indication of accelerated sea-level rise, primarily the result of increased melting of polar icecaps. This computation, however, differs from all other estimates by enough (generally twice as high) to suggest caution in accepting Peltier and Tushingham's conclusions. Pirazzoli (1989), for example, points out that Peltier and others have placed undue emphasis on tide-gauge data from selected parts of the world (notably Europe and North America), and the lack of data from other areas has tended to maximize our estimates of sea-level rise. Pirazzoli concludes that sea-level over the past century rose only 4–6 cm, with a possible accelerated rise from the end of the last century to about 1930 and then a leveling off or even fall in sea-level. This rate of sea-level rise is only about half that cited by Raper and Wigley or Emery and Aubrey and four to six times less than that cited by Peltier and Tushingham.

Raper et al. (1992) have attempted to compute a global balance for sea-level rise based on recent climate change. Although the uncertainty is great, it seems possible to explain a 12 cm rise of sea-level over the past 100 years by assuming a 0.5°C rise in temperature: most of the rise comes from thermosteric change and alpine glacial melting (e.g. Meier, 1984); the impact of Greenland and Antarctic ice melting is not clear. If Peltier and Tushingham's higher value of sea-level rise is accepted, however, the balance becomes more problematic.

At present, then, there are no firm and unambiguous data to suggest that the rate of sea-level rise has accelerated in the 20th century as a result of climatic warming. Clearly, if after exhaustive study the range of conclusions about sea-level in the past century varies by a factor of 4 to 6 and predictions for future sea-level range from accelerated sea-level rise to falling sea-level over the past few decades (Peltier and Tushingham vs. Pirazzoli), the peril in either documenting past change or predicting future change becomes obvious. The bulk of evidence, however, shows a gradual rise of sea-level at between 1 and 2 mm/yr, with most of the eustatic sea-level rise associated with thermal expansion of sea water and melting of alpine ice. Any global trend, however, is still hidden in the tectonic, subsidence, oceanic and atmospheric "noise". Until more sophisticated measurements than tide-gauge data are used, it is doubtful that our understanding of sea-level change in this century will be more insightful.

5 GLOBAL SEA-LEVEL RISE BY 2025

Given a 1.5 to 3.6°C global warming by the year 2025, most researchers expect an accelerated rise of sea-level. These changes will continue to result in thermal expansion of ocean waters and alpine glacial melting, but the response of the polar (Greenland and Antarctica) ice sheets to climatic change is unknown. Depending on the assumptions and models used, the range of predictions is great and increases with time, as the variables used in the different models become more speculative. The range of sea-level-rise scenarios is shown in Fig.2.2.

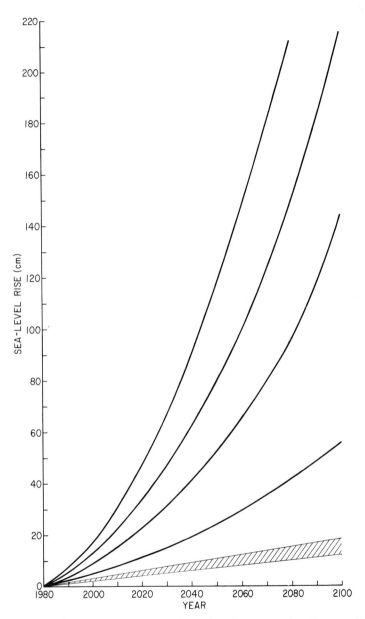

Fig. 2.2. Predicted sea-level rise based on four scenarios discussed by Hoffman *et al.* (1983). The highest rate of sea-level rise assumes large-scale melting of the Antarctic ice shelf, whereas the minimal rise of predicted sea-level rise assumes that the Antarctic ice shelf remains more or less as presently constituted. For comparison, the rate of global rise based on the historic records (past 100 years), if continued into the next century, would be 1–1.5 mm/yr shown in the shaded area. The predicted rise used for the Mediterranean in this volume is 10–20 cm by the year 2025, a figure than lies within the low-range scenario of Hoffman *et al.*

Thermal expansion of sea water is related to temperature change by the coefficient of thermal expansion, which in turn is related to temperature, pressure and salinity. According to the box-upwelling-diffusion model of Wigley and Raper (1987), modified slightly by Raper et al. (1992), approximately half of the global rise in sea-level by the year 2030 (12–18 cm) will be accounted for by the thermal expansion of the ocean, 6–10 cm.

The contribution from melting ice is more difficult to predict, as it also involves absolute changes in weather patterns. For example, even if mean annual temperature increased by 3.5°C, altered patterns of precipitation over Antarctica could cause an increase in snow and ice formation, and thus contribute to a drop in sea-level. Whether Antarctic ice volume would decrease or increase is not known, and therefore the amount of corresponding sea-level change is difficult to predict. Given sufficient time and an adequate increase in temperature, of course, it is possible that the western Antarctic ice shelf could experience significant melting; this was reason for the very high estimates of sea-level rise (348 cm by the year 2100) offered by Hoffman et al. (1983).

One problem is that recent ODP drilling suggests that the western Antarctic ice shelf has been present since the late Miocene, and temperatures during intervening periods were significantly warmer than even the high range of predicted temperatures in 2100. With this in mind, it is highly questionable as to the amount of ice melt in Antarctica. In fact, more moderate estimates for Antarctica suggest that its input to sea-level rise, at least over the next 100 years, may be minimal (e.g. Stewart et al., 1990). Recently Meier (1990) has lowered his estimates of sea-level rise due to glacier and ice-cap melting: according to his new estimate, by 2050 ice build-up may have contributed to a minimal lowering of sea-level. If this is true, then the catastrophic predictions of coastal inundation due to accelerated sea-level rise may turn out to be (in hindsight) highly exaggerated.

6 LAND MOVEMENT IN THE MEDITERRANEAN

Most conservative projections call for a global sea-level rise of 18 cm in the next 40 years, slightly more than 4 mm/yr, or two to three times faster than for the past 100 years. From the arguments cited above, even these presumably moderate estimates, however, may prove to be erroneous. In any event, the effect of tectonic movement may accentuate or minimize the local effects of any sea-level rise.

Much of the Mediterranean coast appears to have experienced sea-level changes within the generally accepted range of sea-level rise (e.g. 1 to 2 mm/yr). It is assumed, therefore, that these areas have remained more or less tectonically stable (Fig. 2.3). Those areas in which the rate of sea-level has been less than 1–2 mm/yr (e.g. the eastern Mediterranean, from Ashdod, Israel to Antalya) are assumed to have undergone slight tectonic uplift, whereas several of the larger river deltas discussed in this volume have experienced sea-level rises substantially greater than the global rise: Nile – 4.8 mm/yr (in agreement with the findings of Stanley, 1989); Thessaloniki – 4.0 mm/yr; Venice (San Stefano) – 7.3 mm/yr. These areas are assumed to have undergone subsidence. Marseille (near the

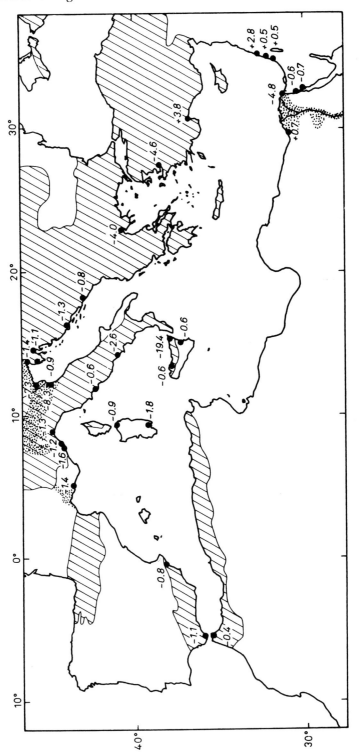

Fig. 2.3. Mean annual rise in sea-level at various acceptable tide-gauge stations in the Mediterranean. Negative numbers indicate falling sea-level, i.e. rising land (after Emery et al., (1988).

Rhône Delta) shows a normal (1.4 mm/yr) sea-level rise, and Alexandria has experienced a slight drop in sea-level (Fig. 2.4). As Emery *et al.* (1988) point out, however, there are essentially no secular tide-gauge measurements west of Alexandria, thus giving us no short-term picture of sea-level change along most of the southern Mediterranean.

Finally, mention should be made of accelerated subsidence resulting from the withdrawal of water and/or petroleum. Removal of water essentially accelerates the compaction (i.e. de-watering) of underlying strata and therefore can increase local subsidence substantially. Perhaps the best example is in the city of Bangkok, where a marked increase in the removal of ground water (in response to the marked urbanization

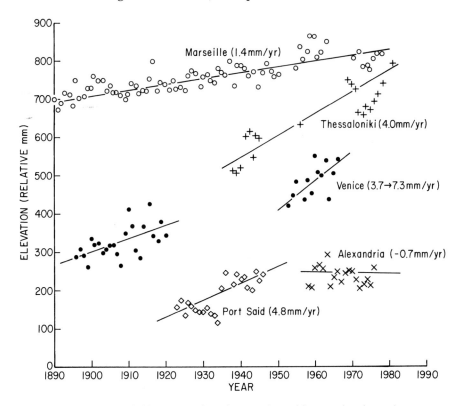

Fig. 2.4. Mean annual tide-gauge elevations and resulting sea-level trends (determined by Eigenanalysis) for selected Mediterranean sites. Data from D.G. Aubrey as reported in Emery *et al.* (1988). Several trends should be noted in this figure: (1) Annual variations in tidal elevation can exceed 5 cm relative both long-term trends and adjacent years. Note also the general fall in sea-level at these stations in the early 1970s, apparently reflecting a regional trend. (2) At three stations (Venice, Port Said and Tessaloniki) sea-level rise (4.0 to 7.3 mm/yr) was significantly greater than the global average (1–2 mm/yr), indicating local subsidence. In contrast, sea-level at Marseille (1.4 mm/yr) was similar to the global average, and at Alexandria (–0.7 mm/yr) it was considerably less, indicative of land uplift. (3) The rate of sea-level rise at Venice increased two-fold between 1890–1920 and 1950–1970, the result of ground-water removal.

54   *Climatic Change in the Mediterranean*

of Bangkok) beginning in 1960 resulted in local rates of subsidence as great as 13 cm/yr, about two orders of magnitude greater than the global rise of sea-level (Fig. 2.5). Once this water is removed and the strata have consolidated, the process cannot be reversed; what has subsided remains lowered relative to sea-level. Although the magnitude has been less dramatic, the enhanced subsidence in Venice in the 1950s and 1960s had similar causes (Sestini, 1990). If climate changes and/or increased population pressures in the next century result in less available water in coastal areas, it is possible that increased removal of ground water could result in accelerated subsidence. Petroleum removal in coastal areas could result in similar subsidence.

7 SUMMARY AND FUTURE STUDIES

Documenting recent sea-level change is an inexact science, and predicting future sea-level change is even more problematical. Sea-level appears to be rising at less than 2 mm/yr, and there is no indication of an accelerated rise over the past 100 years. Although global sea-level rise may increase to as much as 4 mm/yr by the year 2025, the chances seem better than ever (based on the record over the past 100 years) that it will be less. By the year 2025, then, only very low coastal lands will have flooded, although more flooding could occur if storm frequency or magnitude increases as a result of global warming (Emanuel, 1987).

The major problem of sea-level for coastal areas (at least for the next 30–40 years) appears to be related to local subsidence (Fig. 2.2). Those areas with greatest danger appear to be low-lying deltas (such as the Nile), where subsidence rates can be relatively high. Ironically, thickly populated cities are often located on or adjacent to large deltas. By documenting rates of local subsidence, areas with greatest hazard can be identified and protective/mitigating procedures undertaken. Conversely, those areas identified as undergoing uplift may not require substantial modification to existing coastal engineering installations.

As alluded to previously, trying to delineate sea-level change using tide gauges means using 19th century technology to solve 21st century problems. Sea-level rise (or fall) estimated from tide-gauge data requires periods sufficiently long (more than 20 years) that meaningful short-term events cannot be delineated. Moreover, because tide gauges typically are located in ports and harbours, there are large parts of the coast for which no record is available. How representative of the Nile Delta is the tide- gauge record at Alexandria; said another way, what is local rise of sea-level 1, 20 or 100 km from Alexandria?

Because of the slow response time and site-specific nature of tide-gauge records, we need another type of measure of the absolute and local change in sea-level. We need to monitor short-term changes within a broad area, say within the entire coastal zone of the Nile Delta. Such measurements almost certainly will require integration of variations within a local grid of stations. A Global Positioning System (GPS), satellite altimetry or satellite laser ranging (SLR) system could correlate specific changes to a geodetic benchmark, an International Earth Rotational Service (IERS) Terrestrial Reference Frame (TRF), allowing one to monitor local changes in sea-level

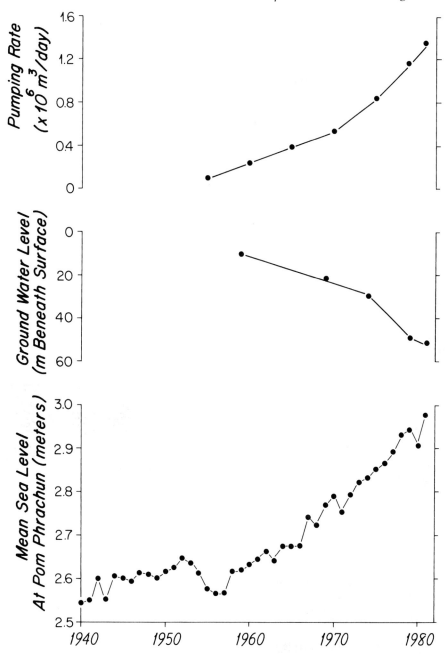

Fig. 2.5. Change in sea-level at the tidal station at Pom Phrachun (near Bangkok), 1940–1982. Increased pumping of ground water led to a drop in the water table and a corresponding rise in relative sea-level (from Milliman et al., 1989). Between 1940 and 1960, the rate of local sea-level rise was about 6 mm/yr (or a local subsidence of 4–5 mm/yr); between 1960 and 1980, the rate of sea-level rise increased to 22 mm/yr (local subsidence of about 20 cm/yr). In central Bangkok subsidence rates were as high as 13 cm/yr.

(Carter et al., 1989) over short time periods. Vertical changes throughout the grid (at relatively short temporal spacing) could provide an image of relative land motion. At present GPS resolution is less than 1 cm, meaning that vertical movement in many areas could be delineated within three to five years, and absolute movement in subsiding areas might be delineated within months.

Such systems are expensive and experimental. For example, the necessary GPS satellites will not be fully operational for several more years, at which time the vertical resolution may be improved. Such a grid, however, should be considered as early as possible to minimize the time required to respond to the adverse effects of local sea-level change.

8 REFERENCES

Aubrey, D.G. and Emery, K.O., 1983. Eigenanalysis of recent United States sea-level. *Cont. Shelf. Res.*, **2**, 21–33.

Carter, W.E., Aubrey, D.G., Baker, T., Boucher, C., LeProvost, C., Pugh, D., Peltier, W.R., Zumberge, M., Rapp, R.H., Schutz, R.E., Emery, K.O. and Enfield, D.B., 1989. Geodetic fixing of tide gauge bench marks. Woods Hole Oceanographic Institution Technical Report, WHOI–89–31 (CRC–89–5), Woods Hole, MA, p. 51.

Emanuel, K.A., 1987. The dependence of hurricane intensity on climate. *Nature*, **326**, 483–485.

Emery, K.O., Aubrey D.G. and Goldsmith, V., 1988. Coastal neo-tectonics of the Mediterranean from tide-gauge records. *Mar. Geol.*, **81**, 41–52.

Flemming, N.C., 1990. Predictions of relatively coastal sea-level change in the Mediterranean based on archaeological, historical and tide gauge data, Chapter 8, this volume.

Haq, B.U., Hardenbol, J., and Vail, P.R., 1988. Mesozoic and Cenozoic chronostratigraphy and cycles of sea-level change. *Soc. Econ. Paleont. Mineral. Spec. Publ.* **42**, 71–108.

Hoffman, J.S., Keyes D., and Titus, J.G., 1983. Projecting future sea-level rise. US Environmental Protection Agency, EPA 230–09–007, p. 121.

Meier, M.F., 1984. Contribution of small glaciers to global sea-level. *Science*, **226**, 1418–1421.

Meier, M.F., 1990. Reduced rise in sea-level. *Nature*, **343**, 115–116.

Milliman, J.D., Broadus, J.M. and Gable, F., 1989. Environmental and economic implications of rising sea-level and subsiding deltas: The Nile and Bengal examples. *Ambio*, **18**, 340–345.

Milliman, J.D. and Emery, K.O., 1968. Sealevels during the past 35,000 years. *Science*, **162**, 1121–1123.

Peltier, W.R. and Tushingham, A.M., 1989. Global sea-level rise and the greenhouse effect: Might they be related? *Science*, **244**, 806–810.

Pirazzoli, P.A., 1989. Present and near-future global sea-level changes. *Paleogeog., Paleoclim., Paleoecol.*, **75**, 241–258.

Raper, S.C.B., Warwick, R.A. and Wigley, T.M.L., 1990. Global sea-level rise: Past and future. In: Milliman, J.D., (ed.), *Sealevel Rise and Coastal Subsidence*, John Wiley, in press.

Sestini, G., 1990. Implications of climatic changes for the Po Delta and Venice Lagoon. Chapter 12, this volume.

Stanley, D.J., 1989. Subsidence in the northeastern Nile Delta: Rapid rates, possible causes and consequences. *Science*, **240**, 497–500.

Stewart, R.W., Kjerfve, B., Milliman, J.D. and Dwivedi, S.N., 1990. Relative sea-level change: A critical evaluation. Final Unpublished Report to Unesco/COMAR, p. 27.

Van de Plassche, O., 1986. Sea-level Research: A manual for the collection and evaluation of data. Geo Books, Norwich (UK), p. 618.

Wigley, T.M.L. and S.C.B. Raper, 1987. Thermal expansion of sea water associated with global warming. *Nature*, **330**, 127–131.

3
Hydrological and Water Resources Impact of Climate Change

G. Lindh
(Department of Water Resources, Lund University, Sweden)

Abstract

Among the societal consequences of climate change, those affecting the hydrological cycle and water resources are expected to be particularly serious. Moreover, salt-water intrusion due to sea-level rise may cause increased salinity in estuaries and aquifers. Excessive concentrations of sea water at water intakes could create public health risks, increase the cost of water treatment, and damage plumbing and machinery. More generally, an increased salt-water intrusion could upset the ecology of the coastal area.

This chapter gives a general overview of these societal and ecological implications followed by an account of the specific situation in the Mediterranean area. Present conditions as well as possible future conditions as regards hydrological characteristics and water resources are presented. Some suggestions are made for research activities needed to provide planners and decision makers with more reliable background material as a base for future actions.

1 Introduction

The regional impact of climate change on hydrology and water resources may have serious societal and ecological implications. The need to understand better climate change and hydrological processes is still more pressing in view of the fact that in many areas water availability is only marginally adequate to meet basic human needs.

Water plays an important role in societal development in the respect that it is needed for drinking purposes, but is also a necessary condition for the development of agriculture and industry. Some of the countries around the Mediterranean are already badly off, e.g. Libya and Tunisia (Hindrichsen, 1986). Also worldwide occurrence of droughts, which may be influenced by natural events, such as the severe El Niño event of 1982–1983, has to be kept in mind when discussing the effects on water resources due to global changes (UNEP, 1987).

2 Societal Consequences of Climate Changes

2.1 Effects of Salt-water Intrusion in Coastal Regions

Among the effects of rising sea-level (Titus, 1986) will be increased salinity in estuaries and aquifers. Excessive concentrations of sea water at water intakes could create public health risks, increase the cost of water treatment, and damage plumbing and machinery. Moreover, an increased salinity of intruding water could also upset the ecology of the coastal area.

Salt-water intrusion is not a new problem, as witnessed by the number of "salt-water intrusion meetings" held since 1968. Unesco has facilitated the analysis of the important contribution of the adverse effects of salt-water intrusion on groundwater in coastal areas (Custodio and Bruggeman, 1987). Today the consequences of the overexploitation of ground water from coastal wells are clearly understood. An interesting parallel to extreme withdrawal of water from wells is the effect of salinization in arid and semi-arid irrigation projects by ground-water extraction. In both examples, it has proved to be extremely difficult to renovate such salt-contaminated wells. Quantitative observations of salt-water depth in porous coastal aquifers were made already at the beginning of this century. From these observations, a formula was deduced for salt-water intrusion, a relationship now known as the Ghyben–Herzberg formula (Fig. 3.1), namely

$$(h_f + z) \gamma_f = z \gamma_s$$

$$z = \frac{\gamma_f}{\gamma_s \gamma_f} \quad h_f = \alpha h_f \tag{1}$$

where
  h_f = fresh-water head (water-table elevation over mean sea-level)
  z = depth of the interface below mean sea-level
  γf = fresh-water density = 1000 kg/m³
  γs = salt-water density = 1025 kg/m³
  α = $\gamma f / (\gamma s - \gamma f)$ = 40
  h_s = salt-water head

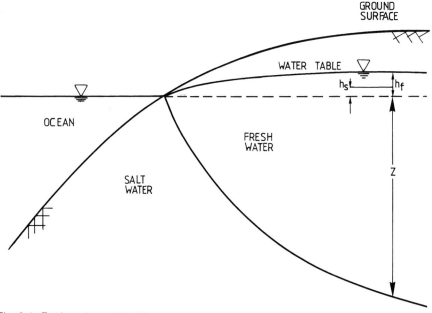

Fig. 3.1 Explanation of the Ghyben–Herzberg principle.

The Ghyben–Herzberg formula is a hydrostatic approach, with in fact some severe limitations (Custodio and Bruggeman, 1987). There have been many attempts to find more appropriate approaches to the salt-water intrusion problem than those presented by Ghyben–Herzberg, mainly in order to correspond to the dynamic processes involved. In order to account for the mixing of fresh and salt water, mainly through microscopic and macroscopic dispersion, a new relationship has been deduced, replacing the older formula. According to this hypothesis, salt-water is dragged along with fresh water towards the sea. In order to conserve the salt-water balance this implies a small landward salt-water flow in a coastal ground-water aquifer under equilibrium. Some considerations then lead to the so-called Hubbert's formula

$$z = \frac{\gamma_f}{\gamma_s - \gamma_f} h_f - \frac{\gamma_s}{\gamma s - \gamma_f} h_s = \alpha h_f - (1 + \alpha) h_s$$
$$= \alpha (h_f - h_s) - h_s \qquad (2)$$

With reference to Fig. 3.1, a sea-level rise may lead to a landward movement of the salt-water wedge. The hydrostatic formula will then lead to an increase in the ground-water table indicating a decrease in water availability. However, the only portion of an aquifer likely to be salty is that part below sea-level. In coastal aquifers the fresh water floats on top of the heavier salt water.

There is an apparent risk that salt water and fresh water mix, thereby causing contamination of the fresh water. Heavy withdrawal from wells also can cause the equilibrium salt-water line to move far inland. If pumping from wells is excessive, wells may be recharged by nearby rivers. As a consequence, the wells will be contaminated by salty water through infiltration into the aquifer. It has proved to be very difficult to wash away salt from such aquifers.

Salt-water intrusion can be predicted from fundamental equations that also consider the special geometry caused by inundation of river reaches (Fig. 3.2, Hull and Titus, 1986). The basic equations describing salt-water intrusion are the continuity equation

$$b \frac{\partial h}{\partial t} + \frac{\partial Q}{\partial x} - q = 0 \qquad (3)$$

the longitudinal momentum equation

$$\frac{\partial Q}{\partial t} + \frac{\partial (QU)}{\partial x} + gA \frac{\partial h}{\partial x} + g \frac{Ad_c \cdot \partial_\rho}{\rho \, \partial \xi}$$
$$+ g \frac{Q|Q|}{AC^2 \cdot R_n} = 0 \qquad (4)$$

the salt-balance equation

$$\frac{\partial (A_t S)}{\partial t} + \frac{\partial (QS)}{\partial x} = \frac{\partial}{\partial x} (A_t E \frac{\partial S}{\partial x}) \qquad (5)$$

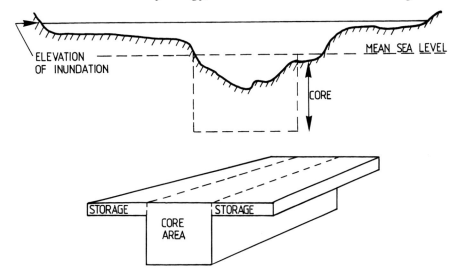

Fig. 3.2 Increase in width resulting from sea-level rise and a schematic sketch of the geometry.

and the equation of state, i.e. the relation between density, salinity and temperature

$$\rho = \alpha_T + \beta_T \cdot S \qquad (6)$$

where
- b = total surface width
- h = depth from water surface to horizontal datum
- Q = cross-sectional discharge
- q = lateral inflow per unit length
- U = longitudinal velocity averaged over the cross-section of the core area
- A = core area
- g = acceleration of gravity
- d_c = distance from the water surface to the centroid of cross-section
- ρ = mass density of water
- R_h = hydraulic radius of core area, including that of storage areas
- S = salinity averaged over the total cross-section
- E = the longitudinal dispersion coefficient

The quantities mentioned are functions of longitudinal location (x), and time (t). The coefficients αT and βT depend only on temperature.

In order to use the above theoretical approach for prediction, one has to start from information already known about the salinity content of a great number of wells in the region in question. The model has been applied for salt-water intrusion in the Delaware Estuary in order to calculate effects of a sea-level rise (Hull and Titus, 1986).

2.2 Influences on the Hydrologic Cycle

One of the most important consequences of the global climate change is the change in the hydrologic cycle, with further repercussions on the water resources. When we talk about consequences for the hydrological cycle we mean changes in the elements that interact within the hydrologic cycle, primarily precipitation, evapotranspiration, infiltration, surface runoff and soil-moisture content.

To study changes in the hydrologic cycle, various global circulation models have been used, particularly the so-called General Circulation Models (GCMs) (Manabe and Stouffer, 1980). However, insufficient knowledge about heat exchange between the atmosphere and the oceans may lead to dubious conclusions. Another part of the hydrologic cycle not well covered by the model is the role of vegetation as well as soil moisture. This is not surprising, because transport processes through the unsaturated and saturated zones are, as yet, not well understood. These rather simple remarks are, however, extremely important for the reliability of the global models, and in general it may be said that they are still rather crude.

Some of the studies using these global models, however, give some indication of changes in precipitation and evaporation that can be estimated from the changes in question. Mitchell (1983), for instance, gives maps showing simulated values of total precipitation, evaporation and soil-moisture on a global scale. In another study, Wilson and Mitchell (1987) have made a doubled CO_2 climate-sensitivity experiment with a global climate model including a simple ocean. The model used is an atmospheric general circulation model coupled to a 50 m oceanic mixed layer and an energy-balance sea ice model. The prescribed heat convergence in the ocean ensures that sea temperatures as well as seasonal variation of ice closely agree with observations. The results presented by Wilson and Mitchell comprise a series of maps showing soil moisture content, precipitation changes for a doubled CO_2 content, etc., from which it may be concluded that the representation of cloud cover is extremely crude and the dynamics of sea ice and changes in ocean dynamics are not taken into account. As a final conclusion, the authors claim that "they consider themselves still some way from predicting the magnitude and distribution of climate change due to increased atmospheric CO_2 with the accuracy required for climatic impact studies".

In a somewhat older study Washington and Meehl (1984) made use of a simplified ocean model, a simple slab of 50 m depth. Seasonal heat storage and release is included in the slab but not heat transport in the ocean. Also in this case a series of maps showing hydrological parameters are shown, with comments on their reliability.

2.2.1 Impacts on water resources

Looking at the impact on available water resources due to the global change, it is certainly a useful approach to consider how the hydrological cycle processes are affected by climate change. The idea behind this approach is explained schematically in Fig. 3.3.

Climatological variables influence the hydrological system, the output of which then serves as the input to the water resources system. This output

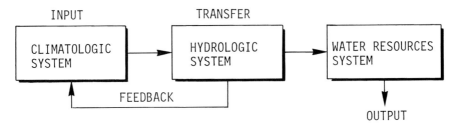

Fig. 3.3 Interaction between climatologic, hydrologic and water resources systems.

hopefully will indicate what changes are to be expected from variations in the climatological variables. If not for other reasons such an interacting process makes clear that there is a conceptual distinction between what we call hydrology and water resources. This fact has become more clear over the years, as water resource activities have been increased due to economics and civil engineering. Moreover, as is pointed out by Beran (1986), "while all natural water is of hydrological interest, only water that is available at the time and place required and in suitable quantity and quality, is of water resources interest".

In order to reveal possible impacts on the hydrological cycle, one could use the interaction systems shown in Fig. 3.3. The most straightforward procedure then would be to look first at the interplay between the meteorological and the hydrological systems. However, we do not have sufficient understanding of this interaction, as has been pointed out several times (cf., for instance EOS, 1988). In general, the GCMs are not yet so fully developed that they can be used as input to the hydrological system. It should be stressed, however, that the global circulation models cannot be developed without efficient cooperation between meteorologists and hydrologists.

For this discussion another approach seems promising, namely to start from the hydrological system itself, assuming certain fixed-input values to the system. This, for instance, could mean that we test the response of the hydrological system to fixed changes in precipitation, etc. This, however, would infer that the hydrological system itself is known. The most basic illustration of this system would then be the hydrological cycle process which schematically can be expressed in Fig. 3.4 (Lindh, 1979). The mathematical background of this cycle process is the water-balance equation (Sokolov and Chapman, 1974):

$$P = Q + E \qquad (7)$$

where
P = precipitation
Q = runoff
E = evapotranspiration

This equation, of course, can be subdivided to depict partial processes.

Several interesting papers confirm that the water-balance model can be used to assess the consequences of the global change impact on the hydrological system. I would like to mention some of them.

64 *Climatic Change in the Mediterranean*

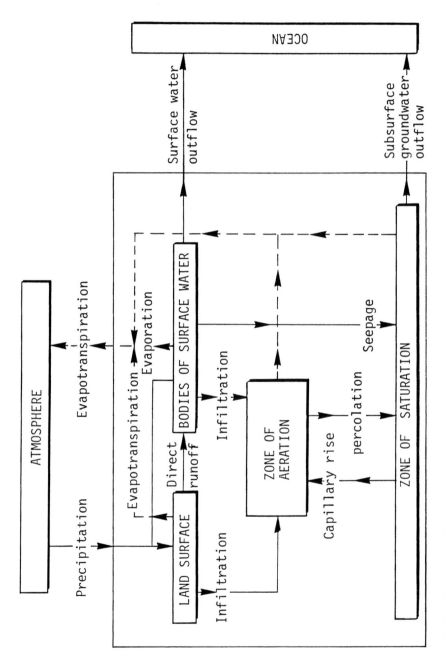

Fig. 3.4 Schematic illustration of the hydrological cycle.

Let me start with a study of the Sacramento Basin (Gleick, 1986, 1987a, 1987b), the first of which makes use of a modified water-balance method to evaluate the regional hydrological impacts of global changes. Gleick begins by quoting the US National Academy of Science from 1979, "At present we cannot simulate accurately the details of regional climate and thus cannot predict the locations and intensities of regional climate changes with confidence. This situation may be expected to improve gradually as greater scientific understanding is acquired." Gleick's comments that the situation is only marginally better in 1986, and he expresses the view that even current climatic assessments are limited to a few global consequences, such as changes in average or zonal temperatures and precipitation. But this is not sufficient knowledge if the intention is to determine a regional or national policy. He also stresses that General Circulation Models (GCM) are of limited value because of their complexity and because the resolution of these models is coarse compared to hydrological events. However, he admits that the models *per se* may indicate important changes in, for instance, precipitation. One such indication is the suggestion of a major reduction in summer soil-moisture patterns in the middle latitudes, a possible consequence of a northward shift in the mid-latitude rain belt and an earlier onset of winter snowmelt and spring runoff (Manabe and Wetherald, 1986). Very important also are the relatively small changes in regional precipitation and evaporation patterns that might result in a significant change in regional availability. Gleick points out that once a region has been characterized in terms of water balances, the effects of climatic change can be evaluated in three ways, namely:

1) After having verified the accuracy of the model by using long-term historical data it should be possible to use historical data to evaluate the effects of historical fluctuations in precipitation and temperatures on historical runoff and soil moisture.
2) After determining the sensitivity of runoff and soil moisture to theoretical changes in the magnitude and temporal distribution of precipitation and temperature, a wide range of hypothetical climatic changes can be assessed to evaluate the hydrological sensitivity of the watershed.
3) By incorporating even rough, regionally disaggregated changes in temperature and precipitation predicted by General Circulation Models, a first estimate can be made of the impacts of future predicted climatic change on regional hydrology.

Following this paper, Gleick (1987a, 1987b) has applied his ideas to the Sacramento basin, California, USA. For this basin he developed a water balance model, and as input to this model he used 18 widely varying climate-change scenarios designed to evaluate the impacts of global climatic changes on runoff and soil moisture in the Sacramento Basin. Four particularly important and consistent changes were observed:
- large decreases in summer soil-moisture levels for all scenarios;
- decrease in summer runoff for all scenarios;
- major shifts in timing of average monthly runoff throughout the year;
- large increase in winter runoff volumes for almost all of the scenarios.

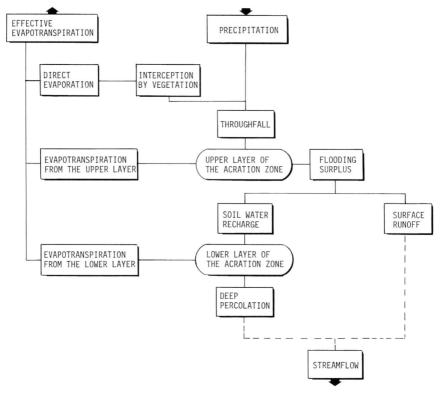

Fig. 3.5 Flow chart of the upper part of the IRMB model (Bultot and Dupriez, 1976a, 1976b).

Gleick concludes that these changes may have serious implications for any aspects of water resources if they materialize.

In a series of papers Bultot and Dupriez (1976a, 1976b) considered a conceptual, hydrological model for an average-sized catchment. This model is a complete description of the hydrological system, including transport process as well as the storing of water at various positions in the system (cf., Fig. 3.5). It aims to calculate the outflow for given inputs. With this model, data from the period 1966–1972 have been used to determine the parameters used in the model description. The model also has been used to forecast the outflow of the system. As a complement to these studies, Bultot *et al.* (1988) have made a special study of the estimated annual regime on energy-balance components, evapotranspiration and soil-moisture for a drainage basin in the case of a CO_2-doubling. This study was a preliminary step to a more detailed impact analysis dealing with three Belgian river catchments with very different features.

In a study by Aston (1984), particular interest has been devoted to the role of vegetation and evapotranspiration to develop a special hydrologic model called SHOLSIM. It simulates a large catchment area with a number of varying land uses. Four computer simulations were undertaken in order to illustrate the maximum effect that CO_2 enrichment can have on large-scale catchment hydrology. Some results seem to be very encouraging for the part of Australia studied.

A good study of modelling catchment response is the paper by Ward (1984), containing a rigorous process-based catchment model. Precipitation and potential evapotranspiration are used as input variables. Outputs consist of hourly or daily values of actual evapotranspiration, soil moisture status and streamflow.

Over the years, a series of conceptual and physically based models have appeared. One very detailed and thoroughly used model is the one by Abbott et al. (1986). The model, "SHE", is a physically based, distributed system jointly produced with the Danish Hydraulic Institute, the British Institute of Hydrology and SOGREAH (France). The model consists of different components:
- the ARAME component, which coordinates the parallel running of other components;
- the interception and evapotranspiration component;
- the overland and channel flow component;
- the unsaturated zone component;
- the saturated zone component;
- snowmelt component.

According to the scientists involved in the project, it is relatively easy to add other components to the model.

Wigley and Jones (1985) made a special study of the hydrological cycle response to changes in precipitation and direct CO_2 effects on streamflow. They also started by considering the fundamental water-balance operation, valid for small non-evaporative losses, namely

$$Q = P - E$$

where Q is runoff and P and E precipitation and evaporation respectively. The runoff coefficient k is related to Q and P through

$$Q = kP$$

In order to estimate changes in runoff due to changes in precipitation and evapotranspiration, the authors denote the change in precipitation from P_o to P_1 due to a doubling of atmospheric CO_2 by

$$P_1 = a P_o$$

where a indicates changes in evapotranspiration as explained in the text below.

Changes in evapotranspiration will occur for two reasons. One is due to a climatic change, the other through direct CO_2-induced change in vegetation. The authors claim that runoff is more sensitive to precipitation changes than to evapotranspiration changes, particularly for higher values of the runoff coefficient. They also state that precipitation changes have

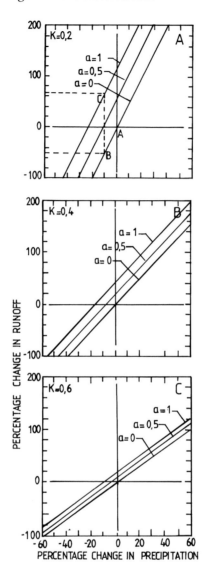

Fig. 3.6 Runoff changes due to changes in precipitation (P) and evapotranspiration (E).
a = corresponds to no change in E.
a = corresponds to 30% reduction in E
(maximum direct CO_2 effect).
Three values of runoff coefficient K are shown.
Precipitation change alone is determined by a = 0.
Thus, the line AB shows the effect of a 10% reduction in P.
The line BC corresponds to a range of E reduction from 0% (point B)
to 30% (point C).
(From Wigley and Jones, 1985.)

an amplifying effect on runoff, especially in arid regions where the runoff coefficient is small. A list of runoff ratios for selected major rivers in the world shows that these values range from 0.1 or less for rivers flowing through arid regions, up to more than 0.6 for some tropical rivers. From the list it is evident that in temperate latitudes runoff coefficients around 0.4 are fairly typical. In Fig. 3.6, we see that when $a = 0$ (no change in evapotranspiration) there is no direct effect. When $a = 1$ (30% reduction in evapotranspiration), the direct effect corresponds to a maximum. Changes in evapotranspiration due to climatic changes can also be read off since the range $a = 0$ to $a = 1$ in fact corresponds to a range of reductions in evapotranspiration from 0 to 30%.

Wigley and Jones show how the curves can be used to calculate changes in future runoff due to changes in precipitation and evapotranspiration given a certain value of runoff for a catchment area. As a result, Wigley and Jones' studies make it possible to make a first estimate of annual mean runoff changes for any river where the runoff ratio is known. Moreover, it is possible to identify regions that most likely will show large runoff changes due to increasing CO_2.

A special study applied to American conditions is presented by Rind (1988) using results from the Goddard Institute for Space Studies' (GISS) General Circulation Model to investigate the consistency of changes in water availability over the United States.

At the same time, the author provides the reader with a critical survey of global models in general. From the result derived by the model it can be mentioned that the climate of the northern and western regions of the United States and southwestern Canada will become wetter, particularly in winter and spring; the southern and eastern parts may become drier. The chapter concludes with the comment that "it is unlikely that climate simulations of sufficient length to differentiate signal from noise will be made in the near future". Rind thinks that at least a decade is needed to develop better ground hydrology and ocean models in order to fit in the General Circulation Models. And he adds: "From this perspective,
'consensus' results from current GCMs may represent risky projections of future water availability. Nevertheless, the problem is of such great importance, and the planning of water resources needs such a long lead time, that every attempt should be made to reduce the uncertain ties as quickly as possible."

Recently another approach in studying details of a general circulation model was published by Koster et al. (1988). The main problem was to investigate lateral distances that water vapour generally travels while in the atmosphere. Or, stated in another way, the problem is to answer what fraction of the precipitation in a given region is derived from water evaporating from the region itself? It is quite evident that answers to these questions cannot be derived from direct measurements in nature. The method used in this study was to incorporate in the GISS model a tracer water isotope model. The three most important water isotopes have the same chemical properties as normal water, but have slightly different physical properties due to their higher mass. The two physical properties affecting water isotope transport are vapour pressure and molecular diffusitivity. The tracer water model was especially applied to determine the evaporative sources of precipitation falling on representative regions in the Northern Hemisphere for the CGM climate. A very interesting result of this study is the apparently larger degree of water recycling over mid-latitude continental regions than is generally estimated in the literature.

A study by Shukla (1982) indicates the importance of the influence of land-surface evapotranspiration on the earth's climate. A numerical model of the atmosphere has been used to show the global fields of rainfall, temperature and motion strongly dependent on the land-surface evapotranspiration. The short paper includes a series of maps showing simulated precipitation, surface temperature and surface pressure on a global scale.

Last but not least, the study of Olejnik (1988) may be mentioned. His studies are based on several models presented earlier in the literature. A series of maps are presented showing variations in the hydrological parameters studied for Europe.

2.2.2 Consequences of runoff in urban areas

Possible effects on the hydrologic system due to climate changes discussed earlier in this chapter have excluded the effects in urbanized areas. The sea-level rise may cause inundations that may render gravity drainage systems useless or even require that they be shut off to prevent sea water from backing up into the community (Titus et al., 1985). Increased precipitation may increase flood frequencies. It is difficult to know if this would mean more frequent storms or more rainfall during the same number of storms. In order to meet an increased precipitation, larger pipes or wider drainage channels can be used. Where drainage systems already are in place, a supplement pipe system can be installed or older pipes may be replaced with larger ones. Communities that rely on gravity drainage systems may have to shift to forced drainage when sea-level rises; If force drainage is already used, larger pumps may be necessary to work against the higher water head. Moreover, a larger capacity is certainly needed to handle more water because less water is taken care of by natural drainage. One other interesting aspect in managing urban and suburban drainage is the increased use of detention basins to control surface runoff as sea-level rises or when precipitation increases.

One example may be taken from the City of Charleston where 70% of the existing storm-water drainage facilities were undersized. Total cost for construction of the recommended improvements was estimated at $135 million (Laroache, and Webb, 1985). One way of meeting the new requirements was to change the return period of design storm from a five-year storm to a 10-year return frequency. In fact, calculations for reconstruction of the storm-water drainage system are perhaps not so difficult. The difficulty lies in making decisions and creating consensus about new plans and being aware of the increased cost of replacing or supplementing the system. This can be expensive, but it is better than waiting too long. One may express it also in the way that the sea-level rise and increased precipitation may involve a degree of uncertainty. One way to reduce the uncertainty is to plan for potential future changes (Waddell, 1985).

3 Implications for the Mediterranean Area

3.1 Introductory Remarks

In this chapter I will look particularly at the consequences of a changing global climate upon the hydrological cycle that might be expected in the Mediterranean area. However, in presenting such consequences I am forced to be more general than specific, because of the lack of detailed data and information. One example of the difficulties is the consequence of increased precipitation on traditional storm-water management. An increase of 10–20% in storm water runoff volume, for example, cannot be discussed without accurate information about the capacity of the drainage system in question. Thus my intention is to indicate where problems may show up in the Mediterranean as well as the scale of such problems.

3.2 Hydrological Characteristics of the Mediterranean Area

3.2.1 Present conditions

The main hydrological features of the Mediterranean region are rather well known from earlier studies. A good source of information is the excellent book, published by Unesco (1978) (Figs 3.7–3.13). From there, the following observations can be made:

1) There is a rather clear topographic dependence on precipitation (Fig. 3.7). A minimum is to be found along the Spanish east coast, whereas rather high values of precipitation occur along the Moroccan, Algerian, Yugoslavian and Greek coasts. One could summarize the precipitation situation by saying that along the Spanish east coast the precipitation is 200–400 mm/yr, along the coasts of the Ligurian Sea 800–1200 mm/yr, and along the Yugoslavian coast more than 1500 mm/year. In the southern part of the Mediterranean region, precipitation generally averages 100–400 mm/year. According to Henry (1977), three to seven months of drought occur along the Spanish east coast (e.g. Fig. 3.8). Along the coasts of the Ligurian Sea and the Italian west coast, two or three months may be dry, along the Yugoslavian coast one or two months, and on the Greek coast four or five months may be dry. In the southern part of the Mediterranean, four to seven months or perhaps more will be dry. In keeping with the Mediterranean climate, precipitation clearly shows a minimum during summer, with the exception of areas north of the 45° latitude, where seasonal precipitation is more evenly distributed.

2) Information about potential evapotranspiration is more sparsely documented because it is very difficult to determine. To do so we must assume sufficient precipitation or ground water to allow an evapotranspiration from the ground as well as transpiration from the vegetation. It is easier to comment on the annual evapotranspiration and its seasonal distribution (Figs 3.9 and 3.10). It varies somewhat along the Mediterranean, but along the northern coast it seems to be around 400–600 mm/yr, with a maximum during the first part of the year.

3) Mean annual runoff varies considerably throughout the Mediterranean (Fig. 3.11). Along the Spanish east coast it amounts to about 100 mm/yr, whereas along the coasts of the Ligurian Sea values increase

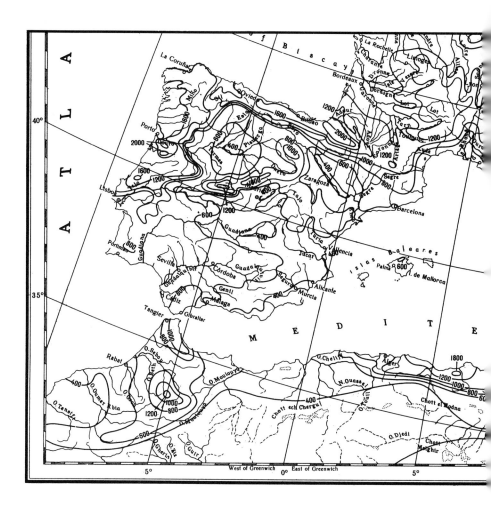

Fig. 3.7 Annual precipitation in mm (Unesco, 1978).

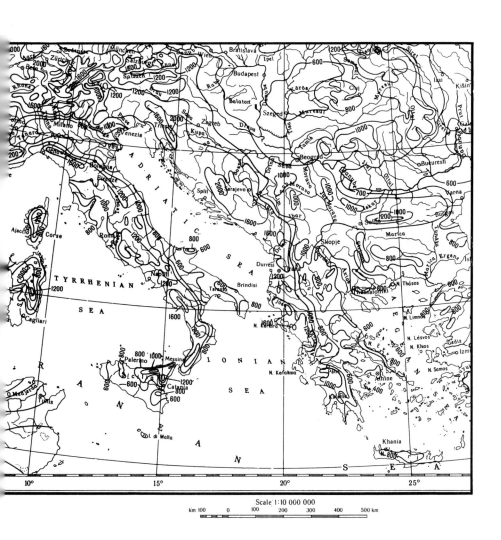

74 *Climatic Change in the Mediterranean*

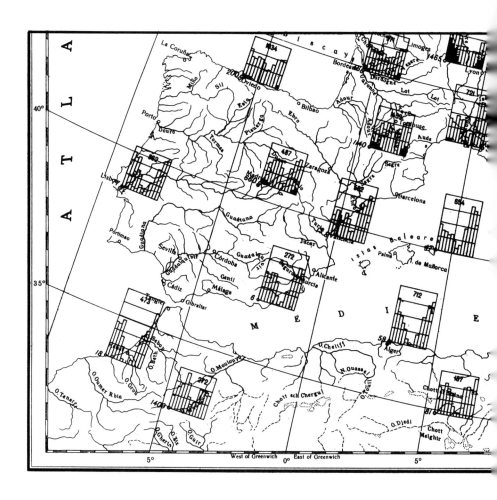

Fig. 3.8 Monthly distribution of precipitation (Unesco, 1978).

Hydrology, water resources and climatic change

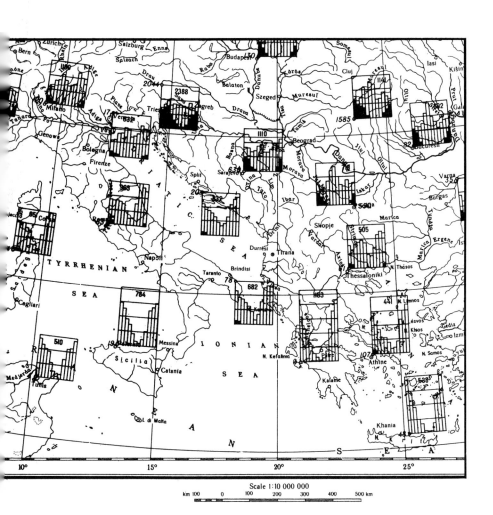

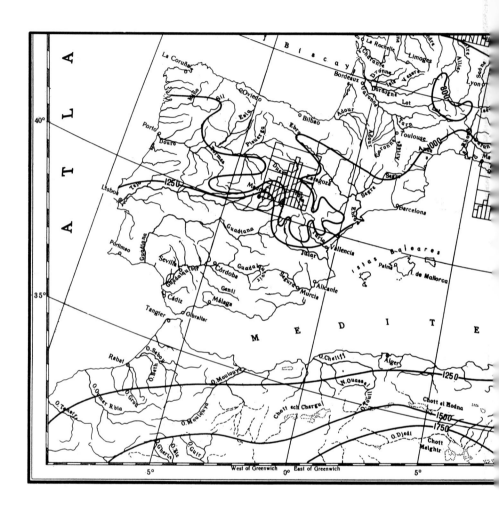

Fig. 3.9 Annual potential evapotranspiration in mm (Unesco, 1978).

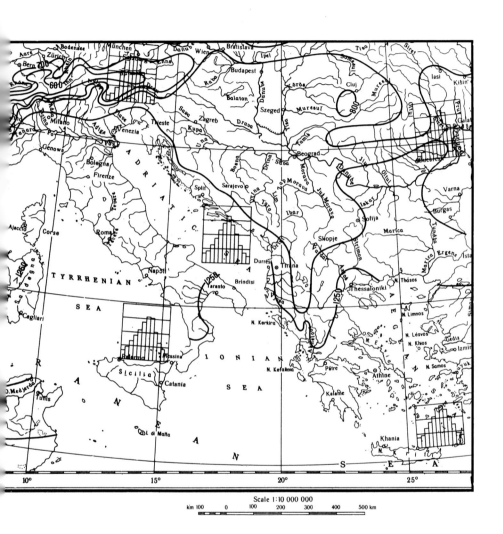

78 *Climatic Change in the Mediterranean*

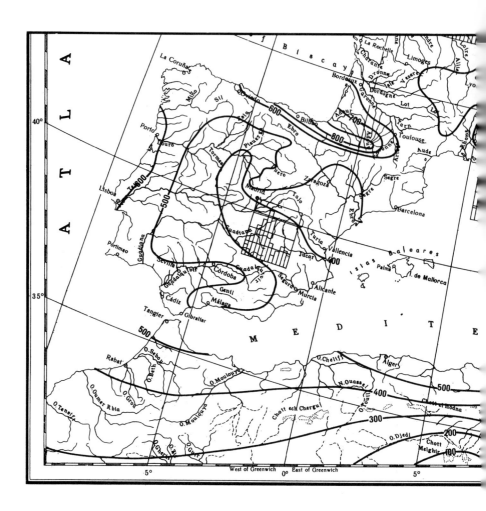

Fig. 3.10 Monthly evapotranspiration and its annual distribution (Unesco, 1978).

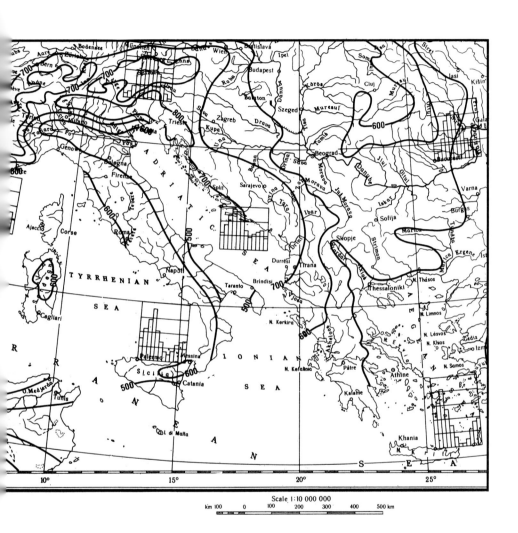

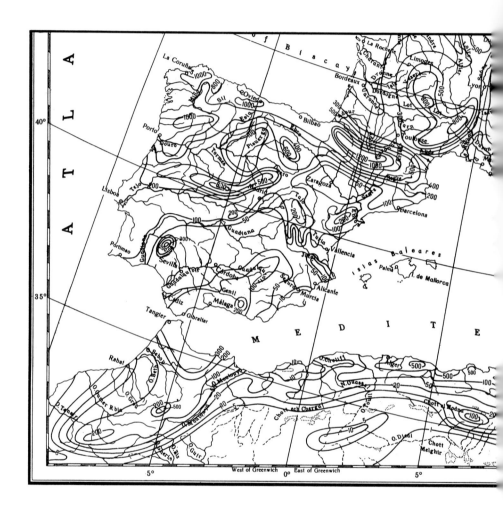

Fig. 3.11 Mean annual river runoff in mm (Unesco, 1978).

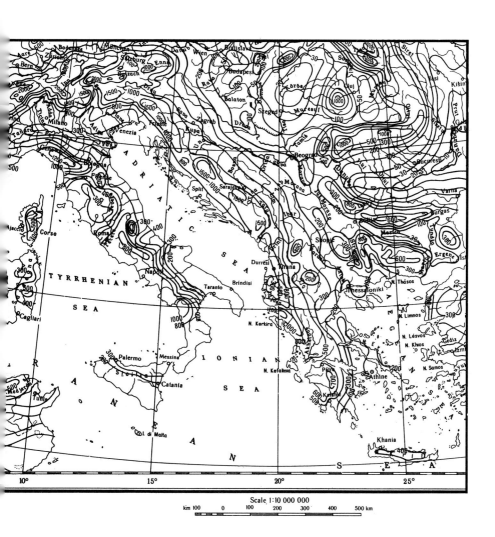

82 *Climatic Change in the Mediterranean*

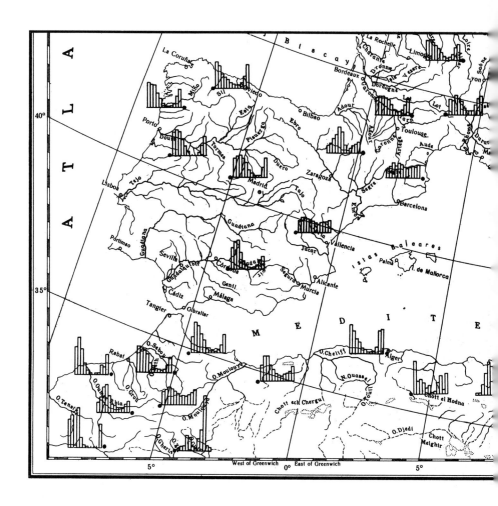

Fig. 3.12 Monthly distribution of runoff (Unesco, 1978).

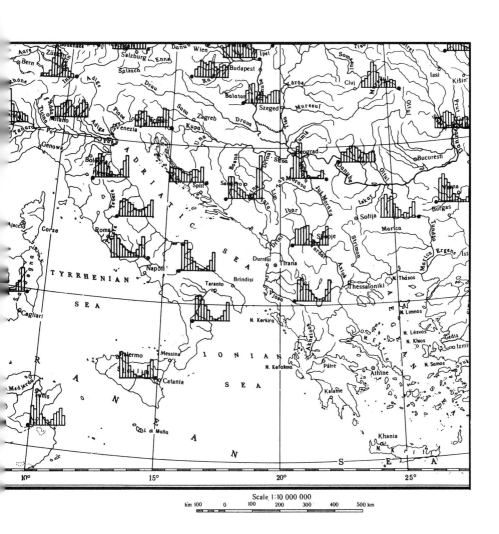

84 *Climatic Change in the Mediterranean*

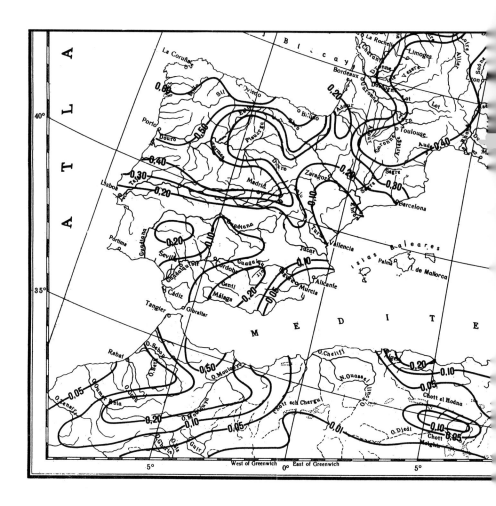

Fig. 3.13 Runoff coefficients (Unesco, 1978).

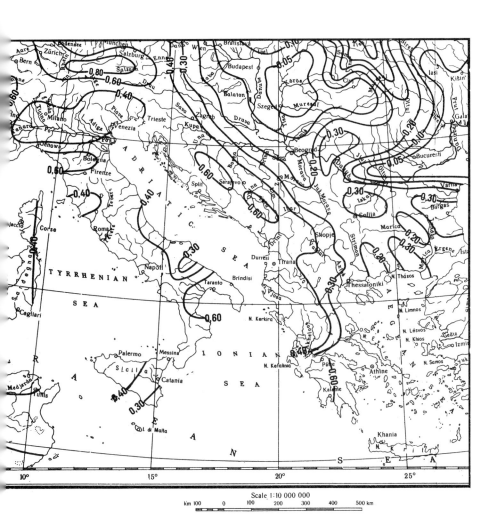

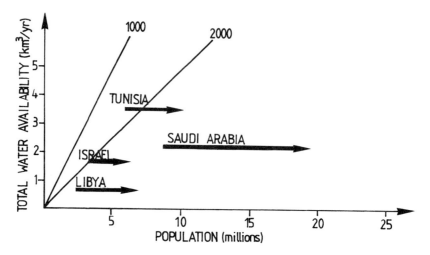

Fig. 3.14 Total water availability per country in relation to population growth from 1975 to 2000. The diagonal lines represent the number of persons to be supplied by one flow unit ($10^6 \times m^3 \times yr^{-1}$ of water; Falkenmark, 1986).

to 400–500 mm/yr. In the Adriatic Sea it is around 300 mm annually, but with somewhat higher values along the western part of the sea. Along the southern coast of the Mediterranean the runoff is highly variable. For instance, near Tunis, it amounts to only 100 mm/yr. A glance of the monthly distribution of runoff reveals a general tendency towards high annual runoff at the beginning and end of the year, coincident with seasonal precipitation (see Fig. 3.12). North toward the 45° latitude, the annual distribution becomes more evenly distributed.

4) Runoff coefficient is defined as the ratio between quantity of water in runoff and quantity of water precipitated. In the Mediterranean this coefficient is about 0.5 to 0.6, except in southern Spain and in Tunisia, where it amounts to only 0.1 on average (Fig. 3.13).

5) Surplus and deficit river-water resources are important parameters as they are measures of water availability. Moisture surplus can be calculated approximately from the difference between precipitation and maximum potential evapotranspiration. According to Unesco (1978) the deficit ranges from 200 to 600 mm/yr in many parts of the Adriatic, southern

Table 3.1 The absolute and relative changes in precipitation (P), evapotranspiration (ET) and runoff (R). (The values refer to data within latitudes 35° to 45°.)

	P	ET	R
Present	726	583	141
Future	807	620	187
Change	81	37	46
	(+11%)	(+6%)	(+33%)

regions of France and along the Spanish east coast. In southern Spain values up to 1000 mm/yr have been calculated.

3.2.2 Future conditions

The information presented here agrees rather well with that given by Oljenik (1988). His study, moreover, is interesting because he also has tried to calculate the future situation, assuming a doubling of atmospheric carbon dioxide. From information about changes in precipitation, based on the GISS model, Oljenik calculates evapotranspiration and runoff for the whole of Europe. In determining the evapotranspiration (ET) he starts from the Morton model (Morton, 1983) with modifications by Kowacs (cf. Oljenik, 1988). In order to determine the runoff, Oljenik uses the water-balance method:

$$R = P - ET$$

for long periods and large areas.

Table 3.2 Utilization of surface water and ground water in some countries of the Mediterranean area in per cent and total available water resources in m³ per inhabitant and year. (Ennabli, 1982)

Country	Withdrawal (%)		Totally available resource m³/inhabitant/year
	Surface water	Ground water	
Italy	87	13	2990
France	50	50	3400
Spain	70	30	2900
Algeria	84	16	1490
Syria	65	35	590
Morocco	76	24	1720
Lebanon	30	70	1480
Cyprus	63	37	1560
Malta	5	95	83
Tunisia			580
Libya			290

Briefly his results are presented and discussed in Table 3.1. In the northern part of the Mediterranean there will be a small increase in precipitation and a rather high increase in evapotranspiration as compared to other parts of Europe. Nearly half the water connected to an increase in precipitation would be evaporated. Although the increase in runoff will be only 46 mm/yr, water deficit in this part of Europe may be reduced, which has strong implications for future agriculture and general land use (Oljenik, 1988). However, we must bear in mind that these conclusions refer to changes on a yearly basis. One must also consider small changes in timing or magnitude of seasonality, which can have important consequences (Gleick, 1987c). Severe stresses, for instance, may be placed upon agriculture if reservoir systems cannot store seasonal precipitation

88 *Climatic Change in the Mediterranean*

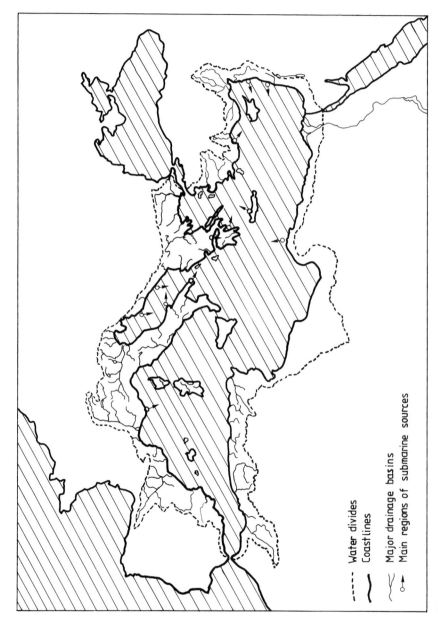

Fig. 3.15 Drainage basins around the Mediterranean (Ambroggi, 1977).

and runoff for later distribution. It should be mentioned, however, that it is difficult to know how useful Oljenik's data may be, as all his results neglect temperature changes.

3.3 Water Resources, Present and Future

As stated earlier, the close relationship between hydrology and water resources means that changes in hydrological parameters directly or indirectly affect influence in water availability for different reasons. One approach is to document those activities that depend on water in the Mediterranean area (cf. WMO 1987). Such activities include:
- water treatment and domestic drinking supply;
- water supply for industry, irrigation and power production;
- agricultural water use;
- sewage, sewage treatment, effluent disposal and dilution;
- land drainage and flood protection;
- navigation;
- fisheries, conservation and recreation.

Regarding domestic water use, Ambroggi (1977) mentions that in the northern Mediterranean there were in 1975 about 2000 m^3 of water per year available for each inhabitant, of which about 400 m^3/year were used. In the southern Mediterranean, for each inhabitant there were about 100 m^3/yr (see Fig. 3.15). Ennabli's (1982) estimated water resources and withdrawals (Table 3.2) do not correspond well with those of Ambroggi, because Ennabli's figures refer to the country as a whole, not just the Mediterranean part.

Several factors may jeopardize domestic water supply. Obviously, one relates to the increased population and degradation of the water resources (Fig. 3.14). Some experts contend that once that number of individuals increases above 2000 individuals per flow unit the country is likely to suffer from inherent water deficit problems. Israel, Libya and Tunisia probably have passed, and by the year 2000 the situation will become worse as almost 4000 individuals will have to share one flow unit.

Another problem concerns sea-level rise and the accompanying salt-water intrusion, which may cause extensive damage to surface as well as ground-water resources. The ground-water reservoirs may have connections with the sea via coastal and submarine springs that can activate salt-water intrusion. It is not possible, without detailed geological information as well as estimated sea-level rise, however, to calculate the effect of salt-water intrusion. In Italy there exist several vulnerable aquifers in Calabria, Puglia, Liguria, Marche, Abruzzo, Sicily and Sardinia, and the extent of the saline intrusion is already quite variable, from a few hundred metres to several kilometres (Zavatarelli, 1988).

Salt-water intrusion also can occur via rivers. The penetration length of such a salt-water wedge can depend on the extent of the sea-level rise, the slope of the river bed as well as other factors such as river discharge and tide. It would not be surprising, however, if a salt wedge could penetrate 10–50 km, at least seasonally. The salt-water intrusion phenomenon is not a new problem in itself for the Mediterranean countries, as seen in the tidal effects. However, a sea-level rise, caused by the greenhouse effect, will involve many secondary effects, such as change in agriculture and agricultural water use.

The sea-level rise in itself will have many deteriorating effects. The primary effect would be an inundation. Not only beaches may be spoiled but also lagoons, salt marshes, wetlands and brackish lakes, such as, for instance, Orbetello lagoon and Burano lake on the Tyrrhenian coast. This will mean that the many productive ecosystems will be damaged or destroyed. The biological production of these areas in the Mediterranean is presently highly utilized by terrestrial animals and coastal fisheries.

Other problems would accompany an increase in precipitation, particularly in coastal cities. An increase in precipitation may result in a more than proportional amount of runoff. This is because green areas may rapidly come to act as impervious areas because of clogging and temporary saturation of the uppermost layer. Whether a storm water drainage system and the treatment plants can manage this, of course, depends on local conditions. Because of the poor condition of coastal waters, the situation may be aggravated.

4 SUGGESTED RESEARCH ACTIVITIES

1) In order to learn more about the importance of evaporation and transpiration from the vegetation due to changes in carbon dioxide and other tracer gases, it is of paramount importance to study in more detail the behaviour of plant stomata in the regulation of carbon dioxide uptake and water transpiration. Since evapotranspiration plays an important role in any hydrological model aimed at depicting the series of processes occurring in the hydrological cycle, the lack of knowledge of the evapotranspiration behaviour will prove an obstacle in combining general circulation models and hydrological models.

It is of utmost urgency to integrate information from different disciplines to assess the role of vegetation in the hydrological cycle. Thus, more efficient cooperation must be established between ecophysiologists, plant ecologists and hydrologists as well as climate modellers to develop better models that describe just the importance of vegetation. Moreover, one has to focus more on the problems of different spatial and temporal scales; scaling issues will be crucial in any effort to adequately describe the role of vegetation in the GCMs (Rosswall and Ojima, 1988). The lack of knowledge is apparently very great, and in this cumbersome situation various ideas have to be tested. One way may be to make use of advances in remote sensing in order to estimate evapotranspiration on a regional basis. However, it is of paramount importance when developing a methodology for regional prediction of hydrological processes that a proper scale of resolution is adopted.

The problem of the role of vegetation is, of course, closely related to the evapotranspiration. However, this phenomenon is also related to precipitation. Consequently, in solving the role of vegetation we also need information about the regional precipitation pattern. Unfortunately, our knowledge of precipitation changes due to the greenhouse effect is not well developed. In fact, the problem of vegetation is clearly more difficult than could be imagined at first sight.

The problems just mentioned are not only characteristic of the Mediterranean region; they have to be solved in cooperation with scientists from other regions.

2) One interesting and important problem would be to investigate how river runoff is affected by changes in precipitation and evapotranspiration (cf. Nemec, 1985). Such studies should be initiated in several river basins in the Mediterranean region.

3) While waiting for more useful regional and local data for hydrological parameters, a series of studies could be initiated as sensitivity tests. For instance:
– How changes in precipitation quantities could affect the storm-water runoff at a particular site, what and when countermeasures should be undertaken. In order to perform such a study one needs to understand how an urban storm-water drainage works. Based on such information one has to apply for instance the Storm Water Management Model or a similar model in order to calculate the effects of an increased amount of precipitation.
– What effects the intrusion of salt-water may have at specific sites in the Mediterranean. One could also investigate the expected effects of a barrier designed to prevent or alleviate the salt-water intrusion, (cf. OMVS, 1986; WW, 1988).
– It would be interesting to estimate the costs incurred by the greenhouse effect. For instance, what will be the economic losses when sea-level rise causes additional erosion, inundation and loss of sandy beaches. This must be an urgent task for planners and decision makers in regions of the Mediterranean where tourism is an important source of income. Another problem to consider would be calculating the costs for reconstruction of harbours and other coastal facilities due to a sea-level rise.

5 REFERENCES

Abbot, M.B. et al., 1986. An introduction to the European hydrological system – systeme hydrologique europeen, "SHE", 2: Structure of a physically based, distributed modelling system. *J. Hydrol.*, **87**, 61–77.

Ambroggi, R. P., 1987. Ressources en eau douce du Bassin Méditerranéen. *Ambio*, **6** (6).

Aston, A.R., 1984. The effect of doubling atmospheric CO_2 on streamflow: a simulation. *J. Hydrol.*, **67**, 273–280.

Beran, M., 1986. Impact of future climate change and variability. In: Titus, J. G. (ed.), *Effects of changes in stratospheric ozone and global climate*. US Environmental Protection Agency/UN Environment Programme.

Bultot, F. and Dupriez, G.L., 1976a. Conceptual hydrologcal model for an average-sized catchment area, 1. Concepts and relationships. *J. Hydrol.*, **29**, 251–272.

Bultot, F. and Dupriez, G.L., 1976b. Conceptual hydrological model for an average-sized catchment area, 2. Estimate of parameters, validity of model, applications. *J. Hydrol.*, **29**, 273–292.

Bultot, F., Dupriez, G.L. and Gellens, D., 1988. Estimated annual regime of energy-balance components, evapotranspiration and soil moisture for a drainage basin in the case of a CO_2 doubling. *Climate Change*, **12**, 39–56.

Custodio, E. and Bruggeman, G.A., 1987. Groundwater problems in coastal areas. *Studies and Reports in Hydrol.*, No. 45, Unesco, Paris.

Ennabli, M., 1982. Etude sur les ressources en eau dans le bassin Méditerranéen. *Planbleu*.

EOS, 1988. Contribution of geophysics to climate changes studies. EOS, May 17, p.602.

Falkenmark, M., 1986. Fresh water: time for a modified approach. *Ambio*, **15**, 192–200.

Gleick, P.H., 1986. Methods for evaluating the regional hydrologic impacts of global climate changes. *J. Hydrol.*, **88**, 97–116.

Gleick, P.H., 1987a. Regional hydrologic consequences of increases in atmospheric CO_2 and other trace gases. *Climate Change*, **10**, 137–161.

Gleick, P.H., 1987b. The development and testing of a water balance model for climate impact assessment: modelling the Sacramento basin. *Water Res. Research*, **23** (6), 1049–1061.

Gleick, P.H., 1987c. Global climatic changes and regional hydrology, impacts and responses. IAHS publication No. 168.

Henry, P.-M., 1977. La Méditerranée: un microcosme menacé. *Ambio*, **6** (6).

Hindrichsen, D. (ed.), 1986. *World Resources 1986*. Basic Books, Inc, New York.

Hull, C.H.J. and Titus, J.G., 1986. Response to salinity increases. In: Hull and Titus, J.G. (eds), *Greenhouse Effect, Sea-Level Rise, and Salinity in the Delaware Estuary*. US Environment Protection Agency/Delaware River Basin Commission, EPA 230–05–86–010.

Koster, R.D., Eagleson, P.S. and Broecker, W.S., 1988. Tracer water transport and subgrid precipitation variation within atmospheric general circulation models. Department of Civil Engineering, M.I.T., Report No. 317.

Laroache, T.B. and Webb, M.K., 1985. Impact of sea-level rise on stormwater drainage systems in the Charleston, South Carolina area. US Environment Protection Agency.

Lindh, G., 1979. Socio-economic aspects of urban hydrology. *Studies and Reports in Hydrology*, Unesco, Paris, **27**.

Manabe, S. and Stouffer, R. J., 1980. Sensitivity of global models to an increase of CO_2 concentrations in the atmosphere. *J. Geophys. Res.*, **85**, C10, 5529–5554.

Manabe, S. and Wetherald, R. T., 1986. Reduction in summer soil wetness induced by an increase in atmospheric carbon dioxide. *Science*, **232**, 626–628.

Mitchell, J.F.B., 1983. The hydrological cycle as simulated by an atmospheric general circulation model. In Street-Perrott, Beran and Ratcliffe (eds), *Variation in the Global Water Budget*, D. Reidel Publishing.

Morton, F.I., 1983. Operational estimates of evapotranspiration and their significance to the science and practice in hydrology. *Hydrol.*, **66**, 1–76.

Nemec, J., 1985. Water resource systems and climate change. In: Rodda (ed.), *Facets of Hydrol., Vol. II*, John Wiley and Sons.

Olejnik, J., 1988. Present and future etimates of evapotranspiration and runoff for Europe. Working Paper WP–88–037, IIASA, Laxenburg.

OMWS, 1986. Miracle or white elephant? – West Africa, 666–667.

Rind, D., 1988. The doubled CO_2 climate and the sensitivity of the modelled hydrologic cycle. *J. Geophys. Res.*, **93**, No. D5, 5385–5412.

Rosswall, Th. and Ojima, D., 1988. Hydrology and climate change – A Need for Transdisciplinary Research. IGBP Secretariat, Royal Swedish Academy of Sciences, Stockholm.

Shukla, J., 1982. Influence of land-surface evapotranspiration on the earth's surface. *Science*, **215**, 1498–1500.

Sokolov, A.A. and Chapman, T.G., 1974. Methods for water balance computations. *Studies and Reports in Hydrol.*, No. 17, Unesco, Paris.

Titus, J.G., 1986. The causes and effects of sea-level rise. In: Tituts, J.G. (ed.), *Effects of Changes in Stratospheric Ozone and Global Climate*, US Environmental Protection Board/UN Environment Programme.
Titus, J.G., Kuo, Ch. Y. and Gibbs, J., 1985. An overview of the possible impacts of the expected greenhouse warming on storm drainage systems in coastal areas. US Environmental Protection Agency.
UNEP, 1987. 1986 Annual report of the executive director. United Nations Environment Programme, Nairobi, 1987.
Unesco, 1978. World water balance and water resources of the earth. *Studies and Reports in Hydrology*, No. 25.
Unesco, 1986. Climate variations and environment impact, Villach conference 1985. *Nat. Res.*, **22**, 3–5, Unesco, Paris.
Waddell, J.O., 1985. Impact of sea-level rise on Gap Creek watershed in the Fort Walton beach, Florida area. US Environmental Protection Agency.
Ward, R.C., 1984. Hypothesis testing by modelling catchment response. *J. Hydrol.*, **67**, 281–305.
Washington, N. M. and Meehl, G. A., 1984. Seasonal cycle experiment on the climate sensitivity due to a doubling of CO_2 with an atmospheric general circulation model coupled to a simple mixed-layer ocean model. *J. Geophys. Res.*, **89**, No. D6, 9475–9503.
Wigley, T.M.L. and Jones, P.D., 1985. Influences of precipitation changes and direct CO_2 effects on streamflow. Nature, 314, 149–152.
Wilson, C.A. and Mitchell, J.F.B., 1987. A doubled CO_2 climate sensitivity experiment with a global climate model including a simple ocean. *J. Geophys. Res.*, **92**, No. D11, 13315–13343.
WMO, 1987. Water resources and climatic change: sensitivity of water-resource systems to climate change and variability. World Meteorological Organization WMO/TD – No. 247.
Zavatarelli, M., 1988. Potential impact of the greenhouse effect on the Mediterranean sea: Overview. International Institute for Applied Systems Analysis, WP–88–76.

4

Implications of Climatic Change on Land Degradation in the Mediterranean

A.C. Imeson and I.M. Emmer
*(Institute for Physical Geography,
University of Amsterdam, Netherlands)*

Abstract

The degree to which reliable statements concerning the impact of climatic change on various forms of land degradation on the basis of available information is questioned. Attention is directed towards climate-sensitive processes that will be influenced by the large effect that a small increase in temperature and evapotranspiration will have on the water and salt balances of the Mediterranean region. Land-degradation problems specific to the Mediterranean region are described.

An important effect of climatic change could be on soil organic matter and on the chemical composition of the soil. A general deterioration in soil structure is to be expected. Particularly vulnerable are silty soils, soils with a duplex character and soils susceptible to dispersion. Fire will have a greater impact due to the negative effect of altered climatic conditions on soil moisture and on the post-fire recovery of vegetation.

Soil degradation or soil erosion can be estimated by using known relationships between erosion (considered at different scales and for different processes) and climate. Rainfall-simulation applications and modelling offer two means of obtaining useful information. Although general comments about future trends can be made, the site-specific nature of the impact can not be evaluated from present information.

A number of important problems have far-reaching implications for the Mediterranean region. These include greater soil aridity, increased soil degradation, a higher risk of erosion, and changes in sediment budgets and river-channel stability. These changes are likely to increase the occurrence of catastrophic floods and lower food production.

1 Introduction

This study presents an analysis of the impact of climatic change on land degradation in the Mediterranean region. The Mediterranean region is characterized by a seasonal climate and specific ecological and pedological conditions that make it particulary vulnerable for land degradation. The impact of CO_2-induced climatic change is potentially serious because of the low rates of primary production in the drier areas, the sensitivity of the soils to degradation and erosion, and the high risk of drought and fire.

A vast body of information on various aspects of Mediterranean environments that can be used to assess the potential impacts of climatic change. However, very little is known about the response time of processes which are considered to alter the susceptibility of landscapes to climatic change. With respect to analogue information from past periods, prehistoric paleoclimatic records only provide coarse and speculative information on intermediate and long time-scales. Historical climatic data allow more detailed study, but suffer from several other limitations. Furthermore, major temperature fluctuations in Europe in the past 1000 years (Medieval Warm Phase and Little Ice Age) involve statistically significant temperature deviations of no more than 0.5–1.0°C. Statements about rainfall and its frequency and soil moisture in these periods (Lamb, 1984) must be regarded as speculative. It also remains questionable whether climatic data from the last century, which make it possible to compare warm and cool periods (see, for example, Berger, 1984), give insight into future climatic response to a CO_2 increase.

Several direct relations between climatic parameters and erosion have been studied in varying depth. The findings are not conclusive because an enormous range of interplays, feedbacks and different types of response have been suggested by research. Because the impact of altered climatic conditions involves complex interactions between biotic and abiotic processes, many temperature- and moisture-dependent mechanisms have to be considered. Because of the many uncertainties concerning the processes involved, a discussion of the implications of climatic change must be limited to generalizations.

Two important aspects of data interpretation must be mentioned here. First, various methods of investigating short-term climatic trends lay emphasis on different types of time series, such as long-term change, continuous variation, cyclical oscillation, abrupt transition, etc. (c.f. Chorley et al., 1984), which create difficulties when trying to relate them to land degradation processes. Second, threshold conditions, which emerge from field data, involve factors varying widely in time and space. The response of a soil system to climate-induced changes will not be the same at different locations. Statistics and probability analyses offer opportunities for reliable predictions, but sufficient data are not yet available for this to be done extensively.

1.1 Scope of this Chapter

Land degradation is a term used here to refer to the deterioration in the physical and chemical properties of the soil that occur as a result of environmental change and which result in soil erosion, problems of sedimentation and flooding, the loss of fertility and sometimes in salinization. Land degradation is thought to be primarily a result of human activities, such as cultivation, deforestation and fire, but phases of land degradation are known to have occurred in the past during periods of environmental stress before modern man is likely to have had an effect on these phenomena. It should be pointed out that degradation can occur as geomorphological systems respond to internal changes, so that erosion can occur periodically as intrinsic geomorphic thresholds are transgressed. In this way degradation can vary independently of present environmental conditions and it can reflect a pulsating delayed response to environmental changes that occurred perhaps hundreds of years ago (see, for example, Schumm, 1977).

In this chapter, the soil system will be approached as an integrated component of the biotic environment and will be described as a body in which several key processes, in connection with land degradation, take place. Then these findings will be evaluated in terms of their significance for the prediction of land degradation upon changes in climate, whether approached qualitatively or by use of mathematical models.

The processes of land degradation considered here for convenience are separated under the headings of soil erosion and river channel changes and sediment supply. These will be discussed further in part 3.

1.2 Climatic Changes Indicated by Global Climate Models

The CO_2 predictions indicated by the two global climate models (BMO and GISS, Jung and Bach, 1986) predict important and relatively uniform changes in climate for the Mediterranean region. Both models suggest a temperature increase in the order of 4°C throughout the year. Precipitation predictions are different for the GISS and BMO models. The GISS model indicates a general precipitation increase for the Mediterranean region. This increase is more than compensated by the increase in potential evapotranspiration. The data suggest a very large increase in the soil-moisture deficit, particularly between the late summer and the early winter.

The BMO model predicts slightly lower increases in temperature but a decrease in precipitation. The predicted summer decreases are sometimes higher than the average precipitation in certain areas. The increase in potential evapotranspiration predicted by the BMO model is less than that predicted by GISS. Both models are in agreement concerning the prediction of very large increases in soil-moisture deficit.

It should be stated that mean annual values of precipitation and evapotranspiration are much less significant for the prognostication of a climatic impact on soil and vegetation than are mean monthly or mean weekly averages. Also, the effect of shifts in recurrence intervals and magnitude of extreme events are of major importance and must not be overlooked. At the best, the above-mentioned climate models give predictions of monthly averages of temperature, precipitation and evapotranspiration for a large area of several hundred square kilometres. Unfortunately, these are the only kinds of climatic data attainable. As an example, the forecasted changes in these parameters are given here for the Madrid region (see Table 4.1).

GISS and BMO models suggest an extension of the period with a rainfall deficit by three and four months, respectively. Moreover, the difference between precipitation and evapotranspiration in the moist period (September to March) is expected to decrease significantly. The BMO model calculates a period with rainfall deficits of 11 months and a decrease in the rainfall excess of 100% in the period November-February. GISS suggests a more modest decrease of about 30%.

To meet at least partly the disadvantages of the global climate models, use should be made of data having high spatial and temporal resolution, based on regional and seasonal information. Forthcoming instrumental data-based scenarios (Wigley, 1986) should be applied to those areas considered to be representative of recognizable Mediterranean sub-

98 *Climatic Change in the Mediterranean*

Table 4.1 Predicted climatic changes by GISS (scenario 1) and BMO (scenario 2) for the Madrid region in Spain at 660 m altitude

		Mean annual average	D	J	F	M	A	M	J	J	A	S	O	N
Actual	temp.	14	6	5	7	10	13	16	21	24	23	20	14	9
	prec.	444	48	39	34	43	48	47	27	11	15	32	53	47
	Epot.	746	11	10	15	30	48	79	113	144	135	87	52	22
	Ep-p	425				30	0	32	86	133	120	55		
Scenario 1 changes	temp.	+4	+4	+4	+4	+4	+4	+4	+4	+4	+4	+4	+4	+4
	prec.	+54				+6	+6	+6				+12	+12	+12
	Epot.	+312	+8	+8	+14	+32	+32	+32	+32	+32	+32	+32	+32	+26
	Ep-p	656				25	38	64	118	165	152	75	19	
Scenario 2 changes	temp.	+3	+2	+2	+2	+4	+4	+4	+3	+3	+3	+2.5	+2.5	+2.5
	prec.	−198	−25	−25	−25	−12	−12	−12	−25	−11	−15	−12	−12	−12
	Epot.	+234	+4	+4	+4	+32	+32	+32	+24	+24	+24	+20	+20	+14
	Ep-p	742			10	31	44	76	135	168	159	87	31	1

regions. For every such sub-region, climatological information (including extremes, variability, etc.), together with data about soil degradation and erosion, crop yield and management should be obtained. Only with this knowledge can the implications of future climatic change on land degradation be given on a regional scale.

1.3 Problems Specific to Mediterranean Regions

Mediterranean environments are characterized by hot dry summers and by mild moist winters. With respect to agricultural production in the Mediterranean, three climatic aspects are important (Rosini, 1984): a marked tendency to water shortage during the whole year, high solar radiation, and the occurrence of extreme weather phenomena (e.g. heavy rainstorms and hail). Today, agriculture is already facing problems through the effect high short-term variability of climate has on plant growth. In the drier Mediterranean areas of North Africa and southern Spain the winter rains are less reliable and the climate can sometimes be described as semi-arid. Due to climatic change, the area of unreliable rainfall might become more extensive and shift northwards. This is important because water is the main limiting factor to plant growth and because net primary production is related to the amount of precipitation during the growing season. Because, in general, the Mediterranean vegetation is adapted to summer drought, which extends to several months, and to the marginal moist periods between summer and winter, these two aspects of climate rank above mean annual values in importance. It is therefore erroneous to consider mean annual values adequate for evaluating the impact of climatic change. Furthermore, an increasing overlap of the growing season and period of drought (see section 1.2) will greatly affect Mediterranean agriculture, especially when thresholds are involved, such as a sustained experienced permanent wilting point. The Mediterranean environment, therefore, is potentially vulnerable to an increase in aridity. For aridity to increase, however, it is not necessary that the yearly amount of precipitation decrease. Increasing aridity could be caused by: (1) higher evapotranspiration; (2) by a change in the frequency and magnitude of rainfall events; and (3) by processes of soil degradation that lower the ability of soil to retain water.

The Mediterranean area cannot be regarded as a homogeneous climatic or climomorphic region. A relatively wide range of marked climatic zones exists between moist and dry. In southwestern Australia, for example, a distinction is made between moderate, dry, extra dry and semi-desert Mediterranean environments (Beard, 1983).

The global climate models suggest that in general the potential evapotranspiration will increase to cause a significant increase in soil-moisture deficits. This general trend should be superimposed upon local environments but to evaluate the environmental impact of these changes requires much more detailed information about the predicted weather patterns and local climates. The relationship of a number of specific environmental processes to, for example, temperature and precipitation in many cases is well understood. To be able to apply such relationships, more information about local weather conditions should be available, as for example, is demonstrated by Garcia (1984) for an agricultural area in south-eastern Spain. The bioclimatic response at any location could be influenced by local winds, dew, sunshine and temperature inversions.

The vulnerability of the Mediterranean environments to aridity takes on many forms. The implications of increasing aridity are many, and these pose many problems that can be traced back to altered vegetational structures, to the lower rates of organic matter production or higher decomposition in the soil and to changes in the salt and water balances. A major chain of problems is associated with the probable deterioration in soil structure and the effect this will have on the partitioning of rainfall between infiltration and runoff. In rivers, low flows may become inadequate to dilute effluents, which may cause pollution, and higher salinity will become a problem with special significance to irrigation. The last problem of salinity will be intensified in shallow coastal aquifers where a sea-level rise of ± 40 cm is predicted.

On the basis of present knowledge and also because of the diversity of Mediterranean landscapes and climatic conditions, in the framework of this study, it is not possible to deal with spatial variability at a local scale. Before this can be done much research will be required in the near future. With respect to existing knowledge, much less information seems to be available for the European Mediterranean area than for areas with Mediterranean climates in Australia and South Africa.

In conclusion, most problems arising from climatic impact are likely to be associated with various aspects of increased aridity. Therefore, this aspect is regarded as the force behind imposed alterations of soil characteristics and hydrological behaviour.

2 SENSITIVITY AND VULNERABILITY OF MEDITERRANEAN SOILS TO CHANGES IN CLIMATE

2.1 Introduction

For the purpose of this study, a distinction should be made between arable soils and soils under a (semi-) permanent forest or shrub cover. This is especially important with respect to organic matter dynamics, element concentrations and tillage. It is also useful to limit considerations to those aspects of soils that are likely to be influenced by relatively small changes in precipitation or temperature. The soil contains elements which vary in age from a few weeks (e.g. nutrient concentrations) to tens of thousands of years (e.g. calcic and oxic horizons). Many soil characteristics, therefore, reflect the operation of soil-forming processes under a wide range of former climatic conditions; because these conditions differed much more from the present climate than the climate predicted for a doubling of the CO_2 concentration, these characteristics are not expected to be influenced to a degree that requires consideration.

A great diversity of soil types is found in the Mediterranean region. This reflects differences in the major soil-forming factors, two of which are climate and parent material. The climatic changes associated with an increase in CO_2 are comparable to differences in climate that occur locally as a result, for example, of relief and exposure. Unless threshold factors are involved, it is not thought likely that CO_2-induced climatic changes will result in major shift in the boundaries between the main soil types. The soil may be considered as an open system, which responds to a number of dynamic processes. These processes are driven by the inputs of material and energy. The processes likely to be influenced by the predicted climatic change are those involving: (1) the input and output of water, gasses and soluble salts; (2) the input and output of calcium carbonate; (3) the input, decomposition and output of organic matter.

The changes in these processes might not be very significant for the long-term morphological evolution of the soil, but they are undoubtedly important with respect to (soil) ecology and land use. This is particularly the case when small changes in climate lead to the occurrence or intensification of certain pedological processes. For example, the response of soil aggregates to wetting by rainfall is extremely sensitive to the chemistry of the soil solution and to microbiological activity. Slight changes brought about, for example, by longer dry seasons or changing atmospheric deposition may lead to changes in the soil structure which dramatically alter the hydrology and nutrient balance of the soil.

2.2 Processes Sensitive to Climatic Change

The envisaged changes in climate are likely to have the greatest short-term (50 years) impact through the effect they have on: (1) the salt balance and salt composition of the soil; (2) the chemical precipitation of Ca/Mg carbonates in the soil; (3) processes associated with the supply and breakdown of organic matter.

In the following paragraphs we discuss each of these aspects.

2.2.1 The salt balance of the soil

Shainberg and Letey (1984) have summarized the problems faced when trying to point out the relations between saline and sodic conditions and soil response, as the inability to indicate threshold values of, for example, exchangeable sodium percentages (ESP) and other cations, the effect of clay content and clay mineralogy and the sensitivity and continuity of certain processes in terms of their response to exchangeable sodium (e.g. clay swelling and clay dispersion).

The transport and distribution of salts within a landscape and in a soil profile reflect the water balance, ground-water levels and transport in throughflow. Therefore, precipitation and evapotranspiration together with soil-profile characteristics are important in this respect. The general level of water-soluble salt concentrations in the soil is also partly influenced by atmospheric inputs that will be sensitive to climatic change.

In general a restricted number of quantifiable inter-relationships between soil and soil solution chemistry and soil physical properties are important. Focussing on the interactions between the solid and liquid phases of the soil, a few general statements can be made. First, the salt concentration in the soil solution is inversely proportional to the thickness of the water layers on clay particles and therefore influences the bonds between these particles. As a result, a change in electrolyte concentration may cause flocculation (if increased) or dispersion (if decreased) of soil particles; the threshold is referred to as the flocculation value. Also with Na as the predominant cation on the external surfaces of clay particles, the water layer is relatively thick (according to the Boltzmann equation) and the sensitivity to dispersion far exceeds that of (for example) Ca clays. Since sodium affects the soil's physical properties, the exchangeable sodium percentage (ESP) of a soil is an important criterion for evaluating soil stability and erodibility in semi-arid regions. High ESP values not only cause a loss of aggregate stability but also can lead to decreasing hydraulic conductivity. Montmorillonitic soils, for example, tend to show a considerable increase in water retention and swelling capacity with increasing ESP. Illitic soils become particularly dispersive. High ESP and low water salinity destabilize soil aggregates (Fig. 4.1); destabilized soil aggregates break down to form surface seals or crusts.

When soil surfaces are sealed by crusts, the infiltration capacity becomes reduced and overland flow occurs at lower threshold amounts of rainfall (see Fig. 4.2). Two other processes that cause sealing are slaking, caused by sudden wetting, and mechanical destruction of aggregates by raindrop impact. These are both more effective when ESP values are high (Agassi *et al.*, 1985).

To illustrate the high sensitivity to changes in salt concentration of the soil properties that regulate the infiltration process, results from research by Shainberg and associates in Israel are shown in Fig. 4.3. The importance of antecedent soil-moisture content with respect to surface response to rainfall has become clear from a study by Ben-Hur *et al.* (1985). They show that surficial crusts become more permeable upon drying due to structure development and cracking, but they also become more sensitive to destruction by raindrop impact.

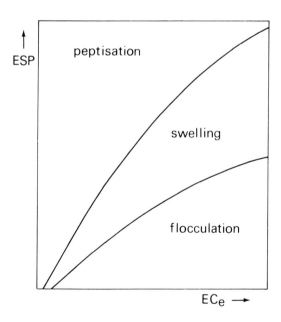

Fig. 4.1 The influence of electrical conductivity (EC) and exchangeable sodium per cent (ESP) on peptisation and flocculation thresholds. (From Bolt and Bruggenwart, 1982.)

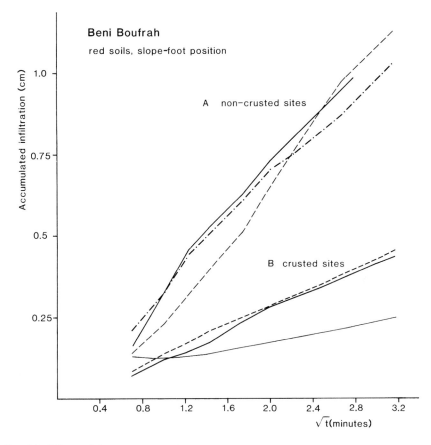

Fig. 4.2 Effect of the presence of a surface crust on cumulative infiltration near Beni Boufrah, Morocco. (From Imeson, 1983).

A general decrease in precipitation or increase in evapotranspiration will cause an increase in the area of soils affected by saline or sodic conditions. This is because in those regions with high evaporation rates capillary rise is accelerated and salts accumulate residually where drainage is nearly absent. Particularly serious would be a decrease in winter precipitation to the extent that seasonally accumulated salts are not flushed from the soil. A change in the duration and intensity of precipitation events would exacerbate salt accumulation where they both decrease and possibly ameliorate the conditions where they increase. A decrease in precipitation or a more adverse rainfall distribution would affect large areas where the annual rainfall is less than about 600 mm. Particularly in Spain and Italy changes in salt accumulation could lead to an increase in the area of soils affected by vertic conditions, in particular by poor physical conditions (low permeability, shrinking and swelling and waterlogging). Irrigation and crop yield management would become more difficult or expensive, and the need for crops with higher salt tolerance would increase.

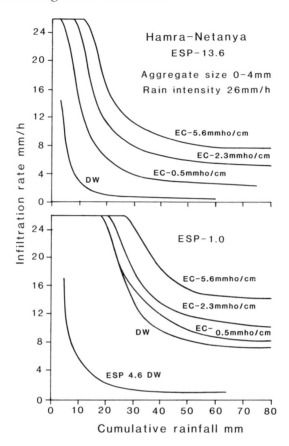

Fig. 4.3 Effect of electrolyte concentration in rain-simulation experiments on the infiltration rate of loess soil. (From Shainberg and Letey, 1984.)

To have an adverse effect, salt concentration need only change slightly in composition or amount. A slight increase in aridity, resulting in a shift in the ESP value at the end of dry season, could make soils sensitive to dispersion and slaking when the electrolyte concentration is lowered at the onset of winter or autumn rainfall. Dispersion and slaking then lead to surface sealing, lower infiltration rates, more runoff, high erodibility of the soil and a potentially catastrophic series of erosion and sedimentation problems during extreme rainfall events.

An increase in precipitation or decrease in evapotranspiration could have a positive effect in areas suffering from sodic or saline conditions in cases where higher rainfall leads to better and deeper drainage. However, it is known from Australia and Kenya that increased rainfall in semi-arid areas can lead to the development of perched water tables, which when they rise into upper soil horizons mobilize salts which are subsequently transported downslope in throughflow. This has led to salinization and erosion at colluvial slope-foot positions and has made runoff unsuitable for irrigation or domestic use.

Temperature increases will increase problems with sodic and saline soils, in as much as they influence the water balance and the mineralization of organic matter. It seems probable that higher levels of organic matter to some extent increases the degree of sodicity required to produce negative soil structural properties (Shainberg and Letey, 1985; and below).

Although it can be concluded that increased aridity and resulting high soil ESP will cause soil physical properties to deteriorate and soil erodibility to increase, the extent to which this will happen at a local scale will depend on various factors controlled by the water balance, soil type, exposition and slope, and by the total salt and sodium inputs. Salinity problems would be most severe in areas receiving between 300 and 600 mm of rainfall and would become less important in areas where precipitation dropped below this level. Such areas with generally irrigated agriculture and the occurrence of natric and sodic horizons in the soil profile are not extensive in European and most other Mediterranean regions today but there is a large area potentially vulnerable to salt problems.

2.2.2 Chemical precipitation of Ca/Mg carbonates

Pedogenic caliche development is restricted to areas with low precipitation (100 to 500 mm) and is often connected with desertification. Caliche accumulation can result from the redistribution of calcium carbonate and some accessory components in the parent material but also from atmospherically supplied material. Under special conditions a distinct caliche layer can develop within several months (Hattman, 1983).

Petrocalcic horizons can have serious implications for hydrology where they are situated at or near to the soil surface. With the increased aridity envisaged, however, present problems probably will become much more serious in the coming decades.

Attention should be paid, however, to the effect that the accumulation of calcium carbonate in the solum has on plant growth and net primary productivity. Primary production might decrease in cases where microcrystalline chalk accumulates and inhibits the uptake of trace elements, such as iron and manganese by plant roots (Finck, 1982). This may especially affect non-irrigated agriculture.

2.2.3 Supply and breakdown of organic matter

The effects of small changes in the supply and mineralization of organic matter could have a large impact on soil structure in certain cases and greatly influence soil and hillslope hydrology. As mentioned above, the Mediterranean area is especially vulnerable to climatic change. In this chapter this will be considered in relation to the low organic matter content in most cultivated Mediterranean soils and to the susceptibility of litter on forest floors to degradation. This is especially significant because of the silty character of many upper soil horizons due to pedological soil development and the deposition of atmospheric dust. Furthermore, many marls weather into silty residual carbonaceous materials.

According to the nutrient status of different soil types, a distinction can be made according to type of substratum and climatic conditions, of soils with varying degrees of leaching and calcium carbonate content, and of nutrient-rich/high pH soils (Specht and Moll, 1983; Quezel, 1981) (see

Fig. 4.4). Unless amended, phosphorus and nitrogen are suboptimally supplied and if pH values are high, trace elements are inadequately available for root uptake.

The ratio Ea/Eo (the actual and the potential evapotranspiration respectively) is a useful index of water availability for plant ecosystems (Specht and Moll, 1983). An increase in aridity may lead to an extended period with low ratios and less favourable conditions for nutrition, with consequential shifts in plant and tree species composition and decrease of organic matter input into the soil. A decreasing area of deciduous thermophyllous forests (e.g. montane beech forests) and steppes can be expected in favour of forests with more sclerophyllous components and maquis, with less vegetation cover. This is confirmed by Huis and Ketner (1987). They anticipate lateral shifts of hundreds of kilometres and several hundreds of metres in altitude in some cases. These lateral and altitudinal vegetational shifts could alter the humus profile of the soil and the soil micro- and meso-fauna quite dramatically.

The importance of a lag time in the response of ecosystems and soil characteristics to changing climate must be stressed. In general, the

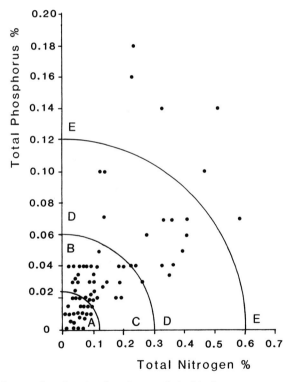

Fig. 4.4 Relative nutrient levels of surface soils in Mediterranean-type ecosystems in southern Australia, based on total soil nitrogen (H_2SO_4 extraction) and total soil phosphorus (HC1) extraction). A: strongly leached soils. B: moderately leached soils (semi-arid zone). C: moderately leached soils (humid, sub-humid zone). D: weakly leached soils. E: nutrient-rich soils. From Specht and Moll, 1983.

organic matter content of the topsoil will adapt to altered vegetation structure and composition quite rapidly. In consequence, the vegetation in forest ecosystems will be the major factor in determining response times to climatic change. Response times for vegetation adjustments are estimated as being at least several decades (Huis and Ketner, 1987). This temporal aspect should be taken into account, together with other external factors such as an increased frequency of forest fires and/or forest clear-cutting that might result from climatic change.

The organic matter and organic carbon contents in soils are closely related to mean annual precipitation. This is due to the positive influence of soil moisture on pedoturbating soil biota (Fig. 4.5). Although the amount of organic matter is very highly correlated with precipitation, other climatic and environmental parameters play a role but are difficult to isolate. Berg *et al.* (1983) and Meentemeyer (1978) have estimated the effect of annual evapotranspiration on organic matter decomposition in non-arid regions. (Fig. 4.6). Annual evapotranspiration encompasses both moisture and energy, which are major factors influencing microbiological activity, and it is therefore a more valuable variable connected with organic matter decay than is precipitation. A factor that very much complicates the picture is the lignin content of the organic tissue. A high lignin content is considered to hinder decomposition and for this reason it should be superposed on decomposition rate-AE relationships. The introduction of more sclerophyllous plant species, with higher lignin content, will probably temper the effect of increasing evapotranspiration on organic matter decomposition, but whether this positively influences the maintenance of soil structure depends on the composition of organic

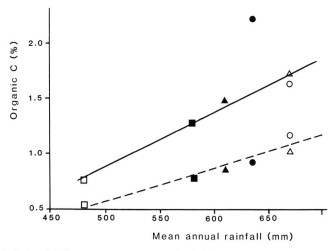

Fig. 4.5 Relationship between mean annual rainfall and organic C content of virgin soils ($r = 0.99$, P 0.001) and adjacent soils cultivated for wheat from 20 years, southern Queensland, Australia ($r = 0.97$, $P < 0.01$). The regression coefficients (% organic C/mm mean annual rainfall) are 0.0048 (—) and 0.0029 (--). From Dalal and Mayer, 1986.

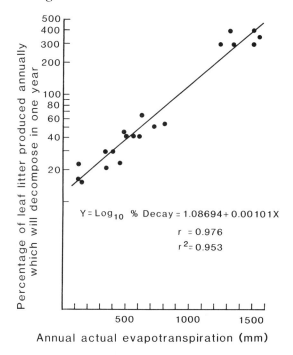

Fig. 4.6 Correlation and regression of estimated annual actual evapotranspiration with measured annual weight loss of leaf litters. At a decomposition rate of 100% all of the available leaf litter is predicted to be decomposed in an average year. At 200% there would be complete decay in six months and 300% four months, etc. (From Meentemeyer, 1978.)

substances. Humic and fulvic acids have proved to have particularly positive effects on aggregate stability (Soong, 1980).

Morgan and co-workers (1986) found that plant cover effects on hillslope runoff and erosion were not only regulated by a simple linear relation between cover percentage and soil detachment by rainfall, but also by increasing leaf-drip action.

Unpublished data from an inventory investigation of relations between vegetational cover, soil erosion and colluviation in north-east Catalonia in Spain show that south exposed slopes are more affected by soil erosion than north exposed slopes. On the latter, a continuous colluvium layer, covered by an Ah horizon, is generally present, but is absent on south exposed slopes (see Fig. 4.7), pointing to higher erosion rates. The percentage of surface cover of the vegetation on north-facing slopes generally exceeds that on south-facing slopes, due to differences in soil moisture availability. Two major factors are considered to be related to this slope dimorphism. First, due to greater dryness, fire events will be more frequent on south exposed slopes, leading to less protection of the soil by the vegetation and, therefore, to an increased susceptibility to erosion. Erosion research after forest fires in the United States confirms this (DeByle, 1981). Secondly, as mentioned before, from the greater

Land degradation and climatic change 109

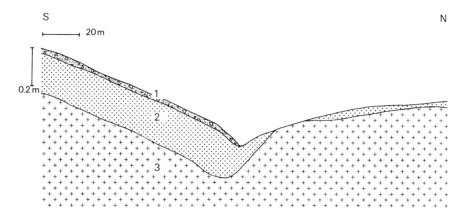

Fig. 4.7 Schematic figure illustrating the different soil horizons present on slopes subject to forest fire. The parent material granite (3) is overlain by BC material (2) and on north-facing slopes, where erosion is less severe, by colluvium (1).

decomposition of organic matter in the topsoil, it may be expected that increased aridity will affect hillslope processes on north-exposed slopes in such way that they become increasingly similar to processes currently influencing south-facing slopes, increasing losses of topsoil material in the area as a whole.

In the same study area, the effect of different types of forest stands, exposition and humus form profile development has been investigated by Sevink (1988). He found that organic profile morphology is closely correlated with vegetational cover evidently influences hillslope hydrology. This means that extended accumulation of organic matter on the soil surface in some situations delays time to ponding and surface runoff during prolonged rainfall events, protecting the soil from erosion. In his inventory research of several forest sites, a clear difference in organic profile morphology was observed between north- and south-facing slopes at about 1200 m height. Chemical analyses show that after an initially high release of nutrients, mineralization is retarded and decomposition mainly involves humification. On the south-facing slopes this process seems to be more pronounced and leads to the development of relatively thick F3/H horizons, compared to north-facing slopes. Due to their high water storage capacity and high root content, these organic horizons may protect the soil from erosion. This will be of major importance in open sclerophyllous forests, where significant transport of litter occurs during rainstorms. The significance of the dependence of soil profile development on exposition needs more detailed study, in view of the postulated increase in aridity in the Mediterranean area.

The effect of increased aridity will depend on the water-retaining characteristics of the soil, the discontinuity of hydraulic processes and on the way in which the frequency of rainfall events of different magnitude is altered.

More research needs to be done in this area before the effects of decline in precipitation can be generalized. In highly calcareous, sandy and silty soils, organic matter plays a vital role in soil fertility because it has a very high exchange capacity and retains nutrients in the soil, and also because it results in a favourable micro-aggregation which promotes relatively high levels of water retention. Lower inputs of organic matter lead to an increased aridity of the soil and to a lower fertility.

It is stressed that at the scale of the Mediterranean basin it is only possible to generalize in a crude way. Local conditions will influence the likely impact of that climatic change. The climatic impact will not depend on average climatic parameters but rather on those parameters that influence the water balance of the soil and the availability of moisture for plants. In detail, a number of critical soil properties are dynamic and highly sensitive to the soil climate. The way in which a cultivated soil will behave when wet reflects the number of wetting cycles it has already experienced since being ploughed and the moisture content at the moment of wetting. This problem is reviewed by Utomo and Dexter (1982) and discussed by Imeson and Verstraten (1986).

On agricultural land organic matter is critical in terms of its effect on soil structure, particularly in the case of highly calcareous or silty soils where bonding forces between mineral particles are low. Several case studies have documented the scale of these effects in Mediterranean areas. The data from southern Spain (Fig. 4.8) clearly indicate how the micro-aggregation of the soil under both forest and cultivation is related to the organic matter content. The data shown are for highly calcareous soils in southern Spain, but similar results would probably be obtained for other soils. A compilation of the data in Figs 4.5, 4.6 and 4.8 may indicate a very dramatic reaction of the soil system to increasing aridity. Figure 4.9 shows that decomposition rates are strongly influenced by soil-moisture adjustments, particularly in warm climates. Undoubtedly, modelling of climatic variables related to soil erosion rates must emphasize these relationships.

Pittock and Nix (1986) have studied the impact of a rise in temperature on net primary productivity (NPP) in Australia. In their preliminary research they use a model which calculates the NPP on the basis of anticipated rainfall amounts and temperatures together with general relations between temperature, precipitation and NPP. Evapotranspiration and soil erosion predictions are not included. Therefore, the results are of restricted value and the model needs to be improved. For the Mediterranean regions in Australia, a decrease in mean annual precipitation is expected and the model estimates a zero increase in NPP. Upon improvement of the model, which besides soil erosion should include soil moisture and nutrient status, a decrease of the NPP may be expected.

2.3 Forest Fires

Evidence suggests that an increase in temperature and dryer conditions will be accompanied by an increase in the frequency of forest fires. In Spain it seems that forest fires tend to be more frequent in distinctly dry years.

Today fires can be usually traced to anthropogenic origins. In Spain fires caused by lightning are of only subordinate importance, but some authors suggest the converse for North America (Kimmins, 1987).

Land degradation and climatic change 111

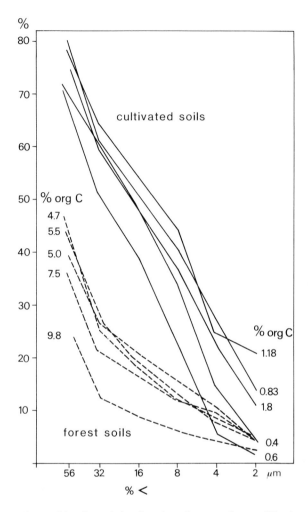

Fig. 4.8 Water-dispersible silt and clay fractions from surface soil horizons from forested and cultivated sites near Teulada, Spain. (From Imeson and Verstraten, 1986.)

Fire can cause a loss of the ecto-organic horizons and a decrease in organic matter content in the upper soil horizon, so that soil nitrogen availability may be reduced significantly. Where erosion does not ensue a forest fire, there is usually a recovery of the forest and a quick regeneration of biomass in the soil. Fires may cause a water-repellent layer to develop, which can lead to increased runoff during rainfall in dry seasons. This layer may also protect the subsoil from extreme evaporation. In a review of clear-cutting and fire in the United States, DeByle (1980) showed that following fire, runoff, erosion and also nutrient concentrations in runoff increase initially, but later decrease.

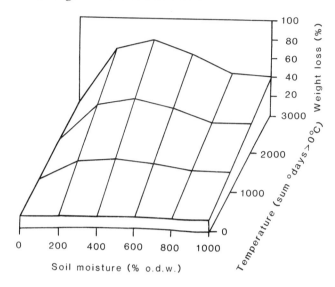

Fig. 4.9 A regression surface showing the interactive effects of soil moisture (% o.d.w.) and thermal environment (temperature sum = degree days at temperatures > 0°C). The regression equation for the relationship of the three factors is $Y = 11.62 + 0.0147.T.W - 0.00289.T.W^2 + 0.000152.T.W^2$ ($r = 0.769$), where Y = percent weight loss, T = temperature sum and W = soil moisture. (From Swift et al., 1979.)

From northeastern Spain it is known that soil losses after forest fires are limited to the upper centimetres and that vegetation recovers rather quickly. Severe erosion, however, is invariably experienced when a dense network of roads is built to make the burned area accessible for forest work. Soil erosion is more pronounced when the organic matter profile has a mor- type profile rather than a moder type (Sevink et al., 1988).

In Sardinia forest fires are the most important cause of severe soil erosion (Aru, 1984).

Another point of interest is the influence that forest fires have on the water balance by decreasing evapotranspiration. This in particular can have a large effect on the salt balance of the soil. Such effects in southwestern Australia are described by Sharma (in press). Removal of native perennial vegetation decreases evapotranspiration significantly and promotes ground water transport on slopes, leading to salt seepage at slopefoot positions and increased salinity in streams. Sommerfeldt and MacKay (1982) show that such salt seepage can enhance soil degradation.

2.4 Desertification

In this chapter, desertification means the diminution of biological potential and primary productivity due to climatic change and excessive exploitation by man. Land degradation in arid and semi-arid Mediterranean regions (rainfall < 600 mm) has great analogies with desertification and to some extent may be considered synonymous. In Europe, large parts of Spain and Greece, the southern part of Italy, Sardinia and Sicily are involved.

Mabutt (1986) summarizes the variables that are understood to be indicators for desertification. Three types (physical, biological and social) of indicators are distinguished. In this chapter organic matter (soil and standing biomass), salinization, alkalization, ground-water and vegetation cover already have been discussed. Surface albedo is also one of the major parameters used to quantify desertification. Frequently, positive feedback mechanisms are involved in desertification, and the ability to inhibit or stop the process is very limited. One such mechanism is the sequence of deforestation, increased soil albedo, lower soil temperature, lower convective activity and decreasing precipitation.

The sensitivity to desertification will be significantly increased by any increase in aridity, and this might accelerate or reinforce present or past degradation processes.

3 LAND DEGRADATION, SOIL EROSION AND RIVER CHANNEL CHANGES

Land degradation in the Mediterranean region dates back to antiquity. Many historians and scientists have attempted to document its occurrence and to relate it to changing environmental conditions or to human occupation. It is not appropriate here to review the findings of the vast body of work on this subject, only to draw attention to a number of conclusions relevant to the potential for further degradation due to the effects of climatic change on climate-sensitive processes.

It should be pointed out that a distinction needs to be made between existing land degradation resulting from past causes and the potential for further degradation under altered conditions. In the past the Mediterranean environment has proved to be sensitive and vulnerable to environmental change, so that for all intents and purposes the landscape in many places has been irreversibly degraded. Large areas of the Mediterranean today have only truncated remnants of the original soil cover and there is relatively little potential for further degradation. Two examples of such areas are the maquis regions in Greece and France. Whereas degradation formerly was probably greatest on slopes, the potential for present degradation is probably greatest on colluvial and valley-fill deposits.

In considering the relationship between climate change and soil erosion in the Mediterranean region, it is convenient to distinguish between the erodibility of materials subject to erosion and the erosivity of the rainfall, runoff or wind supplying the energy for detachment and transport of soil particles. Most of the discussion below refers to the processes of sheet erosion, which includes erosion in shallow channels or rills.

3.1 Rainfall and Runoff Erosivity

Rainfall erosivity, which may be simply defined as the ability of rainfall to cause erosion, is related to properties of falling raindrops, such as size and velocity, that determine the impact force that the drops have upon striking the ground. Raindrop impact increases with drop-size, and the number of large drops increases with rainfall intensity. Because of this it has been possible to calculate indices of rainfall erosivity based on

114 *Climatic Change in the Mediterranean*

rainfall intensity and duration parameters. The most well-known index is the R (rainfall) factor of the Universal Soil Loss Equation. Fig. 4.10 simply reflects the distribution of high-intensity storms in Morocco. Whilst such maps give a general impression, it must be stressed that the index was not developed from reliable observations. Suitable rainfall data need to be available to calculate such indices on a storm by storm basis. Because such data are often absent more simple indices have been applied such as those in Fig. 4.11.

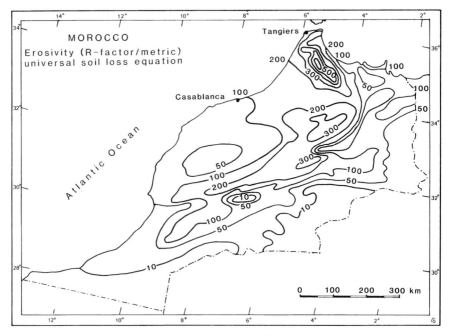

Fig. 4.10 Rainfall erosivity calculated for Morocco (After Arnoldus, 1977).

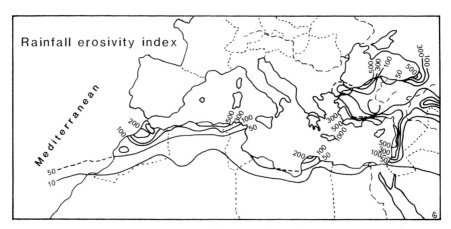

Fig. 4.11 Rainfall erosivity for northern Africa calculated from the index p2/P (p = max. rainfall in any month; P = total yearly rainfall) (From Arnoldus, 1978.)

If it could be established how the relevant properties of rainfall would be influenced by climatic change, it would be simple to examine rainfall erosivity under different climatic scenarios. This might become possible if the precipitation forecasts of the climatic models become more reliable. In a following section the combined effects of erosivity and erodibility are considered further.

3.2 Soil Erodibility

Lowering the organic matter content of cultivated soils will produce smaller, less stable soil aggregates, so that both the rates of water acceptance and the resistance of soil particles to detachment will be lowered. This will be most marked in soils containing low amounts of clay and could lead to higher concentrations of sediment in runoff, to lower thresholds of particle entrainment, and to lower rates of infiltration. Most affected will be the way in which soils respond to rainfall. Whilst soils with a good structure stabilized by organic matter retain their coherence and strength when moistened, soils that contain low amounts of certain necessary types of organic matter, (and/or are dispersible) can respond by slaking, swelling, dispersion, cracking and mellowing, when they are wetted, and because of this can develop surface seals or crusts.

The effect of organic matter on the water-stable micro-aggregation of highly calcareous soils from Alicante has already been illustrated in Fig. 4.8. Soils having low organic matter contents are less stable. The less stable the soils, the lower the infiltration capacity and the more frequently the soils will produce runoff. The frequency with which soils can be expected to produce runoff can be established by relating rainfall intensity and the time to ponding (infiltration envelopes). For example, infiltration envelopes from Alicante (Fig. 4.12) illustrate the effect of crusting on the amount of rain required at different intensities to produce runoff. Relationships between infiltration measurements and soil erodibility parameters (Table 4.2) enable a link with the climate, since the infiltration envelopes can be compared with rainfall-frequency-duration curves.

Comparisons between infiltration envelopes and rainfall-frequency-duration data allow a clear distinction to be made between soils sensitive or insensitive to higher rainfall intensities. Especially highly calcareous soils, lacking either pedogenic iron oxides and/or organic matter (i.e. sensitive to slaking and crust development), and soils that swell or disperse would be sensitive to increased precipitation. An increase in aridity would increase the area of soils having unfavourable infiltration characteristics.

Recently Lavee *et al.* (1988) have studied the infiltration characteristics and soil properties in the Judean desert along a transect from the summit area at about 700 to 100 m below sea-level near the Dead Sea. As rainfall decreased from about 600 mm at the highest sites to 100 mm at the lowest locations, on the same calcareous parent material, so the infiltration capacity of the soil decreased from a final infiltration capacity of about 2 cm/hour at the highest site to about 8 mm/hour at the lowest. As conditions become more arid so infiltration rates become lower and the soil erodibility higher.

To summarize, the results of climatic change will be different for different soils but in general the water-retention characteristics of many

116 *Climatic Change in the Mediterranean*

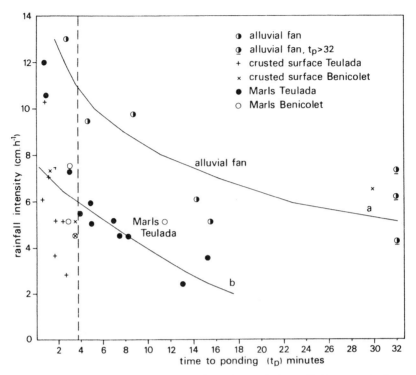

Fig. 4.12 Time required to produce ponding when rain is applied at a particular intensity. Note how little rain is required to pond crusted soils on both the alluvial fan and marl soils (from Imeson and Verstraten, 1985).

soils may deteriorate, the porosity and bulk-density increase, and the CEC decrease so that less nutrients can be retained for plants and that the soil would slake and disperse more readily so that crust-forming becomes a problem. This last problem is important since it influences the amount of infiltration and indirectly the rate or organic matter production and because a greater proportion of rainfall will become overland flow and have a potentially great effect on erosion.

The changes envisaged in the concentrations and composition of water-soluble salts could have a potentially serious effect on crusting. Irrigation practices may have to be modified to make allowance for increased salinity. There is a large literature describing how this can be done.

The erodibility of a soil describes its inherent susceptibility to erosion. The effect of decreasing organic matter contents and salt concentrations on soil erodibility is well documented. Soil erodibility is easily increased by an order of magnitude, particularly when the changes which occur result in the activation of formerly inoperative processes.

3.3 Soil Erosion

From a mountainous area in south-east Spain, Sala (1988) quantified the yearly variations in rainfall, runoff and erosion in catchments on two

Table 4.2 Some characteristics of the soils and infiltration parameters at the infiltration experiment sites in the Lower Zarqa Catchment, Jordan

Site	T_{P_5}	Infiltration experiment data $i = 5.3\,cm\,h^{-1}$			Soil properties			
		$T_{P_{40}}$	T_r	S	SM	DI	AS	BD
A	1.05	3.50	9.35	3.0	20	10	16	1.4
B	0.58	2.36	5.09	2.1	20	3	81	1.1
C	0.24	1.26	1.40	1.4	15	11	8	1.1
D	0.32	0.56	1.15	1.2	17	11	13	1.3
E	3.12	5.46	9.45	2.5	16	11	37	1.4
F	0.39	2.31	7.05	1.8	17	5	37	1.1
G	2.04	3.14	5.30	1.9	14	11	23	1.2
H	1.54	5.22	16.12	2.5	20	11	10	1.1
I	5.51	16.49	n.r.	4.8	18	14	15	1.2
J	2.52	8.3	14.05	2.9	18	9	9	1.0

Notes: T_{P_5} = time to 5% ponding (min., sec.); $T_{P_{40}}$ = time to 40% ponding (min., sec.);
T_r = time to runoff (min., sec.);
S = sorptivity (cm. min.$^{1/2}$);
DI = dispersion index (Loveday and Pyle, 1973);
AS = aggregate stability index (Low, 1954);
SM = volumetric soil moisture at time of test (%);
BD = bulk density
Source: Marijnissen, 1983.

different types of parent material (slates and granite). She measured yearly differences of up to and sometimes more than 100%, with wide seasonal contrasts. Generally, spring and autumn rains were most erosive. Rainfall appeared to be closely correlated with erosion within each area, whereas runoff was not significantly correlated with erosion. This means that the response of the soil surface to rainfall is of greater importance for achieving erosion than is runoff alone, suggesting a marked influence of extreme rainfall events. A great difference existed in the response of the two materials to rainfall. The significance of this kind of research is firstly that it illustrates the variability in erosion at a local and short time-scale and, secondly, that it can be applied to hydrological and soil-erosion modelling. Francis et al. (1986) have approached the variability and the interrelations between several variables at an even smaller scale. They conclude that at the scale of the hillslopes in semi-arid regions, significant relationships exist between plant cover, soil organic matter and soil moisture.

Generally, soil erosion is quantified by estimating sediment yields from surface wash and is classified as rill of sheet erosion. That mass movements (e.g. mud flows) can contribute to erosion on a large scale in the Mediterranean is well known and there is currently much concern about the hazards of landslides and mudflows. Any statement concerning the impact of climatic change on mass-movements will require good information about expected changes in precipitation.

An attempt to model the relationship between global climatic change and regional hydrology in a drainage basin in California has been made by Gleick (1987a,b). He states that climatic changes are expected to generate, for instance, shifts in the timing and magnitude of water availability. This would especially have repercussions for vegetation and hydrology response in the Mediterranean growing seasons.

Climatic change will influence the erodibility of the soil through the effects it has on the organic matter content or in the chemistry of the soil solution as described above. These effects are usually separated into those that primarily influence the ability of a soil to accept water as infiltration, and which therefore influence the partitioning of rainfall between infiltration and overland flow, and those that influence the resistance of soil particles to detachment. The most important climatic characteristics with respect to erosivity are those which determine the energy of impacting raindrops (drop-size, wind velocity and plant cover) and the erosivity of overland flow. Field studies that enable the effects of erosivity and erodibility to be isolated and expressed in rates of soil loss or erosion are few because they require replicated measurements over long periods of time. It is, however, possible to conduct laboratory or field experiments that enable the erodibility of soils to be compared and related to properties that reflect the effects of climate on soil structure. Such studies often consider the aggregate stability or infiltration characteristics as surrogate parameters for soil erosion. For any one soil, erosivity can be studied by comparing soil loss measured on plots, under natural or simulated rainfall, to obtain an evaluation of the erosivity of rainfall and runoff.

In general, the relationship between long-term erosion and rainfall, (and to some extent temperature) can be inferred from sediment yield data. Most authors agree that erosion is highest in areas receiving between 200–600 mm of rainfall per year (Fig. 4.13). Yair and Enzel (1985) have questioned the validity of such relationships in Israel because higher rates of erosion occur in areas with less than 100 mm of precipitation than in more humid areas. The applicability of sediment yield data to this problem in the Mediterranean region is difficult to verify due to limited field data.

A more realistic approach is to consider interactions between climate vegetation and erosion in simulation models. A promising attempt to this is made by Kirkby and Neale (1986) based on vegetation models, models for catchment hydrology and the Universal Soil Loss Equation; it embodies seasonality (e.g. Fig. 4.14). Although the authors have presented results from their model for realistic seasonality and number of rain days, the effect of seasonality must be approached with caution. Fig. 4.15 shows the substantial effect of seasonality on simulated erosion rates. Mediterranean climates are characterized by precipitation low in summer and high in winter, autumn or spring. The picture in Fig. 4.15 is then altered according to curve −50% in Fig. 4.16, which indicates that high rainfall peaks in periods with relatively thin vegetation (early spring) and low soil organic biomass (autumn). This is illustrated in Fig. 4.17 for an arid Mediterranean area in southern Spain.

Disregarding the effect of increasing precipitation, the decrease in organic soil biomass and the increase in soil erosion for a 4°C temperature rise are estimated for two areas (Table 4.3). At present Thornes and co-workers at Murcia (1987) are attempting to collect data to test and validate such models.

Land degradation and climatic change   119

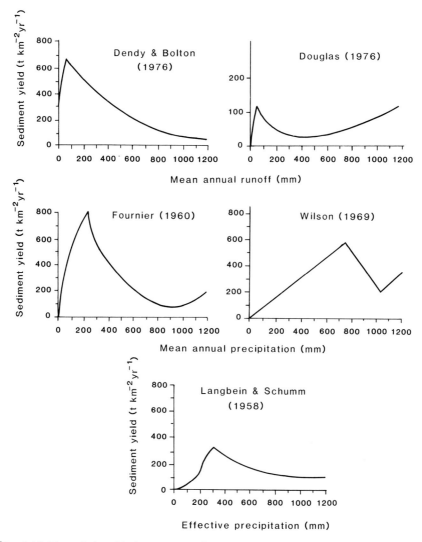

Fig. 4.13 The relationship between sediment yield and annual rainfall (from Yair and Enzel, 1985).

Table 4.3

		Organic soil biomass (kg/m²)	Erosion rate
Crete	500 mm	15 → 19°C	2.8 → 1.5 0.1 → 0.4 (4x)
SW Spain	600 mm	17 → 21°C	2.6 → 1.3 0.12 → 0.8 (7x)

120 *Climatic Change in the Mediterranean*

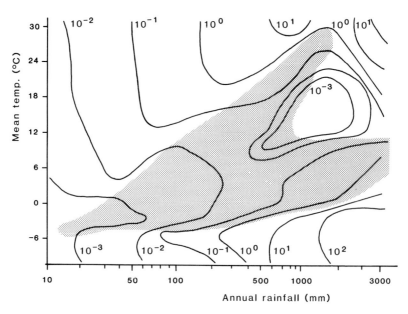

Fig. 4.14 Relationship between forecasted erosion loss and average annual rainfall in summer (from Kirkby and Neale, 1986).

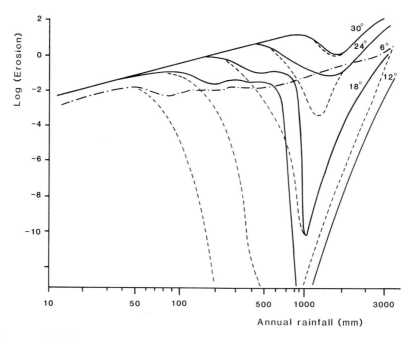

Fig. 4.15 Relationship between forecasted erosion loss and average annual rainfall in summer (from Kirkby and Neale, 1986).

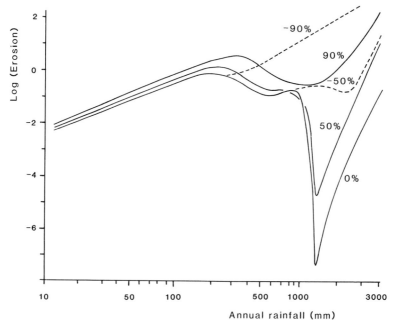

Fig. 4.16 Relationship between forecasted erosion loss and average annual rainfall in summer (from Kirkby and Neale, 1986).

General relationships between climate and erosion need to recognize the importance of strong seasonal contrasts. The erodibility of the soil, for reasons partly explained above is dynamic, often highest after a prolonged dry spell. Consequently, at such times storms, with a relatively low erosivity, often can cause more erosion than general relationships might predict.

It is, therefore, unrealistic to expect good statistical relationships between soil loss measurements and rainfall parameters. Soil loss measurements on plots from various locations in the Mediterranean have been examined for many locations. Poor relationships between rainfall parameters and soil data from Florence, Morocco, Valencia, Catalonia and Murcia for example, illustrate the lack of any simple correlation between soil loss and rainfall. What is nevertheless found in all cases, is that one or two rainfall events each year produce an extremely high proportion of the total yearly soil loss, in many cases upwards of 80%. These rainfall events are not always the most extreme with respect to duration, amount or intensity of rainfall. Rather they are related to the effects of land use and weather on the soil structure and to various erosional thresholds.

Erosion takes on several forms. Sheet or wash erosion caused by overland flow and flow concentrated in shallow channels or rills, occurs where infiltration rates are low or where flow is concentrated in various ways. A large number of models have been developed to predict soil erosion rates by this type of erosion as a function of climatic, soil, vegetation and land-use parameters. These models are difficult to test

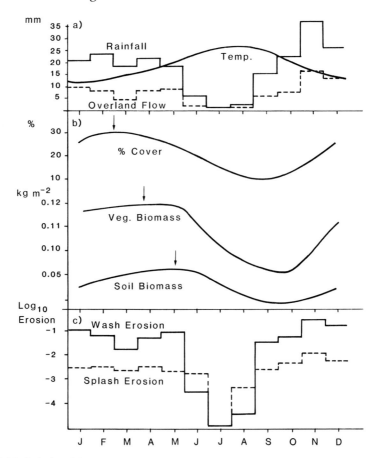

Fig. 4.17 Relationship between forecasted erosion loss and average annual rainfall in summer (from Kirkby and Neale, 1986).

due to a lack of data and are usually used to predict the impact of changes in land use. They could easily be used to estimate the impact of climatic change.

Gully erosion occurs under specific field conditions in various forms. Arroyo-type gullies, which occur in valley fills, are likely to become active as they are related to peak discharges in ephemeral channels. Gully incisions also could be reactivated in alluvial fans. Gully erosion in alluvial fans has in the past been associated in a complex way with climatic change (Harvey, 1987). The types of gully associated with dispersive soils or sediments also are likely to become active or extend their distribution. This is because areas of dispersive soils will increase due to altered salt balances and because the amount of runoff will increase due to higher degree of subsurface and surface runoff concentration. To model the increase in gully erosion that might result from climatic change is feasible but much more difficult than in the case of sheet erosion due to the general lack of research in this area.

The possibility that wind erosion might increase as a result of increasing aridity can not be overlooked. Certainly, the potential for wind erosion (higher soil erodibility) will increase and this can be quantified. However, it is not known how the frequency of erosive wind velocities will change.

3.4 Sediment Budgets

The sediment derived from the more intensive erosional processes postulated above can be considered as a component of the sediment budget. In this budget, sediment is supplied by land surface and channel processes to river systems, where it is either stored or transported as an output to the ocean. Studies of sediment budgets are few and recent because of difficulties in obtaining data. Nevertheless, a number of general statements relevant to the scenario of increased erosion due to higher aridity can be made for the Mediterranean areas.

With increased aridity and higher erosion, sediment will accumulate at slope-foot positions and in small-order drainage basins. Where rivers are ephemeral and have an effluent character as in south-eastern Spain and north-east Morocco, for example, they are characterized by high transmission losses, so that discharges produced by flood events decrease downstream. Particularly in mountainous areas supplying coarse sediment, rivers will develop steep channel gradients as coarse bedload accumulates. Much of the coarse sediment leaving mountainous regions in southern Spain and elsewhere is deposited in alluvial fans at pediment or basin boundaries. Relationships between climate and processes on alluvial fans have been studied by Harvey (1987) and compared for the semi-arid south-west of the USA and for southern Spain. Under present conditions, most alluvial fans in south-eastern Spain are dissected by channels so that extremely high discharge events can carry material across the fan to adjacent river systems. Under the drier conditions prevailing in the semi-arid south-west, and in the driest part of south-west Spain (< 250 mm/yr) channels capable of carrying coarse debris across fan boundaries do not occur and material transported during floods accumulates on the fan. With increasing aridity, therefore, in spite of higher rates of erosion, there will be a greater tendency for any transported material to accumulate on fans and consequently for a lower supply of coarse material to river channels. Since most mountains bordering the drier Mediterranean regions are bordered by coastal plains or lowlands on which fans are developing, there will be very few areas where coarse sediment input into the Mediterranean Sea will be increased.

In the Mediterranean region there are extensive areas of marine marls or shales which weather into a highly erodible regolith. These deposits supply large amounts of silt and clay to river systems, particularly where badlands have developed. Increased erosion in such areas will of course increase sediment and solute levels in rivers receiving water draining from them. Morphological adjustments in river channels in these areas will be different from those where coarse sediments predominate.

A higher sediment input into low order rivers, and the accumulation of this material in the channels of rivers in the dryer parts of the Mediterranean basin, will be a major impact in these regions. The types of adjustment that will occur in the morphology of the river channels is well understood (see Petts, 1984, for example), as are the implications for flooding and river engineering. In more humid regions of the Mediterranean basin, without information on the altered precipitation regimes it is difficult to hypothesize the impact of temperature alone. Nevertheless, in the Mississippi basin Knox (1984) has shown that colder periods are accompanied by lower peak discharges. If these findings can be translated to the Mediterranean, lower sediment inputs into rivers and oceans could be hypothesized.

Because most studies of river-sediment yields in the Mediterranean regions have not been considered as elements of budgets, it is difficult to provide data that could substantiate the above without a substantial analysis of the data. An analysis of available sediment yield data for selected drainage basins could provide useful information.

3.5 Salinity and Water Quality

The altered hydrological regimes described above might be expected to produce a general increase in the solute loads of drainage waters. The trend towards an increase in salinity, although general, will depend very much on the seasonal rainfall and runoff distribution and on the relative contributions of water following different pathways to river systems.

The increase in aridity and the resulting soil degradation will result in a general decline in percolation and eventually in a decrease in the contribution of delayed flow supplying water in equilibrium with the minerals present in the drainage basin; this is often water in equilibrium with calcium carbonate or gypsum. A decrease in the discharge of such water might mean a shift towards streams having perennial or ephemeral regimes.

Evaporation of water present in the sediments of perennial or ephemeral channels will lead to salt accumulation. These salts together with those that seasonally accumulate in the soil, will be transported during relatively short periods of flow. The degree of deterioration in water quality will reflect the balance between dilution and accumulation.

A problem can arise when the runoff has a high sodium adsorption ratio (SAR). In combination with relatively low electrolyte concentrations this can lead to a common situation whereby transported clayey sediments become dispersed. The viscosity of the water is increased by the dispersed clay to such an extent that sand particles remain in suspension and hypersediment concentrations occur, giving the runoff the character of dilute mudflow. This makes the water completely unsuitable for domestic use and causes problems of sedimentation at water intake points and in irrigation channels.

4 Conclusions

4.1 Limitations

Predicting the potential impact of climatic change on soil erosion in the Mediterranean is limited by a number of constraints. These include: (1) the lack of information concerning the nature of the climatic changes that will occur at a regional scale; (2) in the Mediterranean and elsewhere there are very few actual measurements of erosion that can be used to test model hypotheses (such measurements are required to establish relationships between climate or weather phenomena and climatic-sensitive processes); (3) the complexity of the system being considered (see Introduction); (4) the only conclusions that can be made are rather obvious generalizations. Although a scientific basis for most of them is argued, some links are unsubstantiated.

4.2 Major Impacts

An increase in the temperature of 3 to 4°C, as predicted by the GISS model is argued to have the following impacts: (1) an increase in aridity due to higher PE; (2) an increase in soil degradation due to the effect of increased temperature and PE on soil organic matter, highly soluble salts and the moisture balance of the soil; (3) an increase in soil erodibility due to lower rates of water acceptance and a decrease in aggregate stability; (4) higher rates of erosion on slopes and either relatively or absolutely more overland flow during extreme rainfall events; (5) higher sediment concentrations in slope runoff; (6) higher sediment and solute supply from badland and/or marl areas; (7) an increase in the risk of rill and gully erosion; (8) higher sediment and lower runoff inputs into higher order streams; (9) an increase in the discontinuity of runoff and sediment transport in drainage basins; (10) lower rates of sediment transport in the major rivers; (11) adjustment of river channels, especially those carrying coarse load (channel capacity will decrease, braiding will increase, fans will grow and form barriers and flooding will be more frequent); (12) a higher susceptibility to forest fire and a decrease in biodiversity; (13) more arid soil conditions and a decrease in the area suitable for profitable agriculture; (14) increasing management costs in agriculture to maintain crop yields.

4.3 Research Requirements

As has become clear in this review, numerous studies have been made on the effect of (changing) precipitation, temperature and evapotranspiration on the soil system, even if not always carried out in the context of climatic change. Many workers have tried to find general relationships between climatic variables and parameters that are commonly used to assess soil erosion and soil erodibility. Others have attempted to construct models that generate quantitative approximations of, for example, primary productivity or soil erosion rates. The complexity of the systems considered, however, impedes reliable assessments. Knowledge of the behaviour of, and interrelations between, any kind of subsystem involved still needs to be improved considerably. It seems worthwhile to make more use of instrumental-data-based scenarios considering actual climatological and environmental situations, as well as historical instrumental data.

A large number of case studies that integrate climatic, biotic an abiotic aspects in the Mediterranean area are available.

5 REFERENCES

Agassi, M., Shainberg, I. and Morin, J., 1985. Infiltration and runoff on wheat fields in the semi-arid region of Israel. *Geoderma*, **36**, 263–276.

Arnoldus, H.M.J., 1977. Methodology used to determine the maximum potential average annual soil loss due to sheet and rill erosion in Morocco. In: *Assessing Soil Degradation*, FAO Soil Bulletin No. 34, Rome, 127–132.

Arnoldus, H.M.J., 1978. An approximation of the rainfall factor in the Universal Soil Loss Equation. In: de Boodt, M. and Gabriels, D. (eds), *Assessment of Erosion*, John Wiley & Sons, 8–9.

Aru, A., 1984. Aspects of desertification in Sardinia – Italy. In: *Desertification in Europe. Proceedings of the Information symposium in the EEC Programme on Climatology*, Mytilene, Greece, 194–198.

Beard, J.S., 1983. Ecological control of the vegetation of southwestern Australia: moisture versus nutrients. In: *Mediterranean Type Ecosystems; the Role of Nutrients*, Ecological Studies No. 43, Springer-Verlag, 66–73.

Ben-Hur, M., Shainberg, I., Keren, R. and Gal, M., 1985. Effect of water quality and drying on soil crust properties. *Soil Sci. Soc. Am. J.*, **49**, 191–196.

Berg, B., Jansson, P. and Meentemeyer, V., 1983. Litter decomposition and climate — regional and local models. Manuscript work group meeting, Uppsala.

Berger, A., 1984. Desertification in a changing climate with a particular attention to the Mediterranean countries. In: *Desertification in Europe. Proceedings of the Information Symposium in the EEC Programme on Climatology*, Mytilene, Greece, 15–34.

Bolt, G.H. and Bruggenwert, M.G.M., 1982. Soil chemistry A. Basic Elements. Elsevier Scientific Publishing Company.

Chorley, R.J., Schumm, S.A. and Sugden, D.E., 1984. *Geomorphology*. Methuen, London.

Dalal, R.C. and Mayer, R.J., 1986. Long-term trends in fertility of soils under continuous cultivation and cereal cropping in southern Queensland. II, Total organic carbon and its rate of loss from the soil profile. *Aust. J. Soil Res.*, **24**, 281–292.

DeByle, N.V., 1981. Clearcutting and fire in the Larch/Douglas-fir forest of western Montana – A multifaced research summary. General Technical Report INT-99. United States Department of Agriculture.

Elliot, G.L., Lang, R.D. and Campbell, B.L., 1983. The association of tree species, landform, soils and erosion on Narrabeen sandstone west of Putty, New South Wales. *Austr. J. Ecol.*, **8**, 321–331.

Finck, A., 1982. Fertilizers and fertilization. Introduction and practical guide to crop fertilization. *Verlag Chemie*,

Francis, C.F., Thornes, J.B., Romero Diaz, A., Lopez Bermudez, F. and Fisher, G.C., 1986. Topographic control of soil moisture, vegetation cover and land degradation in a moisture stressed Mediterranean environment. *Catena*, **13**, 211–225.

Garcia, C.C., 1984. El potencial agrario de los suelos de torre pachedo. *Excmo. ayuntamiento de Torre Pachedo a.o.*

Gleick, P.H., 1987a. The development and testing of a water balance model for climate impact assessment. Modeling the Sacramento Basin. *Water. Res.*, **23**(6), 1049–1061.

Gleick, P.H., 1987b. Global climatic changes and regional hydrology: Impacts and responses. In: *The Influence of Climate Change and Climatic Variability on the Hydrologic Regime and Water Resources. Proceedings of the Vancouver Symposium*, IAHS Publication, 168, 389–402.

Harvey, A.M., 1987. Alluvial fan dissection: Relationship between morphology and sedimentation. In: Frostick, L. and Reids, I. (eds), *Desert Sediments: Ancient and Modern. Geological Society Special Publication*, **35**, 87–103.

Huis, J. Van and Ketner, P., 1987. Climate sensitivity of natural ecosystems in Europe. Discussion paper, prepared for the European Workshop on Interrelated Bioclimatic and Land Use Changes, The Netherlands.
Imeson, A.C., 1983. Studies of erosion thresholds in semi-arid areas: field measurements of soil loss and infiltration in northern Morocco. In: de Ploey, J. (ed.), *Rainfall Simulation, Runoff and Soil Erosion, Catena Supplement*, **4**, 79–89.
Imeson, A.C. and Verstraten, J.M., 1985. The erodibility of highly calcareous soil material from southern Spain. *Catena*, **12**, 291–306.
Imeson, A.C. and Verstraten, J.M., 1986. Erosion and sediment generation in semi-arid and Mediterranean environments: the response of soils to wetting by rainfall. *J. Water Res.*, **5**, 388–418.
Jung, H.J. and Bach, W., 1985. GCM-derived climatic change scenarios due to an CO_2-doubling applied for the Mediterranean area. *Arch. Met. Geoph. Boicl., Ser. B*, **35**, 323–339.
Kimmins, J.P., 1987. *Forest Ecology*. MacMillan Publishing.
Kirkby, M.J. and Neale, R.H., 1986. A soil erosion model incorporating seasonal factors. In: Gardiner, V. (ed.), *International Geomorphology Part II*, John Wiley & Sons, 189–210.
Knox, J.C., 1984. Fluvial responses to small scale climate changes. In: Costa, J.E. and Fleisher, P.J. (eds), *Developments and Applications of Geomorphology*, Springer-Verlag, 318–342.
Lamb, H., 1984. Climate and history in northern Europe and elsewhere. In: Muerner, N.A. and Karlen, W. (eds), *Climatic Change on a Yearly to Millenial Basis*, Reidel Publishing, 225–240.
La Roca Cervigon, N. and Calvo Cases, A., 1988. Slope evolution by mass movements and surface wash (Valls d'Alcoi, Allicante, Spain). In: Imeson, A.C. and Sala, M. (eds), *Geomorphology in Environments with Strong Seasonal Contrasts, Catena Supplement*, **12**.
Lavee, H., Imeson, A.C., Pariente, S. and Benyamini, Y., 1988. The infiltration characteristics and erodibility of soils along a climatalogical gradient in the Judean desert. Manuscript.
Mabutt, J.A., 1986. Desertification indicators. *Clim. Change*, **9**, 113–122.
Meentemeyer, V., 1978. An approach to the biometeorology of decomposer organisms. *Int. J. Biometeor.*, **22**(2), 94–102.
Morgan, R.P.C., Finney, H.J., Lavee, H., Merritt, E. and Noble, C.A., 1986. Plant cover effects on hillslope runoff and erosion: Evidence From Two Laboratory Experiments. In: Abrahams, A.D. (ed.), *Hillslope Processes*, Allen & Unwin, 77–97.
Petts, G.E., 1984. *Impounded Rivers*. John Wiley & Sons.
Pittock, A.B. and Nix, H.A., 1986. The effect of changing climate in Australian biomass production – a preliminary study. *Climate Change*, **8**, 243–255.
Quezel, P., 1981. Floristic composition and phytosociological structure of sclerophyllous mattoral around the Mediterranean. In: Di Castri, F., Goodall, D.W. and Specht, R.L. (eds), *Ecosystems of the World*, Mediterranean-type Shrublands, Elsevier, **11**, 107–121.
Renard, K.G., 1960. Some hydrological characteristics of arid-land watersheds. *Proc. Joint ARS-SCS Hydr. Workshop*, **36**, 36.1–36.9.
Sala, M., 1988. Hillslope runoff and sediment production in two Mediterranean mountain environments. In: Imeson, A.C. and Sala, M. (eds), *Geomorphology in Environments with Strong Seasonal Contrasts, Catena Supplement*, **12**, in press.
Schumm, S.A., 1977. *The Fluvial System*. John Wiley & Sons.
Sevink, J., 1988. Soil organic profiles and their importance for hillslope runoff. In: Imeson, A.C. and Sala, M. (eds), *Geomorphology in Environments with Strong Seasonal Contrasts, Catena Supplement*, **12**, in press.

Shainberg, I. and Letey, J., 1984. Response of soils to sodic and saline conditions. *Hilgardia*, **52**(2),

Sharma, M.L. Evapotranspiration and stream salinity as a consequence of land-use change in southwestern Australia. Manuscript.

Sommerfeldt, T.G. and MacKay, D.C., 1982. Dryland salinity in a closed drainage basin at Nobleford, Alberta. *J. Hydro.*, **55**, 25–41.

Soong, N.K., 1980. Influence of soil organic matter on aggregation of soils in Peninsular Malaysia. *J. Rubb. Inst. Mal.*, **28**(1), 32–46.

Specht, R.L. and Moll, E.J., 1983. Mediterranean-type heathlands and sclerophyllous shrublands of the world: an overview. In: Kruger, F.J., Mitchell, D.T. and Jarvis, J.U.M. (eds), *Mediterranean Type Ecosystems; the Role of Nutrients, Ecological Studies No.* **43**, Springer-Verlag, 41–65.

Utomo, W.H. and Dexter, A.R., 1981. Tilth mellowing. *J. Soil Sci.*, **32**, 187–201.

Wigley, T.M.L., 1986. Emperical climate studies. In: Bolin, B. *et al.* (eds), *The Greenhouse Effect, Climatic Change and Ecosystems*, SCOPE **29**, 271–322.

Yair, A. and Enzel, Y., 1987. The relationship betwen annual rainfall and sediment yield in arid and semi-arid areas. The Case of the Northern Negev. In: Ahnert, F. (ed.), *Geomorphological Models. Theoretical and Empirical Aspects, Catena Supplement*, **10**,

[5]

Implications of Climatic Change on the Socio-Economic Activities in the Mediterranean Coastal Zones

A. Barić
*(Institute of Oceanography
and Fisheries, University of Split, Croatia)*

F. Gašparović
*(Consultant on Land-use and Environment,
Zagreb, Croatia)*

Abstract

This report gives an estimate of possible impacts of climatic changes on the socio-economic structures and activities, and on the distribution and dynamics of population in Mediterranean coastal areas. Special attention has been given to human settlements and major activities (agriculture, forestry, industry, energy, tourism, transportation, fisheries, aquaculture) and to the possible implications of climatic changes on them. Furthermore, where the available data permit, the study goes into greater detail with the coastline and the coastal belt of the region (the areas influenced by the Mediterranean climate or influencing the Mediterranean environment).

Socio-economic structures and activities are closely dependent upon the existing climatic conditions. It is assumed that the impact on the socio-economic structures and activities will be primarily local, with possibly important common characteristics for small and large areas or regions. The assessment of local and regional impacts presents great difficulties due to a lack of understanding and insufficient knowledge of the local conditions. In the integrated planning component (Blue Plan and Priority Actions Programme) of the Mediterranean Action Plan, impressive amounts of data have been gathered relating to Mediterranean coastal zones, and were used in assessing the impact of climatic changes. A commonly shared confidence is that to date research of the climatic changes has enabled fairly reliable predictions of what is going to happen in the future (the years 2025 and 2100). However, still to be found is how it is going to happen in different areas. Therefore, further research into typical sites is imperative, and without it the assessment of the impacts of climatic changes on various socio-economic situations and activities will remain on a general level.

At present, approximately 133 million people (representing 37% of the total population of the Mediterranean countries) live in the littoral zone (which represents only 17% of the total area of these countries). Sixty-one per cent of them live in urban areas. According to the results of five Blue Plan scenarios, in the year 2025 there will be between 200 and 220 million inhabitants living in the coastal zone. Seventy-five per cent of the population of coastal zones will be located in cities.

The changes of climatic conditions will have a limited impact on the distribution and dynamics of the population of the littoral zones. The natural population growth will not be affected by climatic changes and will continue to follow the present general trends of population, i.e. decrease of natural population growth in the northern Mediterranean countries and a strong increase of population in the countries on the southern and eastern coasts of the Mediterranean. The population influx caused by the present process of littoralization probably will not change, although it could accelerate in the south due to the natural spreading of the deserts. Approximately 5% of the population living in coastal zones will be indirectly affected by the impacts of climatic changes and sea-level rise. A part of this percentage (15–20%) will be directly affected by sea-level rise.

About 54% of the total Mediterranean coastline is rocky, the rest consists of low sedimentary shores. The rise of sea-level will affect primarily those parts of the coast that are alluvial and low-lying.

The climatic changes will affect both the natural and the man-made environment, but their impacts on individual structures and activities will differ. Coastal settlements and harbours, as well as lowlands and tourist establishments, will be affected directly by sea-level rise. Amplifying the static sea-level rise, the changed wave dynamics will increase the threat to the coastal fresh-water resources, waste-water collection and disposal systems and aquaculture.

The temperature rise and its consequences (evaporation, changed wind and ground-water patterns, increased salination of soils which constitute the lowlands, greater danger of forest fires) will be the major generators of changes in agricultural production and forestry. It is to be expected that many present crop cultures traditionally raised in the region will change quality and yield, be replaced by other cultures, and so on. However, in assessing the impact of climatic changes on agriculture and forestry knowing local conditions is imperative.

An inherent quality of the socio-economic and political structures and their complexity is a certain inertness when it comes to reacting to the phenomena, such as the climatic changes, which will not manifest themselves in the very near future. But now, since the existing knowledge of natural systems have stressed that these changes will occur as early as in the next decades, the countries concerned should incorporate the mitigation of climatic changes in their national development strategies and environmental management.

This calls for (1) alerting the public (avoiding unnecessary alarm) and all administrative and economic structures involved in the decision-making process by informing them of possible effects of the climatic changes; (2) studying the local conditions under which these changes will occur; (3) incorporating the knowledge of the climatic implications in the integrated planning process and methods for global development and for the development of individual economic activities; (4) developing appropriate cost-benefit and environmental management; and (5) promoting and developing new technologies for the mitigation of impacts of climatic changes.

Interventions for the mitigation of adverse effects of climatic changes will require considerable funds which should be anticipated in the national, regional and local plans.

Finally, the entire activity of research, mitigation and protection related to the impacts of climatic changes requires efficient coordination at the national level and cooperation of all Mediterranean countries.

1 Introduction

The socio-economic structure and activities of any individual country or region are closely dependent upon climatic conditions. This is so because the climatic conditions impact the natural environment (the soil, water, sea, air, vegetation and the forests) – essentially the basis of all human activities and conditions of life on Earth. Such climate affects the way people live, their activities and the dynamics of society.

In assessing the impact of climatic changes on socio-economic structure and activities, a number of problems and difficulties arise (Clark, 1985; UNEP, 1985; WHO, 1988). Many of these have been fixed by William C. Clark at the International Conference in Villach 1985. Supposing that the appearance and accumulation of the greenhouse gases will be manifested simultaneously both on a general area and locally, we may assume that their impact on the socio-economic structure will be primarily local with possibly important common characteristics for both small and large areas or regions such as the Mediterranean area. Naturally, the climatic changes will progressively affect the general socio-economic structure of the world as a whole as a sum of all local impacts. However, the assessment of both local and regional impacts presents great difficulties due to inadequate understanding and knowledge of the local situations.

Besides the mentioned problems in documenting the local situations in which the socio-economic structures and activities will be impacted by climatic changes, there are other problems with specific and different characteristics of the greenhouse effect with respect to the structure and activities being analysed. For example the climatic changes that impact the environment and the consequent environmental changes are fairly well defined when it comes to the air-temperature rise and sea-level rise. But, changes in other climatic conditions also impacted (e.g. vegetation, soil, water, evapotransportation, etc.) are not so well defined.

Reducing the impact of climatic changes in the next century will mainly have financial repercussions. There will be new investments for reducing the impacts of the sea-level rise as well as for insuring adequate amounts of water for agriculture or for the population of a given area. The correlation between the greenhouse effect and the problems of energy, agriculture, urbanization and environment in general will require that new strategies be developed for socio-economic development, technology, environmental protection on the whole, protection against pollution, etc.

The changes brought about by the greenhouse effect occur slowly, while the socio-political and economic structures are adapted to react only to impending emergencies and dangers. A very complex net of socio-economic structures needs, according to present experience, a period of 20 to 50 years to adapt. Increased investments and expenditures to reduce the negative impacts of climatic change may affect developing countries particularly adversely. This is significant for the southern and eastern Mediterranean countries.

The examination and analysis of the impacts of climatic changes on the socio-economic structures require a realistic assessment of the ways those changes impact the natural systems both regionally and locally. Taking into account the complexity and inertness of socio-economic structures to deal with new social processes brought about by the changes, it is difficult to make a firm appraisal of the ways in which a socio-economic structure could adapt its activities either to a new situation or to a process of continual change. It is also difficult to distinguish the impacts of climatic change on the socio-economic structure from the impacts of human activities on the development of the greenhouse effect. Nevertheless, the assessment of these impacts should serve as a stimulus for preliminary research on the local, regional and sectoral socio-economic levels (and that for scientific research) as well as a stimulus for developing tentative strategies to combat these impacts.

2 BACKGROUND

Although the Barcelona Convention (UNEP, 1982) has defined the Mediterranean area as having common geographic, cultural and climatic characteristics, the area possesses quite different geographic, economic, social and political structures. They are on the one hand interconnected while on the other tied to other political, cultural and social unities, which themselves reach far inland. There are 18 countries located on the coasts of the Mediterranean Sea, and these coastal areas in turn directly impact and are impacted by the adjacent environment of the Mediterranean Sea.

Trying to fix the boundaries of the impact by the Mediterranean Sea is not as easy as it might seem. In part it depends on climatic, cultural, traffic, economic and political criteria. For to assess the impact of climatic factors, the following definitions seem the most important:

1) *The coastline* – an area where land and sea meet, along with that part of the shore on which strong impacts of the sea-level rise are expected. This territory includes both the inhabited and the uninhabited regions, ports, industry, plus roads and traffic ways erected near the shore. This zone can vary in width from a few metres to hundreds of metres.

2) *The immediate coastal belt* – an area that varies in width from 1 to 10 km (and frequently more) on which indirect consequences of the sea-level rise will be felt. This zone also will suffer from possible changes in the rainfall and thus in the supply of drinking water; and whose width depends upon the restructuring and changes along the coastline. Included in the coastal belt are parts of cities further from the coastline, a large part of the industry, harbours located further inland, agricultural areas, traffic ways, etc.

3) *Mediterranean regions*, as defined by the Blue Plan, are regions with strong Mediterranean characteristics and are located within individual administrative units; this allows for the relatively easy determination of important demographic figures such as population, income, etc. In any given country some administrative units reach further inland than others, and in turn some comprise a greater percentage of the national territory than others. Those administrative units will carry out and play a major role in both the organizational and economic activities geared toward alleviating the impacts of climatic change.

4) *The national territories* of the Mediterranean countries – due to the impact of the dynamic growth, migration and urbanization of populations and of the total economic growth as well as the impact of coastal-zone decision-making centres on the country as a whole, it is often necessary to include the entire countries when assessing the impacts of climatic change.

Precise information on the environmental conditions, economic activities, plus infrastructures and superstructures for the first two areas does not exist. An assessment of the number of inhabitants immediately next to the coastline or in the immediate coastal belt may be made with reference to the percentages of people in the Mediterranean administrative regions. But, this only serves as a general guide. Therefore, the number of people living next to the coastline (1) is estimated as 5% of the population of the coastal administrative regions. Likewise, the number of inhabitants of the immediate coastal belt (2) is estimated as 30 – 50% of the population of the coastal administrative regions (3). This, however, is assumed for the Mediterranean area as a whole and will vary greatly in specific localities.

To evaluate the adaptability of the socio-economic structures and activities to change in climatic conditions, the state of the economy in each individual Mediterranean country must be known. Clearly, the climatic changes will tax the national economy and especially the coastal economic structures of any individual country through their mitigation. Therefore facts and figures on the distribution of the Gross Domestic Product (GDP) and of the Gross National Product (GNP) are most valuable (Figs 5.1 and 5.2; Tables 5.1–5.3).

3 INTEGRATED PLANNING

3.1 Scenarios of the Possible Long-term Development of the Mediterranean for the Years 2000 and 2025

Within the framework of the Mediterranean Action Plan (MAP) and its component called "Integrated Planning", an objective of the Blue Plan was to study the long-term relationship between development and the Mediterranean environment. After assessing both the past and present history/status of those relationships (first phase of the Blue Plan), possible future relationships between development and the environment could be discussed. Collecting and applying the data to scenarios for years 2000 and 2025 presented great difficulties for two main reasons:

1) In relation to large global macromodels the Mediterranean is not defined as a macroregion. In most models it is divided into several different regions.
2) Specific figures on the socio-economic structure in coastal zones are hidden in national statistics; at best they may be found on the level of administrative regions. Likewise, data concerning the state of the natural environment often are hidden in local studies, surveys, assessments, etc.

Without precise data on the environment of coastal regions, it is difficult to define the impact of the socio-economic structure on the environment and specially of climatic change. During the initial concept of the Blue Plan in 1978/1980, the need for a special scenario dealing with interruptions, standstills, ruptures, surprises, mutations and catastrophes was examined, but because the time range of Blue Plan's research extends only to 2000 and 2025 years (in which climatic changes still will not be dramatic) this idea was abandoned.

134 *Climatic Change in the Mediterranean*

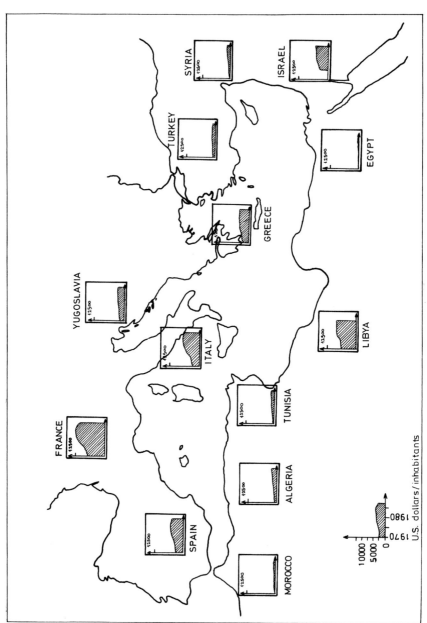

Fig. 5.1. Gross national product per capita, 1976–1983.

Socio-economic activities and climatic change   135

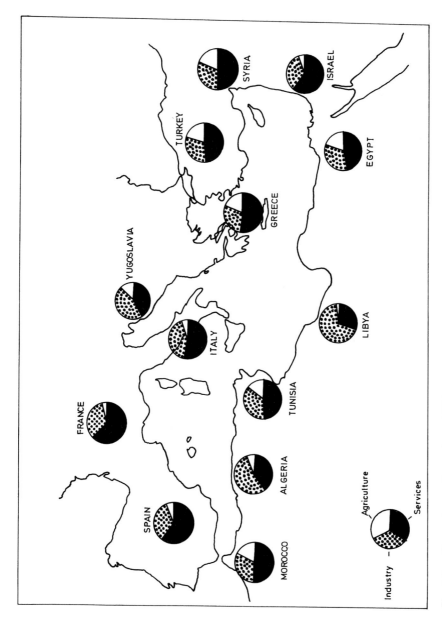

Fig. 5.2. Gross domestic product, 1982.

Table 5.1 Distribution of gross domestic product (percentage)

	Agriculture				Industry				Services			
	1960	1976	1980	1984	1960	1976	1980	1984	1960	1976	1980	1984
Spain	21	9	8	6[a]	39	39	37	34[a]	40	52	55	60[a]
France	9	6	4	4	48	43	36	34	43	51	60	62
Italy	15	8	6	5	38	41	43	40	47	51	51	55
Yugoslavia	24	15	12	15	45	43	43	46	31	42	45	40
Greece	23	18	16	18	26	31	32	29	31	51	52	53
Turkey	41	29	23	19	21	28	30	33	38	43	47	47
Syria	25	17	20	20	21	36	27	24	24	47	53	57
Israel	11	8	5	5	32	43	36	27	57	49	59	68
Egypt	30	29	23	20	24	30	35	33	46	41	42	48
Libya	14	3	2	2	9	68	72	64	77	29	26	34
Tunisia	24	21	17	15	18	30	35	35	58	49	48	50
Algeria	21	7	6	6	24	57	57	53	55	36	37	41
Morocco	29	21	18	17	24	31	32	32	47	48	50	51

Note: [a] = 1982
Source: World Bank, World Development Reports

Table 5.2 Gross National Product Per Capita (US$)

	1978	1979	1980	1981	1982	1983	1984
Spain	3470	4380	5400	5640	5430	4780	4440
France	8260	9950	11730	12190	11680	10500	9760
Italy	3850	5250	6480	6960	6840	6400	6420
Yugoslavia	2380	2430	2620	2790	2800	2570	2120
Greece	3250	3960	4380	4420	4290	3920	3770
Turkey	1200	1330	1470	1540	1370	1240	1160
Syria	930	1030	1340	1570	1680	1760	1620
Israel	3500	4150	4500	5160	5090	5370	5060
Egypt	390	480	580	650	690	700	720
Libya	6910	8170	8640	8450	8510		8520
Tunisia	950	1120	1310	1420	1390	1290	1270
Algeria	1260	1590	1870	2140	2350	2320	2410
Morocco	670	740	900	860	870	760	670

Source: World Bank, World Development Reports

Table 5.3 Gross Domestic Product (US$ million)

	1960	1979	1980	1981	1982	1983	1984
Spain	11403	180800	198320	185080	181250	157880	160930
France	60060	571300	651890	568560	537260		489380
Italy	37190	323600	393950	350220	344580	352840	348380
Yugoslavia	9860	61500	62150	63350	68000	46890	38990
Greece	3110	33370	35650	33390	33950	30770	29550
Turkey	8820	56460	53820	53910	49980	47840	47460
Syria	890	9110	12900	15240	15240	16850	15930
Israel	3040	15300	15340	17440	20490	20660	22350
Egypt	3880	17050	22970	23110	26400	27920	30060
Libya	310	24570	32090	27400	28360	31360	30570
Tunisia	770	6070	7300	7100	7090	7020	6940
Algeria	2740	29810	39870	41830	44930	47200	50690
Morocco	2040	14950	17940	14780	14700	13300	13300

Source: World Bank, World Development Reports

The Blue Plan studied the impact of population and urbanization, agriculture, industry, energy, tourism and transport on the Mediterranean environment (soil and water, the forest, coast and sea) (Blue Plan, 1988b). The aim of the prospective exercise is not to recommend types of development, but to illustrate their effect on the environment. These types of development are greatly affected by population trends, the kind of international economic relations established between countries (particularly by the forms of cooperation between northern and southern countries, or between southern countries) and, at the national level, by the constraints of space and natural resources, and by a country's choice of development/environment strategies.

Compared to a continuation of current trends the Blue Plan reference have been envisioned:

1) **Development with weak economic growth.** If world economic trends are reflected by slow growth in most Mediterranean countries, budgetary constraints could greatly hamper the maintenance and

Table 5.4 Population in the five Mediterranean scenarios (in thousands)

Country	Population 1980	1985	Scenario T1 2000	2025	1985	Scenario T2 2000	2025
Spain	37400	38500	41900	44900	38500	41900	44900
France	53700	54600	55200	52600	54600	53200	52600
Italy	57100	57300	57800	53600	57300	57800	53600
Greece	9640	9880	10100	9560	9880	10100	9560
Yugoslavia	22300	23200	24700	25000	23200	24700	25000
TOTAL REGION A	180000	183000	190000	186000	183000	190000	186000
Turkey	44500	49300	65400	92900	49600	68600	105000
Syria	8800	10500	17000	31800	10500	18300	35500
Egypt	41500	44900	63900	90400	47100	65700	97300
Libya	2970	3610	6080	11100	3620	6240	12500
Tunisia	6390	7080	9430	12900	7130	9830	14000
Algeria	18700	21700	33400	50600	21800	34700	54500
Morocco	19400	21900	29500	40100	22200	31400	45000
TOTAL REGION B	142000	161000	226000	329000	162000	235000	366000
Monaco	24	27	30	34	27	30	34
Malta	369	383	418	459	383	418	459
Albania	2730	3050	4100	5780	3050	4100	5780
Cyprus	629	669	762	902	669	762	902
Lebanon	2670	2670	3620	5220	2670	3620	5220
Israel	3880	4250	5300	6870	4250	5300	6870
TOTAL REGION C	10300	11000	14200	19300	11000	14200	19300
TOTAL MEDITERRANEAN COUNTRIES	333000	356000	430000	533000	354000	439000	571000

Source: Blue Plan, 1988b

Socio-economic activities and climatic change 139

investment operations needed for environment protection. It would be more difficult to enforce regulations (as industrial enterprises would be more vulnerable), and they would be less effective because of lack of new investment.

2) **Development with rapid growth**, but neglectful of the environment. This kind of rapid growth could entail serious, even irreversible, damage to the environment because of the greatly increased pressure on resources and the difficulty in gearing efforts to compensate for the harm done.

3) **Well-balanced development, concerned about the environment**. The combination of certain choices of national strategy (*a priori*) involving environmental regulation policies, among others, and enhanced international cooperation (north–south with a more assertive Europe, and south–south among regional groups) could produce a compatibility between economic growth and protection of the Mediterranean environment.

Scenario T3			Scenario A1			Scenario A2		
1985	2000	2025	1985	2000	2025	1985	2000	2025
38500	42200	46000	38500	42200	46000	38500	43900	51800
54600	57200	58400	54600	57200	58400	54600	58300	63500
57300	58600	57200	57300	58600	57200	57300	60500	63300
9880	10400	10800	9850	10400	10800	9880	10000	12100
23200	25200	26800	23200	25200	26800	23200	25800	29000
183000	194000	199000	183000	194000	199000	183000	199000	220000
49300	65400	91900	48800	62300	81700	48800	62300	81700
10500	17000	33800	10500	17000	28100	10500	17000	28100
46900	63900	90400	46800	62200	85000	46800	62200	85000
3610	6080	11100	3600	5920	9910	3600	5920	9910
7080	9430	12900	7050	9060	12100	7050	9060	12100
21700	33400	50600	21600	32200	46500	21600	32200	46400
21900	29500	40100	21900	28900	39100	21900	28900	39100
161000	226000	329000	160000	218000	302000	160000	218000	302000
27	30	36	27	30	36	27	30	36
383	418	459	383	418	459	383	418	459
3050	4100	5780	3050	4100	770	3080	4260	6500
669	762	902	669	762	902	669	762	902
2670	3620	5220	2670	3620	3220	3720	3830	5950
4250	5300	6870	4250	5300	4870	4350	5720	8120
11000	14200	14300	11000	14200	19300	11300	15000	22000
356000	433000	547000	355000	426000	521000	355000	432000	544000

The results achieved by the application of the methods in the Blue Plan scenario do not show any national "plans" geared toward development and protection of the environment nor "plans" for the entire Mediterranean. They do suggest, however, research for the "possible future" within the framework of what is realistically probable.

Apart from the fact that Blue Plan did not specifically assess the impact of the climatic changes, it is true that the results of the scenarios in general draw a picture of the possible changes in development and their influence on the environmental components. The general picture is best seen through the presentation of the scenario dealing with population migrations (movements) in the Mediterranean countries and their administrative coastal regions (Table 5.4).

Other important general guide-points are the facts and figures covering the economic growth, both the Gross Domestic Product and the Gross National Product.

As stated in the introduction, the impact of the climatic changes on the socio-economic structure and activities will be first felt by the rise of expenditures for the reduction of the negative effects of those changes. This is especially true in the case of keeping back the sea-level rise. Also, it is very difficult or nearly impossible at this time to determine the size of the expenditures for the reduction of the effects that will come about in agriculture and especially in food production.

For those reasons, the following chart showing the "behavioural strategies" for the reduction of the effects of climatic changes has been made for four scenarios (Table 5.5).

On a very generalized scheme, such as this one, showing the strategies for the reduction of the effects of climatic changes for the four scenarios, it is very difficult to elaborate any further. That is, further elaboration in the direction of either the sizes or base orders of magnitude of new expenditures is not possible, because it is both regionally and locally conditioned. However, according to experience so far in reducing the effects brought about by climatic changes (that is, specifically in reducing the dynamic sea-level rise witnessed in Holland), we may conclude that the expenditures take up a relatively acceptable part (1–3%) of the Gross National Product (GNP).

If the action of reducing the climatic changes in terms of the sea-level rise were a national strategic endeavour, then it should not be a great burden. But if such action were left up to coastal regions and coastal authorities, the results might be individual and separate interventions which would not ensure complete protection. Difficulties such as these would probably not be present in the developed countries, but they might occur in the developing countries.

The need for expenditures geared toward the protection from the effects of climatic changes might bring about a reduction of the expenditures for environmental protection. However, in forming internationally coordinated national actions for protection from the effects of climatic changes, a great psychological problem arises: changes are slow in their manifestation and structures that make political, administrative and economic decisions are generally oriented toward solving current problems (that is those which are already present or will arise in the near future).

Table 5.5 Behavioural strategies for the mitigation of climatic changes

	Impact effects in the world	Conditions and reactions in the Mediterranean	Assessment of actions and expenditures
T_2: worse trend scenario	No great impacts of emission due to reduced activity: lower energy consumption	Insignificant reactions	Sporadic reactions – one by one, everyone acts for himself, minimal effects of reducing the consequences, slowest and unsure preventions
T_3: moderate trend scenario	Maximum emission, much activity, high energy consumption	Stronger but delayed reactions (economic growth more or less spontaneous, uncontrolled)	Effects of maximal reaction, local, regional or national. Expenditures very high due to uncoordinated and delayed reactions
A_1: alternative trend scenario	Approximately like T_3, but start of fight for protection, start of research for less emissions, start of technology for reduction of effects (installing natural gas, greater solar energy use)	Strong cooperation north–south, sub-stable financial funds, tendencies for new technologies	Coordinated reaction for fight against consequences, also for on-time initiation of measures (prevention)
A_2: alternative integration scenario	Approximately like A_1, but with the use of solar energy and natural gas	Approximately like A_1; weaker cooperation north–south, stronger south–south	Approximately like A_1, less coordinated, smaller funds

Note: Prepared in consultation with M. Grenon, Scientific Director of the Blue Plan

3.2 Priority Actions Programme

The Priority Actions Programme (PAP) is a part of the Integrated Planning of MAP (MAP, 1985). The task of PAP is to carry out practical actions which are expected to yield immediate results. Its aim is to contribute through the exchange of available experience and cooperation among the Mediterranean countries to the protection and enhancement of the Mediterranean environment, to the strengthening of the local and national capacities which plan and manage the coastal zones, and to contribute to the alleviation of the existing socio-economic inequalities among the countries of the region.

The activity of PAP covers six priority fields selected and agreed upon by the contracting parties. Within these priority fields, PAP is carrying out 11 priority actions:

142 *Climatic Change in the Mediterranean*

- Water resources management on Mediterranean islands and in isolated coastal zones;
- Integrated planning and management of the Mediterranean coastal zones;
- Rehabilitation and reconstruction of the historic Mediterranean settlements;
- Land-use planning in earthquake zones;
- Solid and liquid waste collection, disposal and management;
- Promotion of soil protection as the essential component of the environmental protection of the Mediterranean coastal areas;
- Development of Mediterranean tourism harmonized with the environment;
- PAP-FAO cooperative project on environmentally sound management of Mediterranean aquaculture;
- A Mediterranean network of renewable sources of energy;
- Application of the environmental impact assessment in the development of the Mediterranean coastal zones; and
- A balanced development of the Mediterranean coastal zones and their hinterland.

4 THE COASTLINE AND THE IMMEDIATE COASTAL ZONES

The coastline and the immediate coastal zones are of prime importance in the assessment of the impact of climatic changes, especially with respect to the rise in the sea-level.

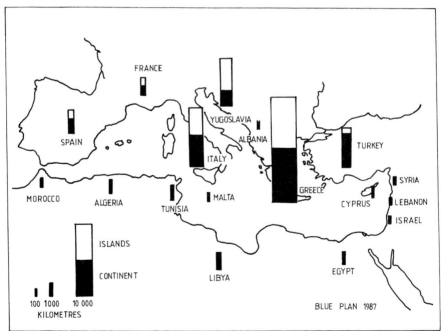

Fig. 5.3. Mediterranean coastline, by coastal country, in 100 km and nation continent/islands.

The Mediterranean coastline is very ragged and indented. Of the 46,000 km of coastline, 75% is taken up by just four countries: Greece, Yugoslavia, Italy and Turkey. Forty per cent of the coastline consists of islands found mainly in Greece (7700 km), Yugoslavia (4024 km), Italy (3766 km), with lesser amounts in Spain (910 km) and France (Corsica 802 km) (Fig. 5.3).

According to accessible data (Table 5.6) some 25,000 km or 54% of the total coast length is rocky, while the other 21,000 km is considered to be sedimentary (sandy and rocky beaches, dunes, wetlands, lagoons, estuaries and deltas). Recently, some of the coast has been turned artificially into beaches by depositing sand or flattening out rocky platforms. This rearranging or reconstruction is local, but in Spain already it has reached up to 5% of the coastal length (Couder, 1989).

Climatic changes would primarily impact the sedimentary parts of the coast. This impact would not be just mechanical (both static and dynamic, a consequence of the sea-level rise) but also chemical due to increased salinization of ground waters and aquifers.

There are no figures on the coastal gradients, but the main impact of climatic changes or the sea-level rise probably will be felt in the cities and tourist establishments, most of which are located on the coast. Therefore, harbours will be affected as well. The total area covered by cities and settlements is 14,213 km^2. This is 1.11% of the total area of the Mediterranean administrative regions, but urban inhabitants account for 87 million people, 65% of the Mediterranean coastal population.

Table 5.6 Distribution of rocky and sedimentary coasts within the Mediterranean

| Country | Type of coasts | | | |
| | Rocky coasts | | Coasts created by sediment | |
	%	km	%	km
Spain*	3	80	92	2370
France	64	1090	36	613
Italy	40	3181	60	4772
Malta	100	180	0	0
Yugoslavia	80	4893	20	1223
Albania	30	125	70	293
Greece	70	10500	30	4500
Turkey	60	3115	40	2076
Cyrpus	50	391	50	391
Syria	65	119	35	64
Lebanon	65	146	35	79
Israel	5	10	95	190
Egypt	5	50	95	900
Libya	5	90	95	1680
Tunisia	20	260	80	1040
Algeria	50	600	50	600
Morocco	50	256	50	256
TOTAL	54	25086	46	21047

Note: *plus 5% of artificial coastal beaches

144 *Climatic Change in the Mediterranean*

The knowledge received from case studies and from other studies plus empirical knowledge of the Mediterranean coast, allows us to accept the following ideas. First, in addition to the primary impact on coastal cities, harbours and tourist beaches, a secondary and equally strong impact would be felt in the lowlands and swampy coastal regions located mostly in the eastern Mediterranean. Secondly, this impact also would affect the estuaries in the western Mediterranean, especially the Po Delta (Ravenna, Venice, Trieste), parts of the Albanian coast, the southernmost parts of the Yugoslav coast (Ulcinj, Lake Skadar), the Neretva and Mirna deltas in Yugoslavia, Tunis (Biserrta), the mouth of the Ebro River, and especially the regions around the mouth of the Rhône in France.

5 Distribution and Dynamics of Population in Littoral Zones

At the present time approximately 133 million people, 37% of the population of the Mediterranean countries, live in the littoral or coastal zones, which only represent 17% of the total area of these countries. About 87 million people live in the urban zones (Table 5.7; Blue Plan, 1988b). The Blue Plan scenarios predict between 200 and 220 million inhabitants living in the coastal zones in the year 2025; 150 to 175 million of these inhabitants in the cities, 75% of the coastal zone population.

The process of littoralization will continue but with much lesser intensity than urbanization. This latter process will be especially intense in the southern and eastern parts of the Mediterranean, where today 40 to

Table 5.7 Selected data for the Mediterranean coastal zones

Country	Urbanized coastal zones (km^2)	Population in coastal zones (1000) inhabitants	Percentage of urban population in the coastal zone	Population in the coastal zones (inhabitants/km of coastline)	Total length of coastline (km)	Coastal Islands
Spain	2794	13860	80.64	5372	2500	910
France	1203	5496	87.52	3227	1703	82
Monaco	2	27	100.00	6750	4	
Italy	4981	41829	66.76	5260	7953	3766
Malta	13	383	85.38	2128	180	
Yugoslavia	351	2582	54.38	422	6116	4024
Albania	52	3050	34.10	7297	418	
Greece	1315	8862	59.37	591	15000	7700
Turkey	371	10000	53.00	1926	5191	
Cyprus	20	669	49.48	855	789	
Syria	17	1155	35.93	6311	183	
Lebanon	86	1668	80.15	11867	225	
Israel	154	2886	90.35	21250	200	
Egypt	236	16511	35.73	17300	950	
Libya	85	2284	62.17	1290	1770	
Tunisia	168	4965	67.47	3819	1300	
Algeria	276	11500	48.00	9583	1200	
Morocco	91	3390	44.89	6621	512	

Source: Blue Plan, 1988b

50% of the population lives in the urban areas. According to the Blue Plan scenario this figure is expected to reach 70 to 80% by the year 2025.

The changes in the climatic conditions will have a definite but limited impact, especially in the first years of accelerated sea-level rise, on the distribution and dynamics of the population in the littoral zones. This premise assumes that adequate measures for the protection of the inhabited parts immediately next to the coastal line already will have been taken.

According to all assessments, the dynamics of natural growth will not be affected by sea-level rise and other impacts of climatic changes, but rather will continue to follow present trends. These trends are: decreased population growth in the countries on the northern Mediterranean coast, compared to a strong increase in population in southern and eastern coasts. The natural population growth is primarily affected by other factors such as the increase in standard of living, the growth of economic vitality, the level of medical care, cultural and religious customs, etc.

Population growth caused by littoralization probably will not change with initial climatic changes. Therefore, the total population growth in the coastal zones will continue to follow present trends. However, in the future people living immediately next to the coast may experience manifestations of climatic changes, depending on a local coastal configuration and individual settlement structures.

The expected climatic changes probably will not diminish the predicted speed of littoralization in either the northern or southern Mediterranean. In fact, littoralization probably will increase speed in the south due to northward growth of the desert zone.

Urbanization probably will be slower than predicted by the Blue Plan. Decreased water supplies, poorer water quality, the need for more purified drinking water, transport of water from further sources, plus a greater level of waste-water treatment and consequent reuse in cities will require greater expenditures for necessary infrastructures. This will raise the cost of urban living which in turn will destimulate urbanization. The temperature rise will make living conditions less pleasant during the summer months in crowded city spaces. The hot exhausts of air-conditioning systems, which probably will be increasingly used, also will contribute to the temperature rise. A reduction of their use might only occur with a drastic rise in energy prices.

It will be necessary to establish better methods of water management for both private and communal use because of the expected reduced precipitation. Because of the temperature rise and corresponding increased evaporation, we can expect increased use of water, unless this problem can be solved by the recycling of waste water.

By evaluating the assessments explained in section 2, we can conclude that no more than 5% of the population would be affected by the impacts of sea-level rise, or some 6,700,000 inhabitants. Part of this number may be directly affected by the sea-level rise, but most likely this will not be greater than 10–20% of the total number affected. The repercussions to the impacts may be felt in the need for the removal of housing units and communal infrastructures tied to these units and eventually for the relocation of those inhabitants. This need will probably arise only in later years of the manifestations of climatic-change impacts.

6 ACTIVITIES

6.1 Agriculture and Forestry

All agriculture is based upon a few natural factors, primarily climate (temperature, precipitation and its distribution, water availability) and soil. Within the limits of these factors, man produces goods for his personal use or sale.

Agricultural production in the Mediterranean countries is generally oriented toward food production. The major exceptions are tobacco and cotton in a few countries (Blue Plan, 1988b). Agricultural areas comprise less than 50% of the total area of each country (Table 5.8). In north African countries, these areas are not more than 10% of the total area. Generally speaking, in recent times Mediterranean agricultural regions have remained unchanged although in some highly developed countries (such as Italy and France) they have decreased in size (Blue Plan, 1988a, b).

In the northern Mediterranean countries and Turkey, most agricultural production does not take place in the coastal zones, while in the eastern and southern Mediterranean countries most agriculture is located in coastal zones. The exception is Egypt, where production occurs along the Nile River as well as in the Nile Delta.

Since 1970 agricultural production has become more modern and intense. The use of fertilizers has increased by 50%, while the number of tractors has risen by 40%. But, these changes have been greatest in the developed countries where most of the agricultural lands are being irrigated and fertilized with the use of modern mechanization (Table 5.9). In less developed countries these advances have been insignificant. The consequences are major differences in crop yields. Thus, the yield of cereal crops in the countries of the Maghreb amounts to 10 quintal per hectare, while in the northern Mediterranean countries it is well over 50 quintal per hectare. The agricultural production of some southern and eastern Mediterranean countries cannot fulfill their present needs, and the prognosis of FAO is that the gap between need and production will increase regardless of projected increases in agricultural productivity.

The expected climatic changes in the Mediterranean regions could have far-reaching consequences on agriculture. Sea-level rise will impact directly and primarily the lowlands, the wetlands and the estuaries. These lands constitute a very important but generally not the main part of the agricultural areas due to the quality of their soils. However, ultimately they could be strongly affected by other factors such as the temperature rise, salinization and the changes in the amount and distribution of precipitation. These factors also could cause changes in the geographical distribution of individual crop cultures, their quality, and so on.

Lengthening of the summer dry period may affect the existence of crops or plantations, the incomes of farmers, and the commercial values of products. The temperature rise in many ways will directly impact agricultural production. It will affect growth cycles and will affect harvest times. It also will have an effect on the quality of produce. As well, temperature rise will have indirect effects, such as increased evaporation, lower moisture levels of the soil, and increased erosion. The

Socio-economic activities and climatic change   147

Table 5.8 Land use in the Mediterranean region and countries according to national statistical yearbooks (1980s) (in 1000s of ha)

Country / Med. region	Utilized agricultural land			Arable land			Other land			Other non-agricultural land		Total area
	Total area (UAL)	Temporary crops and meadows, fallows	Gardens (trees and vineyards, market gardens)	Total arable land	Permanent meadows	Range land	Non-utilized agricultural land	Forests and other wooded land	Water areas	Unproductive land	Built areas	
Spain, 1984	39898.1	10847.2	4916.3	20507.7	19390.4	7233.1	3343.8	11792.0				50.475
	6881.6	1445.2	1767.4	3981.4	2900.2	1868.6	800.2	1974.8				9.550
France, 1982	31649.0	17350.0	1677.0	19027.0	12622.0	—	2747.0	14369.0	124	5809	560	54.703
	1711.0	344.0	580.0	924.0	787.0	—	827.0	1411.0	45	684	214	4.683
Italy, 1982	17252.1	9050.5	3209.9	12260.4	4991.7	2775.7	3706.5	6393.4				30.122
	13786.9	7220.0	2934.7	10156.7	3360.2	2025.0	2476.7	4382.0				22.668
Yugoslavia, 1982	10019.0	7218.0	625.0	7843.0	2176.0	4349.0	2488.0	8746.0				25.602
	3241.0	1961.0	263.0	2224	1017.0	2202.0	1324.0	4450.0				11.115
Albania												2.875
Greece												13.195
Turkey, 1983	47751.0	25590.0	1063.0	26653.0	21098.0	643.0	1042.0	23478.0	1102	3208(1)		77.784
	9642.0	7112.0	448.0	7560.0	2082.0	34.0	555.0	10560.0	250	1030		22.294
Cyprus												925
Syria	5655.0	5655.0	—	5655.0	—	8317.0	514.0	498.0	113	3040	381	18.518
	215.0	215.0	—	215.0	—	4.0	16.0	116.0	3	40	26	420
Lebanon												1.040
Israel							205.0	110.0				267
												2.077
Egypt, 1980	2992.0											477
												100.145
Libya												40.372
												175.954
Tunisia, 1984	3677.0	2101.0	1576.0	3677.0	—	3358.0	3260.0	4505.0				16.361
	1968.0	1002.0(2)	966.0(2)	1968.0	—	1627.0(2)	545.0	965.0				4.571
Algeria, 1984	7510.0	6868.0	625.0	7493.0	17	31661.0	346.0					238.174
	2740.0	2279.0	452.0	2731.0	9.0	700.0	174.0					6.829
Morocco, 1984	8353.0	7834.0	519.0	8359.0	—	28269.0	—	7852.8	—	—	180.0	44.655
	880.0	816.0	64.0	880.0	—	1696.0	—	1591.4	—	—	28.0	4.195
TOTAL												

Notes: ¹ Salty soil, river beds, coastal dunes, rock soil
² Including the Governorate of Tataouine; in Tunisia, ranges include fallow

Source: Blue Plan, 1987b

Table 5.9 Irrigation in the Mediterranean Basin (1975–1984)

	Irrigated area (1000 ha)		Rate of variation 1975–1984 (%)	Irrigated area in % of arable land & permanent crops			Use of micro-irrigation (ha)	
	1975	1984		1965	1974–1976	1984	1974	1981–1982
Spain	2818	3145	+12	10	14	15	—	
France	805	1160	+44	—	4	6	—	22000
Italy	2717	2970	+9	16	22	24		10300
Malta	1	1	0	6	8	8		
Yugoslavia	130	161	+24	2	2	2		
Albania	330	390	+18	45*	49	55		
Greece	887	1030	+16	14	23	26		
Turkey	1983	2140	+8	5	7	8		
Cyprus	94	94	—	22	22	22	160	6600
Syria	547	617	+13	9	10	11		
Lebanon	86	86	—	11	26	29		
Israel	182	220	+21	35	43	50	6070	81700
Egypt	2799	2474	−14	100	100	100		
Libya	200	232	+16	5	10	11		
Tunisia	123	210	+71	2	3	4		
Algeria	244	298	+22	4	4	4		
Morocco	426	520	+22	3	6	6		

Note: * 1970
Source: Blue Plan, 1988b

temperature rise may have significant effects in the development of various parasites and insects, which may directly affect agricultural productivity and income.

Most plants require the greatest amounts of water during the spring or summer, the period of least precipitation in the Mediterranean. The absence of showers in late spring or early summer may significantly reduce productivity in those areas with no irrigation. Simultaneously with the increased need for irrigation, the available amounts of water will decrease. To satisfy the needs it will be necessary to use whenever possible closed systems of water distribution and ways of supplying water that can maximally reduce water evaporation. Likewise, it may be necessary to reuse waste waters.

To prevent the mentioned consequences it will be necessary to invest significant funds to build irrigation systems, to take protective measures against erosion and to erect dams for protection from the sea. These actions will make agricultural production much more expensive, which could have far-reaching socio-economic consequences. The economically richer countries in the northern Mediterranean should be able to solve these problems more easily. The economically poorer countries in the south, which even today have problems with supplying their inhabitants with food, will be faced with greater problems.

As spring comes earlier in northern Europe, the need for early fruits and vegetables produced in the Mediterranean will decrease.

The forest areas in the immediate Mediterranean coastal belt have limited capacities for timber production (Table 5.10). This timber is for the most part abandoned since it is unsuitable for most constructional uses. These forests have limited capacities even though wood also has been abandoned as a heating fuel. However, forest areas constitute the basis of the plant and animal ecosystem as well as a strong influence on microclimatic conditions. Also, forests and trees are one of the typical characteristics of the Mediterranean landscape.

The sea-level rise probably will not affect directly the forest areas since there are practically none located on the coastline. However, other changes (temperature, amount and distribution of precipitation, salinization) may greatly affect forest growth. Their effects may be intensified since the forests already have been attacked by pollution from countless sources. In addition to this, the temperature rise and the smaller precipitation in the summer months will surely enhance the forest fires. A more precise statistical survey of forest fires is presented in Table 5.11.

To minimize the impacts of climatic changes on waters, on precipitation and its distribution, and on agricultural and forest soils will require complex reforms in agrarian techniques and technology. Since some hypothetical examinations of the effects of changes in climatic conditions on agriculture already exist (e.g. individual crops, individual plantings, fruits/vineyards and vegetables) and since those studies point out the need for further research, it follows that more study of the local conditions of the existing climate, soil, land configuration, precipitation and soil permeability (water retention) is imperative.

Table 5.10 Forests, wooded land and other covered areas (1980)

	Total land area (waters excluded) (1000 ha)	Forest Areas Natural forests and wooded lands (*)		Closed forests		Open forests (1000 ha)	Other wooded land (1000 ha)	Plantations (all species) end 1980 (1000 ha)
		Total area (1000 ha)	% of land	Total area (1000 ha)	% of land			
Spain	49954	12511	25.0	9506	19.0	3005		3387
Med. region	18883	1975						
France	54564	15075	27.6	13875	25.4	792	408	2800
Med. region	6746	3293	31.3	2102		1091		
Italy	29402	8063	27.4	6363	21.6	1700		
Med. region	22514	4192	19.3					
Yugoslavia	25402	10500	41.1	9100	35.6	1400		
Med. region		4450	40.0					
Albania	2740	1202	44.0	1090	39.8	192		
Greece	13000	5754	44.0	2512	19.2	3242		
Turkey	77076	20199	26.2	8856	11.4	8843	2500	1100
Med. region	22294	10569	67.2					
Cyprus	924	173	18.7	153	16.8	19	1	34
Syria	18422	159	0.9	60	2.2	50	9	40
Med. region								
Lebanon	1023	30	2.9	0	0	20	10	19
Med. region								
Israel	2033	109	3.4	70	3.4	28	11	
Egypt	99545	0	0	0	0	0	0	40
Med. region		0	0	0	0	0	0	
Libya	1755594	636	0.4	78	0	112	446	159
Med. region								
Tunisia	15536	601	3.9	143	1.0	134	304	127
Med. region								
Algeria	230174	3894	1.6	1021	0.4	746	2127	431
Med. region	4829							
Morocco	4430	4397	9.9	687	1.5	2549	1161	322
Med. region	4195	1191	27.9					

Note: * Plantations are not included from Turkey to Morocco inclusive; they are included from Spain to Greece

Source: Blue Plan, 1987b

Table 5.11 Fire damage: area of forest and other land burnt (ha)

	1961	1965	1970	1975	1980	1981	1982	1983	1984	1985
Spain	34507	38018	87324	187314	265954	298436	151644	117599	164546	412426
France (Med. region)	43986	59694	61230	25849	22176	27711	55145	53729	27203	57400
				17600	13980	20052	45775	47627	12086	46628
Italy	43000	—	91176	71425	144302	242218	130239	223728	75272	190640
Yugoslavia	14146	2114	3139	7851	4033	12170	19358	20585	9010	15000
Greece	10646	27030	3189	30955	32965	81417	27372	19613	33655	105450
Turkey (Med. region)	8989	3954	15019	17515	10546	5470	4018	3554	7358	26007
Cyprus	279	2706*			754	371	7512	3718		
Syria	1500 (ha/year on average)									
Lebanon	1200 (ha/year on average)									
Israel	71	600	1400	900	1805	2395	3441	4788	1740	1476
Libya								1	3	43
Tunisia						376	1613	4139	1287	396
Algeria					19700	33516	9381	221367	4731	4668
Morocco						1707	1818	17730	1423	1888

Note: * in 1964
Source: Blue Plan, 1987b

6.2 Industry and Energy Production

There are great advantages in locating industry in the coastal regions, particularly because the coastal zone represents the confluence of two transport systems (land and sea). This is one reason why in the last few decades littoralization has included a migration of industry to the coastline. Particularly affected have been energy-generating plants (both thermal and nuclear power-plants), basic chemical industry, oil refineries, and cement and steel-producing factories. These installations have located themselves to coastal regions both for access to transportation and large water supplies.

These basic "heavy" industries have developed primarily on the north-eastern coasts next to large harbours where they form harbour-industrial zones. Some separate industrial zones have been built on their own harbours, as is the case with Marghera near Venice and Foss near Marseille. Where possible, these industrial zones were built on flatlands next to the coast, so the impact of the sea-level rise could be strong. They will consequently require protective structures, such as embankments or dams. They will also require the reconstruction of systems dealing with waste waters in places where they empty into the sea.

Because of the recent development of industrial capacities in the north, it is assumed that most future development will occur in south (e.g. Table 5.12). In addition to the plants listed here, there are 73 petrochemical plants existing and 31 planned, 28 metallurgic plants existing and 8 planned or being built, plus 56 chemical industry plants existing and 28 planned or being built in the coastal regions of the Mediterranean.

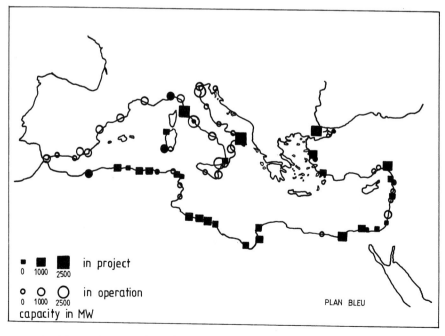

Fig. 5.4. Thermal power-plants located in the coastal zones.

Table 5.12 Energy production plants, existing and planned

		North	Turkey	South	Total
Refineries	existing	32	3	15	50
	planned	0	1	10	11
Thermal	existing	28	11	23	62
power-plants	planned	4	5	23	32
Total no.	existing	60	14	38	112
of plants	planned	4	6	33	43

Source: Blue Plan, 1988b

There are no figures on the number of industrial plants located at sea-level which may be threatened directly by the sea-level rise, nor the percentage of plants that will have difficulties in adapting their systems of waste-water disposal to new standards. The manufacture of sea salt will find itself in a critical state with a sea-level rise of 20 cm. Greater difficulties may arise for those industries forced to use mainly fresh-water resources. Any assessment will be hard to make since this situation will vary locally. More importantly, the repercussions of climatic changes on the amounts of available water are basically unknown and also are locally conditioned.

According to the assessments of Blue Plan, the growth of the need for more energy in the future will lessen the exploitation of crude oil after the year 2000. This need will require that 100–250 units of 1000 megawatt power, which corresponds to 200–375 million tons of coal, be built in addition to the existing energy-production plants.

Keeping in mind that industry requires and produces large capital and that it is usually concentrated, we may conclude that projected climatic and sea-level changes should not create insurmountable difficulties in the protection of industrial plants.

6.3 Tourism

Mediterranean tourism is by far the strongest of any region in the world. For over 30% of the tourists who vacation outside their own country, their final destination is one or more of the Mediterranean countries. In 1984 there were about 100 million visitors in the Mediterranean (Table 5.13), of which 45 million were domestic tourists (Blue Plan, 1987a). This number is probably too low, since most of the countries do not accurately register their domestic tourists.

Mediterranean tourism is primarily oriented on swimming and sun-bathing, recreational activities that directly use the seashore. Therefore, the immediate coast, its relief, climate and the quality of land and sea are of prime importance to Mediterranean tourism. However, tourism also utilizes the attractiveness of Mediterranean life style and architecture, landscapes and cultural heritage.

Mediterranean tourism fluctuates seasonally. The number of visitors from June to September is from 50% to as much as 70–80% greater than other periods of the year. The high-peak season lasts from the end of June to end of August.

Table 5.13 Number of international tourists (thousands)

	1970	1975	1980	1981	1982	1983	1984
Spain	15320	19800	22500	23800	25300	25583	27100
France	18130	25710	30100	31340	33467	34018	34812
Italy	14188	15500	22087	20036	22297	22140	22855*
Malta	171	335	729	706	511	491	480
Yugoslavia	4749	5834	6410	6616	5955	5947	7224
Greece	1407	2840	4796	5034	5033	4778	5523
Turkey	446	1201	865	997	1026	1178	1533
Cyprus	127	47	353	421	547	621	737
Syria	409	678	1204	1043	831	836	976
Lebanon	900	1555	118	–	–	–	–
Israel	419	559	1116	1090	949	1043	1095
Egypt	348	730	1253	1376	1423	1498	1560
Libya	77	341	126	126*	126*	126*	126*
Tunisia	411	1014	1602	2151	1355	1439	1580
Algeria	236	296	290	320	280	285	410
Morocco	747	1242	1425	1567	1815	1877	1936
TOTAL	58085	77582	94974	96623	100915	101860	107947

Note: * Blue Plan Assessment
Source: Blue Plan, 1987a

Most tourist facilities, such as hotels, camps and youth hostels, are located within 200–300 m of the coast. Facilities further from the shore are found mainly in the developed and luxurious tourist areas. In 1984 the tourist accommodations covered an area of 2169×10^6 m^2 of which hotels took up 203×10^6 m^2. The hotels can accommodate 5,082,000 guests while the other establishments can accommodate 28,223,000 guests (Table 5.14).

Tourism should continue to develop. According to some Blue Plan scenarios, the expected number of tourists for the year 2000 is 139 to 184 million, and for the year 2025 this number is 219 to 341 million (Table 5.15). Tour operators and airline companies believe that the number of tourists will be even higher.

Tourism is a great user of water and a producer of both liquid and solid wastes that go directly into the sea. Furthermore, tourism will suffer the greatest perturbations from the impacts of climatic changes regardless of changes in sea-level. Even if the tourist accommodation facilities are not affected by the sea-level rise, it is most certain that a large number of the beaches would disappear or would require substantial investments for reconstruction. The impacts of both the sea-level rise and temperature rise on water needed for tourism and on waste-water drainage are just some of the consequences that will come about in other economic activities and cities. Islands may experience even graver consequences than those on land.

A special type of tourism, nautical tourism, has stimulated the building of a large number of marinas, sport harbours and berths. They should be harder hit than the large freight harbours.

Socio-economic activities and climatic change 155

Table 5.14 Capacity and appropriation of soil of hotel and other lodgings (1984, in thousand beds and m²)

Country	Number of hotel beds (× 1000)	Appropriation of soil (× 1000m²)	Number of other beds (× 1000)	Appropriation of soil (× 1000m²)
Spain	840	33.600	8.923	624.610
France	1.590	63.600	10.894	762.580
Italy	1.598	63.920	5.854	409.780
Malta	14	560	41	2.870
Yugoslavia	319	12.760	1.127	78.890
Greece	323	12.920	324	22.680
Turkey	68	2.720	182	12.740
Cyprus	27	1.080	137	9.590
Syria	23	920	71	4.970
Israel	64	2.600	162	11.340
Egypt	48	1.920	218	15.260
Libya	9	360	24	1.680
Tunisia	72	2.880	77	5.390
Algeria	27	1.080	37	2.590
Morocco	59	2.360	152	10.640
TOTAL	5.082	203.280	28.223	1.975.610

Source: Blue Plan, 1988b

Table 5.15 Coastal tourism in 1984 and prognoses for the years 2000 and 2025

		Tourists (millions)			Nights (millions)		
		LT	NT	IT+NT	LT	NT	IT+NT
1984		51	44.7	95.7	418.2	983.4	1.401.6
Scenario 2000	T_1	85.4	53.9	139.3	743	1256	1999
	T_2	76.4	45	121.4	664	1048	1712
	T_3	94.	64.1	158.1	817	1493	2310
	A_1	97.7	71.4	169.1	850	1663	2513
	A_2	107	77.3	184.3	930	1801	2731
Scenario 2025	T_1	147	72	219	1205	1584	2789
	T_2	125	48	173	1025	1056	2081
	T_3	162	98	260	1328	2156	3484
	A_1	168	130	298	1377	2860	4237
	A_2	193	148	341	1582	3256	4838

Note: IT – International Tourism
NT – National Tourism
Source: Blue Plan, 1988b

Expected climatic changes also may lengthen the tourist season by creating more favourable weather conditions. However, the rise in air temperature and sea-level may enhance the attractiveness of the continental lakes and of the coastal tourist areas of other seas (Baltic, North Sea), thus negating the need for many northern Europeans to vacation in the south.

Beaches with rocky inclines only a few metres from the sea would be first to disappear. There also would be problems with beaches where large infrastructures located only a few metres from the sea-limit beach expansion. But, the problems here would deal with major relocation of these city roads and promenades (Nice–Cannes) or with the relocation of streets and railway lines (coastal stretch between Nice and Cannes, parts of the Italian and Yugoslav coasts).

Any attempts to relieve the impacts of climatic changes on tourism will be faced with great organizational and financial difficulties for at least two reasons. First, tourism is a typical form of dispersional concentrated capital (a great number of small or large tourist accommodation and catering establishments). Secondly, any mitigation of the adverse impact of climatic change asks for very high investment into traffic and other infrastructures of tourist complexes.

6.4 Traffic

Until the end of the last century and the beginning of this century, traffic in the Mediterranean region was maritime traffic. Except for the stretch of land from Spain through southern France into Italy, connections between various Mediterranean countries as well as with other parts of the world was by sea.

This changed at the end of the 19th century with construction of roads and railway lines in the western Mediterranean, that connected areas both along and interior from the coastline. In the south-western Mediterranean, where traffic already existed immediately next to the coast, fast roads further from the coast were built. In some harbours traffic became concentrated. These harbours grew and expanded their industrial complexes. However, most harbours located in small or average-size cities and settlements remained virtually unchanged, continuing to serve as city ports for local passenger traffic and for local sport activities.

Maritime traffic continued to connect the Mediterranean countries, but mostly in terms of freight transport; passenger traffic increasingly has become airborne. Redevelopment of air traffic has emphasized north–south connections, whereas longitudinal connections have been neglected. Most longitudinal trips along the Mediterranean require travelling inland and taking a connecting flight.

There are no figures on the length of roads and railway lines immediately next to the coastline. According to Blue Plan's assessment roads and railways take up 10–20,000 km^2 of land (Blue Plan, 1988b). A few airports have been built immediately next to the coast (Barcelona, Athens, Split, Tunis, Malta) or on artificial addition (Nice, France; Tivat, Yugoslavia; and others).

The impacts of the climatic changes will be very strong and more significant on the harbours than on the roads, railway lines and airports. Bearing in mind that maintaining traffic is crucial to existing social, political and economic structures, there should be no difficulty in raising funds to combat climatic changes, especially the sea-level rise.

6.5 Fisheries and Aquaculture

6.5.1 Fisheries

Foods harvested from the sea historically have provided an important staple in the diets of the inhabitants throughout the Mediterranean, especially small islands. With the development of food-preservation tech-niques and faster transport, the consumption of fresh sea foods has spread inland.

The production of sea foods in the Mediterranean countries is about 4 million tons a year. The needs of individual Mediterranean countries for sea food are basically satisfied through national fishing. However, the annual catch on the Mediterranean is approximately 1 million tons, the rest comes from fishing in other water bodies such as the Atlantic (Table 5.16) (Blue Plan, 1988b).

Even though this amount has remained constant for about the last 10 years, there have been significant changes in the consumer structure. Generally speaking, the change in structure of the catch in the past years has been related to the overexploitation of some species, the demands of the market, and the development of specific new techniques in fishing. Looking at individual countries, we find that major differences exist in the catch structure. In more developed northern countries, there has been a significant decrease in the catch of average-quality fish (sardines, mackerel) and an increase in the catch amount of high-quality fish (mullet, perch, dorade). In the less developed countries there has been no real change in the catch amount of poorer-quality fish. According to some assessments the present level of fishing is somewhat lower or is already past the allowable catch, thus endangering the existing fish stock. While regionally individual fish species may be caught in amounts significantly lower than those that would endanger their survival, some species in some areas are clearly overfished. The consequences of this are reductions in the amounts of individual species and their eventual disappearance (e.g. mackerel in the Adriatic Sea). For every individual region and for the entire Mediterranean, proper stock assessments, established systems of catch monitoring, and further development of fishing techniques could enable an optimal catch without the danger of overexploitation or the disappearance of some species.

Some experts warn that the increase in coastal pollution will result in reduction of the total stock as well as individual species by forcing them into farther regions that will make fishing more difficult. Both of these elements may cause a reduction of the catch.

It is very likely that the expected climatic changes and their direct consequences will cause changes within the ecosystem. Existing ecological balances and chains will be broken, and new ones will be formed on significantly different levels. The Mediterranean region is typical of this problem because it contains species that are heterogeneous in every individual area, the result of the vastly different biotypes.

Generally speaking, the Mediterranean Sea can be considered poor, with a weak primary production that generally is the limiting factor of

Table 5.16 Marine catches of Mediterranean countries: fish, crustaceans and molluscs from all marine zones and the Mediterranean, 1970–1984

Country	All marine zones (tonnes)					In the Mediterranean (tonnes)					Proportions of Mediterranean marine catches (US$) 1984
	1970	1975	1980	1983	1984	1970	1975	1980	1983	1984	
Spain	1522200	1497359	1231780	1227345	1243581	124200	139410	149606	379168	168240	13.5
France	782500	784495	765676	774142	738813	45600	46828	46394	52356	46423	6.3
Italy	383200	386672	413236	437040	451223	320300	351976	352631	398635	409440	90.7
Malta	1200	15067	1054	993	1216	1200	1506	1054	993	1216	100.0
Yugoslavia	26700	32251	34968	53238	48420	26700	32251	34968	53238	48420	100.0
Albania	4000	4000	4000	4000	4000	4000	4000	4000	4000	400	100.0
Greece	91500	86482	95131	89454	91380	59100	62666	73069	80893	82635	90.4
Turkey	165300	103226	394600	518359	520437	55660	25876*	64831	105297	83788	16.1
Cyprus	1300	919	1305	1933	2205	1300	919	1305	1933	2205	100.0
Syria	1000	826	976	923	923	1000	826	976	923	923	100.0
Lebanon	2200	2400	1700	1300	1200	2200	2400	1700	1300	1200	100.0
Israel	10300	9615	12053	9500	9710	3200	3200	3702	4173	4596	47.3
Egypt	27200	25910	32249	26146	26146	12000	5384	17466	12538	12538	48.0
Libya	5500	4803	5200	7500	7800	5500	4803	5200	7500	7800	100.0
Tunisia	24400	44500	60154	67145	74944	24400	44500	60154	67145	74944	100.0
Algeria	25700	37693	48000	70000	75000	25700	37693	48000	70000	75000	100.0
Morocco	250200	224895	329680	452529	466130	10300	15202	27316	32314	41788	9.0
Japan							1263	121	683	1057	
Other						3700	4652	994	845	1553	

Note: * 1974
Source: Blue Plan, 1987a

fish stock. Cyclonal activity directly affects the dynamics of water-mass movements, especially in shallower coastal regions.

A general decrease of precipitation is predicted for the entire year and for the summer period. However, in some regions of the Mediterranean precipitation may increase and/or receive increased inflow of fresh water. Non-uniform changes in precipitation could cause very complex alterations in the physical characteristics of the Mediterranean Sea, especially the shallow coastal areas. Assuming, however, that the entire Mediterranean region will experience a decrease in precipitation and increase in evaporation, then salinity and sea temperature should rise. These changes could affect the inflow of nutrients from river runoff or from "upwelling" of deep-sea waters. The changes in the physical characteristics also could affect the oxygen solubility in the sea.

The changes in the physical characteristics could speed-up the physiological processes of marine organisms. They also could increase the speed of the mineralization process of organic matter.

The rise in the sea temperature in the shallow coastal waters could create subtropical or even tropical conditions. That will probably stimulate a greater immigration of numerous plant and animal species from the Red Sea through the Suez Canal. Some plant and animal species indigenous to the Red Sea already have naturalized in the eastern Mediterranean, but in the future they will spread westward. In contrast, boreal species will be endangered or might even disappear. All these things could bring about major changes in the qualitative and quantitative composition of the Mediterranean marine flora and fauna. Economically important populations of marine organisms also will be affected. Climatic changes may cause subsequent changes in the migratory habits of fish species that constitute the largest part of the annual catch. Likewise, a shift in the distribution of niches of some other economically important species may require significant modification of existing fishing techniques. This may include new devices and equipment for detecting fish, larger boats capable of fishing in farther and deeper waters, and new fishing equipment. All these changes will require significant financial expenditures that may depress further development of fishing in the economically weaker countries.

6.5.2 Aquaculture
Some countries have tried to satisfy the demand for high-quality fish by artificial rearing. At present, the amounts of produced fish in comparison to the amounts caught are negligible: in 1987 approximately 26.5 thousand tons of high-quality fish were produced; 1992 this number may reach 44 thousand tons (Blue Plan, 1988b). Ninety per cent of the present aquatic farming in the Mediterranean occurs in lagoons. Other techniques, for example rearing in cages, are being introduced in some Mediterranean countries. The Mediterranean coast provides excellent conditions for further development of aquaculture. However, uncontrolled development often endangers the areas suitable for aquaculture. If the production of fish through artificial rearing is to be notably increased, measures for the protection of suitable habitats must be taken immediately. Within the framework of the PAP programme in cooperation with FAO, ecological

160 *Climatic Change in the Mediterranean*

criteria for the rational development and protection of aquaculture in the Mediterranean coastal regions are being developed, so that the Mediterranean countries in time could legislate the proper protective measures.

Since marine aquaculture is located mainly in the coastal zones, climatic changes and their direct consequences (increase of sea temperature and salinity, sea-level rise) will greatly impact its development. Keeping in mind the fact that marine culture in the Mediterranean comes from lagoons whose average depths are only 50 cm, expected changes may make the lagoons completely unfit for production. It is most likely that new lagoons will need to be created or existing ones greatly modified.

A rise in both the salinity and temperature of the sea will result in a decrease of oxygen solubility and increased organic matter decomposition. This may enhance the oxygen depletion and may even create anoxic conditions. On the other hand, the sea-temperature rise will accelerate the growth of marine organisms in the colder periods of the year. This will make the time necessary for rearing shorter and thus more effective.

The changes in climatic conditions may alter the effects of pollutants on certain organisms. In other words, the temperature change significantly affects the metabolic functions of the marine organisms. The expected temperature rise, therefore, may create major changes in the bioaccumulation of certain pollutants. This could have a negative impact on some commercially important species. In turn, criteria on the acceptable levels of sea-water pollution in areas intended for rearing may need to be amended. Moreover, some existing rearing areas will need to be relocated.

7 COMPONENTS

7.1 Water Resources

Because water is scarce throughout much of the Mediterranean, its availability is one of the key conditions for development in the region. Data on water sources and water consumption are shown in Table 5.17. In many individual areas there is a conflict between different consumers, as well as between the consumers and the natural needs.

Of the total current annual water consumption of 154,000 million m^3, about 72% is used for irrigation, 10% for drinking water distributed through municipal waterworks, and 16% for industry not connected to the communal waterworks (Blue Plan, 1988b). The level of exploitation of available sources is high in most countries. In some countries it even exceeds 100% (Israel, Libya), which means that some water is reused.

Blue Plan scenarios have estimated water consumption in the individual Mediterranean countries. A significant increase in the water consumption is predicted for the entire region and for the individual countries. Three groups of countries will have different problems with respect to future water supply:

1) In countries where available water supplies will remain abundant until 2025 and beyond, a comfortable margin should remain for higher per capita consumption. Countries with low demographic growth (France, Italy, Greece, and Yugoslavia) and with high growth

Table 5.17 Water: supply and demand in the Mediterranean drainage basin, 1980

Country	(a) Estimated population (millions)	SUPPLY (resources)		DEMAND			RATIO	
		(b) Total water resources Gm³/year	(c) Stable or stabilized Gm³/year	(d) Water distribution Gm³/year	(e) Net consumption Gm³/year		(f) Exploitation index in relation to[b] d/b	(g) Exploitation index in relation to[c] d/c
Spain	15.5	28.3	24.1	12.7	6.9		45	53
France, Monaco	11.2	74	39	12.7	2.2		17	33
Italy	57.3	187	30	56.2	22		30	112
Malta	0.37	0.03	0.03	0.023	0.02		77	77
Yugoslavia	6.9	62	12	1.5	0.3		2.5	12
Albania	2.7	21.3	7	0.2	0.04		1	3
Greece	9.6	62.9	10	6.9	3.6		11	70
Turkey	11.0	77	23	7.5	3.8		10	33
Cyprus	0.6	0.9	0.4	0.54	0.3		60	135
Syria	1.0	4.4	2.8	2.3	1.3		57	90
Lebanon	3.1	4	3.3	0.6	0.4		15	18
Israel	3.9	1.0	0.3	1.4	0.8		140	463
Egypt	41.2	57.3	53.8	43	41		79	81
Libya	3	0.6	0.23	1.0	0.63		167	435
Tunisia	6.3	3	2.9	1.2	0.65		40	41
Algeria	14.6	13.2	3.1	2.8	1.4		21	90
Morocco	1.7	4.0	1.5	1.0	0.6		25	67
TOTAL	190	572	—	152	86			

Source: Blue Plan, 1988b

(Albania, Turkey, Lebanon) should have sufficient water, provided that efforts are made to harness and control water supplies (including preservation of quality), especially in countries with high population growth rates.
2) In countries with presently satisfactory water supplies and high demographic growth (Spain, Morocco, Algeria, Cyprus) will reduce the amount of available water. But, overall water demands can be met until 2025, mainly by new development schemes and by large interregional transfers of water in countries with very uneven distribution of resources, provided that per capita consumption remains near present levels. Marked increases in consumption, however, would quickly place these countries in the critical position of those in the following group, requiring solutions other than conventional water schemes.
3) In countries where exploitation presently (or will by 2000) exceeds supply, demographic growth rates ranges from low (Malta) to average (Israel, Tunisia) to high (Egypt, Syria, Libya). Particularly in these latter countries, per capita consumption from conventional resources probably will have to be reduced or the country will have to resort either to exploiting non-conventional resources (fossil water, desalinized sea water) or importing water.

Expected climatic changes will significantly affect the water balance in the Mediterranean region. This is quantitatively impossible to express without more detailed research. However, it is envisaged that decreased precipitation and increased evaporation will reduce the amounts of running surface water and ground water.

Since the distribution of rain water depends upon a number of simultaneous processes that will be affected by the climatic changes, we will briefly describe the impacts of these climatic changes. Part of the precipitation volume after reaching the ground infiltrates the soil and part stays on the surface as runoff. The level of infiltration and the depth of penetration depend on the characteristics of the rain showers (force, duration, frequency) and on the characteristics of the soil (also vegetation density). Of the numerous soil characteristics, porosity and permeability are the most important; both are dependent on the soil structure and texture. The speed of infiltration markedly decreases as the intensity of rainfall increases. Total infiltration of rain water per unit area also can be increased by maintaining high soil infiltrability and by increasing surface storage.

The decrease in the infiltration speed by rain is the result of decreased hydraulic gradient and crust formation. The term crust refers to a compact layer that forms on the surface of the soil by the aggregation of fine soil particles, mainly clays around and between larger particles. The permeability of such a layer is significantly less than the permeability of the layer just below. Crust formation is conditioned by the joint action of the kinetic energy of each raindrop and the dispersion effect of rain.

By examining the formation of the crust, it was found that the electrolyte content of water and the soil sodicity have a significant effect on the rate of infiltration, and likewise on the crust formation itself. The speed of infiltration markedly decreases with a decrease in electrolyte concentration. At the same time, when there is a steady electrolyte concentration, a greater vegetation cover will cause a faster reduction of the infiltration speed and a significantly lower final speed (Agassi *et al.*, 1986).

The impact of raindrops has two consequences: breaking down the aggregates and enhancing the rate of the chemical dispersion by stearing the soil surface. Crusts formed by raindrop impact of saline or distilled water are irreversible when high energy rainfall changes to low energy rainfall. Changing high-energy rain to low-energy, more saline rain does not lead to an increase in the infiltration rate. However, crust formed by high-energy or low-energy rain made of distilled water was found to be reversible to some extent when the salinity of the rain water was increased. These findings are important for arid and semi-arid regions, where intermittent application of saline irrigation water with relatively low-impact energy occurs simultaneously with natural rainfall.

The expected climatic changes will impact the process of infiltration and in turn the amounts of ground waters. A rise in soil aridity and the resulting soil degradation will reduce the volume of water which infiltrates the ground. The temperature rise and the evaporation increase also will cause a decrease in water volume. This will directly cause a rise in salinity levels, locally making the water unfit for home use as well as for other purposes.

Owing to the sea-level rise there may be an intrusion of sea water into the aquifers of the coastal areas, thus increasing the salinity of the ground waters. The reduction of the amount of both surface and ground water also will alter water character in rivers, streams, lakes and aquifers. Small rivers will have a minimal flow-through for most of the year and some will dry up in summer months. Seasonal streams will move farther north. In the small and shallow lakes there may be an increase in the primary production due to the temperature rise, creating anoxic conditions. Salinity levels also will rise. The water volumes in the aquifers will decrease significantly. Due to overextraction of these waters there may be an intrusion of sea water and a corresponding salinity rise. Many instances (e.g. Malta) are known where the overextraction/exploitation of these aquifers has resulted in the intrusion of sea water and salinity increase to the point of rendering the water unusable for household consumption (PAP, 1987a, c).

A decrease in fresh-water sources will be accompanied by a greater demand for water consumption which will be the result of developmental needs as well as of climatic changes. The greatest water consumption rise will be in agriculture, which presently consumes the largest quantities of water.

To satisfy the growing demands with decreasing quantities of available water, it will be necessary to alter greatly the existing management of these waters. Water management must ensure the following:
- that available water will be protected from any pollution;
- that specific water consumption in individual areas will be reduced; and
- that more water will be reused.

The realization of these conditions will require that substantial funds be invested for the construction of waste-water treatment plants and facilities needed to protect existing waters from pollution. It also will require investments for the development of technology that will enable a smaller consumption per produced unit. In agriculture, for example, specific consumption may be reduced by constructing closed systems of water distribution which will bring water to the very root of a plant. This greatly decreases evaporation and allows for a more effective water use.

7.2 Soil

The Mediterranean coastal zone displays a marked variety in geomorphic character as indicated by the following coastal types:
- large river deltas (Po, Rhône, Nile);
- narrow plains with mountainous hinterlands;
- large plains and depressions frequently containing lakes;
- hills and mountain ranges touching the coastline;
- coastal dunes.

These geomorphological elements can be found throughout of Mediterranean, but in each zone they are represented in a different proportion.

The parent substrates in the Mediterranean coastal zones are very diverse, including limestone, marl, volcanic rock and tuft, clastic sediments, various alluvial deposits, bentonite clay, etc. The percentage of soil types in each Mediterranean country is shown in Table 5.18 (Blue Plan, 1987b).

Water erosion is an accute problem in the entire Mediterranean region (Table 5.19). The major contributing factors include:
- periodically heavy rains in an arid climate;
- high erodibility of some substrates (marl, volcanic tuft, aeolian deposits and some unconsolidate Pleistocene sediments);
- improper land use, such as inadequate soil cultivation and crop rotation;
- bad forest management; and frequent forest fires;
- sodification on arid areas, reducing the infiltration ability of the soil.

To date, assessments of the erosion processes and risks have been made only in a few Mediterranean countries (PAP, 1987b). However, some types of erosion are specific to certain locales. For example, in the arid regions of the eastern Mediterranean, wind erosion is an acute problem. In river deltas and sections of the coast composed of soft sediments, beach erosion is a serious problem (Egypt, Israel, Tunisia and Italy). Landslides are a specific problem in some Mediterranean countries (Cyprus, Greece, Turkey). Terraces as traditional measures of soil protection from water erosion are hardly acceptable, because they make mechanization and access by vehicular transport impossible.

Salinization and sodification of soils is a problem closely related to water resources use. The main form of secondary salinization is the increased salinity of subsoil water used for irrigation. Aquifers become saline because of excessive pumping for irrigation and the intrusion of sea water.

Abandonment of agricultural land has become a widespread phenomenon in the Mediterranean historically and appears to be increasing today. The reasons are various:
- decreased soil fertility due to erosion, salinization, coastal subsidence (e.g. North Africa and Middle East);

Table 5.18 Soil resources in the Mediterranean countries

Country/ Med. region	Total area area (1000 ha)	Soil groups (in percentage of total area)								
		I	II	III	IV	V	VI	VII	VIII	IX
Spain	50475	87.9	—	1.0	1.3	7.3	—	1.0	1.5	—
	12192	91.5	—	1.0	—	4.1	—	—	3.4	—
France	54396	—	—	—	—	—	—	—	—	—
	5884	64.6	—	2.9	—	32.4	—	—	—	—
Italy	31028	75.0	0.4	—	1.3	8.0	6.1	9.2	—	—
	24407	80.5	0.4	—	—	5.3	2.4	11.3	—	—
Yugoslavia	25580	74.1	3.4	—	4.0	17.4	—	1.1	—	—
		62.4	—	—	—	35.4	—	2.2	—	—
Albania	2870	92.7	—	—	—	6.2	—	1.1	—	—
Greece	13071	70.5	0.9	—	—	16.8	—	11.8	—	—
Turkey	77481	—	—	—	—	—	—	—	—	—
	13431	64.9	10.4	—	—	22.5	2.1	—	—	—
Cyprus	920	77.9	10.9	—	—	11.2	—	—	—	—
Syria	18520	1.4	6.0	—	—	8.7	—	—	83.8	—
Lebanon	1040	57.6	7.8	—	—	18.3	—	—	16.3	—
Israel	2080	28.6	7.4	1.1	—	19.1	—	8.6	35.2	15.5
Egypt	99545	3.5	0.1	6.9	1.0	21.7	—	18.6	32.6	13.4
	6644	24.6	—	11.8	—	0.9	—	17.3	32.0	22.4
Libya	175954	5.7	—	1.4	—	11.8	—	9.2	49.3	0.4
	20337	14.2	—	4.1	—	4.9	—	0.5	75.8	9.1
Tunisia	15536	20.7	3.0	9.1	—	29.3	—	2.4	26.2	—
	5215	29.6	4.7	6.6	—	18.9	—	—	40.2	—
Algeria	238174	6.3	0.2	2.2	—	24.7	—	9.3	31.6	25.7
	9024	85.7	2.4	0.3	—	0.7	—	0.5	10.3	—
Morocco	46630	31.1	2.2	4.5	0.7	25.9	—	8.9	26.6	—
	6780	51.1	2.4	0.8	—	0.5	—	0.4	44.8	—
Med. Total	296095									

Source: Blue Plan, 1987b

Table 5.19 Soil erosion intensity, in area and for selected countries

Country	Importance of erosion	Eroded area (1000 ha)	as % of total area	Estimated deposit (1000m³)	Area of arable land lost through erosion, in ha/yr (% of total land)
Spain (4,971,900 ha)	none slight	27804	55.9		
	moderate	13744	27.6		
	severe	8171	16.5		
Mediterranean Spain⁽¹⁾ (15,884,000 ha)	none	6035	38.0		
	slight	1250	7.9		
	moderate	3354	21.1		
	severe	5243	33.0		
Yugoslavia⁽²⁾ (20,415,000 ha)	slight	6626	26.0	28213	
	moderate	7264	29.0	54790	
	severe	4585	18.0	50568	
	excessive	1938	7.0	37510	
Greece (13,144,400 ha)	severe and excessive	4700⁽³⁾	35.6		
Turkey⁽⁴⁾	none or slight	5884	28.5		
	moderate to severe	6461	31.5	8,750,000 t/yr	54237 (0.3%)
	very severe	7445	36.3		
	excessive	114	0.6		
Mediterranean Turkey⁽⁵⁾	none or slight	1738	10.0		
	moderate to excessive	15534	90.0		
Tunisia					18000 (1.1 T)
Algeria					–
Morocco					2200 (0.5%)

Notes: (1) Concerns the following regions: Andalusia, Murcia, Valencia and the Balearic islands. *Medio Ambiente en Espana*, 1984, 1985.
(2) *The State of the Environment and the Environmental Policy in Yugoslavia*, 1983
(3) In addition, 300 million hectares threatened with salinization and alkalization (from *Review of Environmental Policies in Greece*, OECD, 1981).
(4) *Environmental Profile of Turkey*, 1981.
The areas studied at national level only represents part of total land surface (78058 thousand ha).
(5) Figures refer to the three Mediterranean provinces: Marmara, Aegean, Mediterranean (whose total area is 22294 thousand ha)

Source: Blue Plan, 1987b

- lack of water for irrigation (e.g. Yugoslav islands);
- fragmentation of agricultural holdings below the actual limits of profitability;
- discrepancy between the crops raised and the technical and economic requirements of contemporary production;

- cultivation on steep slopes inaccessible to machinery;
- countryside and forest fires; and
- migration of labour to employment in tourism and industry.

The inequitable distribution of natural resources in some Mediterranean countries resulted in high concentration of the population in their coastal zones (Egypt, Tunisia, Cyprus). The development of tourism, particularly intensified in the last decades, has generated large migration flows in all the Mediterranean countries. The result of such a trend has been enormous urban expansion and the irreparable loss of hundreds of thousands of hectares of fertile land. This loss of fertile land also represents both a biological and an aesthetic degradation of the environment.

The expected climatic changes brought about by temperature rise will significantly affect the soil. Although direct change may be slight, the indirect impact may be diverse and great, depending on the local soil characteristics and climatic conditions.

Increase in evaporation (together with decreased precipitation) will result in decreased soil moisture. This will increase the areas in which desertification occurs (Table 5.20), leading to intensified wind erosion.

The climatic changes also will alter the effects of water erosion. Studies on the effects of erosion conducted in arid and semi-arid areas of Israel show that blocking or reducing runoff directly reduces water erosion.

Surface runoff, and therefore erosion, can be controlled by increasing the surface storage of the soil, by leaving small rocks and plant remains on the ground, and by dispersing electrolites onto the surface of the soil. The vegetation cover of the ground greatly reduces surface runoff. Water erosion would be less intense in the changed climatic conditions since a decrease in rainfall is expected. However, if the number of rain storms increases in summer months, the period when the agricultural lands are least covered by vegetation, then water erosion might be greatly intensified. That this form of erosion probably also will be intensified as a consequence of the increased number of forest fires and to the decreased number of areas covered by vegetation. According to the existing data at FAO, large areas in North Africa and the Middle East are endangered by

Table 5.20 Desertification in three major Mediterranean regions, early 1980s

| | Total productive drylands | | Rangelands | | Productive dryland types | | | |
| | | | | | Rain-fed croplands | | Irrigated lands | |
	Area (M ha)	Number desertified	Area (M ha)	Number desertified	Area (M ha)	Number desertified	Area (M ha)	Number desertified
Mediterranean Europe	76	39	30	30	40	32	3	30
Western Asia	142	82	116	85	18	85	8	40
Mediterranean Africa	101	83	80	85	20	75	1	40
For comparison purposes: Sudan-Sahel region	473	88	380	90	90	80	3	30

Source: Blue Plan, 1988b

salinization. It is estimated that 50% of the irrigated land in the Euphrates valley is threatened by salinization; in the Nile valley in Egypt the value is 30% and in Greece 33%. Salinization results from the inefficient use of water, the lack of adequate drainage, and improper as well as inadequate management. This process of soil degradation may become even worse in the expected warmer conditions where it will be imperative to intensify irrigation in order to ensure the necessary quantities of food.

A probable proliferation of insects due to changed climatic conditions would necessitate increased use of pesticides. This will enhance soil degradation and thus soil erosion. Significant reduction of the soil degradation can be achieved through the dispersion of chemical compounds of short duration. After achieving their purpose they would quickly degrade to simpler compounds that would not be detrimental to the soil.

7.3 Specially Protected Areas

The Mediterranean region is rich both in areas that have an important and diverse biological and ecological value, and also in sites that have scientific, aesthetic, historical, archeological, cultural and educational importance. Many of them have been preserved through the establishment of national, regional or local protected areas (national parks). Today there are approximately 100 protected areas located in the Mediterranean coastal zones, about 70% of which are found in five northern Mediterranean countries (Table 5.21, Fig. 5.5). Currently, there is increased pressure in many countries for the establishment of new protected areas or for the expansion of existing ones. Many of these actions are internationally supported

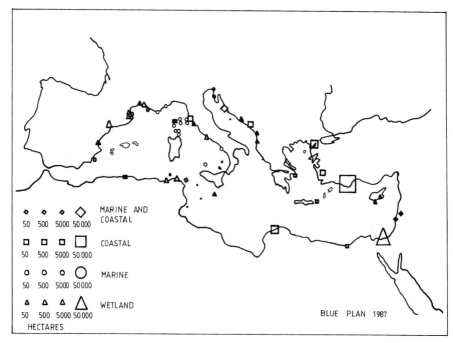

Fig. 5.5. Protected areas on the Mediterranean coast, 1986.

Table 5.21 National Parks: Number and surface area as recorded in recent national publications

Country and Med. region	Date	National parks Number	Surface area (Ha)
Spain	1985	9	112689
		—	—
France	1986	6	124670
		4	75678
Italy	1984	5	273400
		3	66400
Yugoslavia	1976/1983	15	301704
		10	82756
Greece	1983	10	350000
Turkey	1985	17	271032
		8	170390
Cyprus		—	—
Syria	1980	—	—
Lebanon	1985	—	—
Israel	1985	2	15140
Egypt	1980	—	—
Libya	1985	1	30000
Tunisia	1985	4	36102
		4	36102
Algeria	1985	7	170000
		6	160000
Morocco	1980	2	36580
		1	580

Source: Blue Plan, 1988b

through coordination with the Special Protected Areas Protocol, one of the protocols of the Convention on the Protection of the Mediterranean Sea against Pollution (Blue Plan, 1988b).

The expected climatic changes will have an impact on existing as well as on future protected areas. The impact and its magnitude, however, will depend on the characteristics of each area, that is on the characteristics of the biotype and biocenose.

It is very probable that the temperature rise will have greater impact on land biocenosis than marine biocenosis because the resultant changes (e.g. decreased precipitation, increased evaporation) will be felt more on the land. The temperature rise will enhance the development of that species which favour higher temperatures. This may cause changes in the existing ecological balances, and in turn create new ones. As mentioned previously, the land ecosystems also will be more threatened by fires which will be more frequent in the changed climatic conditions.

The sea-level rise will shift of the border between the land and the sea biocenoses. Depending on the physical characteristics of the biotype, the sea-level rise may cause major erosion which in turn will bring about numerous environmental and sociological changes. The rise of the sea-level also will affect deltaic regions (e.g. the protected deltaic areas of the rivers Ebro, Nile, Po and Rhône).

It is necessary to gather sufficient data on the individual regions in order to predict the specific impacts of the expected climatic changes. These data should include the following:
- topographic maps for every protected region,
- facts on tidal exchanges,
- sedimentary facts on the biotype, and
- facts on the existing biocenoses and their distributions in the protected areas.

7.4 Communal Infrastructure

In this chapter we restrict the discussion to communal infrastructure, fresh water supply and the drainage, treatment and disposal of communal waste waters. There are few available technical data on the communal infrastructures in the Mediterranean region. In spite of this, we can predict the possible impacts on these structures by the expected climatic changes. However, every settlement has its own specific communal infrastructure upon which the impact of the expected changes will depend. Clearly, this problem needs proper national attention.

When discussing the development of the communal infrastructure in the Mediterranean, we must distinguish between the northern, eastern and southern areas. Within each area, however, there are other divisions between the urban and rural units. Infrastructures are generally most numerous in the cities of the more developed northern countries, less in the rural areas of the north, even less in the city areas of the south, and least in the rural areas of the south. This means that while all northern cities have systems of waterworks, some parts of some of the southern cities do not even have running fresh water. Furthermore, most country settlements in the north have waterworks, whereas most in the south do not.

Until recently most sewage systems in the Mediterranean released their wastes directly to the sea by the shortest possible route. A large number of small systems released wastes immediately under or just at the surface of the sea. It was difficult to control these releases, and consequently coastal waters became polluted. Only recently has there been a change in both conception and practice. The general belief now is that waste waters should be collected, treated to an acceptable degree of contamination, and then released through a long pipeline far into the sea. This method of waste disposal also allows for the re-use of waste waters, a point of a great importance to areas with inadequate water sources. For some time now, for example, in Israel there has been no release of waste waters into the sea. After proper treatment, all communal waste waters are re-used, mainly for irrigation of agricultural crops.

The expected climatic changes will have an impact on the existing communal infrastructure and will require the creation of new infrastructures. Basically, the reduction of available water will only exacerbate present-day problems in many areas. It will be necessary to supply some cities and settlements with water from far away sources since the local sources will have a lessened capacity and poorer quality.

Due to the temperature rise, the rate of growth of various microoganisms in some water sources will increase, thus increasing possible water-borne environmental health risks. Eutrophication may be intensified in some sources and rivers. All this will require the construction of equipment and facilities for fresh water purification that will significantly increase the price of fresh water.

The sea-level rise may endanger the sewage systems in some coastal cities, particularly those whose sewage systems also serve as cisterns. The sea-level rise may make the drainage of waste waters more difficult. After rainstorms, waste waters may flood the lower parts of these cities. However, the shortage of water will require the re-use of large quantities of waste water, and this will in turn require the construction of completely new sewage systems and facilities for waste-water treatment.

7.5 Historic Settlements

A large number of important cities and settlements are located in the coastal regions of the Mediterranean, with different types of typology (typography, morphology). Moreover, the diversity of architectural and urban types is great, covering a wide range of human history. There is evidence of great civilizations of antiquity, medieval cities belonging to West European and Byzantine cultures, North African cities belonging to Islamic cultures, more recent cities that have served as forts or as trade centers, and modern settlements built for the development of tourism and industry.

In many Mediterranean cities all these histories and uses can be found located next to (or on top of) one another. This gives a specific value to these towns, while at the same time it creates major problems in their protection and worth. Despite unique local characteristics and history, the problems related to the protection of these historic settlements have much in common (PAP, 1986):

- deterioration of cultural values, especially those that have not been preserved or those that are not very important. Most values are unprotected ones;
- economic degradation that is characteristic of the historic zones in larger towns and some entire smaller settlements;
- social degradation of (primarily) old cores that have lost social balance either due to abandoned "living" functions or to overcrowded activities. This can be seen in the general neglect and deterioration of historical zones, architectural and public utilities, housing and hygienic conditions and infrastructure. The functions of the historical zones are not adequate to their importance: old cores and other zones, in many cases, have lost any semblance of vital functions;
- disorganized traffic, especially vehicular, deforms the historical and ambient values of old zones and becomes a physical menace to them;
- industrial construction surrounding historical settlements means visual degradation and is a cause of difficult ecological problems in the wider historical environment;
- construction of tourist facilities means the degradation of historical-ambient values of the old settlements unless it is in accordance with the best interest of cultural heritage;
- some Mediterranean countries also have specific problems requiring special attention, such as preventive protection of the cultural values in earthquake zones.

To protect their historical heritage (monuments, culture, etc.), all Mediterranean countries have established organizational and legal systems. However, each country has a system that is significantly different from that in other countries.

In the coastal areas of the Mediterranean a considerable number of urban units (as well as individual structures) are registered with the World Cultural and Natural Heritage. Some of the urban units and monuments are of such importance that major international help is requested for their restoration and rehabilitation. One such example is Venice, a city located on a lagoon, which has enormous cultural, artistic and ambiental values (see Sestini, Chapter 12).

The expected climatic changes and their direct consequences will have an impact on the historic settlements and monuments. The predicted sea-level rise will endanger all those monuments and settlements located just above sea-level. Venice may find itself in a particularly difficult situation.

The expected temperature rise and resulting climatic changes along with air pollution may accelerate the erosion of monuments and buildings. The serious damage to the Acropolis brought about by air pollution is well known.

However, the greatest impacts on the protection and reconstruction of most historic settlements and monuments will come from logical socio-economic deterioration. The negative impact may be manifested both through the reduction of funds needed for the reconstruction of these sites and through the development of activities that may directly or indirectly cause negative effects on the historic sites. Funds for the protection of the historic sites will be particularly difficult in the poorer countries, since available moneys will be spent on the alleviation of the negative impacts brought about by the climatic changes (e.g. the construction of irrigation systems in agriculture).

8 Conclusions

1) Judging by the present level of research on the global phenomena and case studies presented in this volume, the estimates of impacts of climatic changes in 2025 on the socio-economic structure, activities and population of the Mediterranean coastal zones suggest that there will be no significant changes in the distribution of population and that population growth will follow present trends. This is also true for urbanization, although a percentage of the population of coastal settlements is expected to be affected (5% indirectly, and up to 1% directly) by climatic changes. The national economies and economic structures in the region, however, will be afflicted by increased expenditures incurred by the attempted abatement of adverse impacts of climatic changes. The effect of these expenditures probably will be greater on developing countries than in developed ones.

Changes of temperature, wind patterns, evaporation and the resulting change of ground-water character in lowlands probably will affect agricultural production and forests. Some crops probably will be abandoned and the new ones introduced, as the traditional Mediterranean vegetation belt moves northwards. Sea-level rise and related conditions (e.g. winds) locally will endanger the coastline and the immediate coastal belt, the coastal towns and harbour installations, the coastal transportation systems (roads and railways), water supply and sewerage systems, beaches and tourist establishments. These, of course, are only general estimates, and a clearer picture will be possible only if local factors and their impacts are known.

2) Although the negative consequences will be gradual and thus not expected before the first half of the next century, the communities and governments of the Mediterranean countries should start thinking about climatic change and appropriate measures for alleviating their possible adverse effects. These measures should be incorporated in the national development plans, environmental protection and management strategies.

In addition to setting into gear the machinery of social and political decision-making regarding what will happen, it is necessary to begin local or regional studies as to how it will happen. Great attention should be given to local studies of the physical conditions, the organizational problems and the application of the most appropriate technology for the alleviation of the impacts of climatic changes. However, the fact that these impacts are linked to the problems of energy, industry, agriculture and the environment as well as the uncertainty with respect to their effect on the existing socio-economic structures, emphasizes the difficulty in making firm assessments and proposals of how existing socio-economic structures can mitigate these impacts. In any case, the first measures should be tentative and flexible so that they can be readjusted at a later point.

Taking into account what has been discussed in this chapter, there are a number of activities that should be part of national plans and strategies:
- Information should be disseminated to the public as well as to all levels of economic and political decision-making structures about both the possible consequences of gradual changes in the climatic conditions and the need for steps leading to their mitigation.
- This should be a national action, not left to the local authorities or to coastal zones.
- Local inventories should be made of the coastal zones and data collected on the local impacts by the sea-level rise and the temperature rise on water, soil, precipitation and individual socio-economic activities.
- An intervention strategy should be developed that can react to changing climatic conditions, keeping in mind the continuity of these changes. Calculating expenditures for a given change or given time ignores any additional changes in subsequent years.
- The effects of changed climate should be incorporated into the methods of integrated planning (integrated, economic, land-use, environmental, town planning and management) process.
- Cost-benefit analyses of Environmental Impact assessment should be used in order to evaluate the feasibility of every expenditure for alleviating the impacts of climatic changes.

- Developing technology for the alleviation of these impacts on a local, regional and general level.
- Those activities ultimately should not be localized, but integrated and coordinated with the entire Mediterranean region.

9 REFERENCES

Agassi, M., Benyamini, Y., Morin, J., Marish, S. and Heukin, E., 1986. The Israeli concept for runoff and erosion control in semi-arid and arid zones in Mediterranean Basin. Submitted to PAP/RAC Dec., 1986.

ASSA, 1986. *Sustainable Development of the Biosphere*, Cambridge University Press, Cambridge.

Coudert, E. 1989. Amenagement de l'espace et regions cotiers Mediterranneanes. Blue Plan, Sophia Antipolis.

Blue Plan, 1987a. Data base of the Mediterranean, Sophia Antipolis.

Blue Plan, 1987b. Mediterranean Basin environmental data (natural environment and resources), Sophia Antipolis.

Blue Plan, 1988a. Futures of the Mediterranean Basin – Executive Summary and Suggestions for Action, Sophia Antipols.

Blue Plan, 1988b. Futures of the Mediterranean Basin: Environment Development 2000–2025, Sophia Antipolis.

Clark, W., 1985. On the practical implications of the Greenhouse Question – 3 C. Presented at the International Conference on the assessment of the role of carbon dioxide and of other Greenhouse gasses in climate variations and associated impacts. Villach, 9–15. Oct., 1985.

Mediterranean Action Plan, 1985, UNEP, Athens.

PAP, 1986. Rehabilitation and reconstruction of Mediterranean historic settlements, Split.

PAP, 1987a. Specific topics related to water resources – development of large Mediterranean islands, Split.

PAP, 1987b. Syntheses report of the case studies on soil erosion problems, Split.

PAP, 1987c. Water resources development of small Mediterranean islands and isolated coastal areas, Split.

UNEP, 1982. Convention for the Protection of the Mediterranean Sea against Pollution and its related protocols. United Nations, New York.

UNEP/ICSU/WHO, 1985. Report of the International Conference on the assessment of the role of carbon dioxide and of other greenhouse gases in climate variations and associated impacts (Villach, 9–15 October 1985), WMO, 1986 (UNEP) (OCA)/WG.2/Inf.4).

WMO, 1988. Developing Policies for Responding to Climatic Change. A summary of the discussions and recommendations of the workshops held in Villach (28 September–2 October 1987) and Bellagio (9–13 November 1987), World Climate Impact Studies, April, 1988.

WHO/UNEP/UNESCO, 1988. Interaction of the oceans with greenhouse gases and atmospheric aerosols. UNEP Regional Seas Reports and Studies No. 94, UNEP, 1988 (UNEP (OCA)/ WG.2/Inf.5).

6

Vegetation and Land-Use in the Mediterranean Basin by the Year 2050: A Prospective Study

H. N. Le Houérou
(Centre d'Ecologie Fonctionnelle et Evolutive Montpellier, France)

Abstract

As a result of the cumulative consequences of the greenhouse effect, temperature in the Mediterranean region is predicted to rise by 3 ± 1.5°C by the mid- 21st century. The resulting increase in evapotranspiration would be 180–200 mm, presumably more or less evenly spread throughout the year.

The impact of increased evapotranspiration on natural vegetation and crops would be moderate, but it could be significant in areas where climatic or soil conditions are marginal for a particular type of vegetation. One may also expect a fair shift in vegetation belts due to increasing aridity and expansion of desertization on the margins of the Sahara and the near-eastern deserts.

Assuming that the temperature increase is constant throughout the year, the change in winter temperature would have a very significant impact in areas where this factor is limiting to plant growth, that is most continental areas of northern Africa and the Near East. There would be a moderate upward and northward shift of cold-sensitive crops such as olives, citruses and vegetables. The economic impact could be quite significant; if, for instance, citruses could be grown over much larger areas in southern Europe (and therefore closer to the ultimate markets) northern Africa and the Near East would have to shift to growing more tropical crops that could not be grown in Europe.

Finally, the change in vegetation and crop patterns induced by the rise of temperature will be slight compared to the change induced by the exponential population growth in the south and east of the Mediterranean basin where population has doubled every 25 years since 1950. If the present growth remains unabated, the total population of these areas will be 1.5 billion by the year 2050. Any foreseeable change in temperature or evapotranspiration would have an almost negligible impact on the environment compared to this demographic explosion. Man has thus become a major geological agent in this part of the world. He is increasingly responsible for a phase of *Biostasis* to the north of the basin due to forest and shrubland encroachment and for an acute state of *Rhexistasis* to the south, as a result of the intense vegetation and ecosystem degradation as well as accelerated anthropic erosion.

1 Introduction: The Study Area and its Climate

The Mediterranean basin is assumed to include areas neighbouring the Mediterranean Sea, but not necessarily adjacent to it, that have a Mediterranean climate and Mediterranean types of vegetation (Figs 6.1–6.4; Table 6.1 and 6.2. The Mediterranean climate is characterized

Table 6.1 Surface area of the countries and provinces under Mediterranean climate in the basin

Area/country	Percentage of country area with Med. climate	Surface area ($10^3 km^2$)	Area
Albania	18	5	Coastal plain
France	16	87	Coastal strip 50 km wide and Corsica
Greece	62	81	Except Macedonia
Italy	40	118	South of Florence and coastal strip North
Portugal	62	57	Southern 2/3
Spain	64	317	Except north-east corner and Pyrénées
Yugoslavia	10	25	Coastal strip and Adriatic Islands
EUROPE	9	690	
Algeria	50	1191	Except central and south Sahara
Egypt	50	500	Except central Sahara
Libya	50	880	Except central Sahara
Morocco	100	620	
Tunisia	100	156	
AFRICA N. OF THE TROPIC OF CANCER	46	3347	
Cyprus	100	9	
Iran	80	1300	Except the Lut and Hyrcanian zones
Iraq	100	434	
Israel	100	20	
Jordan	100	97	
Lebanon	100	10	
Saudi Arabia and Emirates	50	1075	North of the Tropic of Cancer
Syria	100	144	
Turkey	30	231	Except the Euxinian zone and interior mountains
ASIA S. OF 45° LAT. N. AND W. 70° LONG. E.	45	3320	
Grand total	–	7357	

Vegetation, land-use and climatic change 177

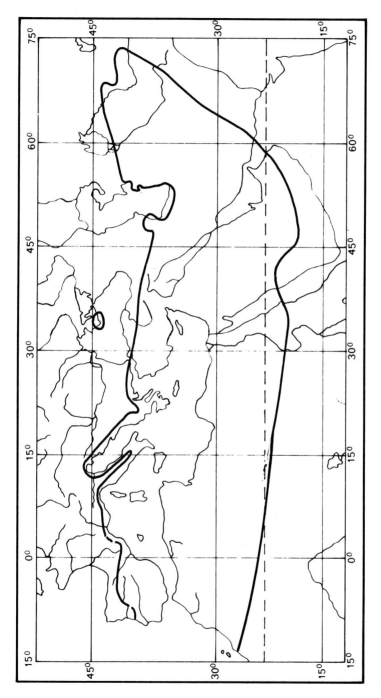

Fig. 6.1. Geographical distribution of Mediterranean climates in the basin.

178 *Climatic Change in the Mediterranean*

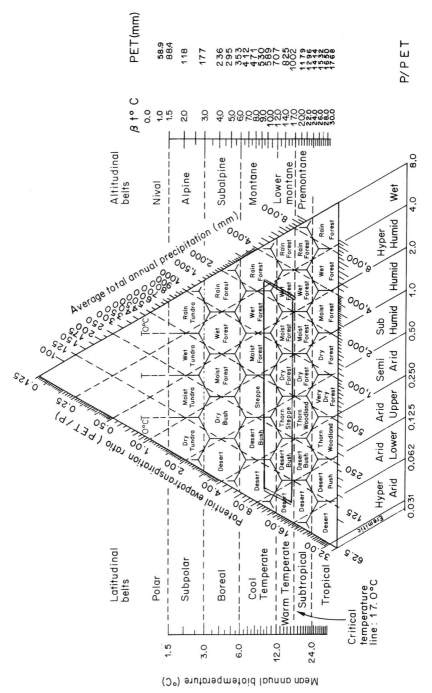

Fig. 6.2. The Mediterranean area in Holdridge's life zone climagram.

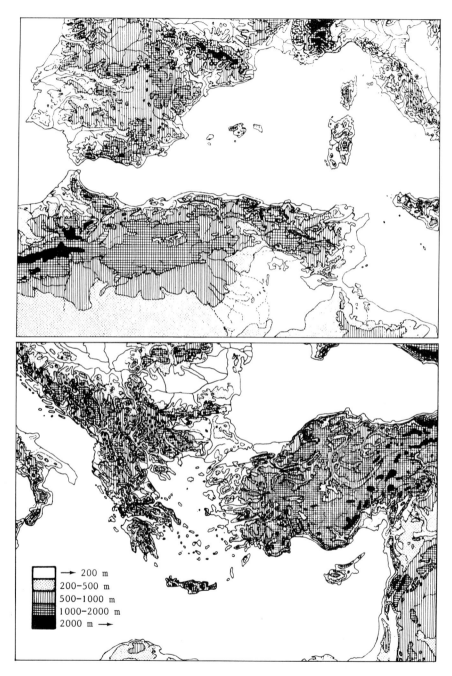

Fig. 6.3. Elevation in circum-Mediterranean countries (Tomaselli, 1976).

180 *Climatic Change in the Mediterranean*

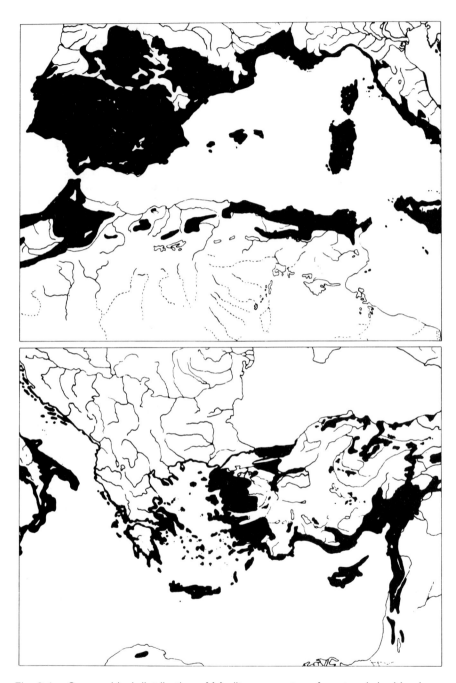

Fig. 6.4. Geographical distribution of Mediterranean-type forest and shrubland around the basin (Tomaselli, 1976).

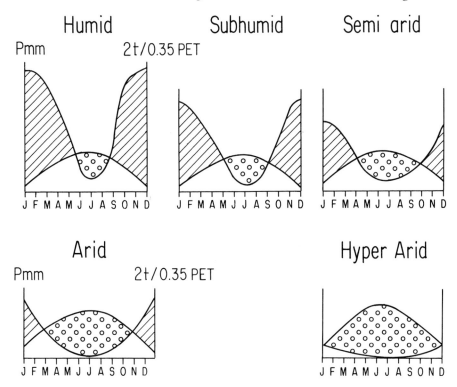

Fig. 6.5. Various degrees of aridity in Mediterranean climates as a function of monthly rainfall, temperature and potential evapotranspiration.

by a prolonged and intense summer drought of at least two to three months and up to 11 months. Rains occur during the winter season (Fig. 6.5). A dry month is empirically considered as one where precipitation expressed in mm is less than twice the mean temperature expressed in °C (Bagnouls and Gaussen, 1953a,b; Walter and Leith, 1967). A scientifically better definition is: a month where precipitation is equal to or smaller than 0.35 PET (Potential Evapotranspiration) (Le Houérou and Popov, 1981).

According to the respective length of the dry and rainy seasons and to the total amount of rainfall, one can differentiate Hyper-arid, Arid, Semi-arid, Sub-humid, Humid and Hyper-humid Mediterranean sub-climates to which correspond specific types of vegetation (Figs 6.9–6.12; Tables 6.1–6.4). Mean annual rainfall can vary from 20–25 mm in the Mediterranean deserts to 2000–2500 mm on mountain slopes or locally in maritime areas exposed to rain-bearing winds.

Temperature varies widely with latitude, altitude and continentality: the annual mean may drop below 10°C or rise above 25°C; the mean daily minimum of January may vary from −10 to +15°C; and the mean daily maximum of July/August from 25 to 45°C.

Another most important attribute of the Mediterranean climate from the biological viewpoint is winter cold stress. It can be non-existent, light, mild or intense and prolonged, depending on latitude, elevation

Table 6.2 Geographical location of bioclimatic zones in various Mediterranean countries

Country	Hyper-arid	Arid Lower	Arid Middle	Arid Upper	Semi-arid	Sub-humid	Humid	Hyper-humid	High mountain
MEDITERRANEAN BASIN									
Portugal	–	–	–	–	+	+	+	+	(+)
Spain	–	(+)	+	+	+	+	+	+	+
France	–	–	–	+	+	+	+	+	+
Italy	–	–	–	(+)	(+)	+	+	+	+
Yugoslavia	–	–	–	–	+	+	+	+	+
Greece	–	–	–	(+)	(+)	+	+	+	+
Turkey	–	–	+	+	+	+	+	+	+
Syria	(+)	–	+	–	+	+	+	+	–
Lebanon	–	+	–	–	+	+	+	+	+
Israel	+	+	+	+	+	+	(+)	–	–
Jordan	+	+	+	+	+	+	(+)	–	(+)
Iran	+	+	–	+	+	+	+	(+)	(+)
Iraq	–	–	–	(+)	+	+	+	(+)	(+)
Cyprus	–	–	–	(+)	+	–	(+)	–	–
Malta	–	–	–	–	+	(+)	–	–	–
Egypt	+	+	+	+	–	+	–	–	–
Libya	+	+	+	–	+	+	(+)	+	(+)
Tunisia	+	+	+	+	+	+	+	+	+
Algeria	+	+	+	+	+	+	–	–	–
Morocco	+	+	+	+	+	+	+	+	(+)
OTHER MED. ZONES									
South Africa	(+)	+	+	+	+	+	+	?	?
Australia	–	+	+	+	+	+	+	(+)	–
California	+	+	+	+	+	+	–	+	+
Mexico (Baja Calif.)	+	+	+	+	+	–	–	–	(+)
Chile	+	+	+	+	+	+	+	+	+

Notes: + = present; – = absent; (+) = pin-point; ? = dubious

Vegetation, land-use and climatic change

Table 6.3 Bioclimatic zoning in the Mediterranean basin

Main zones (Water stress)	Mean annual rainfall (mm)	P/PET ratios (Penman)	Very warm $m>9$	Warm $9>m>7$	Mild $7>m>5$	Temperate $5>m>3$	Sub-zones based on Winter Cold Stress			
							Cool $3>m>1$	Cold $1>m>-2$	Very cold $-2>m>-5$	Extreme cold $-5>m>-0$
Hyper-arid	$100>P$	$0.05>R$	+	+	+	+	+	+	–	–
Arid: Lower	$200>P>100$	$0.12>R>0.05$	+	+	+	+	+	+	–	–
Middle	$300>P>200$	$0.20>R>0.12$	+	+	+	+	+	+	+	+
Upper	$400>P>300$	$0.28>R>0.20$	+	+	+	+	+	+	+	+
Semi-arid	$600>P>400$	$0.43>R>0.28$	+	+	+	+	+	+	+	+
Sub-humid	$800>P>600$	$0.60>R>0.43$	+	+	+	+	+	+	+	+
Humid	$1200>P>800$	$0.90>R>0.60$	+	+	+	+	+	+	+	–
Hyper-humid	$P>1200$	$R>0.90$	+	+	+	+	+	+	–	–

Notes: + combination present. – combination absent
Source: Le Houérou, 1975a or b

184 *Climatic Change in the Mediterranean*

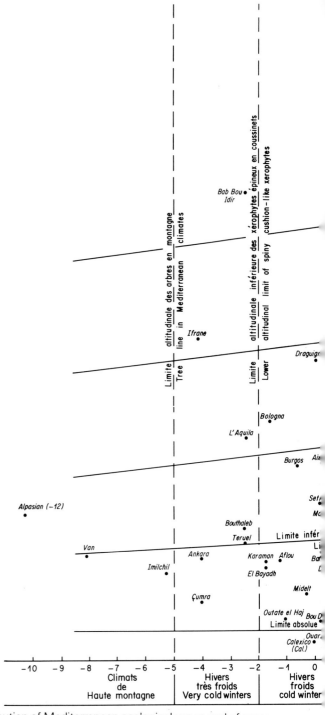

Fig. 6.6. Geographical distribution of Mediterranean ecological zones and of some critical bioclimatic and agroclimatic thresholds under the present-day conditions (Le Houérou, 1973).

$$\text{OOP} \simeq \frac{3.43 P}{M-m}$$
$$Q_2 \simeq Q_3$$

						Etage	Hyper–humid
			Albany			méditerranéen	mediterranean
						perhumide	zone

Limite thermique absolue de la culture des citrus
Limite des Hautes plaines Nord-Africaines
Cold limit of Acacia, Prosopis
Cold limit of commercial cultivation
grape fruits commercial cultivation
Upper thermal limit of commercial cultivation of apples & pears
Limite thermique de la culture des Orangers et des Clémentiniers
Limite thermique des cultures tropicales perennes
Upper thermal limit of tropical crops
Limite thermique inférieure de la flore tropicale spontanée
Lower thermal limit of native tropical species

Ain Draham
Djidjelli
Genova
Zakinthos
Tanger
Kerkira (Corfou)
Durazzo
Limite supérieure des arbres a pépins
Limite supérieure de la culture du blé
Upper limit of rainfed commercial cropping of wheat & barley

San Francisco Monaco
Alanya Bejaia
El Kalla

						Etage	Humid
		Grasse	Coimbra · Tabarca			méditerranéen	mediterranean
			Antalya	Skikda		humide	zone
				Perth	Beyrouth		

Extreme cold limit of citrus cultivation
Lower limit of highlands in Northern Africa
Cold Limite thermique inférieure du maïs sans irrigation
Lower limit of rainfed cropping of Maize

Napoli Annaba
Blida
Istanbul Roma Ajaccio Livorno Cherchell Iskanderun Cape Town
Nice Toulon Cadiz Alger
Perpignan Barcelona Lisboa
Salt Jerusalem Rabat
Bari Brindisi Shahat (Cyrene)
Ancona Izmir Hania

						Etage	Sub–humid
						méditerranéen	mediterranean
						sub–humide	zone

ntpellier
Culture Lower limit of
Limite inférieure des cultures d'été en culture séche (Sorgho, Melon, Pois chiche etc.)
Lower limit of rainfed summer crops (Sorghum, Millet, Beat, Chickpeas etc...)

Nîmes Marseille
Narbonne Beja Faro Oran Naxos
Avignon Perpignan Adana Taranto Casablanca
Meknès Sevilla Thibar Los Angeles Hiraklion
Waga Waga Fez Tel Aviv
Constantina Amman Valencia
Thessaloniki Mazagan Tunis
Moktar San José (Calif) Athenai Safi
Homs (Syrie) Sacramento Famagusta
Thala Badajoz Katanning Berkane Sousse
Denliquin
Limite supérieure de la céréaliculture productive en dry farming et des cultures fourrageres en sec. Lower limit of sclerophyllous strees & shrubs
Benghazi Lower limit of rainfed commercial cropping of cereals
Tripoli (Libya) Upper limit of steppic végétation

						Etage	Semi–arid
						méditerranéen	mediterranean
						semi–aride	zone

Limite supérieure de la végétation steppique
Agadir

						Etage	Mediterranean
ragozb El Asnam Oujda Alicante Nicosia						méditerranéen	arid
rine Hama Murcia Sfax						aride	zone
Aleppo Riverside Sidi Bouzid El Djem Kalgoorlie Almeria Alexandria							
Feriana Fresno Marrakech Matmata Suez							
Damascus Gabès							
Laghouat Ben Gardane Medenine Jericho							
Bakersfield Palmyra Gafsa Biskra							

cultures non irriguées (Céréales, Olivier) Limite supérieure de la végétation "contractée" sur les regs

Béchar Figuig Ghardaia Absolute limit of rainfed cropping upper limit of contrated vegetation
Touggourt Kebili Tozeur Kufra Aqaba-Eilath Etage méditerranéen Hyper–arid & desert
Tindout Kharga Reggane saharien ou désertique mediterranean zone

2	3	4	5	6	7	8	9	10	m
Hivers frais	Hivers tempérés		Hivers doux		Hivers chauds			Variante à Hivers très chauds	
ol winters	Temperate wint.		Mild winters		Warm winters			Very warm winters	

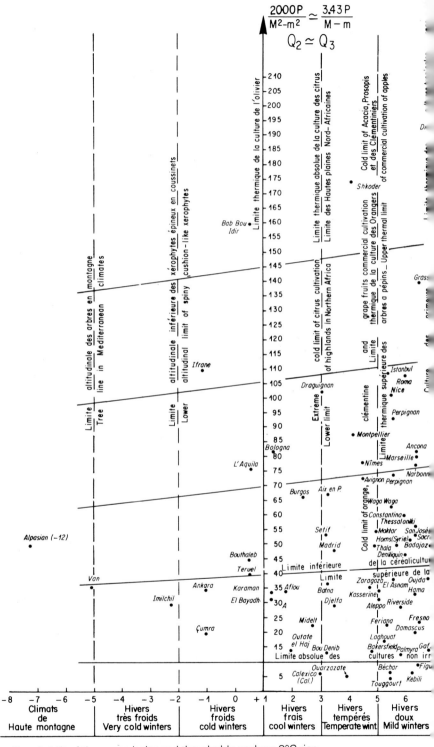

Fig. 6.7. Zonal shift of the same belts and thresholds under a 3°C rise of temperature.

Limite inférieure de la flore tropicale spontanée
thermal limit of native tropical species

· Albany

· Djidjelli
· Genova
· Zakinthos Tanger
· Kerkira (Corfou)
· Durazzo

Limite supérieure de la culture du blé
Lower limit of rainfed commercial cropping of wheat & barley

Etage méditerranéen perhumide — Hyper-humid mediterranean perhumide zone

· San Francisco · Monaco
· Alanya · Bejaia
· Coimbra · Tabarca · El Kalla

Etage méditerranéen humide — Humid mediterranean zone

· Antalya · Skikda
· Napoli · Annaba · Perth · Beyrouth
· Blida

Limite inférieure du maïs sans irrigation
Lower limit of rainfed cropping of Maize

· Iskanderun
· Cherchell · Cape Town
· Alger
Cold · Livorno
·accio
· Barcelona
· Lisboa
· Cadiz

Etage méditerranéen sub-humide — Sub-humid mediterranean zone

· Jerusalem · Brindisi · Rabat
· Bari · Shahat (Cyrene)
· Izmir · Hania

Limite inférieure des cultures d'été en culture sèche (Sorgho, Melon, Pois chiche etc.)
Lower limite of rainfed summer crops (Sorghum, Millet, Beat, Chickpeas etc...)

· Beja · Naxos
·dana · Taranto · Faro · Organ
· Sevilla · Thibar · Casablanca
· Los Angeles · Hiraklion
· Tel Aviv
· Valencia
· Mazagan · Tunis
· Athenai · Famagusta · Safi

Etage méditerranéen semi-aride — Semi-arid mediterranean zone

Lower limit of sclerophyllous strees & shrubs
·anning · Berkane · Sousse
productive en dry farming et des cultures fourrageres en sec.
· Benghazi Lower limit of rainfed commercial cropping of cereals
gétation steppique · Agadir Upper limit of steppic végétation
· Alicante · Tripoli (Libya)
·cia · Nicosia
·Kairoun · Kalgoorlie · Almeria
· El Djem · Sfax · Alexandria
·zid · Matmata · Suez
· Marrakech
Gardane · Gabès
· Medenine
· Jericho
· Biskra

Etage méditerranéen aride — Mediterranean arid zone

(Céréales, Olivier) Limite supérieure de la végétation "contractée" sur les regs
Absolute limit of rainfed cropping upper limit of contrated vegetation

·hardaia
· Tozeur · Kufra
· Tindouf · Kharga · Reggane · Aqaba-Eilath

Etage méditerranéen saharien ou désertique — Hyper-arid & desert mediterranean zone

| 8 | 9 | 10 | 11 | 12 | 13 | m |

Hivers chauds / arm winters | Variante à Hivers très chauds / Very warm winters

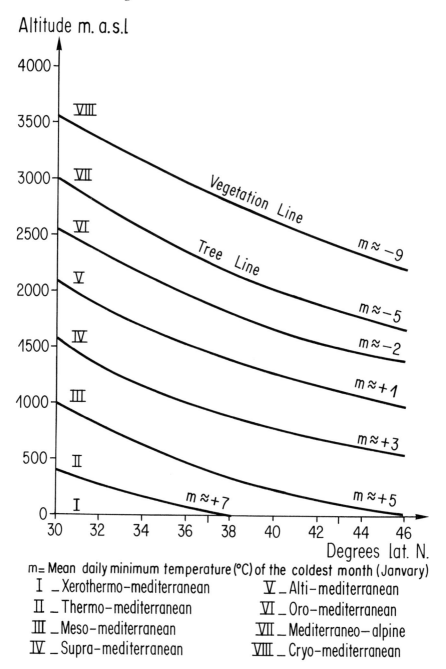

Fig. 6.8. Distribution of altitudinal belts of natural vegetation as a function of elevation and latitude (see also Table 6.4).

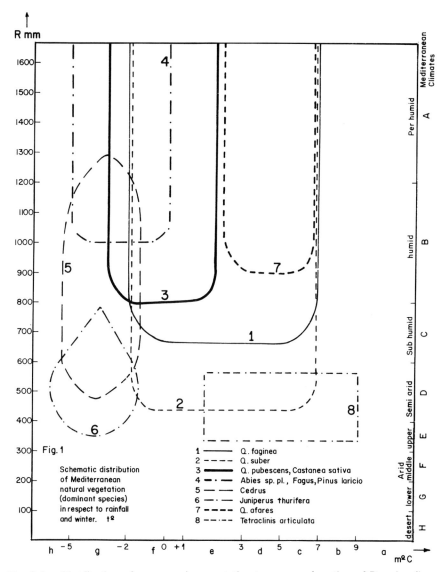

Fig. 6.9. Distribution of some main vegetation types as a function of P and m (Le Houérou, 1973b).

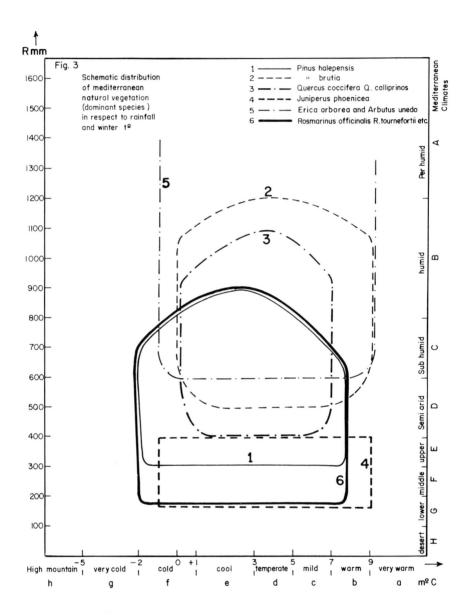

Fig. 6.10. Distribution of some main vegetation types as a function of P and m (Le Houérou, 1973b).

Vegetation, land-use and climatic change 191

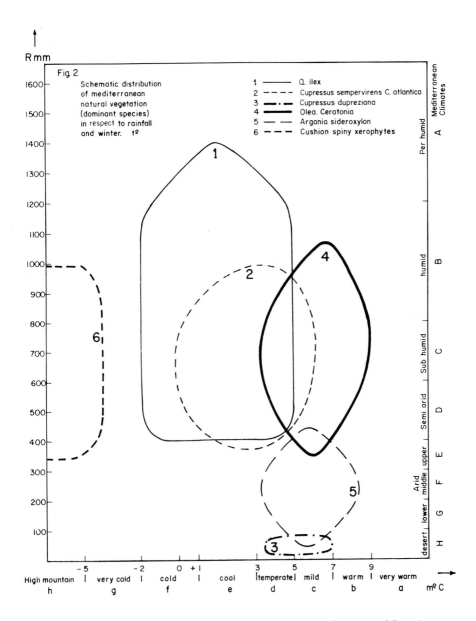

Fig. 6.11. Distribution of some main vegetation types as a function of P and m (Le Houérou, 1973b).

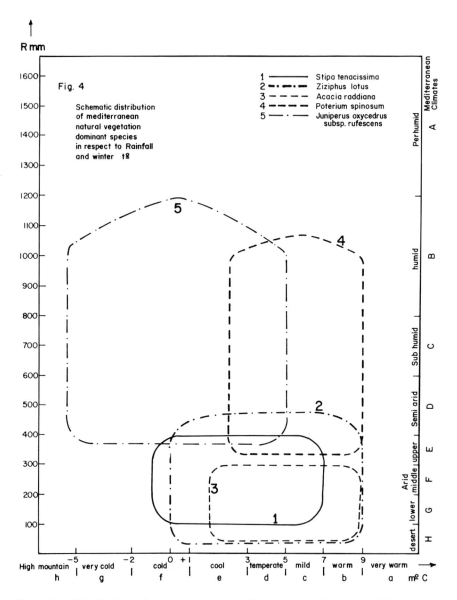

Fig. 6.12. Distribution of some main vegetation types as a function of P and m (Le Houérou, 1973b).

Table 6.4 Vegetation zonation in Mediterranean mountains as a function of altitude and latitude, used as a basis for Fig. 6.8

Latitude (°N)	Mountains	Maximum altitude (m.a.s.l.) (rounded)	Country
30–32	Anti Atlas/Sarhro	2560	Morocco
28–35	Zagros	4550	Iran
31–32	High Atlas	4165	Morocco
32–33	Ksour Mountains	2240	Algeria
32–33	Liban	3090	Lebanon
32–34	Mid Atlas	3340	Morocco
35	Jebel Amour	1980	Algeria
35	Rif	2450	Morocco
35	Hodna & Aurès	2330	Algeria
35	Mount Ida	2500	Crete, Greece
35	Troodros	1950	Cyprus
35	Alborz	5600	Iran
35	Chambi	1540	Tunisia
35	Mount Alaoui	1560	Syria
35.5	Babor	2310	Algeria
34–38	Kurdistan	3610	Iran, Iraq, Turkey
37	Sierra Nevada	3480	Spain
37	Taygete	2410	Greece
37–38	Taurus	3920	Turkey
38	Etna	3260	Italy
38	Aspromonte	1960	Italy
38.5	Parnassus	2510	Greece
40	Olympus	2920	Greece
41	Sierra de Gredos	2590	Spain
42	Sierra de Guadarrama	2470	Spain
42	Mte Terminillo	2210	Italy
42	Gran Sasso d'Italia	2910	Italy
43	Eastern Pyrénées	2300	Spain, France
44	Ventoux and southern Alps	2000	France
44.2	Aigoual (Cevennes)	1570	France
44.5	Mt Lozère (Cevennes)	1700	France

and continentality. Cold stress becomes a potent discriminating factor of vegetation distribution patterns, crop selection and land use (Emberger, 1930). A sketch of vegetation zonations as a function of latitude and elevation in the Semi-arid to Hyper-humid zones is shown in Fig. 6.8. Arid and Hyper-arid zones are differentiated on the basis of two main criteria: (1) A drought index which may be the Precipitation to Potential Evaporation ratio (P/PET), or Emberger's pluviometric quotient (these two indices are tightly correlated); (2) A cold stress index (the mean daily minimum temperature of the coldest month which is a much more sensitive criterion than the more commonly used parameter of mean temperature of the coldest month) (Le Houérou, 1973a). A general classification of Mediterranean climates on the basis of these criteria is shown in Figs 6.6 and 6.7. Fig. 6.6 shows the present-day situation whereas Fig. 6.7 shows the shift of ecological zones under an assumed 3.0°C rise of temperature.

As suggested in the above remarks, the Mediterranean basin is extremely diverse in terms of climate and vegetation. In accordance with our definitions, the study area includes the 19 countries shown in Tables 6.1, 6.2 and 6.5. As mentioned above, this huge area (about the size of

Table 6.5 Countries with Mediterranean climate

Area/country	Percentage of country with Mediterranean climate	Surface area (10^3 km^2)	Area
EUROPE	9	690	
Albania	18	5	Coastal plain
France	16	87	Coastal strip 50 km wide + Corsica
Greece	62	81	Except Macedonia
Italy	40	118	South of Florence + Coastal strip north
Portugal	62	57	Southern 2/3
Spain	64	317	Except north-east corner and Pyrénées
Yugoslavia	10	25	Coastal strip and Adriatic Islands
AFRICA N. OF TROPIC OF CANCER	46	3347	
Algeria	50	1191	Except central and south Sahara
Egypt	50	500	Except central Sahara
Libya	50	880	Except central Sahara
Morocco	100	620	
Tunisia	100	156	
ASIA S. OF 54.5° LAT. N. & W. 70° LONG. E.	45	3320	
Cyprus	100	9	
Iran	80	1300	
Iraq	100	434	
Israel	100	20	
Jordan	100	97	
Lebanon	100	10	
Saudi Arabia and Emirates	50	1075	
Syria	100	144	
Turkey	30	231	
Grand total	–	7357	

Australia or conterminous United States) is climatically very diverse and, as a consequence, bears a large number of vegetation types: desert and sub-desert steppes, evergreen and deciduous Mediterranean shrubland and forest. Land-use also varies widely, depending on climatic, edaphic, historical and socio-economic conditions. The study area has experienced an almost 10,000-year history of land use, both farming and livestock husbandry.

The 7.3 million km^2 of the study area can be subdivided as follows:

1) Semi-arid to Hyper-humid climatic zones (400 mm$<P$)
1.6×10^6 km^2 = 22%

2) Arid steppes, rangeland and cropland ($50 < P < 400$ mm)
2.7×10^6 km^2 = 37%

3) Desert wasteland (25 <P< 50 mm)
     3.0×10^6 km^2 = 41%

TOTAL 7.3×10^6 km^2 = 100%

Of the 1.6 million km^2 of Semi-arid to Hyper-humid zone, about 48% is cropland, 35% shrubland and forest, and 17% rangeland (Le Houérou, 1981). Of the 2.7 million km^2 of Arid steppes, around 50% is periodically cropped to cereals whenever autumn and early winter rains permit. But those 1.35 million km^2 of cereals usually do not show up in official statistics, such as published by the FAO, since only harvested crops are taken into consideration and since crop expectancy in these areas does not exceed 20 to 25%.

In considering land use in the basin as a whole two quite different and almost opposite (in terms of both vegetation and use) areas are distinguished: the northern part (Euro-Mediterranean countries) of the basin and the southern and eastern parts (Afro-Asian Mediterranean countries).

2 Present Vegetation, Land Use and their Evolutionary Trends

2.1 In the Northern Mediterranean

General vegetation belts are laid out along latitudinal and altitudinal gradients, as both dominantly control precipitation and temperature patterns which, in turn, largely determine vegetation and land use. Precipitation increases with latitude and altitude, whereas temperature follows the opposite pattern (Fig. 6.8). The increase of precipitation with altitude and latitude can be modified by such factors as exposure to rain-bearing winds, rain shadows, the pattern of interface between sea and land masses in connection with dominant winds and frontal depression pathways. But, roughly speaking, lowlands in the Tropic of Cancer have mean annual rainfalls approaching zero, whereas towards the 40–45° Lat. N, mean annual precipitation tends towards 800–1200 mm. That is a northwards increase of 40–60 mm of precipitation for each degree of latitude, or 0.35–0.55 mm per km.

The gradient of mean annual precipitation increase with elevation is more reliable: about 10% increase for each 100 m (Le Houérou, 1959; Le Houérou et al., 1975). In other words, precipitation tends to double each kilometre increase in elevation, although, as mentioned above, local departures are common.

The temperature gradient is similar to that in other parts of the world; a decrease of 0.55°C per 100 m of elevation. This rate is also subject to local and seasonal variations, but much less so than precipitation gradients; temperatures can be predicted from elevation in a fairly reliable fashion provided some caution is used regarding site-specific conditions (temperature inversion and the like).

The latitudinal gradient of temperature at any given altitude is a northwards decrease of about 0.6°C per degree of latitude, or 0.0055°C per km.

2.2 Natural Vegetation and its Distribution Pattern

The present-day vegetation is the result of a complex interactive network of a large number of various factors:

1) *Floristic factors*: The flora is the result of geologic, paleoclimatic and paleogeographic history and biological evolution. The presently estimated number of flowering plants in Mediterranean study area is about 25,000 species. This flora is a mixture of recent paleartic and inherited palaeotropical elements (Quézel, 1976).
2) *Climatic*: The governing climatic factors are:
   - rainfall amount and distribution, drought stress;
   - temperature and cold/frost stress;
   - the P/PET ratio.
3) *Edaphic*: Geology, petrology, physiography, topography, hydrology, erosion, soil, water and nutrient budgets.
4) *Anthropozoic activities*: The pressure of man and animals (particularly livestock) on ecosystems.

Broad climatic vegetation types can be correlated with climatic matrices using rainfall, temperature and PET. Three such matrices are shown in Figs 6.2, 6.5, 6.6, 6.7 and 6.8. These matrices have given birth to various classification systems: those of Bagnouls and Gaussen and of Holdridge are general whereas Emberger's is specifically designed for Mediterranean climates.

Other systems are based on elevation (Bagnouls and Gaussen, 1953a; Ozenda, 1976; Quézel, 1976; Tomaselli, 1976; Rivas-Martinez, 1982); but these are only valid within narrow latitudinal limits. Fig. 6.8 is an attempt to show these altitudinal limits as a function of latitude.

For the Mediterranean basin as a whole one may consider the following very simplified sketch in order of decreasing aridity and temperature and of increasing elevation. This sketch, however, does not take into consideration the arid steppes or the Mediterranean deserts.

1) *Xerothermo-Mediterranean zone*. This lowland, low-latitude zone is predominantly semi-arid to sub-humid with mean annual precipitation ranging from 300 to 800 mm. The rainfall regime is Mediterranean but the temperature regime is tropical. Frost virtually never occurs and a number of tropical species are present: *Acacia Sp.Pl., Prosopis, Hyphaene, Argania, Cactoid Euphorbia Sp.Pl., Phoenix, Maytenus, Nannorhops, Lycium Sp.Pl., Periploca Sp.Pl., Rhus Sp.Pl., Juniperus phoenica* and others.
2) *Thermo-Mediterranean zone*. This is a mild winter zone with occasional light frosts in December/January. Rainfall varies from less than 400 to over 1500 mm. Vegetation may be classified as from semi-arid to hyper-humid with dominant sclerophyllous shrubs and trees:

Ceratonia siliqua	Olea europaea var oleaster
Pistacia lentiscus	Tetraclinis articulata
Chamaerops humilis	Periploca laevigata
Rhus pentaphyllum	Myrtus communis
Rhus tripartitum	Ziziphus lotus
Euphorbia dendroides	Quercus coccifera
Juniperus oxycedrus	Quercus calliprinos
Juniperus phoenica	Pinus pinaster mesogeensis
etc.	Pinus pinea

3) *Meso-Mediterranean zone*. This is perhaps the most typical Mediterranean zone, occurring over huge areas around the whole basin (Figs 6.3 and 6.4). Winter temperatures are cool to cold (-2 < m < 3). Rainfall can vary from 350–400 to 1000–1500 mm; vegetation qualifies as semi-arid to hyper-humid. Dominant species are sclerophyllous or acicular, particularly four species of oaks and two pines and a large number of shrubs: *Quercus ilex s.l.* (including *Q. rotundifolia*) *Q. coccifera, Q. suber, Q. calliprinos, Pinus halepensis, Pinus brutia.*

Among the numerous sclerophyllous shrubs: *Arbustus Sp.Pl., Phyllirea Sp.Pl., Erica Sp.Pl., Rhammus Sp.Pl., Ulex Sp.Pl., Pistacia Sp.Pl., Genista Sp.Pl., Rosmarinus Sp.Pl., Styrax, Zelkhova*, etc.

4) *Supra-Mediterranean zone*. Rainfall is rather high: 600–2500 mm and winter temperatures cool to cold. Natural vegetation is a deciduous forest: *Quercus faginea, Q. afares, Q. trojana, Q. cerris, Q. toza, Q. frainetto, Q. pubescens, Q. aegylops, Ostrya carpinifolia, Carpinus orientalis, Pinus brutia, Pinus pinaster mesogeensis.*

5) *Alti-Mediterranean zone*. Rainfall varies from 800 to 1500 mm, winter temperatures are very cold with three to four months of hard frost and snow cover. Vegetation is predominantly coniferous and occasionally deciduous (chesnut, beech). *Pinus nigra s.l., P. silvestris, Fagus sylvatica, F. orientalis, Castanea sativa, Cedrus libani, C. brevifolia, C. atlantica, Abies Sp.Pl. (A. pinsapo, A. numidica, A. cilicica, A. maroccana, A. nebrodensis, A. cephalonica).*

6) *Oro-Mediterranean zone*. Rainfall decreases with respect to the previous zone and winters are very cold. Vegetation consists of scattered Juniper trees: *Juniperus thurifera* in the western basin; *J. excelsa* in the eastern basin.

Occasionally other Junipers may occur: *J. drupacea, J. sabina, J. communis*. Among scattered trees is a more or less continuous layer of cushion-like spiny xerophytes: *Bupleurum spinosum, Vella mairei, Alyssum spinosum, Erinacea anthyllis, Cytisus purgans, Astragalus Sp.Pl.,* etc.

7) *Mediterranean-Alpine zone*. Elevation varies from 1800 m in the north and over 3000 m in the south of the basin. There are over six months of hard frost. Vegetation is made of spiny pulvinate xerophytes with meadows and moors in depressions.

8) *Cryo-Mediterranean zone*. This high-mountain zone has virtually no flowering plants above 2500 m in the north and 3500 m in the south.

9) *The Arid zone* vegetation is made up of various kinds of steppes:
   a) Grass steppes dominated by perennial bunch grasses such as *Stipa tenacissima* in northern Africa and Spain.
   b) Dwarf shrub steppes dominated by shrublets such as *Artemisia Sp.Pl., Hammada Sp.Pl.*
   c) Crassulescent steppes of halophytic *Chenopodiaceae: Suaeda, Salsola, Atriplex, Salicornia, Arthrocnemum, Halocnemon*, etc.
   d) Succulent steppes made up of glycophytic cactoid species: *Euphorbia Sp.Pl., Caralluma, Kleinia*, etc.
   e) Tall shrub steppes located in favourable edaphic conditions: *Ziziphus, Rhus, Retama, Calligonum, Nitraria, Acacia, Pistacia, Amygdalus, Pistacia*, etc.

2.3 Land Use

As mentioned above, land use is quite different in the northern and southern parts of the basin, due to differential anthropozoic pressures.

2.3.1 Northern basin

Farmland represents some 36% of the overall land area while forest and shrubland occupy 29%, rangelands 22%, and wasteland and non-agricultural land 13% (Le Houérou, 1987b).

In Mediterranean Europe, land use is fairly uniform in broad terms since forestation values vary between 20% (Greece) and 40% (Portugal) for an average 29%. The trend is towards an increasing uniformity via farmland and rangeland abandonment in marginal areas, compensated by a sharp increase in forestation and urbanization. In the seven Euro-Mediterranean countries, the hectarage of forest and shrubland increased by 14% from 1965 to 1985, that is 0.7% per annum. The major increase was in France, Spain and Portugal with 23, 19 and 15% respectively. During the same 20-year period, rangelands and cropland receded by 7 and 9% respectively, while non-agricultural land (industry, urban areas, communication networks, etc.) increased by 15%. Albania, however, behaved like a southern Mediterranean country: the forestation rate decreased by 18% and rangelands by 43% over the same 20 years while cropland expanded by 43% (12% in cereals).

We shall examine further the consequences of such trends and what could be the likely evolution in the next few decades with due consideration to expected trends in EEC agricultural policies.

2.3.2 Southern basin

The opposite trend is witnessed in the southern and eastern basin over the same period. According to official statistics, forest and shrublands receded by 3% while cropland expanded by 5%. This trend has been going on since the end of the Second World War (Le Houérou, 1973a), as the mean hectarage of cereals increased by 50% between the 1948–1957 and 1978–1986 periods. (Given the variability in annual hectarage and rainfall only five- to ten-year averages are meaningful.) The increase was 28% from 1948 to 1967 and 22% between 1968 and 1986. Between 1979–1981 and 1983–1986, the average increase was 8%; that is 1% per annum (36.9 vs 39.9 million ha of cereals).

But in fact official statistics do not reflect the real situation since only the hectarage harvested is taken into consideration. In the 14 countries examined, cereals are mainly cultivated in rain-fed conditions (except for Egypt and Saudi Arabia), including over large areas of arid lands. In these arid lands, crop expectancy does not exceed 20 to 25%, which means that the area actually tilled and sown each year is probably 1.5 to 3 times larger than the area harvested and shown in official statistics.

The actual surface area cultivated for cereals is thus probably 3.5 times what was used in the 1950s, that is 70 to 80 million hectares, of which 40 million are being harvested as an annual average (1979–1986). As a matter of fact, many remote sensing surveys with ground truth control confirm this trend. They show that probably some 50% of the arid steppe rangelands between the 100–300 mm annual rainfall isohyets have been cleared over the past 30 years and are now cropped more or less regularly for cereals

Vegetation, land-use and climatic change 199

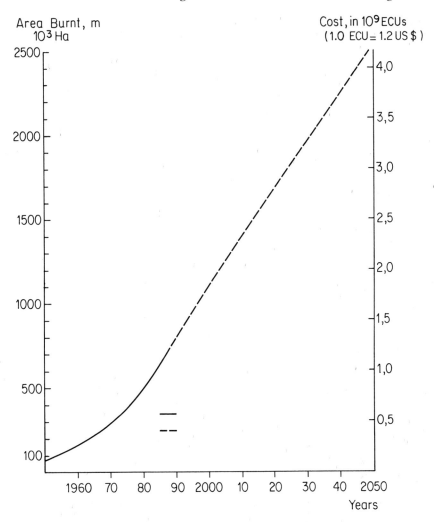

Fig. 6.13. Evolution of the annual area of forest and shrubland burnt in the basin between 1960 and 1986 and projections to 2050 and costs of rehabilitation.

(wheat and barley) under subsistence farming conditions; average yields are very low (100–300 kg/ha/yr), with the kind of crop expectancy mentioned above. The overall annual expansion of cereal cultivation over the past 30 years was thus an exponential 1.2%.

Two exceptions to that model are Saudi Arabia and Egypt where cereal cultivation is carried out entirely under irrigation, with mean annual yields of 3000–4300 kg/ha; but the area concerned is less than 2.5 million ha, only 6% of the total hectarage in the 14 countries.

The official overall forestation for the 14 Afro-Asian countries mentioned is 6.6%, with a maximum of 26% in Turkey (close to the Euro-Mediterranean figure) and a minimum of 0.4% in Jordan. According

200 *Climatic Change in the Mediterranean*

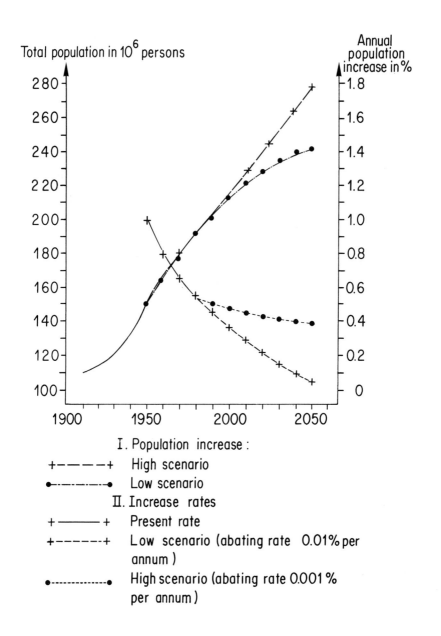

I. Population increase:
+ – – – + High scenario
• – · – • Low scenario

II. Increase rates
+ ——— + Present rate
+ – – – + Low scenario (abating rate 0.01% per annum)
• ········ • High scenario (abating rate 0.001% per annum)

Fig. 6.14. Evolution of the human population numbers in the Euro-Mediterranean countries since the beginning of this century and projections for the year 2050, according to various growth-rate scenarios.

Vegetation, land-use and climatic change 201

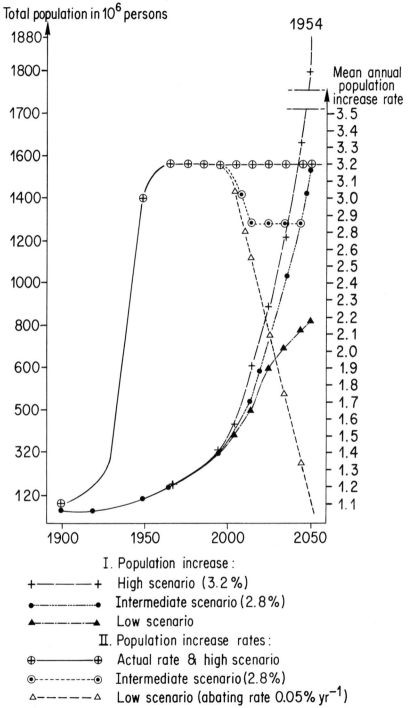

Fig. 6.15. Evolution of the human population numbers in the Afro-Asian Mediterranean countries from 1900 to 1986 and projections to 2050, according to various growth rate scenarios.

to official statistics, the overall forest and shrubland area would have decreased by 17% between 1965 and 1976 (Le Houérou, 1981) and increased by 2.4% between 1965 and 1984 (Le Houérou, 1987a). The subject is complex as the figures are reliable for some countries and quite unrealistic for others. A number of local studies in various countries of the region show that, in spite of very expensive and more or less successful reforestation policies, the overall forested area has decreased by 1 to 2% per annum since the 1960s. This trend is only clearly shown in the official statistics of four countries: Tunisia (-17.7% between 1965 and 1984), Jordan (-67%), Lebanon (-13%), Saudi Arabia (-29%). Other countries show stability or expansion of forested areas, which is obviously unrealistic.

Among the 14 countries under study, three have land-use trends closer to those observed in Euro-Mediterranean countries; that is Israel, Cyprus and Turkey. The two former show a typical Euro-Mediterranean pattern whereas the latter lies in an intermediate situation between north and south.

The ever increasing clearing of arid steppe rangeland in the other countries is the major cause of desertization in northern Africa and the Near East (Le Houérou, 1968). Given trends of the past 40 years, it is safe to predict that in a proximate future, virtually all arid lands able to be tilled will be cultivated for the cropping of cereals. Most likely, all land except rock outcrops and salt marshes down to the 100 mm isohyet of annual rainfall will be tilled episodically. All valuable rangelands will thus be cleared (most already are). Livestock will thus have to rely on stubble and fallow and, increasingly, on concentrated feed.

At the present time, sheep husbandry in the southern Mediterranean depends for at least 50% of its annual diet on concentrated feed, whether subsidized or not. Stocking rates have increased three- to four-fold over the past 30 years (Le Houérou, 1981, 1982, 1987c) and are approaching the overall figure of one sheep per hectare in the arid lands between the 100–350 mm isohyet of annual rainfall; that is about four times what the carrying capacity was some 30 years ago and perhaps as much as six to eight times the present carrying capacity of the depleted arid rangelands (which decreased to ¼–⅕ of pristine condition over the past 30 years (Le Houérou, 1987a, c). Overall stock numbers increased by a factor of 2.5, i.e. an exponential 2.33% per annum between 1950 and 1985 (Le Houérou, 1987c).

In the southern and eastern Mediterranean countries the stock numbers and the tilled areas tended to grow at a pace comparable to the rural population growth, i.e. slightly less than the overall demographic growth. This trend has remained unchanged over the past 20 years (Le Houérou, 1973). As a matter of fact, the figures are consistent with the long-term trends (Le Houérou, 1973a, 1975a), mean an increasing reliance on imported food, made possible from oil or mining revenues. A number of countries that used to be net exporters of agricultural products before the 1960s presently import over 80% of their food.

The gap between local supply and demand of food products grows nearly as fast as the demographic expansion of the human population. The latter grew by an overall average exponential 3.2% per annum between 1965 and 1984 while cereal production grew by only 0.2% per annum (except for

Saudi Arabia). The mean yield increase for the region as a whole (excluding Saudi Arabia) has been 0.9% per annum (i.e. 6 kg/ha/yr) while the figure in Euro-Mediterranean countries has been 6% per annum (i.e. 63 kg/ha/yr) (Le Houérou, 1987b). In other words, cereal yields increased six times faster in the northern Mediterranean, but population increased four times faster in the south (3.2% vs 0.8% per annum; see Figs 6.14 and 6.15 and Tables 6.6 and 6.7). The gap between north and south in terms of food production per inhabitant is thus increasing very fast.

3 Consequences of a Hypothetical Warming of the Atmosphere by $3.0 +/- 1.5°C$

3.1 Likelihood of a Possible Temperature Rise

Although there are some discrepancies between various workers, the Global Circulation Model (GCM) predicts an average increase of $3.0 +/- 1.5°C$ for a doubling of CO_2 in Mediterranean latitudes (30–45°N) and 6 to 9°C in higher latitudes (Charney, 1979; NRC, 1979, 1982; Kellog and Schware, 1981; WMO, 1981). We shall therefore work on the assumption of a 3.0°C increase by the year 2050.

This prediction, however, calls for some words of caution:

(1) Model scenarios are not actually predictions because, although they are based on the best available knowledge, major uncertainties exist regarding nearly every physical parameter entering the model (see Wigley, Chapter 1, for discussion).
(2) No global temperature increase since the beginning of industrial times is problematic (Van Loon and Williams, 1976; Kukla et al., 1977; Jones et al., 1982; Hare, 1985; Jäger, 1988; Wigley, this volume).
(3) Nuclear fusion, expected to be used for power between 2000 and 2020, might considerably reduce the burning of fossil fuel.
(4) An important but totally unpredictable cooling factor is the emission in the stratosphere of aerosol from volcanic explosions as well as man-induced aerosols and dust from deserts and desertizing areas. According to some physicists, for instance, increased desert aerosols would increase rainfall and decrease temperature; according to others they would contribute to temperature rise.
(5) An increasing albedo in desertizing zones might compensate its decrease in polar and sub-polar regions.
(6) A number of leading scientists (climatologists and atmospheric physicists) question the concept of a warming Earth; for instance, "In the global and annual mean, the net cloud effect is a cooling of the planet because the albedo effect ($54 Wm^{-2}$) dominates over the greenhouse effect ($31 Wm^{-2}$) by $23 Wm^{-2}$" (Rockner et al., 1987).
(7) Glaciers are expanding in the Alps and other mid-latitude high mountains.

3.1.1 Other climatic consequences of CO_2 increase in the atmosphere

Present available models do not provide any reliable information on the possible seasonal change of temperature: whether, for instance, the increase would be larger in winter or in summer, which, of course, would have quite different effects on vegetation and crops. We shall

Table 6.6 Land use in the northern and southern Mediterranean countries: human population numbers and projections (in 10⁶ persons)

A.

Country	1950	1960	1970	1980	1985	1990	2000	2010	2020	2030	2040	2050
Albania	1.2	1.6	2.1	2.7	2.8	3.0	3.3	3.6	3.9	4.2	4.5	4.7
France	41.7	45.7	50.8	53.9	55.2	56.5	59.3	62.9	66.0	68.6	70.7	72.8
Greece	7.6	8.2	8.8	9.6	10.0	10.4	11.2	12.0	12.7	13.5	14.2	14.8
Italy	46.8	49.5	53.7	56.4	57.1	57.8	59.0	60.2	63.4	64.7	66.0	67.3
Malta	0.3	0.3	0.3	0.4	0.4	0.4	0.4	0.4	0.4	0.4	0.5	0.5
Portugal	8.4	8.8	8.6	9.6	10.2	10.5	11.5	12.4	13.3	14.1	14.8	15.4
Spain	27.9	30.0	33.6	37.2	38.5	39.9	42.7	45.7	48.4	50.8	52.8	54.4
Yugoslavia	16.3	18.3	20.4	22.3	23.1	24.2	25.4	26.7	28.0	29.1	30.0	30.6
TOTAL	150.2	162.4	169.8	192.1	197.3	202.7	212.8	223.9	236.1	245.4	253.5	260.5

B.

Country	1900[1]	1950	1960	1970	1980	1985	1990[2]	2000	2010	2020	2030	2040	2050
Algeria	4.0	8.7	11.0	13.7	18.6	21.7	25.4	34.0	45.6	61.1	81.9	109.7	147.0
Egypt	8.0	20.0	25.9	33.3	42.0	46.9	52.4	65.0	80.6	99.9	123.9	153.6	190.5
Libya	0.7	1.0	1.2	2.0	3.0	3.6	5.2	6.3	9.4	14.1	21.2	31.8	47.7
Morocco	3.8	7.5	11.6	15.6	20.3	21.9	26.4	34.3	44.6	58.0	75.4	98.0	127.4
Tunisia	1.5	3.5	4.2	5.3	6.4	7.1	7.9	9.8	12.9	14.8	18.2	22.4	27.6
Cyprus	0.30	0.49	0.56	0.63	0.65	0.67	0.70	0.73	0.75	0.77	0.79	0.81	0.83
Iran	8.0	16.0	20.2	28.7	38.1	44.6	52.2	70.0	93.8	125.4	168.4	265.7	302.4
Iraq	2.6	5.2	7.1	9.7	13.1	15.9	17.7	23.9	32.3	43.6	58.9	79.5	107.3
Israel	0.3	1.4	2.1	2.9	3.9	4.2	5.3	7.1	9.6	13.8	17.5	23.6	31.3
Jordan	0.6	1.2	1.7	2.2	3.2	3.5	4.2	5.5	7.1	9.2	12.0	15.0	20.3
Lebanon	0.8	1.5	1.8	2.7	3.2	2.7	3.1	4.0	5.2	6.7	8.7	11.3	14.7
Saudi Arabia	1.7	3.2	4.0	5.3	8.4	11.5	16.1	22.5	31.5	44.1	61.7	86.4	121.0
Syria	1.9	3.4	4.7	6.1	8.6	10.5	12.8	17.9	25.1	35.1	49.2	68.9	96.4
Turkey	10.0	20.7	27.6	35.2	45.3	49.3	56.6	70.8	88.5	110.6	138.3	172.9	216.1[3]
TOTAL	44.2	93.8	123.7	163.3	214.8	244.1	286.0	371.8	486.0	636.7	836.1	1100.2	1451.0

Notes: [1]Estimate. [2]Projections from 1990 onward are made on the basis of an unabated growth rate identical to the 1950–1985 period; all attempts to curb this rate have failed so far (Egypt, Tunisia). [3]Overall exponential annual growth rate: 2.8% between 1950 and

Vegetation, land-use and climatic change   205

Table 6.7 Land use in the Mediterranean basin: agricultural population in 10^3 and in % of total population

Country	1950 nr	1950 %	1960 nr	1960 %	1970 nr	1970 %	1980 nr	1980 %	1985 nr	1985 %
Algeria	6500	75	7800	71	8279	60	9335	49	5720	26
Egypt	14200	71	14760	57	17848	54	21145	50	20407	43
Libya	800	80	720	60	634	32	467	16	519	14
Morocco	5550	74	8240	71	8601	57	10301	51	8965	40
Tunisia	2630	75	2860	68	2553	50	2578	41	2032	28
NORTHERN AFRICA	29680	73	34380	64	37915	55	43826	48	37643	36
Cyprus	290	60	310	55	233	39	212	34	153	23
Iran	9600	60	11510	57	12761	45	14319	38	13620	30
Iraq	3120	60	3550	50	4362	47	5260	40	3885	24
Israel	420	30	378	18	288	10	265	7	213	5
Jordan	600	50	560	33	776	34	836	26	272	7
Lebanon	900	60	990	55	487	20	263	10	304	11
Saudi Arabia	2650	80	2880	72	3794	66	5389	60	5136	43
Syria	2210	65	2585	55	3200	51	4265	47	2915	27
Turkey	16974	82	20980	76	23926	68	24641	58	24217	48
NEAR AND MIDDLE EAST	36674	69	43743	63	49827	53	55448	44	50715	35
Albania	800	80	1200	80	1416	66	1647	60	1606	52
France	12000	30	10000	25	6961	14	4639	9	3310	6
Greece	4000	53	3800	50	4041	46	3574	37	2363	24
Italy	18000	44	20000	44	10077	19	6363	11	4318	8
Portugal	3000	40	3500	42	2875	33	2577	26	2108	21
Spain	11900	51	13200	48	8475	25	6185	17	4861	13
Yugoslavia	10606	76	12000	73	10145	50	8348	37	5493	24
EURO-MED. COUNTRIES	60306	40	63700	39	34889	20	33333	17	24059	12

therefore assume that temperature increase would be uniformly spread over the whole year. Furthermore, we assume no substantial change in annual and seasonal temperature variability, although some variability may be expected.

GC models do not provide any significant indication on change of rainfall in Mediterranean latitudes. Some indicate a slight upward trend, others a downward trend. Wigley (1987) and Petrivanov (1987) indicate a slight increase in the north of the basin and a lessening to the south. We shall therefore assume no significant change in the overall amount of rainfall, seasonal pattern, or variability.

Currently available models do not provide any indication on Potential Evapotranspiration (PET). But, from the relationship between temperature and PET (Griffiths, 1972), one may assume that an increase of 3.0°C × 58.92 t_β^1 = 175 mm/yr (Holdridge, 1947) to 3°C × 68.64 t = 206 mm/yr (Le Houérou and Popov, 1981). We shall therefore assume an annual increase of 200 mm in PET. But, where rain-fed crops are concerned, PET increase through the growing season, assuming a potential six-month growing period, would be approximately half (100 mm). If the growing season is only 120 to 150 days, as for barley and wheat respectively, the increased water demand would be 65.7 and 82.5 mm.

Substantial yield decrease occurs when actual ET (ET_0) drops below 0.5 PET. The minimum additional water demand for barley and wheat would thus be 33 and 42 mm. These amounts are quite significant, but can be overcome with improved agricultural techniques (surficial tillage, tight weed control, crop-rotation patterns, cultivar selection, more timely operations, better use of fertilizers, etc.). In marginal water supply areas (i.e. between mean annual isohyets of 350 to 450 mm), however, crop yield expectancy could be significantly reduced; and, naturally, still more so in subsistence farming zones below an average annual rainfall of 350 mm in northern Africa and the Near East. Increased water demand in permanently irrigated farming would be of the order of 160–180 mm (0.8 PET); for the whole Mediterranean basin (23.5 million ha) at 7000 m³/ha/yr amounts to 165 × 10⁹ m³/yr, or a fictitious continuing discharge of 5 × 10⁶ m³ per second.

3.2 Effect on Native Vegetation

A temperature increase of 3.0°C would, in principle, provoke an average upward shift of vegetation belts of

$$3°C \times \frac{100m}{0.55°C} = 545 m$$

In other words each vegetation zones shown in Fig. 6.8 and Table 6.3 would move about one step upwards. The Xerothermo-Mediterranean Zone would thus reach the 46°N lowlands; the Thermo Mediterranean Eco-Zone would reach 700 m of elevation at 40°N and 1500 m at 30°N; tree-line would reach 2500 m at 46°N and 3500 m at 30°N, etc.

Annual biomass productivity might slightly, though significantly, decrease because of a lower P/PET ratio. The loss of productivity in the Semi-arid to Sub-humid Zones would be 150–300 kg DM ha⁻¹/yr⁻¹, that is about 10%, reckoning a six-month potential growing season and a rain-use efficiency of 3 kg DM ha^{-1}/yr^{-1}/mm^{-1} (Le Houérou, 1984). In high rainfall, higher elevation, higher latitude zones, productivity would increase as a result of a longer growing season and reduced winter cold stress. Primary productivity would be more seriously affected in the Arid Zone, as feedback effects would most likely occur between permanent plant cover, biomass and productivity as a result of the worsening of an already critical P/PET ratio. Under the combined effect of a worsened soil water budget, erosion and of a sharply increased pressure by man and livestock, most of steppic rangeland vegetation would give way to deserts. The northern limit of deserts may correspond by 2050 (or earlier) to the present lower limit of the Semi-arid Zone, i.e. the present 350–400 mm isoline of mean annual precipitation, the foothills of the High, Mid and Tell Atlas and Tunisian Dorsal in northern Africa and of the main mountain ranges of the Near-Middle East (Taurus, Lebanon, Alaoui, Kurdistan, Zagros, Alborz).

Desertization of the North African and Near Eastern steppes would have significant consequences on the livestock (sheep) industry in these regions. Range production, which has receded by an estimated 60 to 80% over the past 40 years (Le Houérou and Aly, 1982; Le Houérou, 1985b, 1988a), may be reduced to 10% of its pristine condition, with a very short growing season as vegetation is reduced to tiny ephemerals (Tachytherophytes) and ephemeroids of very low productivity (*Hordeum leporinum, Stipa capensis, Aegylops Sp.Pl., Trachynia distachya, Cynosurus coloratus, Elymus orientalis, E. delileanus, Schismus arabicus, S. perennis, Poa bulbosa, Poa sinaica, Carex pachystylis, C. physodes*, etc.). As probably over 50% of the sheep industry in the southern Mediterranean is traditionally in the Arid Steppic Zone, sheep-production systems would inevitably shift to a cereal grain/concentrate-based feeding system.

This move already began in the mid-1970s and at present, probably over 50% of sheep diet in this region consists of concentrate feed (Le Houérou and Aly, 1982; Le Houérou, 1985a, 1988b). This trend should accelerate in the future as range deteriorates further. Thus the sheep industry increasingly will rely on imported grain and concentrate feed and therefore on the international cereal market. As a result the cereal trade may become a strategic issue advantage for cereal-exporting countries, as the overall cereal production per inhabitant of the region has declined or (at best) stagnated over the past 40 years (Tables 6.7–6.21).

In the northern Mediterranean countries, natural vegetation will expand as a result of the abandonment of marginal farmland. Over the past 30 years, forest and shrubland expanded by almost 1% annually (Le Houérou, 1981, 1987a and c), as shown in Tables 6.9–6.13). The reduction in cultivated land area in Euro-Mediterranean countries is statistically consistent with the evolutionary trend in range and farmland hectarages. This issue will be further addressed later.

In the southern Mediterranean, forest and shrubland may disappear altogether by mid-21st century because of ever-increasing anthropogenic pressure. Although it does not appear in official statistics, a number of field surveys shows that in many areas, forest and shrubland have receded by 1–2% per annum in spite of reforestation programmes (Le Houérou, 1973a, 1981; Bourbouze, 1982, 1986; Donadieu, 1985; Tomaselli, 1976). Again, this issue will be examined later.

Table 6.8 Land use in the southern mediterranean basin: Land area, forest and shrublands

Country	Land area	Forest and shrublands, 10^3 Ha					
		1948/1952	1961/1965	1969/1971	1974/1976	1979/1981	1984/1986
Algeria	238174	3050	2549	2424	4122	4384	4385
Egypt	100145	2	2	2	2	2	2
Libya	175954	462	491	533	560	600	650
Morocco	44505 (62000)	5385	5302	5164	5182	5200	5200
Tunisia	16415	980	674	470	505	540	557
Cyprus	925	171	171	171	171	171	171
Iran	164800	18000 ?	18000 ?	18000 ?	18000 ?	18000 ?	18000?
Iraq	43492	1770	1953	1940	1930	1910	1900
Israel	2070	75	94	109	116	116	116
Jordan	9774	525	125	56	60	63	69
Lebanon	1040	92	94	95	90	85	80
Saudi Arabia	214969	400	1688	1630	1601	1200	1200
Syria	18518	432	446	468	445	466	516
Turkey	78058	10854 ?	20100	20170	20170	20199	20199
TOTAL	1126334	42198	51689	51232	52954	52936	53044
Afforestation rate % land area		4	5	5	5	5	5

Table 6.9 Evolution of land use in the Mediterranean basin between 1965 and 1984 (areas in 10^3 ha)

Countries	Area	1965	Forest and shrubland						Cropland			
			Aff. rate	1984	Aff. rate	Change	Change %	1965	Cultivation rate	1984	Cultivation rate	Change	Change %
Albania	2875	1266	44.0	1038	36.1	−228	−18.0	501	17.4	713	24.8	+212	+42.3
France	54703	11905	21.8	14603	26.7	+2698	+22.7	20542	37.6	18812	34.4	−1730	−8.4
Greece	13080	2545	19.5	2620	20.2	+75	+3.0	3854	29.5	3974	30.4	+120	+3.1
Italy	31023	5984	19.9	6410	21.3	+426	+7.1	5258	50.6	2233	40.6	−4125	−27.0
Portugal	9208	3165	34.4	3641	39.5	+476	+15.0	4370	47.5	3545	38.5	−825	−18.9
Spain	50478	13160	26.1	15625	31.0	+2465	+18.7	20594	40.8	20540	40.7	−54	−0.3
Yugoslavia	25580	8744	34.2	9294	36.3	+550	+6.3	8266	32.3	7758	30.3	−508	−6.2
SUBTOTAL	186047	46769	25.1	53231	28.6	+6462	+13.8	73452	39.5	67575	36.3	−5877	−8.0
Europe	978919	98355	10.0	106066	11.0	+7711	+8.0	148535	15.3	146103	14.9	−4432	−3.0

Source: FAO Production Yearbooks

Table 6.10 Land use in Euro-Mediterranean countries: evolution and trends between 1965 and 1984 (areas in 10^3 Ha)

1965

Country	Land area	Arable	Cereals	Permanent cropland	Permanent pastureland	Forest and shrubland	Misc. non-agric.
Albania	2740	497	321	55	736	1266	241
France	54475	21056	9160	1774	13221	11905	8382
Greece	13080	3800	1780	859	5100	2545	1635
Italy	29402	15454	6060	2763	5096	5984	2868
Portugal	9164	4332	1860	600	530	3165	1137
Spain	49978	20709	6830	4583	12300	13160	3809
Yugoslavia	25540	8349	5150	693	6472	8744	1975
TOTAL	184379	74208	31161	10727	43455	46769	20047

1984

Country	Land area	Arable	Cereals	Permanent cropland	Permanent pastureland	Forest and shrubland	Misc. non-agric.
Albania	2740	713	360	121	400	1038	589
France	54563	18812	9707	1344	12385	14603	8763
Greece	13080	3974	1482	1026	5255	2620	1231
Italy	29402	12233	4810	3133	4930	6410	5829
Portugal	9164	3545	1124	585	530	3641	1448
Spain	49940	20540	7485	4920	10640	15625	3135
Yugoslavia	25540	7758	4230	740	6379	9294	2109
TOTAL	184435	67575	29198	11869	40519	53231	23101
Difference (Ha) 1984–1965	+56	−6633	−1963	+1142	−2936	+6462	+3054
Difference % of 1965	–	−9	−6	+11	−7	+14	+15

Source: FAO, Production Yearbooks 1965 and 1985

3.3 Consequences on Crop Distribution Patterns and Yields

As for natural vegetation, cold-sensitive crops will expand upwards and northwards. Olive cultivation zones will considerably shrink in the northern Mediterranean in spite of more favourable climatic conditions, because of the increasing cost of labour and the difficulty in developing mechanized harvesting. This trend has been developing over the past 20 years. Olive cultivation thus will be restricted to intensive production irrigated systems in lowlands, using dwarf clones and mechanical harvesting. Most of the present olive groves will return to the wild: pasture[2], rangeland, shrubland and forest.

In the present situation, commercial crops of citruses are restricted to Sicily, southern Italy, south and east coastal Greece, eastern, southeastern and southern Spain, south of 40°N. A 3°C increase of winter temperature would render the crop commercially feasible over large areas of lowlands in Greece, Italy, France, Spain and Portugal between 40 and 46°N, where they are presently excluded by winter frost.

The area presently cultivated to citruses in Europe is some 250,000 ha, producing 8.6 million metric tons per annum. This hectarage could be increased easily by a factor of three with a temperature rise of 3°C. The

Table 6.11 Cereal cultivation in Euro-Mediterranean countries: Evolution and trend of cropped areas, 1965 and 1984 (areas in 10^3 Ha)

1965

Country	Wheat	Rye	Barley	Oats	Maize	Rice and Misc.	Total
Albania	125	10	9	23	150	–	321
France	4520	221	2430	1070	871	170	9159
Greece	1258	16	188	120	144	8	1774
Italy	4288	48	197	367	1028	–	6056
Portugal	628	316	126	271	484	–	1860
Spain	3991	393	1374	502	478	6	6826
Yugoslavia	1683	146	405	321	2550	25	5153
TOTAL	16493	1150	4729	2674	5705	209	31149
TOTAL %	53	4	15	9	18	1	100

1984

Country	Wheat	Rye	Barley	Oats	Maize	Rice and Misc.	Total
Albania	190	11	13	20	95	31	360
France	4832	90	2255	411	1858	261	9707
Greece	898	9	310	43	203	19	1482
Italy	3032	9	468	184	910	207	4810
Portugal	310	163	90	165	360	36	1124
Spain	2024	222	4155	465	516	103	7485
Yugoslavia	1348	45	264	151	2401	21	4230
TOTAL	12634	549	7555	1439	6343	678	29198
TOTAL%	43	2	26	5	22	2	100
Difference Ha 1984–1965	−3859	−601	+2826	−1235	+638	+469	−1951
Difference % of 1965	−23	−52	+60	−46	+11	+224	−6

Source: FAO, Production Yearbooks 1965 and 1985.

additional production could cover far beyond the European consumption, thereby closing the European market to the 5 million tons of citruses presently produced (2% of the presently irrigated 15 million hectares in the southern basin). A large proportion of the citrus groves of the southern Mediterranean should then be converted to other crops that are too cold-sensitive to be commercially grown in southern Europe, even with a 3°C increase of temperature. These include: avocado, mango, banana, paw-paw, sugar cane and other tropical species. Vegetable crops would be little affected as large proportions of those are already grown under controlled conditions (greenhouses, etc.). This proportion will grow in the future to include virtually all the vegetable production both to the south and north of the basin.

Rain-fed agriculture will be affected in the Semi-arid and Arid Zones, but very little in the Sub-humid to Hyper-humid climates (with mean

Table 6.12a Yields of main cereal crops: evolution and trends between 1965 and 1984 (in kg of grain per hectare)

Country	1965			1984		
	Wheat	Barley	Maize	Wheat	Barley	Maize
Albania	800	990	1030	2785	2756	4211
France	3270	3040	3980	6008	5066	6372
Greece	1650	1660	1730	1996	2106	8867
Italy	2280	1530	3230	2808	3479	6979
Portugal	980	570	950	1242	1033	1583
Spain	1120	1380	2390	2631	2570	6455
Yugoslavia	2060	1680	2320	3605	2667	4120

Table 6.12b Yields increase in main cereal crops between 1965 and 1984 (in kg/ha and % of 1965)

	Wheat		Barley		Maize	
	Kg/ha	%	Kg/ha	%	Kg/ha	%
Albania	1985	248	1766	178	3181	309
France	2738	84	2026	67	2392	60
Greece	346	21	446	27	7137	413
Italy	528	23	1949	127	3749	116
Portugal	262	27	463	81	633	67
Spain	1511	135	1190	86	4065	170
Yugoslavia	1545	75	987	59	1800	78

Table 6.12c Mean annual yield increase between 1965 and 1984 (in kg of grain per hectare and in % of 1965)

	Wheat		Barley		Maize	
	Kg/ha	%	Kg/ha	%	Kg/ha	%
Albania	99	15.5	88	8.9	159	15.1
France	137	4.2	101	3.3	120	3.0
Greece	17	1.0	22	1.3	357	20.6
Italy	26	1.1	97	6.3	187	5.8
Portugal	13	1.3	23	4.0	32	3.3
Spain	76	6.7	60	4.3	203	8.5
Yugoslavia	77	3.7	49	2.9	90	3.9

annual rainfall above 600 mm), where water supply is ample during the growing season and will remain so even with the increased PET envisaged. The situation may be quite different in the Semi-arid and Arid Zones, where increasing PET will reduce water availability during the growing season where it already is a critical factor of production. Most of the red Mediterranean oxysols overlying a shallow Pleistocene lime-crust will become inappropriate for cereal growing between the 350–500 mm isohyets of annual rainfall. These soil types cover huge areas in Spain, southern Italy and Greece, most likely over 8 million ha. Cereal growing on these soils is already quite marginal[3]; it will become untenable with a

Vegetation, land-use and climatic change 213

Table 6.13 Land use in the southern mediterranean basin: cultivated land (in 10^3 ha)

Country	1950	1960	1970	1980	1985
Algeria	6784	6820	6200	7500	7610
Egypt	2780	2610	2840	2450	2490
Libya	2509	1070	1725	1750	1790
Morocco	7858	8560	7500	8000	8400
Tunisia	4334	2970	4480	4700	4920
Cyprus	432	430	430	430	430
Iran	11593	16850	15700	13710	14830
Iraq	7496	5460	4990	5450	5450
Israel	996	410	410	410	420
Jordan	1140	890	370	400	420
Lebanon	296	280	325	310	300
Saudi Arabia	373	210	870	1110	1180
Syria	6130	6010	5910	5680	5620
Turkey	26384	25360	27380	28480	27540
TOTAL	79105	77930	79130	80380	81400
Cultivated rate % of land area	7	7	7	7	7

Note: Change 1950 to 1985: 81400–79105 = 2295 = 3%

worsened water balance, particularly if, as expected, the price of cereals in the EEC countries progressively declines to the World Market price (it is about 50% above the World Market's at the present time). These estimated 8 million hectares would then have to be used differently; most likely in part for reforestation for timber production and in part for the extensive grazing of sheep and game (*Cervidae*) under low input conditions, in combination or not with planted fodder shrubs and with cereals on the deeper soils having a favourable water budget.

To the south of the basin, the gamble of growing cereals in the Arid Zone, with harvest expectancies of ⅕ or less, will continue and expand to all soils able to be tilled (i.e. all except rock outcrops and salt marshes). Rangeland will be restricted to stony hills and saline swamps. This will result from the demographic explosion (see Table 6.6, Fig. 6.15). But the worsening of the P/PET ratio will make the arid crops even more risky: the P/PET ratio in the Mediterranean Arid Zone varies presently from 0.25 to 0.08; an increase of PET by 200 mm will drop this ratio to 0.20–0.06.

In the Semi-arid North African and Near Eastern Zones, crop expectancy will likewise decrease as a result of the worsening of the P/PET ratio. This ratio is at present 0.45 to 0.25 between the 350–600 mm isolines of annual rainfall; it will drop to 0.40–0.20. Crop expectancy, which is now about 70 to 80%, may drop to 50–60%. In other words, the lower part of the Semi-arid Zone will become similar to the present upper part of the Arid Zone (Figs 6.6 and 6.7). Similar to the Arid Zone, most of the natural vegetation will be cleared for crops wherever the soil is soft and deep enough to be tilled, no matter what the slope or erosion hazards. Given the tremendous demographic pressure, one cannot see how forestry regulations could be enforced any more. Most natural

214 *Climatic Change in the Mediterranean*

Table 6.14 Evolution of land use in the Mediterranean basin between 1965 and 1984 (areas in 10³ ha)

Countries	Area	Forest and shrubland						Cropland				
		1965	Aff. rate	1984	Aff. rate	Change	Change %	1965	Cult. rate	1984	Cult. rate	Change	Change %
Algeria	238174	2549	1.1	4384	1.8	+1835	+72.0	6863	2.9	7744	3.2	+881	+12.8
Libya	175954	491	0.2	640	0.3	+149	+30.0	2509	1.4	2115	1.2	−394	−15.7
Morocco	44655	5302	11.9	5200	11.6	−102	−1.9	7066	15.8	8331	18.7	+1265	+17.8
Tunisia	16361	674	4.1	555	3.4	−119	−17.7	4406	26.9	4687	28.6	+281	+6.4
Iran	164800	18000	10.9	18020	10.9	+20	+0.1	15353	9.3	14840	9.0	−528	−3.4
Iraq	43492	1953	4.5	1900	4.4	−53	−2.7	4810	11.1	5450	12.5	+640	+13.3
Israel	2077	94	4.5	116	5.6	+22	+23.4	401	19.3	437	21.0	+36	+9.0
Jordan	9774	125	1.3	41	0.4	−84	−67.2	391	4.0	415	4.2	+24	+6.1
Lebanon	1040	94	9.0	82	7.9	−12	−12.8	276	26.5	298	28.7	+22	+8.0
Saudi Arabia	214969	1688	0.8	1200	0.6	−488	−28.9	705	0.3	1156	0.5	+451	+64.0
Syria	18518	446	2.4	498	2.7	+52	+11.7	6523	35.2	5654	30.5	−869	−13.3
Turkey	78058	20170	25.8	20199	25.9	+29	+0.1	25775	33.0	27411	35.1	+1636	+6.4
SUBTOTAL AFRICA AND ASIA	792872	51586	6.5	52835	6.6	+1249	+2.4	75083	9.5	78528	9.9	+3445	+6.4

Source: FAO Production Yearbooks.

Table 6.15 Land use in the Mediterranean basin: irrigated land (in 10^3 ha)

Country	1950	1960	1970	1985	Differ. 1985–1970	Differ. % of 1970
Algeria	200	220	240	340	100	+42
Egypt	2500	2600	2840	2490	−350	−12
Libya	100	130	180	230	50	+28
Morocco	100	200	340	520	180	+53
Tunisia	40	70	90	220	130	+144
NORTHERN AFRICA (intercountry averages)	2940	3220	3690	3800	110	+3
Cyprus	70	80	100	100	0	0
Iran	3800	4600	5200	5740	540	+10
Iraq	600	800	1480	1750	270	+18
Israel	100	140	170	270	100	+59
Jordan	50	50	30	40	10	+33
Lebanon	70	70	70	90	20	+29
Saudi Arabia	100	120	370	420	50	+14
Syria	300	400	450	650	200	+44
Turkey	1300	1700	1800	2150	350	+19
NEAR AND MIDDLE EAST (intercountry average)	6390	7960	9670	11210	1540	+16
Albania	60	140	280	390	110	+39
France (× 0.8)	20	30	600	940	340	+57
Greece	340	410	730	1100	370	+51
Italy (× 0.8)	125	140	2050	2400	350	+17
Portugal	35	40	622	632	10	+2
Spain	130	210	2380	3220	840	+35
Yugoslavia (× 0.4)	40	60	50	70	20	+40
MED. EUROPE	740	1020	6720	8750	2030	+30
OVERALL MED. BASIN	10070	12200	20080	23760	3680	+18

vegetation will disappear, either cleared for farming, eaten by livestock or used for fuel. These processes have been going on for several decades already; they should only grow with the same exponentiality as human and livestock populations.

To summarize: in the northern basin, forest and shrubland will expand steadily due to the abandonment of marginal farmland and the moving of people from cities. But forest and shrubland wildfires will grow exponentially (4.7% annually), putting a heavy burden on Mediterranean communities (650,000 ha burned as an annual average from 1980 to 1986 around the basin with a global cost of 1 billion ECUs [US$1.2 billion]; see Fig. 6.13). Cropland will shrink, particularly cereal cultivation; most marginal farmland would have to be converted to extensive grazing, game and timber production (using highly productive new clones of timber species).

Table 6.16 Land use in the southern mediterranean basin: cereal cultivation (in 10³ ha; yields in kg/ha)

Country	Hectorage					Yields		
	1950	1960	1970	1980	1985	1969–1971	1979–1981	1984–1986
Algeria	2960	3160	3230	2970	2990	614	693	867
Egypt	1780	1870	1940	2010	1904	3847	4052	4471
Libya	229	604	603	538	407	301	430	616
Morocco	4080	4180	4640	4410	4820	985	811	1145
Tunisia	1550	1550	1280	1420	1930	634	828	808
Cyprus	133	143	96	64	72	1109	1780	1610
Iran	3152	4480	7150	7800	8520	832	1149	1185
Iraq	2090	2500	3220	2160	2980	1079	832	1020
Israel	90	135	130	129	107	1445	1840	1679
Jordan	266	232	260	165	139	696	511	542
Lebanon	108	100	72	34	20	870	1307	1225
Saudi Arabia	94	160	190	388	630	1352	820	3355
Syria	1490	2220	2500	2640	2710	588	1156	899
Turkey	8250	12290	13240	13570	13750	1357	1860	196
TOTAL	26272	33624	38551	38298	40979	Mean 1122	1290	1527

Notes: Change: 1950–1985: 40979–26272 = +14707 = +56%.
Increase: 1970–1985: 1527–1122 = +405 = +36%.

In the southern basin, high-risk growing will extend to the Semi-arid Zone. Destruction of the forest and shrubland will result in heavy wind and water erosion. Presently, water erosion averages 5–10 t/ha/yr on medium- to large-size watersheds (200–20,000 km²) (Le Houérou, 1969); it might increase by a factor of five. Some large watersheds on marls and shales, where most vegetation has been destroyed, have erosion rates of 30–60 t/ha/yr (Le Houérou, 1969) and up to 200–300 t/ha/yr on particularly sensitive substrates like the gypsopherous Miocene marls of south-eastern Spain (Lopez-Bermudez et al., 1984). The clearing of a forest and its replacement by annual crops may increase runoff by a factor of 5 and erosion by a factor of 50 (Cormary and Masson, 1964).

The amount of sediments reaching the southern Mediterranean at present is estimated to be around 6.0×10^8 t/yr (assuming an average erosion rate of 5 t/ha/yr over 1,200,000 km² of catchment). This figure might be 3.0×10^9 t/yr by the mid-21st century. The thickness of the soil layer annually eroded would then shift from a present 0.4 mm to 2.0 mm over these 1,200,000 km². The state of *Rhexistasis* (Erhard, 1956) caused by the destruction of virtually all natural vegetation thus becomes a phenomenon of major geological magnitude, even if the erosion rate were only half of the above suggested figures.

In the north of the basin, very few sediments will reach the sea due to the expansion of forest and shrubland over the watersheds. Even if 2 to 5% of the forest and shrubland are burned each year, the increased sediment load would be small. In addition, major hydraulic systems have been dammed by chains of reservoirs that trap the silt load. The Rhône River, for instance, which in the 19th century carried some 50×10^6 t/yr of sediment, dropped to 5.5×10^6 t/yr by 1957 and 2.2×10^6 t/yr in 1977 (Corre et al., this volume). Similar facts are reported from the Ebro, the Po and other major river systems.

Table 6.17 Land use in the southern mediterranean basin: cereal production (in 10^3 m tons)

Country	1949/51	1959/61	1969/71	1979/81	1984/86	Change 1950–1985	
Algeria	1953	1770	2060	2055	2920	+997 =	+51%
Egypt	4101	6080	7470	8130	8860	+4759 =	+116%
Libya	97	129	152	225	235	+138 =	+142%
Morocco	2740	3150	4230	3580	5320	+2580 =	+94%
Tunisia	690	603	615	1150	1270	+520 =	+75%
Cyprus	98	151	165	114	113	+15 =	+15%
Iran	3090	4560	5790	8950	10280	+7190 =	+233%
Iraq	1410	1850	1970	1803	2204	+794 =	+56%
Israel	75	216	200	239	175	+100 =	+133%
Jordan	194	250	165	91	63	−131 =	−68%
Lebanon	99	92	62	41	29	−70 =	−71%
Saudi Arabia	106	230	255	300	1090	+984 =	+928%
Syria	1200	1800	3300	3070	2390	+1190 =	+99%
Turkey	9060	14830	15990	25230	27390	+18330 =	+202%
TOTAL	24913	35709	42424	54978	62339	+37426 =	+150%
Production per inhabitant (kg/yr)	261	289	260	256	255	−6 =	−2%

Table 6.18 Land use in the southern mediterranean basin: wheat cultivation (in 10^3 ha)

Country	1948/52	1961/65	1969/71	1974/76	1979/81	1984/86
Algeria	1597	1969	2214	2240	1943	1625
Egypt	605	557	551	583	577	500
Libya	124	149	163	210	251	270
Morocco	1287	1578	1952	1843	1673	1991
Tunisia	917	1002	908	967	887	824
Cyprus	75	69	74	64	15	12
Iran	2085	3580	5370	5840	5894	6156
Iraq	936	1595	1216	1513	1515	1045
Israel	35	58	111	100	96	82
Jordan	182	268	155	158	111	55
Lebanon	70	68	46	47	26	17
Saudi Arabia	36	92	57	73	71	519
Turkey	4770	7959	8732	9140	9265	9434
TOTAL	13713	20340	22828	24385	26407	23687

Change 1950 to 1980: 26407–13713 = +12694 = +93%; annual increase: 2.3%

The situation prevailing in the north by the mid-21st century thus would be exactly opposite to that in the south: the north would be in a state of *Biostasis* due to its good vegetative cover; erosion would be restricted mainly to chemical processes (dissolution of carbonates, etc.) according to Erhard's theory of the *Biorhexistasis*. In other words, the north will be subject of chemical sedimentation in relatively clear waters (assuming anthropic pollution is controlled), whereas the south will be undergoing heavy detrital sedimentation in highly turbid and polluted waters due to the intense state of overland *Rhexistasis*, resulting from an extreme anthropogenic pressure.

The impact of CO_2 increase on crop yields would be negative in the rain-fed agriculture of the Arid and Semi-arid Zones (100–600 mm) because of the lessening of the P/PET ratio as suggested above. In Sub-arid Zones (600–800 mm of mean annual rainfall), the impact probably would be negligible or slightly positive, because water availability is not a major limiting factor to crop production and productivity may be enhanced by "carbon fertilization".

In the Humid (800–1200 mm) to Hyper-humid (P > 1200 mm) Zones, the combined effect of lowered P/PET ratio, a prolonged growing season and the increased CO_2 content in the atmosphere, would boost photosynthesis, hence primary production significantly. The increase might be as much as 10%, if all other conditions than temperature, CO_2 and PET remain equal to present conditions, including agricultural techniques. But the combined effect of improved climatic production parameters on higher yielding varieties and advanced farming techniques might yield even more positive results in terms of productivity.

The same scenario would apply to irrigated agriculture which presently amounts to 23.8 million ha in the basin (see Table 6.15). Over the past 20

Table 6.19 Land use in the southern mediterranean basin: wheat cultivation (in 10³ m tons)

Country	1948/52	1961/65	1969/71	1974/76	1979/81	1984/86	1948/52	1984/86
							Production per inhabitant kg/yr	
Algeria	996	1254	1359	1523	1270	1374	113	63
Egypt	1111	1458	1598	1859	1754	1762	54	42
Libya	11	36	41	72	125	164	11	47
Morocco	675	1335	1718	1762	1599	2717	86	124
Tunisia	452	446	520	867	837	855	129	121
Cyprus	48	63	75	75	20	16	97	25
Iran	1879	2873	3946	5438	6100	6430	122	144
Iraq	448	849	1080	1162	854	992	86	62
Israel	23	90	160	241	200	142	17	34
Jordan	128	180	127	120	67	47	101	13
Lebanon	51	64	39	60	32	22	34	9
Saudi Arabia	49	132	101	126	160	1846	15	160
Syria	761	1093	763	1657	1878	1584	221	151
Turkey	4770	8585	11423	14163	17058	17756	230	360
TOTAL	11513	18461	22962	29346	31957	25827	120	147
Surface	13713	20340	22828	24385	26407	23687	—	—
Yield, kg/ha	840	910	1010	1200	1210	1510	—	—
Production per inhabitant, kg/yr	120	129	140	154	149	147	—	—

Table 6.20 Land use in the southern mediterranean basin: wheat yields (in kg/ha)

Country	1948/52	1961/65	1969/71	1971/75	1976/80	1981/85
Algeria	620	640	560	600	650	660
Egypt	1840	2620	2580	3100	3670	3560
Libya	90	240	250	400	540	560
Morocco	610	850	860	970	970	860
Cyprus	640	910	1240	1210	1320	1330
Iran	900	800	910	890	1110	1000
Iraq	480	530	540	730	930	870
Israel	660	1540	1600	2360	1680	2030
Jordan	700	670	660	730	490	790
Lebanon	730	940	860	1090	1290	1230
Saudi Arabia	1370	1430	1460	1400	1360	2780
Syria	770	780	700	860	1010	1350
Turkey	1000	1080	1200	1400	1810	1890
Arithmetic average	744	931	959	1124	1202	1351

Table 6.21 Land use in the southern mediterranean Bbsin: sheep numbers (in 10^3 heads of mature animals)

Country	1947/52	1960/65	1966/71	1971/75	1976/80	1981/86
Algeria	4567	4622	6710	8246	11044	14285
Egypt	1254	1697	776	2024	1774	1982
Libya	1390	1376	1778	2697	5044	4974
Morocco	11249	10957	13785	16220	14755	13400
Tunisia	2462	3125	3433	3220	3929	5274
Cyprus	298	418	397	418	494	510
Iran	18000	30320	32500	35500	33260	34440
Iraq	9072	10245	11400	15610	11538	10210
Israel	53	190	196	189	230	247
Jordan	235	677	907	730	861	1041
Lebanon	55	200	208	237	254	139
Saudi Arabia	1098	2288	2940	3010	2990	3533
Syria	2968	4035	5930	5570	7577	11516
Turkey	24282	32863	35315	38534	43110	46227
TOTAL	76983	103013	116275	132205	136860	147778

years, the hectarage of irrigated land increased by an average of 1.27% per annum (20 million ha in 1970 and 23.8 million in 1986). This growth rate cannot continue for very long. Probably in less than 25 years, by the year 2015, most water resources (surface and deep aquifers) will be fully tapped. In many areas, deep aquifers are being already overexploited (i.e. discharge exceeds recharge). The total irrigated area in the Basin would then reach 30 to 35 million hectares, i.e. an agricultural water use of some 2.6×10^{11} m³ per annum, or a fictitious permanent discharge of 8×10^6 m³ per second

Vegetation, land-use and climatic change 221

(reckoning an average discharge of 7500 m³/ha/yr) versus 1.4×10^{11} and 4.3×10^6 m³ per annum and per second respectively at the present time.

However, the expansion of drip irrigation and of greenhouse farming may increase considerably both water use efficiency and yields. Irrigated agricultural production thus may increase by a factor of 5 to 10 with the doubling of the present water discharge before the mid-21st century, not to mention genetic improvement of cultivars.

4 DEMOGRAPHIC AND SOCIO-ECONOMIC SITUATIONS AND THEIR CONSEQUENCES ON NATURAL VEGETATION AND LAND USE

4.1 Demographic and Socio-Economic Situations

In the northern parts of the basin, human population growth is presently about 0.5% per annum (see Table 6.6; Fig. 6.14), a doubling period of 140 years. Growth is projected to decline to 0.3% by 2050, or before, with a doubling period of 233 years. The percentage of rural population is declining steadily (see Table 6.8). To the south, population grows by an average 3.2% per annum, that is an exponential rate of 2.8% and a doubling period of 25 years, 5.6 times faster than in Euro-Mediterranean countries. Several countries of the region have present growth rates between 3.2 and 7.0% per annum: Algeria (3.2), Morocco (3.2), Libya (7.0), Saudi Arabia (5.8), Syria (4.9), Iran (3.7), Iraq (3.5). As mentioned above, all attempts made to curb population growth in the region (i.e. Tunisia, Egypt) have failed.

The total population of the 14 countries under study was about 44 million at the beginning of the present century; it will reach 372 million by the year 2000 if no corrective action is taken in the interim. Projections (see Table 6.6; Fig. 6.15) show a high of almost 2 billion and a low of 800 million by 2050. The most likely figure is an intermediate figure of 1.45 billion. For this projection, we have used the present mean growth rate of the countries having the slowest growth (i.e. Turkey, Tunisia and Egypt).

The agricultural population in the south remained constant between 1970 and 1986, 87.7 *vs* 88.3 million (see Table 6.7). But in relative terms, it decreased from 55% of the total population in 1970 to 36% in 1986. In Euro-Mediterranean countries, the situation is different: The agricultural population decreased from 35 million in 1970 to 24 million in 1986, a 31% reduction in 16 years. The actual percentage of agricultural population in 1970 was 20% versus 12% in 1986, with a high 52% in Albania and a low 6% in France. As a matter of comparison, the proportion of agricultural population in 1986 was as follows: UK 2%; USA 3%; Switzerland and West Germany 4%; Sweden 5%; East Germany 9%; Bulgaria 13%; Hungary and Ireland 14% and Poland and Romania 21%.

The Gross Domestic Product per capita (GDPPC) (in constant 1980 US$[4]) between 1980 and 1987 decreased by −22% in northern Africa (= −2.75% per annum), by −11% in the Near East (= −1.38% per annum), and increased by 48% (= +6% per annum) in the Euro-Mediterranean countries (Table 6.23). The decrease in GDPPC in real terms in the south of the basin was largely due to the fact that population grew faster than the economy. Naturally, the important drop in oil revenues played a role,

Table 6.22 Land use in the southern mediterranean basin: meat production (in 10³ m tons)

Country	Production in 10³ tons						Production per inhabitant Kg/yr	
	1948/52	1961/65	1969/71	1973/75	1979/81	1984/86	1948/52	1984/86
Algeria	56	84	108	120	173	204	6.3	9.3
Libya	8	18	46	46	128	154	8.0	42.7
Morocco	115	156	201	212	288	316	12.5	14.4
Tunisia	30	45	55	75	96	118	8.6	16.7
Egypt	173	283	386	385	431	564	8.5	12.0
Cyprus	8	12	32	25	41	50	16.2	74.7
Iran	141	251	350	455	647	729	9.2	16.3
Iraq	46	110	117	170	168	262	8.9	16.5
Israel	2	83	121	166	187	197	4.0	46.5
Jordan	9	17	28	18	45	56	7.1	15.9
Lebanon	4	48	51	56	70	81	3.0	30.4
Saudi Arabia	48	61	70	88	171	210	13.9	20.0
Syria	13	44	51	82	157	396	4.1	34.3
Turkey	89	521	650	685	792	977	4.3	19.8
TOTAL	742	1733	2266	2583	3394	4314		
Production per inhabitant	9.5	14.0	13.9	14.2	15.8	17.7	9.5	17.7

Note: Increase from 1962 to 1985: 4314−1733 = 2581 = 150%; exponential annual increase: 3.5%

Table 6.23 Land use in the Mediterranean basin: evolution of Gross Domestic Product per capita (GDPPC) (in 1980 US$)

Country	1980	1987	Change in % of 1980
Algeria	980	1620	+65
Egypt	280	492	+76
Libya	6000	3400	−47
Morocco	500	390	−22
Tunisia	770	720	−6
NORTHERN AFRICA (intercountry averages)	1706	1325	−22
Cyprus	1620	2600	+66
Iran	1930	2060	+7
Iraq	1400	1500	+7
Israel	2580	4000	+55
Jordan	630	730	+16
Lebanon	850	720	−15
Saudi Arabia	4480	4020	−10
Syria	800	960	+20
Turkey	770	780	+1
NEAR AND MIDDLE EAST (intercountry averages)	1670	1483	−11
Albania	530	530	0
France	6600	9300	+41
Greece	2550	2760	+8
Italy	3130	7800	+149
Portugal	1370	2040	+49
Spain	2880	4320	+50
Yugoslavia	1720	1050	−39
EURO-MEDITERRANEAN (intercountry averages)	2680	4450	+48

but not necessarily the major one, as some oil/gas-producing countries, like Algeria, exhibited a substantial increase of GDPPC, due largely to increasing gas production. Iran and Iraq showed a slight increase despite eight years of continual war. Libya underwent a considerable drop of US$2600 due to both the drop in oil price and the Chad war. Saudi Arabia and Lebanon exhibited a slight decrease, for different reasons (oil prices and civil war, respectively).

The socio-economic conditions have considerably worsened over the past few decades in the south of the basin as a result of the extremely high demographic growth. GDPPC stagnated or declined, unemployment is high (often 30–50%), and the proportion of the population engaged in agriculture is still over 35% in Afro-Asian Mediterranean countries compared to 12% in Euro-Mediterranean countries.

224 *Climatic Change in the Mediterranean*

To the north of the Basin, GDPPC is growing steadily in most countries (except Albania and Yugoslavia). At the same time, the agricultural population decreased from 40% in 1950 to 12% today (6, 8 and 13% in France, Italy and Spain, respectively). This situation, in turn, creates new problems tied to marginal land abandonment in the countryside.

The abandonment of the countryside, combined with a 7% annual increase in tourism and a sharp decrease in extensive grazing, provoked a rapid increase in the area of forest and shrubland burned annually by wild-fire (see Fig. 6.13): from an average annual mean of 200,000 ha in the 1960s to an average 650,000 ha in the 1980s. The cost in 1987 was estimated about US$1.2 billion (1.0 billion ECUs) (Le Houérou, 1987a). New methods of fire prevention will have to be applied, in particular prescribed burning combined with the reintroduction of extensive grazing by browsing animals (goats, cervidae, camelideae) in order to avoid fuel build-up and thus reduce fire hazards.

5 CONCLUSIONS

5.1 In Euro-Mediterranean Countries

The impact of a possible doubling of the CO_2 content in the atmosphere to 700 ppmv by the mid-21st century should be moderate in the northern Mediterranean Basin, and most likely, generally beneficial, as a result of enhanced primary production. An order of magnitude of 10% increase seems a conservative estimate considering the present state of the art of GCMs. "Carbon fertilization" is known to enhance net photosynthesis, particularly in the species having a C_3 carbo-xylation pathway (most Mediterranean species belong to the C_3 type). Both growth chamber and greenhouse experiments show a 30–65% (and sometimes much more) biomass production increase for a CO_2 increase in the atmosphere up to 600–700 ppmv or above (Lemon, 1983; Mortensen, 1983; Bolin, 1983; Crane, 1985; Shugart and Emanuel, 1985; Morison, 1985; Houghton, 1986; Trabalka and Reichle, 1986). However, the rate of increase of utilizable products, such as cereal grain, may be much less than the overall biomass, as increased photosynthates are not evenly allocated to the various plant tissues. The factors governing this allocation are poorly understood. Moreover, "carbon fertilization" also increases water-use efficiency, perhaps up to 30–50% (Morison, 1985), a fact that may compensate for the higher PET resulting from the temperature rise. The latter would furthermore accelerate soil organic matter oxidation, hence nutrient turn-over and soil fertility.

Table 6.24 Evolution and projections of human population (n 10^6 persons)

Time	1900	1925	1950	1975	2000	2025	2050
North	115	135	150	180	215	241	260
South	44	70	94	190	372	736	1450
Total	159	205	244	370	587	977	1710
North/south ratio	2.6	1.9	1.6	0.9	0.6	0.3	0.2

Table 6.25 Land use (10^6 ha)

Time	1965	1984	Difference	1984/1965
North				
Forest and shrubland	46.8	53.2	+6.4	1.14
Cropland	73.4	66.6	−6.8	0.91
South				
Forest and shrubland	51.6	52.8	+1.2	1.02
Cropland	75.1	78.5	+3.4	1.05

Table 6.26 Land use: cereals

Time	1900	1950	1985	1985/1950	1985/1900
North					
Area (10^6 ha)	34.3	31.0	29.2	0.94	0.85
Prod. (10^6 m.)	41.1	40.3	117.0	2.90	2.85
Yield (kg/ha/yr)	1200	1300	3780	2.91	3.15
Production per capita	358	2678	594	2.22	1.66
South					
Area	–	26.3	41.0	1.56	–
Production	–	24.9	62.3	2.50	–
Yield	–	1222	1527	1.36	–
Production per capita	–	261	255	0.98	–

Table 6.27 Sheep number and meat production

		South	
Time	1950	1985	1985/1950
Sheep (10^6 heads)	77	148	1.92
Meat production (10^3 m.t.)	742	4314	5.81
Meat production per capita (kg/yr)	9.5	17.7	1.86

Table 6.28 Gross Domestic Product (in 1980 US$) per capita per year)

Area	1980	1987	Difference %
North Africa	1706	1325	−22
Near East	1670	1484	−11
Mediterranean Europe	2680	4450	+48

Cold-sensitive tropical and subtropical crops, such as citruses, avocados, bananas, paw-paws, sugar cane, etc., would expand. Cereal cultivation would be eliminated from 8 to 10 million ha presently farmed in Mediterranean Europe. Those lands would have to be converted to low-input rangelands for livestock breeding, game management, hunting, forestry, tourism, etc.

Forest and shrubland would grow from the present 53 million to about 76 million ha in Euro-Mediterranean countries, if the present rate of expansion (0.7% per annum) is sustained over the next 62 years (see Tables 6.8 and 6.9). Reforestation would thus grow by some 68%, from a present 28% to 47% in 2050. If the present rate of shrinking of farmland continues unabated (0.4% per annum), cropland would decrease to some 51 million ha versus the present 67.5 million – nearly a 25% loss.

The coastal areas devoted to tourism will undergo a severe impact if the present 5.6% annual increase in visitors (Barić and Gosparovic, Chapter 5) continues. It reached some 106 million visitors in 1986: 30 million in Spain, 36 in France, 25 in Italy, 8 in Yugoslavia and 7 in Greece. National tourism represents about 45% of the visitors. The number of tourists to the northern shores of the Mediterranean basin will reach 220 million by the year 2000 if the current growth rate continues. In contrast, tourism in southern represents less than 10% of the number of visitors in the north: in 1986, Morocco 2.2 million, Turkey 2.0, Egypt 1.7, Tunisia 1.7, Israel 1.2, Syria 1.1, Cyprus 0.9 million. Not only is the impact of tourism very strong along the narrow (<5 km) coastal belt, but it also contributes significantly to the increasing forest and shrubland fires each summer (Le Houérou, 1973b, 1987a).

Vegetation should evolve towards more thermophilous and more sclerophyllous types. The various vegetation belts shown in Fig. 6.8 will move about one step upward and northward; that is 550 +/− 200 m upward in elevation and 550–660 km northward at any elevation[5]. In other words, temperature presently prevailing on the southern shores of the Mediterranean Sea would occur along its northern shores. Similarly, xerothermo-Mediterranean and thermo-Mediterranean vegetation, now rare on the northern shores, will considerably expand to the mountain foothills of the Appenines, Southern Alps, Cevennes and Pyrénées. The limits of the Mediterranean region, however, are not expected to change as long as seasonal rainfall distribution remains unchanged.

5.2 In Afro-Asian Mediterranean Countries

The situation strongly differs in the Afro-Asian Mediterranean countries in terms of land-use trends, and demographic and socio-economic situations. The impact of CO_2 and temperature increase will be extremely important, not because of their direct consequences but because these consequences will worsen to the major change provoked by the human population explosion. The population will have multiplied 8.5-fold during the 20th century; it will multiply another 24.5 times by the year 2050, a 33-fold increase in 150 years.

The Afro-Asian Mediterranean countries have a non-desert area of some 1.51 million km^2, only 13% of the total land area. The population density on non-desert land grew as follows: 29 inhabitants/km^2 in 1900, 62 in 1950,

108 in 1970, 142 in 1980, 182 in 1988, and should increase to 189 in 1990 and 246 in 2000. It would reach 960 inhabitants/km² by the year 2050 if the present growth rate remains unabated, as it has for the past 40 years. This is an urban density incompatible with the maintenance of any natural vegetation. For the past three decades, forest and shrubland vegetation has been receding by 1 to 2% per annum. Natural steppe vegetation in the arid rangeland has been destroyed to a very large extent, over the past 30 years, giving way to desert encroachment of more than 2% per annum.

By the year 2050, only Turkey may have remaining forests in its Euxinian (non-Mediterranean) region, Iran in its Hyrcanian region (non-Mediterranean) and perhaps Morocco on the high and middle Atlas Mountains exposed to Atlantic moisture. Virtually all land except true desert, rock outcrops and salt marshes will be subjected to high-risk growing of staple cereals.

Erosion, sedimentation and flooding will accelerate freely, and catastrophic events will become more and more frequent as the environment deteriorates. Any climatic change resulting from rising temperature will then be trivial compared to the population-related catastrophe. Unless the demographic explosion is controlled shortly, the scourge is doomed to happen. As a matter of fact, it should be kept in mind that any attempt to curb demographic growth takes at least 25 years to produce a significant impact on population numbers (see Fig. 6.15).

This suggests that unless strong and determined action is taken between now and the year 2025, the catastrophe is arithmetically unavoidable. Social unrest and all sorts of extremisms would flourish. European countries would undergo extreme pressure from the hungry multitudes to the south. Such pressure could, in turn, lead to social unrest, racism and, it is to be feared, totalitarian political regimes. The symptoms already are present in some countries.

NOTES

1. t_B = Biotemperature = mean of total annual temperature above 0°C.
2. Pasture = sown permanent grazing land. Rangeland = extensively grazed natural vegetation.
3. Average yield of barley and wheat: 800–1000 kg/ha/yr for a cost of cultivation equal to the market value of 600-800 kg at 1987 EEC prices, which are expected to drop by some 40% by 1995 to be in line with the world markets.
4. The average altitudinal and latitudinal lapse rates being about 0.55°K per 100 m and 100 km respectively

6 REFERENCES

Abi-Saleh, B., 1982. Altitudinal zonation of vegetation in Lebanon. *Ecol. Medit.*, **VII**, 355–364.
Arrhenius, S., 1896. On the influence of carbonic acid in the air upon the temperature of the ground. *Philos. Mag.*, **41**, 237.
Assadollah, F., Barbero, M. and Quézel, P., 1982. Les écosystèmes préforestiers et forestiers de l'Iran. *Ecol. Medit.*, **VII**, 365–380.

Bach, W., 1987. Scenario analysis – 9 p. mimeo. *Proc. European Workshop on Interrelated Bioclimatic and Land-Use Changes*. Noordwijkerhout, The Netherlands.

Bach, W., 1988. Development of climatic scenarios from General Circulation Models. In: Parry, M.L. et al. (eds), *The Impact of Climatic Variation on Agriculture*, Sect. 4, 125–158.

Bagnouls, F. and Gaussen, H., 1953a. Saison sèche et indice xéro-thermique. *Labor. Forestier, Fac. Sce.*, Univ. de Toulouse, p. 47.

Bagnouls, F. and Gaussen, H., 1953b. Période de sècheresse et végétation. *Cptes. Rend. Acad. Sces*, Paris, **236**, 1076–1077.

Bagnouls, F. and Gaussen, H., 1957. Climats biologiques et leur classification. *Ann. de Géogr.*, **355**, LXVI, 193–220.

Barić, A. and Gasparovic, F., 1988. Implication of climatic changes on the socio-economic activities in the Mediterranean Coastal Zone. Mimeo, UNEP, Split/Athens, p. 88.

Bolin, B., Degens, E.T. and Kletner, P. (eds), 1979. The global carbon cycle. SCOPE, Study No. 13, ICSU, Paris, p. 491.

Bourbouze, A., 1982. L'élevage dans la montagne marocaine: Organisation de l'espace et utilisation des parcours par les éleveurs du Haut Atlas. INA, Paris-Grignon, p. 345.

Bourbouze, A., 1986. Adaptation à différents milieux des systèmes de production des paysans du Haut Atlas. *Techniques et Cultures*, **7**, 59–94.

Chamberlain, T.C., 1989. An attempt to frame a working hypothesis on the cause of glacial periods on a atmospheric basis. *J. Geol.*, **7**, 545.

Charney, J.G., 1979. Carbon dioxide and climate: A Scientific Assessment. *Nat. Acad. of Sci.*, Washington, DC, p. 22.

CEE, 1985. Perspectives d'une Politique Agricole Commune: Le Livre Vert de la Commission. CEE, Bruxelles.

Cormary, Y. and Masson, J., 1964. Etude de conservation des eaux et du sol au Centre de Recherches du Génie Rural de Tunis. Application à un projet-type de la formule universelle de perte en sol de Wischmeier. *Cah. ORSTOM, Sér. Pédologie*, **II(3)**, 1–26.

Corre, J.J. et al., 1988. Implication des changements climatiques dans le Golfe du Lion. PNUE, Athènes/Split, p. 137, multigraph.

Crane, A.J., 1985. Possible effects of rising CO_2 on climate. *Plant, Cell and Environment*, **8**, 371–379.

Donadieu, P., 1985. Géographie et écologie des végétations pastorales Méditerranéennes. *Ecole Nat. Sup. du Paysage*, Versailles, p. 324, multigraph.

Emanuel, W.R., Shugart, H.H. and Stevenson, M.P., 1985. Climatic change and the broad scale distribution of terrestrial ecosystem complexes. In: *The Sensitivity of Natural Ecosystems and Agriculture to Climatic Change*, Parry, M.L. (ed.), Kluwer, Dordrecht, The Netherlands, 29–43.

Emberger, L., 1930. La végétation Méditerranéenne. Essai de classification des groupements végétaux. *Rev. Gen. de Bot.*, **42**, 641–622; 705–721.

Emberger, L., 1939. Aperçu général de la végétation du Maroc. *Bull. Soc. Sc. Phys. et Nat. du Maroc*, Mem. H.S., p. 157.

Emberger, L., 1955. Une classification biogéographique des climats. *Trav. Inst. Bot. Fac. Sc. Montpellier*, **7**, 3–43.

Erhard, H., 1956. La genèse des sols en tant que phénomène géologique. Masson, Paris, 90 pp.

Gates, W.L., Han, Y.J. and Schlesinger, M.E., 1984. The global climate simulated by a coupled Atmosphere-Ocean general circulation model: Preliminary Results. *Climatic Research Institute*, Report No. 57, Oregon State University, Corvallis, OR.

Gentile, S., 1982. Zonation altitudinale de la végétation en Italie Méridionale et en Sicile (Etna exclu). *Ecol. Medit.*, **VII**, 323–338.

Griffiths, J.G., 1972. Report on the Agroclimatic conditions in S.W. Spain. FAO, Rome, 20 pp.
Hansen, J., Johnson, D., Lacis, A., Lebedeff, S., Lee, P., Rind, D. and Russell, G., 1981. Climate impact of increasing atmospheric carbon dioxide. *Science*, **213**, 957.
Hare, F.K., 1985. Climatic variability and change. In: *Climate Impact Assessment*, Ch 2, Kates, R.W., Ausubel, J.H. and Berberian, M. (eds), SCOPE No. 27, John Wiley & Sons, New York, 37–68.
Holdridge, L.R., 1947. Determination of World Plant Formations from simple climatic data. *Science*, **105**, 367–368.
Holdridge, L.R. and Tosi, J.A., 1967. Life zone ecology. Tropical Science Center, San Jose, Costa Rica, p. 206, (revised edition).
Houghton, R.A., 1986. Estimating changes in carbon content of terrestrial ecosystems from historical data. In: Trabalka, J.R. and Reichle, D.E. (eds), *The Changing Carbon Cycle, A Global Analysis*, 175–193.
Idso, S.B., 1980a. The climatological significance of a doubling of Earth's atmospheric carbon dioxide concentration. *Science*, **207**, 1462.
Idso, S.B., 1980b. Carbon dioxide and climate (reply to Schneider *et al.*, 1980 and Leovy, 1980). *Science*, **210**, 7.
Imai, K. and Murata, Y., 1978. Effect of carbon dioxide concentration on growth and dry matter production of crop plants. *Jap. J. Crop. Sc.*, **47**(2), 330–335; **47**(4), 587–595.
Jäger, J., 1988. Development of climatic scenarios: Background to the Instrumental Record. In: Parry, M.L. *et al.* (eds), *The Impact of Climatic Variation on Agriculture*, Sect. 4, 159–181.
Jones, P.D., Wigley, T.M.L. and Kelly, P.M., 1982. Variations in surface air temperatures. Part I, Northern Hemisphere, 1881–1980. *Monthly Weather Review*, **110**, 59–70.
Keeling, C.D., Bacastow, R.B., Bainbridge, A.E., Ekdahl, C.A., Guenther, P.R. and Waterman, L.S., 1976. Carbon dioxide variation at Mauno Loa Observatory, Hawaii, *Tellus*, **28**, 538.
Kellog, W.W., 1977. Effects of Human Activities on global climate. Tech. Note 156, p. 47, WMO publ. No. 486, WMO, Geneva, Switzerland.
Kellog, W.W., 1978. Is mankind warming the Earth? *Bull. of the Atomic Scientists*, **34**, 17.
Kellog, W.W. and Schware, R., 1981. Climate Change and Society. Consequences of Increasing Atmospheric Carbon Dioxide. Westview Press, Boulder, CO, p. 178.
Kukla, G.J. *et al.*, 1977. New data on climatic trends. *Nature*, **270**, 573–580.
Lamb, H.H., 1977. Climates: Present, Past and Future. 2 vols, Methuen, London.
Legg, B.J., 1985. Exchange of carbon dioxide between vegetation and the atmosphere. *Plant Cell and Environment*, **8**, 409–416.
Le Houérou, H.N., 1959. Recherches écologiques et floristiques sur la végétation de la Tunisie méridionale, *Mem. H.S. Inst. Rech. Sahar.*, Alger., 2 vols., p. 510.
Le Houérou, H.N., 1962. Les pâturages naturels de la Tunisie aride et désertique. *Inst. Sces Econ. Appl.*, Paris, p. 106.
Le Houérou, H.N., 1968. La désertisation du Sahara septentrional et des steppes limitrophes. *Ann. Algér. de Géogr.*, **3(6)**, 1–27.
Le Houérou, H.N., 1969. La végétation de la Tunisie steppique (avec référence aux végétations analogues de l'Algérie, de la Libye et du Maroc). p. 645, *Ann. Instit. Nat. Rech. Agron. de Tunisie*, **42**, 5.
Le Houérou, H.N., 1970. North Africa: Past, Present, Future. In: Dregne, H. (ed.), *Arid Lands in Transition*, Amer. Assoc. for the Advanc. of Science, Washington, DC, 227–278.

Le Houérou, H.N., 1973a. Ecologie, démographie et production agricole dans les pays méditerraméens du Tiers-Mode. *Options Méditerranéennes*, **17**, 53–61.

Le Houérou, H.N., 1973b. Fire and vegetation in the Mediterranean Basin. *Proc. Ann. Tall Timbers Fire Ecology Conference*, **13**, 237–277, Tall Timbers Research Station, Tallahassee, FL.

Le Houérou, H.N., 1975a. Problèmes et potentialités des terres arides de l'Afrique du Nord. *Options Méditerranéennes*, **26**, 17–36.

Le Houérou, H.N., 1975b. Le cadre bioclimatique des recherches sur les herbages méditerranéens. *I Georgofili*, **XXI**, 7, 57–67.

Le Houérou, H.N., 1977a. Fire and vegetation in North Africa. In: Mooney, H.A. and Conrad, C.E. (eds), *Proc. Internat. Sympos. on the Environmental Consequences of Fire and Fuel Management in Mediterranean Ecosystems*, Report WOZ, USDA Forest Service, Washington, DC, 334–341.

Le Houérou, H.N., 1977b. Plant sociology and ecology applied to grazing lands research, survey and management in the Mediterranean Basin. In: Krause, W. (ed.), *Application of Vegetation Science to Grassland Husbandry*, Handbook of Vegetation Science, XIII, Junk Publications, The Hague, 213–274.

Le Houérou, H.N., 1981. Impact of man and his animals on Mediterranean vegetation. In: Di Castri, F., Goodall, D.W. and Specht, R.L. (eds), *Mediterranean-type Shrublands*, ch. 25, pp. 479–521, Ecosystems of the World, vol. 11, Elsevier, Amsterdam.

Le Houérou, H.N., 1982. The arid bioclimates in the Mediterranean isoclimatic zone. *Ecologia Mediterranea*, **VII**, 115–134.

Le Houérou, H.N., 1984. Rain-use efficiency: a unifying concept in arid-land ecology. *J. Arid Envir.*, **7**, 1–35.

Le Houérou, H.N., 1985a. The impact of climate on pastoralism. In: Kates, R.W., Ausubel, J.H. and Berberian, M. (eds), *Climate Impact Assessment*, SCOPE No 27, John Wiley & Sons, New York. Ch. 7, 155–186.

Le Houérou, H.N., 1985b. La régénération des steppes algériennes. *Inst. Nat. Rech. Agron. (Relat. Exter.) et Minist. Relat. Extér.*, Paris, 45 pp. multigraph.

Le Houérou, H.N., 1987a. Vegetation wildfires in the Mediterranean Basin: Evolution and Trends. *Proc. Workshop on the Ecological Aspects of Forest Fires in the Mediterranean Basion*. Europ. Sces. Foundation & Forest Ecosystems Res. Network. *Ecologia Mediterranea*, **XIII**, 4, 13–24.

Le Houérou, H.N., 1987b. Ecological guidelines to control land degradation in European Mediterranean countries. *Proc. Internat. Conf. on Policies to Combat Desertification in Europe*, Valencia, Spain, July 1987, EEC, Brussels, p. 20, in press.

Le Houérou, H.N., 1987c. Agro-forestry and Sylvo-pastoralism to combat land degradation in the Mediterranean Basin: Old Approaches to New Problems. *Agr. Ecosyst. 8 Envir.*, 33 (1990), 99–109.

Le Houérou, H.N., 1988a. Considerations biogéographiques sur les steppes Nord-Africaines. *Cpte-Rend. Ve Coll. Intern. de l'Assoc. Franç. de Géogr. Phys.: Biogéographie, Environnement, Aménagement*; CNRS, Univ. Paris VII, p. 23.

Le Houérou, H.N., 1992. The grazing lands of the Mediterranean Basin. In: Coupland, R.T. (ed.), *Natural Grassland Ecosystems, Ecosystems of the World*, Elsevier Publ., Amsterdam, Vol. 8, Chapter 7, 171–196.

Le Houérou, H.N., and Aly, I.M., 1982. *Perspective and evaluation study on agricultural development: The Rangeland Sector*. FAO, Tripoli, Libya, p. 77, mimeo.

Le Houérou, H.N., Haywood, M. and Claudin, J., 1975. Etude phyto-écologique du Hodna. FAO, Rome, 154 pp. , 3 colour maps 1/200,000th.

Le Houérou, H.N. and Popov, G.F., 1981. An ecoclimatic classification of intertropical Africa. Plant Production, Paper No. 31, FAO, Rome, p. 40, 3 maps.

Le Houérou, H.N., Pouget, M. and Claudin, J., 1977/79. Etude bioclimatique des steppes algériennes. *Bull. Soc. Hist. Nat. de l'Afrique du Nord*, **68(3–4)**, 33–74.

Lemon, E.R., 1983. CO_2 and Plants: *The Response of Plants to Rising Levels of Atmospheric Carbon Dioxide*. Westview Press, Publ., Boulder, CO, p. 280.
Lopez-Bermudez, F., Romero-Diaz, A., Fisher, G., Francis, C. and Thornes, J.B., 1984. Erosión y ecología en la España semi-arida (cuenca de Mula, Múrcia). *Cuadernos Investig. Geografica*, X, **1–2**, 113–126.
Manabe, S, and Stouffer, R.J., 1979. A CO_2-climate sensitivity study with a mathematical model of the global climate. *Nature*, **282**, 491.
Manabe, S. and Stouffer, R.J., 1980. Sensitivity of a global climate model to an increase of CO_2 concentration in the atmosphere. *J. Geophys. Res.*, **85**, 5529.
Manabe, S. and Wetherald, R.T., 1975. The effects of doubling the CO_2 concentration on the climate of a general circulation model. *J. Atmos. Sci.*, **32**, 3.
Manabe, S. and Wetherald, R.T., 1980. On the distribution of climate change resulting from an increase in CO_2 content of the atmosphere. *J. Atmos. Sci.*, **37**, 99.
Mass, C. and Schneider, S.H., 1977. Statistical evidence on the influence of sunspots and volcanic dust on long-term temperature records. *J. Atmos. Sci.*, **34**, 1995.
Mitrakos, K., 1982. Winter low temperature in Mediterranean-type ecosystems. *Ecologia Mediterranea*, **VII**, 95–102.
Morison, J.I.L., 1985. Sensitivity of stomata and water-use efficiency to high CO_2. *Plant Cell and Environment*, **8**, 409–416.
Mortensen, L.M., 1983. Growth response of some greenhouse plants to environment. X-long-term effect of CO_2 enrichment on photosynthesis, photorespiration, carbo-hydrate content and growth if *Chrysanthemum morifolium Rawat*. *Medlinger fra Norges Land-bruskshløle*, **62**, 12, 1–11.
National Research Council, 1979. Carbon dioxide and climate: A Scientific Assessment. *Nat. Acad. of Sci.*, Washington, DC, p. 22.
National Research Council, 1982. Carbon dioxide and climate: A second assessment. *Nat. Acad. of Sci.*, Natl. Acad. Press, Washington, DC, p. 72.
Ozenda, P., 1970. Sur une extension de la notion de zone et de l'étage sub-méditerranéen. *C.R. Somm. Séances Soc. Biogéogr.*, **47**, 92–103.
Parry, M.L., Carter, T.R. and Konijn, N.T., (eds), 1988. The impact of climatic variation on agriculture. Kluwer, Dordrecht, Vol. 1, p. 876, Vol. 2, p. 764.
Penman, H.L., 1948. Natural evaporation from open water, bare soil and grass. *Proc. Roy. Soc. London*, **193**, 120–145.
Pitrovanov, S.E., 1987. A climatic scenario based on the Vinnikov/Groisman's approach. *Proc. European Workshop on Interrelated Bioclimatic and Land-Use Changes*, Noordwijkerhout, The Netherlands, p. 23, mimeo.
Poli-Marchese, 1982. Zonation altitudinale de la région de l'Etna comparée avec celle des autres volcans. *Ecol. Medit.*, **VII**, 339–354.
Quézel, P., 1976. Les forêts du pourtour méditerranéen. *Notes Techniques du MAB*, No. 2, UNESCO, Paris, 9–33.
Quézel, P. and Barbero, M., 1982. Definition and characterization of Mediterranean-type ecosystems. *Ecologia Mediterranea*, **VII**, 15–27.
Quézel, P., Gamisans, J. and Gruber, M., 1960. Biogéographie et mise en place des flores Méditerranéennes. *Naturalia Monspeliensia*, Actes du Colloque sur la mise en place et la caractérisation de la flore et de la végétation circum-Méditerranéennes, organisé par la Fondation L. Emberger à l'Institut de Botanique de Montpellier, les 9–10 Avril 1980, 41–51.
Ramanathan, V., 1975. Greenhouse effect due to chlorofluorocarbons: Climatic implications. *Science*, **190**, 50.
Ramanathan, V., 1980. Climatic effects of anthropogenic trace gases. In: Bach, W., Pankrath, J. and Williams, J. (eds), *Interactions of Energy and Climate*, D. Reidel, Boston, MA, p. 269–280.

Ramanathan, V. and Coakley, J.A., Jr., 1978. Climate modeling through radiative-convective models. *Rev. Geophys. Space Phys.*, **16**, 465.

Rivas-Martinez, S., 1982. Etages bioclimatiques, secteurs chorologiques et séries de végétation de l'Espagne Méditerranéenne. *Ecologia Mediterranea*, **VII**, 275–288.

Rockner, E., Schless, U., Biercamp, J. and Loewe, P., 1987. Cloud optical depth feedback and climate modeling. *Nature*, **329**, 138–140.

Rotty, R.M., 1979. Uncertainties associated with global effects of atmospheric carbon dioxide. Orau/IEA Report No. 79–6l, Oak Ridge Associated Universities.

Schneider, S.H., 1975. On the carbon dioxide-climate confusion. *J. Amer. Sci.*, **32**, 2060.

Schneider, S.H., Washington, W.M. and Chervin, R.M., 1978. Cloudiness as a climatic feedback mechanism: Effects on Cloud Amounts of Prescribed Global and Regional Surface Temperature Changes in the NCAR.GCM. *J. Atmos. Sci.*, **35**, 2207.

Shugart, H.H. and Emanuel, W.R., 1985. Carbon dioxide increase: The Implications at the Ecosystem Level. *Plant Cell and Environment*, **8**, 381–386.

Slade, D.H., 1979. Summary of the carbon dioxide effects. Research and Assessment Program. US Dept. of Energy, Washington, DC, p. 36.

Theys, M., 1986. L'Agriculture française et l'environnement dans les 20 prochaines années: Vers un Nouvel Équilibre? Ministère de l'Equipement et de l'Environnement, Paris, p. 59.

Tomaselli, R., 1976. La dégradation du maquis Méditerranéen. *Notes Techniques du MAB*, No. 2, UNESCO, Paris, p. 35–76.

Trabalka, J.R. and Reichle, D.E. (eds), 1986. *The changing carbon cycle: A Global Analysis*. Springer-Verlag, Heidelberg, p. 592.

Tyndall, J., 1863. On radiation through the Earth's atmosphere. *Philos. Mag.*, **4**, 200.

Walter, H. and Lieth, H., 1967. Klimadiagram weltatlas. G. Fisher, Verlag, Iena, p. 200.

Wigley, T.M.L., 1987. Climate scenarios. *Proc. European Workshop on Interrelated Bioclimatic and Land-Use Changes*, Noordwijkerhout, The Netherlands, p. 14.

Wigley, T.M.L., Jones, P.D. and Keely, P.M., 1980. Scenario for a warm high-CO_2 world. *Nature*, **283**, 17–21.

WMO, 1986. Report of the International Conference on the Assessment of the Role of Carbon Dioxide and of Other Greenhouse Gases in Climatic Variation and Associated Impacts. p. 78. Villach, Austria, 9–15 October, 1985. WMO No. 661, WMO, Geneva, Switzerland.

7

Aspects of the Response of the Mediterranean Sea to Long-Term Trends in Atmospheric Forcing

M. Gačić *
(*Institute of Oceanography and Fisheries
Split, Croatia*)

T.S. Hopkins
(*North Carolina State University,
Department of Marine, Earth and Atmospheric Sciences, Raleigh, USA*)

A. Lascaratos
(*University of Athens, Department of Applied Physics,
Athens, Greece*)

Abstract

Successful prediction of the response of the Mediterranean to climatic changes requires detailed understanding of both forcing and response over the entire frequency range. Here we describe briefly our present knowledge of wind forcing and identify lack of representative data for either local or basin-wide regional studies. We then present aspects of the barotropic response of the Mediterranean to the atmospheric forcing. Variations of the barotropic flow through Mediterranean straits at the subinertial time-scales appear to be related to atmospheric pressure changes. Lower-frequency exchange through straits at the subinertial time-scales appear to be related to atmospheric pressure changes. Lower-frequency exchange through straits is mainly driven by the atmospheric buoyancy fluxes as reflected in the production of the dense water within the basin. Finally, a preliminary assessment of the consequences of an atmospheric warming is presented.

1 Introduction

We present in this chapter a very brief discussion of some of the aspects concerning the physical response of the Mediterranean system to atmospheric forcing in the light of predicted climatic changes in the next 35 to 100 years. As a qualifying preface we should make it clear that these comments are neither intended to be comprehensive nor completely rigorous in a scientific sense. We would request the reader to bear this in mind and to consider this only as a first treatment of the subject.

Historically the Mediterranean has had its relative share of oceanographic attention, but unfortunately much of the data collection has

---

*Temporarily at: Istituto per lo Studio della Dinamica delle Grandi Masse, Venice, Italy.

been haphazardly distributed in space and time, perhaps due to the complicated territorial division of the sea itself. This was not conducive to understanding the sea as a unique physical system responding to its various forcing functions. A significant change took place with the Barcelona meeting of February 1975, which inspired the birth of the UNEP Mediterranean Action Plan, a multinational, multifaceted programme dedicated to focusing research on specific aspects of Mediterranean marine problems. In the last decade this has been complemented by several international programmes dedicated to elaborating our understanding of physical oceanographic processes in specific sub regions of the Mediterranean, e.g. Mediterranean Alpine Experiment (MEDALPEX), Western Mediterranean Circulation Experiment (WMCE), Physical Oceanography of the Eastern Mediterranean (POEM), and Programme de Recherche Internationale en Méditerraneé Occidentale (PRIMO). These and other programmes have greatly augmented our knowledge of the regional and global responses of the Mediterranean to its physical forcings. Much of this material is yet to be published but will certainly have a great relevance to the topic of the Mediterranean's response to climatic change.

Predicting the response of the Mediterranean to possible climatic changes requires a detailed understanding of both forcing and response over the entire subtidal frequency range. General prediction of climatic trends indicates major changes in atmospheric pressure and wind fields and in the hydrological cycles. For example, a 300 to 500 km shift northward (Kühn, 1987) is expected in the atmospheric circulation, which will undoubtedly affect the trajectories and frequency of mid-latitude cyclones over the Mediterranean. Quantitative estimates suggest mean air temperature increases of 1.5 to 4.5°K accompanied by a sea-level rise of up to 1.5 m within the next century. Apart from the straightforward implications, these trends have numerous indirect consequences that will require the best possible monitoring and modelling efforts on the part of Mediterranean marine scientists.

2 Some Remarks on the Mediterranean Sea

We begin by mentioning several aspects of the Mediterranean situation that make it both unique in the world's ocean and that will make it particularly susceptible to climatic change. The Mediterranean is a small longitudinal ocean surrounded by continental landmasses. The very restricted connection to the Atlantic (~ 4 km^2) acts to strongly uncouple the Mediterranean circulation from the Atlantic. This acts as a low-pass filter such that, to the first approximation, we can say that the Mediterranean responds only to local meteorological forcing. On the other hand, the atmospheric circulation is not uncoupled by a physical restriction from the influence from the North Atlantic; instead the atmospheric disturbances of North Atlantic origin are gradually modified in their eastward propagation. The influence of smaller-scaled cyclones generated within the Mediterranean also increases to the east. Furthermore, the continental boundaries of the Mediterranean are complex in configuration and in topography resulting in significant energy transfer from large to smaller-scaled meteorological features. The steep coastal topography limits the areal extent of the water catchment for the Mediterranean. The presence of the continents significantly influences the relative humidity of the local air masses. Finally, the Mediterranean lies within the boundary region between sub-tropic and mid-latitudes atmospheric patterns, making it particularly sensitive to a shift in that boundary.

The physical response of the sea to atmospheric forcing occurs through the processes of momentum flux (wind-driven circulations), buoyancy flux (heat and water-vapour exchange), and through the effect of atmospheric pressure on the sea surface (barotropic movements). None of these responses have the same length scale as has its corresponding atmospheric forcing. The fact that these scales often are strongly mismatched is the cause of much of the complexity encountered in our attempts to understand the response of the sea to its atmospheric forcing. In terms of atmospheric forcing, we will distinguish in the text between spatial scales as follows: (1) *Large scale*, to refer to the large synoptic atmospheric disturbances, as for example, the cyclones that are advected in from the North Atlantic or the anticyclones that form over the adjacent continental land masses. (2) *Basin scale*, to refer to the atmospheric disturbances generated within the Mediterranean of dimension approximately 500 km, for example, the Ligurian cyclones, the Bura and the Etesian winds. (3) *Local scale*, to refer to length scales of approximately 50–100 km, e.g. sea-breeze systems. Thus, in discussing the oceanographic response we will refer to a corresponding set of three spatial scales: 'global' referring to the entire sea, 'basin' referring to one of the Mediterranean sub-basins, and 'local' referring to coastal and meso (~10–50 km) scaled phenomena.

We note that there is less of a mismatch between the oceanic and the atmospheric time-scales than there is with length scales. We will identify the major periodicities of the forcing and/or the response as: "interannual" to signify changes of energy occurring over more than a year, "annual" those occurring within a year, "seasonal" within one to four months, "subinertial" within several days to several weeks, "diurnal" within 16–30 hours, and "semi-diurnal" around 12 hours.

The predicted climatic trends are at the interannual scale, and consequently a coupling to a response in the sea at an interannual scale is obvious. We expect that an interannual climatic change will result in modification of the atmospheric forcing at the seasonal and even subinertial scale. For example, the extent to which Ligurian cyclogenesis would be reduced with less frequent outbreaks of alpine air over the Ligurian Sea. This would be the topic of atmospheric modelling research. The central question under discussion is how the changes at each of the atmospheric time-scales will affect changes in the response in the Mediterranean, that is, each atmospheric energy scale can evoke a response, generally at a shorter time-scale, in the sea.

3 Aspects of Wind Forcing and Response

The major wind systems, such as the Mistral, the Sirocco etc., have been known since antiquity. Accurate numerical data for modern scientific needs are seriously lacking. Local coastal data, even if they exist in numerical time-series format, often are irrecoverably contaminated by local boundary layer effects, e.g. as caused by steep local topography, etc. Weather stations are commonly located in large cities or their airports which have a micro-meteorology not representative of their offshore regions. Apart from a need of a few projects (the Bouée Océanographique and 'CNR' Buoys), offshore meteorological data acquisition programs are seriously lacking. Even the ARGOS drifting meteorological data buoys, so extensively placed in the world's ocean, are missing in the Mediterranean. In order to drive the Mediterranean circulation models (c.f. Malanotte-Rizzoli and Hecht, 1988; and Le Provost and Hopkins, 1986 for a summary), the atmospheric forcing functions have been generated from the wind data composites (e.g., that of May, 1982, obtained from ship's sea-surface observations) or from depth-temperature-salinity data composites (e.g. the GDEM data, Davis et al., 1986). Another consequence of this lack of data is the corresponding lack of the baseline data (and the modelled response to this baseline data) needed to assess interannual trends is correspondingly restricted.

An important aspect of the Mediterranean atmospheric forcing is a marked winter-to-summer shift in the energy distribution from the seasonal to subinertial scales. During winter, the penetration of large-scale and locally generated basin-scale cyclones dominate the energy spectrum. During summer, this aspect is nearly absent, and the major atmospheric energy peak is shifted to the subinertial scales. Most of the winds are locally generated by more stable pressure systems instead by propagating ones. The land–sea pressure gradient systems plays an important role. Near the coasts strong sea-breeze systems are established and, in the case of overlapping systems, reach basin-scale dimension, e.g. in the Aegean with the Etesian winds.

Another important aspect concerns the very strong subinertial scale wind events that occur only when certain large–scale meteorological conditions are met, i.e. the intense Mistral, Bura, and Sirocco. Because the buoyancy flux is not linear and because these winds are differently directed, their contribution to the annual buoyancy flux is significantly underestimated in the lower frequency averaging (monthly or seasonal) commonly used to drive Mediterranean circulation models. Unlike the sea-breeze systems, these meteorological events depend on a particular sequence of atmospheric events such that the probability of their occurrence varies from year to year. We suggest that this may make their occurrence even more susceptible to interannual trends. Evidence was presented by Zore-Armanda (1974) that the incidence of the Bura may be correlated with the position of the major pressure systems over the Atlantic (Fig. 7.1).

3.1 Barotropic Response

Sea-level variations at the subinertial scale are mostly generated by the atmospheric forcing and represent an energetic part of the Mediterranean sea-level spectrum. A number of researchers (Kasumović, 1958; Mosetti, 1971; Papa, 1978; Godin and Trotti, 1975; Gačić, 1980) have reported that sea-level variations at time-scales from one to ten days are primarily due to surface pressure changes related to synoptic atmospheric disturbances.

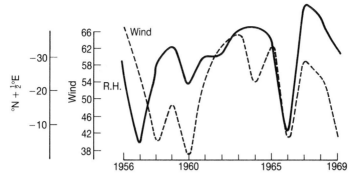

Fig. 7.1 Time series of the uniform atmospheric pressure fluctuations over the Mediterranean Sea (upper plot, dotted line), and of measured subinertial barotropic transports through Gibraltar (lower plot, thick line) in Sverdrups, i.e. 1 Sv = 10^6 m^3/s^{-1} (from Candela et al., 1989). The thin line superposed on the lower plot corresponds to the model "prediction" for transport that is presented in section 6 of the same paper.

Therefore, changes in sea-level at this periodicity depend on large- and basin-scale disturbances passing over the area. On the other hand, sea-level variations at time-scales of ten days to several weeks have been explained in terms of the influence of atmospheric planetary waves (Orlić, 1983).

Lacombe (1961) pointed out that an isostatic response of the Mediterranean sea-level to an atmospheric pressure different from that over the adjoining Atlantic would require, by continuity, a flow through the Strait of Gibraltar. Crépon (1965) found a high negative correlation between the cross-strait sea-level difference, which represents the geostrophic surface inflow, and the mean atmospheric pressure over the western Mediterranean. He concluded that in the western Mediterranean the ratio of atmospheric pressure to the sea-level was consistent with an inverted barometric response. Both he and Lacombe et al. (1964) have commented on the apparent contradiction with the conservation of mass condition, that in the western Mediterranean (WMed), the sea-level rather than the rate of change of the sea-level is related to the inflow at Gibraltar. Garrett (1983) has explained this fact by a simple model in terms of the barotropic response to eastward-propagating cyclones and the exchange of water with the eastern Mediterranean (EMed) through the Strait of Sicily. He also predicted that the EMed should react non-isostatically to the moving atmospheric perturbations. Garrett and Majaess (1984) showed from one station in the EMed (Katakolon, Greece) that the non-barometric response of the sea-level to atmospheric pressure occurs at the time-scale of about four days. For a period longer than four days an isostatic response can occur because there is enough time for the sea-level to adjust by importing water through the Strait. At less than three days, an isostatic response is also possible because the associated spatial scales are less than the basin-scale and the water for adjustment can be supplied internally without requiring a flow through the Strait.

Garrett and Majaess (1984) also computed the cross-spectrum between the atmospheric pressure variations at one station in the EMed (Andravidha, Greece) and at another station in the WMed (Cagliari, Italy). A significant coherence was shown only for time-scales of more than eight days. Rickards (1986) also reported a lack of coherence in the sea-level variations between the EMed (Adriatic) and the WMed. At the same time she found that the sea-level variations in the Adriatic were correlated with the local passage of perturbations. These results imply that within the subinertial time-scale the EMed and WMed are uncorrelated in their isostatic response to atmospheric pressure variations, even though individually they may be well correlated with the local forcing.

Departures from isostatic response are due either to local winds (Palumbo and Mazzarella, 1982) or to the fact that the Straits of Gibraltar and Sicily restrict the supply of water, through the mechanism of geostrophic control (Garrett, 1983; Garrett and Majaess, 1984). Wright (1987) showed that in a number of cases the flow is also restricted by friction to the same or greater degree than by geostrophy. Crépon (1976) showed that the response of a rotating fluid is never completely isostatic. The larger the bottom friction the more isostatic is the response. Also, he showed that coastal boundaries that can support the existence of Kelvin waves make an isostatic response easier.

An intensive study of the sea-level variations at the subinertial time-scale was undertaken by Lascaratos and Gačić (1990). The study area was north-eastern Mediterranean (Adriatic, Ionian and Aegean Seas). At these time-scales the influence of the atmospheric forcing should prevail and thus the air pressure was also included in the analysis. Most of the variability was found to be related to the in-phase sea-level or atmospheric pressure variations over the entire area, while a smaller fraction of the variance represents variations for which the Adriatic Sea is out of phase with respect to both the Ionian and Aegean Seas. The two spatial patterns explain more than 90% of the total variance from both atmospheric pressure and sea-level.

Sea-level variations are closely related to atmospheric pressure changes in both space and time. Departures from the isostatic response evidenced in the low-frequency range are not due to a geostrophic control in the Straits. It was also shown that the sea-level variability, caused by the Adriatic Sea being out of phase with respect to the rest of the area, is generated by the cyclonic basin-scale atmospheric pressure forcing. On the other hand, in-phase sea-level variations over the entire area are induced by air-pressure changes associated with the atmospheric planetary waves of the global scale (6000–8000 km). The location of the zero-crossing of the atmospheric pressure variability over the area of the Strait of Otranto was explained in terms of the prevalent cyclone trajectories over the north-eastern Mediterranean region (Oszoy, 1981). Therefore, any appreciable shift of the cyclone path over the Mediterranean will result in a change of the position of the zero-crossing.

The major fraction of the variance of the sea-level slope between the Adriatic and the Ionian Seas, i.e. on the basin-scale, is associated with the atmospheric pressure forcing due to passages of mid-latitude cyclones. Thus, the water exchange between different basins of the area varies depending on the cyclone frequencies. On the other hand, the variance of the sea-level slope within the area related to the atmospheric planetary- wave time-scale is very small since the typical length scale of the forcing function becomes larger than the entire Mediterranean. Evidences on the relationship between sea-level slope and water-flow variations at the synoptic time-scale within the area were presented by Zore-Armanda (1985) for the Adriatic Sea. On spatial scales of the entire Mediterranean (global scale), the observations analysed by Candela *et al.* (1989) suggest that the in-phase atmospheric pressure variations over the entire Mediterranean are a principal forcing for the subinertial barotropic flow through the Strait of Gibraltar (Fig. 7.2). The flow appears to be restricted by friction in the Strait. It was also shown that the Strait of Sicily do not impose any serious restriction to the subinertial water exchange between eastern and western Mediterranean.

3.2 Baroclinic Response

While the atmospheric pressure variations can cause large barotropic flows through the Straits at the higher end of the subinertial scale, it is the atmospheric buoyancy flux that drives the lower frequency exchange through the Straits. Defant (1961) made the point that the mean exchange is driven by the density differences between adjacent water bodies connected by a strait. Hopkins (1989, in press) has elaborated this concept, demonstrating that the low-subinertial exchange through a strait can be determined by an assessment of the pressure gradient force created by the difference in weights of the water columns on either side of a strait. In this way the inflow is primarily a barotropic compensating

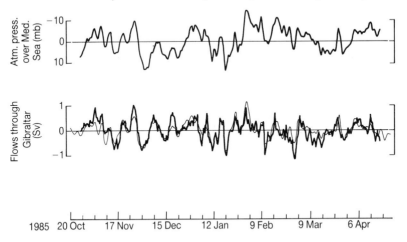

Fig. 7.2 Winter position of the High over southern Europe (Russian High) expressed as the sum of degrees of latitude and longitude and considering Irkutsk as a zero position (solid line). Broken line represents the north and north-east wind index (product of the strength and frequency) for the island Hvar in the Adriatic Sea (from Zore-Armanda, 1974). (From J Candela, C D Winant and H L Bryden, *Journal of Geophysical Research*, **94**: 12667–79. Copyright 1989 American Geophysical Union. Published with Permission)

flow for the outflow. Variability in the baroclinic driving force is converted into a variability in the exchange, although the conversion relationship is not linear due to the effects of frictional and hydraulic control within the strait. Central to understanding this concept is that the density (or water mass) differences exist because of the inefficiency of the strait to exchange waters with the external water body, i.e. the two-way exchange continues as long as the production of more dense water exceeds the capacity of the Strait to expel it. This then makes the vigour of the exchange a function of the production of denser waters inside the basin and ultimately then to the atmospheric buoyancy flux over the basin.

4 Dense-Water Formation

The production of dense water is primarily dependent on the atmospheric buoyancy flux. The wind speed and direction can also be important by directly affecting the latent heat and conductive heat losses. The general circulation of the sea itself, both wind and thermohaline driven, also enters importantly in the dense water production cycle. The latter effect is often referred to a preconditioning phase of dense water production (MEDOC Group, 1970) in which the general circulation first acts to import denser waters, already exposed to some atmospheric buoyancy-loss, to the ultimate site of production. Then other aspects of the circulation expose the more dense waters to extreme buoyancy loss events.

The Adriatic Sea is considered as the most important source of the EMed deep water (Pollak, 1951; Buljan and Zore-Armanda, 1976; Hopkins, 1978). Violent outbreaks of cold dry Asian air over the Adriatic (Bura Winds) have been shown to cause such intense evaporation losses that a sudden and deep convection is immediately triggered (Hendershott and Rizzoli, 1976). These Bura Wind events have also been shown to be correlated with the presence of Adriatic Deep Water at the surface of the centre of the South Adriatic cyclonic gyre. The effects of the Bura are also notable, generating basin-wide circulation patterns (Zore-Armanda and Gačić, 1987; Fig. 7.3), and the density gradients set up by the dense water production contribute to the general wintertime circulation pattern (Hendershott and Rizzoli, 1976).

The dense water production in the Ligural-Provencial basin has been well documented by a series of observational campaigns, commencing with the MEDOC 1969 (MEDOC Group, 1970; Anati and Stommel, 1970; Gascard, 1974; Killworth, 1976). These studies greatly expanded the knowledge of the processes of atmospheric buoyancy exchange processes, and the mechanisms of deep convection (cf. Hopkins, 1978 for a summary). The WMed deep-water production is about 5000 km^3 (Sankey, 1973) or approximately the same order of magnitude as EMed water produced in the Adriatic (Zore-Armanda and Pucher-Petković, 1977; Hopkins, 1978).

The Levantine Sea is recognized as the primary location for the production of the Levantine Intermediate Water, which after its formation spreads westward through the entire Mediterranean, becoming not only the major component of the Gibraltar outflow but also the critical component in the formation of the WMed Deep Water. In the northern Levantine Sea, the atmospheric buoyancy loss associated with cold dry outbreaks of

Fig. 7.3 21 January (upper) and 22 January, 1987 (lower) NOAA9 infrared satellite images of the Adriatic (SATMER, 1987). Large cyclonic gyre of the northern Adriatic dimensions is easily seen due to large thermal contrast between sea water and the Po River water. Inspection of synoptic maps from the same period suggests that the gyre was transient wind-induced feature. It was noted that, e.g. on 22 January at 7.00 a.m. the Bura wind speed was over 18 m/s^{-1} at some locations in the northern Adriatic.

continental air and the local circulation producing the Rhodes gyre are the determining factors (e.g. Lacombe and Tchernia, 1960; Wüst, 1961; Hopkins, 1978, Oszoy et al., 1981). Morcos (1971) presented strong evidence for additional production over the southern Levantine, apparently a result of the very dry winds blowing off the North African land mass.

242 *Climatic Change in the Mediterranean*

The production of these dense waters is vital to the general Mediterranean thermohaline circulation, forcing a vertical circulation necessary for the oxygenation of the deep water. The deep-water production in the EMed and the WMed is relatively small compared to the volume that they renew, giving large renewal times: approximately 250 and 200 years for the EMed and WMed, respectively. However, a reduction in the production may result in more than a linear increase in these renewal times because a reduction in volume probably will be correlated with a reduction in the density of the product, making it convect to lesser depth than the deep-water products from previous years.

The Levantine Intermediate Water (LIW) acts to ventilate a shallower depth range 200–700m, and a reduction of its production would have effects other than those on the deep waters. Primarily, this water provides the east-to-west pressure gradient at the depth of the Sicily and Gibraltar sills, thereby providing the main baroclinic forcing of the Mediterranean thermohaline exchange. Hopkins (in press) has suggested that the supply of the dense waters, particularly the LIW, to the immediate vicinity of these major sills is a decisive factor in the magnitude of the strait exchange. The supply is controlled ultimately by the production, but more immediately by the internal circulations, e.g. by that in the south-western Ionian or in the Western Alboran. Wüst (1961) stated that the summer flow through the Strait of Sicily was weaker than the winter flow. Manzella *et al.* (1988) report a winter flow of more than double that of the summer. This led to an important finding reported by Manzella and La Violette (1990) that the volume of LIW in the WMed has a strong seasonal fluctuation, which then effects the supply of LIW available for the exchange with the Atlantic.

5 A Preliminary Assessment

Hopkins (1989) has used the Strait of Gibraltar to demonstrate some of the consequences that an atmospheric warming might have on the exchange through the Strait and, consequently, on the entire Mediterranean circulation. Associated with an atmospheric warming, we might expect the sensible heat loss to decrease and to result in warmer surface waters and a warmer winter dense-water product. Much less obvious would be the net result of evaporative exchange, which has a more complicated dependence on temperature due to the non-linear dependence of water vapour on temperature. According to Stewart (pers. comm.) the hydrological cycle would speed up, that is, both evaporation and precipitation would increase. However, the Mediterranean atmospheric water balance is not closed, in the sense that, although evaporative loss is restricted to the Mediterranean surface, the precipitation resulting from it may not return to the Mediterranean and significant portions would come from Atlantic evaporation. A likely implication of atmospheric warming might be that the net evaporation and deep-water production would decrease as a result of an increase in the surface-water buoyancy.

The result of increasing the surface buoyancy on both sides of Gibraltar would have only little effect on the exchange through the Strait, because the internal pressure gradient (see section 3.2) would remain nearly the same. If the Mediterranean surface waters warmed more quickly than

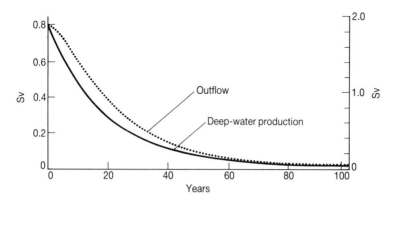

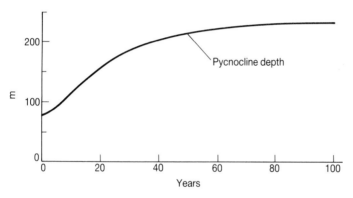

Fig. 7.4 The decrease of Gibraltar outflow and of the Mediterranean pycnocline depth as a result of a 5% yr⁻¹ decrease in the deep-water production (from Hopkins, 1989).

those of the Atlantic, the resultant effect on the exchange would still be small. For example, Hopkins calculates that if the surface Atlantic waters were warmed by 1°K to 75 m, whereas the Mediterranean surface waters were warmed by 1°K to 200 m, the reduction in the annual outflow would be about 5%. An equivalent increase in buoyancy due to freshening would require the addition of 2 m of fresh water, or a salinity decrease of 0.35 ppt.

However, a much more pronounced effect would be generated if the pycnocline were lowered in simulation of a reduction in the amount of deep-water produced. Figure 7.4 illustrates the effect if, for example, the deep water formation were decreased at a rate of 5% per annum from an initial value 25,000 km³/yr. The result shows that the outflow would decrease slightly less rapidly than the deep-water production, reaching one-half its historical value in about 20 years. More realistically, the production of the LIW would continue even though the production of the deep water might be significantly reduced. However, even if the LIW was produced at the same volume but not as dense, the thermohaline circulation would still slow down in the same fashion as depicted in Fig. 7.4, because of the internal pressure gradient through the Strait would be correspondingly reduced. The cessation of several warm winters would noticeably slow down the Gibraltar exchange, and among the many side effects would be a lesser import of Atlantic water with its nutrients and less ventilation of the deep waters.

6 Conclusions

To assess the response to climatic change in the Mediterranean the careful analysis of the existing long time series of data is needed. This especially concerns data on atmospheric forcing functions, sea-level and other oceanographic data if available.

Acquired knowledge by international on-going projects in the region will shed more light on the general circulation of the Mediterranean as well as on the forcing mechanisms.

General circulation models with better spatially and temporally resolved wind forcing are needed for an accurate prediction. For that purpose more representative wind data are necessary. Finally, coupled ocean-atmosphere regional models should offer the most reliable answer to questions about implications of the expected climatic changes in the Mediterranean.

7 References

Anati, D.A. and H. Stommel, 1970. The initial phase of deep water formation in the northwest Mediterranean during MEDOC '69 on the basis of observations made by "ATLANTIS II". *Cahiers Océanographiques*, **22**, 343–351.

Bormans, M., Garrett, C. and Thomson, K.R., 1986. Seasonal variability of the surface inflow through the Strait of Gibraltar. *Oceanologica Acta*, **9**, 403–414.

Buljan, M. and Zore-Armanda, M., 1976. Oceanographical properties of the Adriatic Sea. In: Barnes, H. (ed.), *Oceanogr. Mar. Biol. Ann. Rev.*, Aberdeen University Press, **14**, 11–98.

Crépon, M., 1965. Influence de la pression atmosphérique sur le niveau marin de la Méditerranée occidentale et sur le flux à travers le détroit de Gibraltar. *Cahiers Océanographiques.*, **17**, 15–32.

Crépon, M., 1976. Sea level, bottom pressure and geostrophic adjustment. *Mémoires de la Sociéte Royale des Sciences Liège*, **6**(10), 43–60.

Candela, J., Winant, C.D. and Bryden, H.L., 1989. Meteorologically forced subinertial flows through the Strait of Gibraltar. *J. of Geophys. Res.*, **94**, C9, 12667–12679.

Davis, T.M., Countryman, K.A. and Carron, M.J., 1986. Tailored acoustic products utilising the NAVOCEANO GDEM (A Generalized Digital Environmental Model). *Proc. 36th Naval Symp. on Underwater Acoustics*, NOSC, San Diego.

Defant, A., 1961. Physical Oceanography. Vol. 1., Pergamon Press, New York.

Gačić, M., 1980. Some characteristics of the response of the Adriatic Sea coastal region to the atmospheric forcing. *Acta Adriatica*, **21**(2), 239–240.

Garrett, C.J.R., 1983. Variable sea-level and strait flow in the Mediterranean: A Theoretical Study of the Response to Meteorological Forcing. *Oceanologica Acta*, **6**, 79–87.

Garrett, C. and Majaess, F., 1984. Nonisostatic response of sea-level to atmospheric pressure in the Eastern Mediterranean. *J. of Physical Oceanography*, **14**, 656–665.

Gascard, J.C., 1974. Deep convection: deep water formation in the northwest Mediterranean Sea. Procés-Verbaux N., 13, IAPSO Special Assembly, Melbourne, 1974.

Godin, G. and Trotti, L., 1975. Trieste-water levels 1952–1971: A Study of the Tide, Mean Level and Seiche Activity. Environment Canada, Fisheries and Marine Services, Miscellaneous Special Publication 28, p. 23.

Hendershott, M.C. and Rizzoli, P., 1976. The winter circulation in the Adriatic Sea. *Deep-Sea Research*, **23**(5), 353–370.

Hopkins, T.S., 1978. Physical Processes in the Mediterranean Basin. In: Kjerfve, B. (ed.), *Estuarine Transport Processes*, University South Carolina Press, Columbia, 269–309.

Hopkins, T.S., 1989. The effect of thermohaline variability on the exchange through Mediterranean Straits. In: *Proceedings of the 8th Congress of the Italian Association of Oceanology and Limnology*, 11 June 1990, **16**(11), and Saclantcen Memorandum SM–225, p. 22.

Hopkins, T.S., in press. On the variability of the thermohaline forcing through Gibraltar Strait. Submitted to *J. Geophys. Res.*

Kasumović, M., 1958. On the influence of air pressure and wind upon the Adriatic sea-level changes (in Croatian). *Hidrografski godišnjak* 1956/57, 107–121.

Killworth, P.D., 1976. The mixing and spreading phases of MEDOC. *Progress in Oceanography*, **7**, Pergamon Press, 59–90.

Kühn, M., 1987. Consequences of changes in temperature, precipitation, evaporation. Shift of global circulation pattern. In: *Impact Analysis of Climatic Change in the Central European Mountain Range*, European Workshop on Interrelated Bioclimatic and Land Use Changes, Noordwijkerhout, 17–21 October 1987.

Kuzmić, M., Orlić, M., Karabeg, M. and Jeftić, Lj., 1985. An investigation of wind-driven topographically controlled motions in the Northern Adriatic. *Estuarine, Coastal and Shelf Science*, **21**, 481–499.

Lacombe, H., 1961. Contribution à l'étude du régime du détroit de Gibraltar. L'étude dynamique. *Cahiers Océanographiques*, **13**, 2, 73–107.

Lacombe, H., Tchernia, P., Richez, C. and Gamberoni, L., 1964. Deuxiéme contribution à l'étude du régime du détroit de Gibraltar (travaux de 1960). *Cahiers Océanographiques*, **16**, 283–314.

Lacombe, H. and Tchernia, P., 1972. Caractères hydrologiques et circulation des eaux en Méditerranée. In: Stanley, D.J. (ed.), *Mediterranean Sea: A Natural Sedimentation Laboratory*, Hutchinson and Ross, Stroudsburg, PA, 25–36.

Lascaratos, A. and Gačić, M., 1990. Low frequency sea-level variability in the northeastern Mediterranean. *J. of Physical Oceanography*, **20**, 4, 522-533.

Le Provost, C. and Hopkins, T.S., 1986. Modelling of physical processes. In: *Review of the Physical Oceanographic Processes Influencing the Distribution, Transfer, and Fate of Pollutants in the Mediterranean Sea*, Chapter IV, IOC/UNEP Mediterranean Action Plan Report, UNEP Athens, Greece.

Malanotte-Rizzoli, P. and Hecht, A., 1988. Large-scale properties of the Eastern Mediterranean: A Review. *Oceanologica Acta*, **11**, **4**, 323–336.

Manzella, G., Gasparini, G. and Astraldi, M., 1988. Water exchange through the Strait of Sicily. *Deep-Sea Research*, **35**, 1021–1035.

Manzella, G.M.R. and La Violette, P.E., 1990. The relation of the transport through the Strait of Gibraltar and the seasonal transport of LIW through the Strait of Sicily. *J. of Geophys. Res.*, **95**, C2, 1623–1626.

May, P.W., 1982. Climatological flux estimates in the Mediterranean Sea. Part I: Winds and Wind Stresses. Report 54, NORDA, NSTLStation, MS39529, p. 56.

MEDOC Group, 1970. Observation of formation of deep water in the Mediterranean, 1969. *Nature*, **227**, 1037–1040.

Morcos, S.A., 1972. Sources of Mediterranean Intermediate Water in the Levantine Sea. In: Gordon, A.L. (ed.), *Studies in Physical Oceanography*. Vol. 2, Gordon and Breach, New York, 185–206.

Mosetti, F., 1969. Oscillazioni del livello marino in rapporto con le oscillazioni di pressione atmosferica. *Bolletino di Geofisica Teorica ed Applicata*, **11**, (43–44), 264–278.

Mosetti, F., 1971. Considerazioni sulle cause dell' acqua alta a Venezia. *Bolletino di Geofisica Teorica ed Applicata*, **13**, 169–184.

Orlić, M., 1983. On the frictionless influence of planetary atmospheric waves on the Adriatic sea-level. *J. of Physical Oceanography*, **13**, 1301–1306.

Oszoy, E., 1981. On the atmospheric factors affecting the Levantine Sea. European Center for Medium Range Weather Forecasts, Tech. Rep. 25, p. 29.

Oszoy, E., Latif, H. and Unluata, U., 1981. On the formation of Levantine Intermediate Water. *Rapp. Comm. Inter. Mer. Medit.*, **27**, 51–65.

Ovchinnikov, I.M., 1974. On the water balance of the Mediterranean Sea. *Oceanology*, **14**, 198–202.

Ovchinnikov, I.M., Zac, V.I., Krivosheya, V.G. and Udodov, A.I., 1985. Formation of Deep Eastern Mediterranean Water in the Adriatic Sea. *Oceanology*, **25**(6), 911–917.

Palumbo, A. and Mazzarella, A., 1982. Mean sea-level variations and their practical applications. *J. of Geophys. Res.*, **87**, 4249–4256.

Papa, L., 1978. A statistical investigation of low-frequency sea-level variations at Genoa. Istituto Idrografico della Marina, Universita' degli Studi di Genova, F.C. 1087, *Grog.* **6**, p.13.

Pollak, M.I., 1951. The sources of deep water of the eastern Mediterranean Sea. *J. of Marine Research*, **10**, 128–152.

Rickards, L.J., 1985. Report on sea-level data collected during the MEDALPEX experiment from September 1981-September 1982. Institute of Oceanographic Sciences, Report No. 209, p. 170.

Sankey, T., 1973. The formation of deep water in the northwestern Mediterranean. In: Waren, B. A. (ed.), *Progress in Oceanography*, 6, Pergamon Press, Oxford, 159–179.

SATMER, 1987. Bulletin Mensuel, Janvier 1987, Le Centre de Météorologie Spatiale, Lannion, 41.

Wright, D.G., 1987. Comments on "Geostrophic Control of Fluctuating Barotropic Flow through Straits". *J. of Physical Oceanography*, **17**(12), 2375–2377.

Wüst, G., 1961. On the vertical circulation in the Mediterranean Sea. *J. of Geophys. Res.*, **66**, 3261–3271.

Zore-Armanda, M., 1968. The system of currents in the Adriatic Sea. *Stud. Rev. Gen. Fish. Coun. Medit.*, **34**, p. 48.

Zore-Armanda, M., 1972. Formation of Eastern Mediterranean Deep Water in the Adriatic. *Colloques Internationaux du C.N.R.S.* No. 215, 127–133.

Zore-Armanda, M., 1974. Some aspects of air-sea interaction in the Eastern Mediterranean. *Proc. Colloq. Physics of Sea*, Trieste, Oct. 13–16 1971. Accademia Nazionale dei Lincei.

Zore-Armanda, M., 1985. Sea water climatic characteristics of middle Adriatic (in Croatian), *Geofizika*, **2**, 179–193.

Zore-Armanda, M. and Pucher-Petković, T., 1977. Some dynamic and biological characteristics of the Adriatic and other basins of the Eastern Mediterranean Sea. In: Huilings, N. (ed.), *Symposium on the Eastern Mediterranean Sea, IBP/PM – UNESCO*, p. 396., *Acta Adriatica*, **2B**, 15–28.

Zore-Armanda, M. and Gačić, M., 1987. Effects of bura on the circulation in the North Adriatic. *Annales Geophysicae*, **5B**(1), 93–102.

8

Predictions of Relative Coastal Sea-Level Change in the Mediterranean Based on Archaeological, Historical and Tide-Gauge Data

N.C. Flemming
*(Institute of Oceanographic Sciences,
Deacon Laboratory, Wormley, Godalming, UK)*

Abstract

Prediction of relative sea-level change on the Mediterranean coast over the next 50–100 years requires an estimate of the global change of absolute sea level which may be caused by climate change, and the local vertical displacement of the level of the land at each point on the coast relative to absolute co-ordinates. Vertical land movements, which must be added to eustatic sea-level changes, can be defined by archaeological data (100 to 2000 years ago), and tide-gauge data (10 to 100 years ago); 335 coastal archaeological sites have been identified, but only 32 tide-gauge records are available. The tide gauge data are more accurate, and refer to the appropriate time-scale for extrapolation into the future. Data on earth movements at discrete points can be interpolated laterally for only a few tens of kilometres between data sites, and thus the widely spaced tide-gauge sites can only be used to indicate expected earth movements in limited regions along a coast of 45,000 km in length. By comparing tide-gauge and archaeological data from those sites where both are available, a correction rule is derived to compensate for the aliasing which occurs when noisy data are sampled over different time periods. As a result it is possible to suggest statistically probable estimates for the rate of vertical movement to be expected in the next 50–100 years for most sectors of the Mediterranean coast.

Field data from archaeological sites suggest that many sites are experiencing rapid erosion. Submerged ruins represent a unique record of ancient maritime technology and are being destroyed by wave action and human activities, with marked collapse and scattering of structures observable over 1–10 years. Assuming that most sites cannot be protected, they should at least be surveyed and photographed accurately as soon as possible. Older sites from the Neolithic and Palaeolithic are usually in deeper water, 8–30 m, and so are more protected from wave action.

1 Introduction

The tidal amplitude throughout most of the Mediterranean is of the order of 20–30 cm, with the exception of the immediate region of Gibraltar, where it is approximately 1 m. Wind forcing and variations in barometric pressure cause fluctuations in the level of the order of 30 cm over periods of several days and 10–12 cm in monthly mean sea-level during the year (Striem, 1974;

Rickards, 1985). Structures built close to the water, such as jetties, landing docks, boat slipways, etc., are naturally designed to resist the local wave conditions, but are not designed to resist an extreme sea-level change of more than 50 cm; in practice the expected annual range of extremes is of the order of 30–40 cm, being the sum of the seasonal and multi-day cycles. Similarly, natural coastal processes, the formation of stable beach gradients, marshes and wetlands, deltaic alluviation, etc., are adjusted to a near constant sea-level which fluctuates by only 30–50 cm.

Anthropogenic climate change caused by the release of so-called greenhouse gases at present is predicted to cause a global rise in sea-level of the order of 20 cm by 2030 and 44 cm by 2070 (Warwick and Oerlmans, 1990, p. 277); or 2–4 cm/decade (Houghton et al., 1992), from thermal expansion of the ocean alone. Even at the lower estimate, such a change would bring wave action consistently into contact with structures or natural shoreline features that never previously had been exposed to wave forces. The upper estimate would cause massive inundation of low-lying structures and coastal landforms in many parts of the Mediterranean. However, no general prediction can be made for the Mediterranean coast because local vertical earth movements are of approximately the same magnitude, 10–50 cm/century (Pirazzoli, 1976a, 1976b; Flemming, 1978; Flemming et al., 1978; Flemming and Webb, 1986; Flemming and Woodworth, 1988). Tectonic earth movements may be vertically up or down, and thus in some places the relative rise of sea-level over the next 100 years may be as much as twice the absolute rise of sea-level, whereas in others the local relative sea-level may even drop. From the point of view of planning defensive action, it is important to know the likely areas of maximum and minimum relative rises of sea-level.

2 Method

The Mediterranean lies across a major global plate tectonic boundary (Le Pichon and Angelier, 1981; Dewey et al., 1973; McKenzie, 1972, 1978), and many parts of the coast are subject to regular earthquake activity. Volcanism is also active in south-west Italy, the Lipari Islands, Sicily and the central Aegean. These processes cause sections of the coast to be displaced vertically at varying rates (Schmiedt, 1972; Flemming, 1969, 1972, 1978; Pirazzoli, 1976a, b, 1987b; Vita-Finzi and King, 1985; Flemming and Webb, 1986; Flemming and Woodworth, 1988). Although parts of the Pacific rim (Indonesia, Japan and Alaska) have higher rates of movement, it is reasonable to say that the Mediterranean coast is amongst the most active in the world.

In addition to the up-and-down displacements due to tectonism and volcanism, the coast is liable to downwards movement due to the accumulation of alluvium in coastal plains and on deltas. The net subsidence may be caused both by the isostatic response to the weight of the alluvium and by compaction within the sediment column. Several areas of such subsidence have been documented both by identification of archaeological remains and through coring (Venice, Pirazzoli, 1987a; Messenia, Kraft et al., 1975; Kraft and Aschenbrenner, 1977; Larnaca, Gifford, 1978).

In the most active regions of the Mediterranean sites separated by only 10–20 km can show different directions and rates of vertical movement. In the western Mediterranean, the variability is less pronounced. Nevertheless, the mechanisms being proposed for the tectonic movements in the crust suggest that spatial variability on the scale of 10–20 km would be expected. A description of the vertical movements to be expected in the future on the coast of the Mediterranean therefore would ideally be based on a data set which provided reliable data at 10–20 km intervals along the whole coast length of 45,000 km.

Such a data set is plainly unobtainable. Possible sources of data include tide gauges, erosional and solution formations, biological formations such as vermetid terraces or lithodomus borings, sedimentary accumulations with layers containing organic carbon which can be dated by radiocarbon methods, and archaeological structures on the coast (Van de Plaasche, 1986).

All the above types of data can be obtained for the last 2000 years in some parts of the Mediterranean coast, but the most widely spread data type is the archaeological, which provides sea-level estimates at 335 locations (Flemming and Webb, 1986). Other data sources could be integrated with the archaeological data over the same time-scale of 100–2000 years, but the present author does not have experience of the errors and corrections that would need to be applied to create a homogeneous data set. For the moment it seems better to rely on the internally consistent archaeological data for the time-span longer than that of tide gauges.

For periods of a few decades, tide gauges are extremely accurate, provided that the calibration and reference datum of the instrument is properly maintained. Standards for this procedure are maintained by the Permanent Service for Mean Sea Level (PSMSL), and the data referred to in this study have been obtained by various authors from the PSMSL. Pirazzoli (1987b) studied 22 tide-gauge records within the MAP area of which 16 provide reliable records; Flemming and Woodworth (1988) studied 16 sites in Greece. Emery and Aubrey (1985, p. 250) deduce rates of earth movement for northern and north-west Europe adjacent to the present study area.

Tide-gauge accuracy is of the order of millimetres (Rossiter, 1967; Pugh, 1987; Emery et al., 1988; and others). The principal source of errors arises from undocumented changes in datum, or corrections or adjustments to the instrument calibration. Additionally, data may be "corrected" by authors before they are published to allow for presumed local earth movements. If this has occurred it is pointless to try to detect earth movements by studying the difference between adjacent tide gauges, which is the principal method. Various methods have been used to separate global or regional sea-level change from earth movements on the basis of statistical treatment of large numbers of tide-gauge records (Graaf, 1980; Gornitz et al., 1982; Emery and Aubrey, 1985; Pirazzoli, 1987b; Emery et al., 1988; Douglas, 1991), including regression analysis and eigen analysis. It is not the purpose of this chapter to discuss the virtues of the different methods. Suffice it to say that there is no reservation about the capability of modern well-maintained tide gauges to produce excellent records: the problem is simply that there are not sufficient instruments in the Mediterranean to give an accurate picture of the tectonic variability of the complex coastline. Furthermore, in the absence of existing records of a decade or more at a desired location, an instrument installed now will only provide a suitable time series in 10–20 years time. Several hundred tide gauges would be needed to show the true land movement at a sufficient number of sites.

2.1 Archaeological Methods

2.1.1 Site selection

The deduction of relative land/sea-level changes at a number of archaeological sites was attempted by several 18th and 19th century authors (cited by Flemming, 1969). Gnirs (1908) was the first author to attempt a complete synthesis for the whole Mediterranean, using field data from eight sites, referenced data on a further 34 sites, and very rough generalizations for another 25 sites based only on inspection of charts. Observations and deductions were based on the assumption of no earth movements. Very little attempt was made to estimate errors, and the probable errors at the sites where data were observed were of the order of 0.5–1.0 m or more. There was a strong tendency to force data into the simple model of constant rate of eustatic change. Data were concentrated in the Adriatic and Aegean.

A problem with early works is that the authors searched for "submerged ruins", and found them. It is a simple matter to list over 100 sites in the Mediterranean where masonry can be found to a depth of 1.0 m or more. The masonry may be part of a harbour wall built at that depth, but before the invention of inexpensive diving equipment, that was difficult to ascertain. On the principle of "no smoke without a fire", authors naturally assumed that the only question at stake was the measurement of the size of the sea-level change since Roman times, which was assumed to be at least 2 m. Hafemann (1960) lists well-known submerged sites and makes calculations accordingly.

Lehman-Hartleben (1923) provided an inventory of 303 sites in the Mediterranean, with descriptions of harbours derived from classical literature and archaeological sources. There was no estimate of sea-level change, but some useful site maps. In some studies the authors have tried to separate stable and unstable coastal areas, and use the "stable" areas to measure eustatic sea-level change. Schmiedt (1972), in a series of brilliant observations on fish tanks on the west Italian coast, deliberately avoided the areas around Naples and Mount Etna. And yet most of the Italian coast is liable to some seismic activity, and it is arbitrary as to what level of low seismicity can be classified as stable.

The present study is designed to measure earth movements, and therefore all possible sites with good data are included. The data set includes all valid sites, whether the data show relative submergence, constant relative sea-level or relative uplift. When this method is used (Flemming, 1968, 1969, 1972, 1978; Flemming et al., 1973, 1978) it is apparent that a large number of sites are relatively stable or uplifted, and that the samples used in the papers prior to 1969 were biased towards selection of submerged sites.

Pre-1969 studies attempted to measure the sea-level at 2000 BP, an objective also assumed by Flemming (1969). This results in a partial data set, since many sites provide data for the Mediaeval, Norman, Arab, Byzantine, Minoan and other periods, but not the classical Roman. Given the limitations of computer statistics, earlier workers did not attempt to analyse the general problem, assuming random earth movements and eustatic sea-level change, utilizing a data set dispersed in space and time. However, by the early 1970s suitable statistical packages were available, and this generalized method has been used by the author since 1972 (Flemming, 1972).

To construct the MEDSITE database all known sites were identified from the multi-site sources already quoted, together with reports and data on single sites from the archaeological literature and atlases (Frost, 1972; Le Gall, 1981; Negris, 1904; Raban, 1981, 1983; Vrsalovic, 1974; Yorke and Dallas, 1968; Yorke and Davidson, 1969; and York et al., 1973). So far as possible all sites were labelled with an ancient name and a modern name, and recorded with their latitude and longitude. A mid-19th century folio of Admiralty Mediterranean charts was used to locate many of the ruins that have not been visited, since early chart-surveyors were well-versed in classical geography and noted the existence of ruins with accuracy and detail. Of the sites listed by Lehman-Hartleben (1923) 18 named sites could not be located on any chart, and there has not been time to check the early German archaeological references which would be needed to locate the sites. Some discrepancy may be due to variations in spelling.

Of the 1053 sites, sea-level estimates and age were derived at 335 (Table 8.1). The sites for which data were not obtained constitute an unexploited reservoir of data that could be utilized to extend or improve the present study.

Large-scale site maps exist for most sites with data, and most sites are documented with land and underwater photographs. These files can be accessed separately from the MEDSITE database.

2.1.2 Observational methods

Field observations use echo sounding, snorkel diving, scuba diving, wading, aerial photography, underwater photography, and a wide range of simple and sophisticated survey methods depending upon time and budget. Methods and accuracy are discussed fully by Flemming (1969, 1978, 1979), Blackman (1973, 1982), Schmiedt (1972) and Guery et al. (1981). During the preparation of Flemming (1969) it became clear that certain ancient structures, such as ship slipways, mooring bollards and mooring rings, fish cultivation tanks, salt pans, docks, quays, sluice gates, etc., could provide a link to the sea-level at the time of construction with greater accuracy than previously obtained.

Data for over 270 of the present sites with conclusive data are based on field surveys by the author. For the majority of the other sites employed in the calculations, the present paper uses the same estimate of relative sea-level changes as cited by the original authors. However, the data of other published authors have been used as data only, and the conclusions with regard to eustatic or tectonic change may be different in the present paper. Additionally, the present data base contains estimates of the accuracy of

Table 8.1 Distribution of valid and invalid data records by region

Region number	Name	SIT	DTA	NON	A/S %	DDD	MLT	ONE	X km	Y km	AREA km²	CST km	DSP km	SSP km
1	Spain	64	24	40	38	24	2	2	1780	280	498400	2200	92	34
2	Balearics	5	2	2	40	4	2	0	310	130	40300	6500	325	130
3	Morocco	52	24	28	46	26	2	2	1600	280	448000	2200	92	42
4	Sardinia	16	3	13	19	3	9	9	180	599	90000	1400	467	88
5	W. Italy	102	42	60	41	45	4	0	1100	270	297000	1800	43	18
6	Adriatic	71	3	68	4	6	2	0	890	330	293700	3100	1033	44
7	Sicily	43	9	34	21	9	0	0	470	610	286700	900	100	21
8	Sirte	13	5	8	38	5	0	0	980	260	254800	1550	310	119
9	W. Greece	38	1	37	3	1	0	0	360	190	68400	600	600	16
10	Peloponnese	71	30	41	42	32	2	0	240	300	72000	1000	33	14
11	Crete	72	43	29	60	47	3	0	330	130	42900	650	15	9
12	N. Aegean	185	5	180	3	8	1	0	680	420	285600	–	–	–
13	S. Aegean	36	5	31	14	6	1	3	410	240	98400	–	–	–
14	S.W. Turkey	66	36	30	55	47	9	0	380	310	117800	–	–	–
15	Rhodes	25	17	8	68	20	3	0	200	80	16000	300	18	12
16	Cyrenaica	36	4	32	11	5	1	0	830	120	99600	950	238	26
17	E. Turkey	32	23	9	72	23	0	0	530	130	68900	700	30	22
18	Cyprus	36	17	19	47	20	4	0	250	170	42500	700	41	19
19	Egypt	4	0	4	0	0	0	0	450	110	49500	500	0	125
20	Syria/Israel	83	42	41	51	75	20	6	710	110	78100	650	15	8
21	Islands	3	0	3	0	0	0	0	–	–	–	–	–	–
	Total/means (mn)	1053	335	718	32 (mn)	406	56	13			162430 (mn)	993 (mn)	164 (mn)	36 (mn)

Notes: Codes as follows: SIT = number of sites; DTA = number of sites with valid data; NON = sites with no data; A/S = percentage of sites with data; DDD = number of valid data records; MLT = number of sites with more than one valid period; ONE = sites with one-sided records; X and Y = dimension of the region in km; AREA = area of region km²; CST = length of coastline in km; DSP = mean spacing between sites with valid data; SSP = mean spacing of all sites (note that regions 12, 13 and 14 have such complex coastlines that the length cannot be measured accurately); mn = mean value.

data, defined by a probable error bar. The error bar has been defined on the basis of the type of sea-level indicators used, and the inherent limits on accuracy described in the next section.

In the Bronze Age, (5000–3000 BP), and later archaeological periods, specialized structures were built in relation to the contemporary sea-levels. In earlier periods there do not appear to be identifiable water-line structures, although Aghios Petros (site number 964 in Table 8.2 and MEDSITE database) is a neolithic village facing directly onto a small estuary which was probably a harbour, and the data from the submerged site at Franchthi Cave (965) also suggest a waterside dwelling area. In general, data earlier than 5000 BP only indicate that an occupation site was dry land at a given date, but the sea-level may have been several metres lower, unless geomorphological evidence can be found for linking shoreline features and human artifacts into a contemporaneous assemblage. Sites that can be dated by artifacts, but can only be proven to be above sea-level, are termed "one-sided" in the present study. In statistical studies the "one-sided" data points sometimes were included and sometimes excluded in order to test different hypotheses. There are 14 one-sided sites, all older than 5000 BP.

The method of observing solution notches and comparing the present notch with higher or lower notches, has been used by many authors (e.g. Spratt, 1865; Günther, 1903a, 1903b; Hafemann, 1965; Pirazzoli et al., 1981; Flemming et al., 1973). In the open sea a solution notch is often 0.5–1.0 m in height, and the mean sea-level point can only be determined to an accuracy of about 0.2–0.3 m. Even in sheltered bays, the error is of the order of 0.2 m. Notches occasionally cut across structures, as at the Roman quays in Marseille (32), the harbour entrance at Phalasarna (365), or the Casa degli Spiriti at Posilipo (75). Because of wave exposure and unevenness of the substrate, the width of the notch is such as to leave uncertainty of exact mean sea-level of the order of 0.2 m (Guery et al., 1981).

A rock-cut storage tank or fish tank can act as a stilling pool, with the long entrance channel acting as a filter to remove wave action. This occurs at Dor (199), Lambousa (382), and Caesarea, Israel (203/1051) where the tanks are not submerged below the level of the outer walls. In these circumstances the calm water surface, combined with the tidal range of 20–30 cm, results in a double-solution notch, with a small indentation at mean high tide, and a second at mean low tide. The W-shaped indentation has a profile that can be matched very precisely as between ancient and modern notch system, with an accuracy of 0.1 m. Since the notch system records mean sea-level averaged over several decades, this observation requires no correction for seasonal or barometric factors, provided that there is a similar double notch at present sea-level.

2.1.3 Limits to accuracy
Once the accuracy of individual observation approaches 0.3–0.2 m further limitations occur. Assuming that the tidal cycle is known, the relative change of level can be corrected, but often the state of the tide is not known in relation to the time and date of observation. The Mediterranean tides are of the order of 0.2–0.5 m, and so are not regarded as important for shipping and navigation. The only reliable way to obtain tidal data is

Table 8.2 List of sites used to compute earth movements and sea-level changes. The sites are those with valid data analysed by Flemming and Webb (1986)

1[a]	2[b]	3[c]	4[d]	5[e]
1	Belo	1	2.00	0.00
3	Carteia	1	2.00	0.00
13	Hemeroskopeion	1	2.50	0.00
17	Tarraco	1	2.00	0.00
21	Emporiae	1	2.00	1.00
25	Leuxos	1	7.00	−10.00
26	Narbo	1	2.00	0.00
29	Artemis	1	2.50	−1.00
30	Arelate	1	2.00	−1.00
31	Fossae Marinae	1	1.80	−4.00
32	Massilia	1	1.80	−0.40
37	Olbia	1	2.40	0.00
38	Forum Iulii	1	2.00	0.00
39	Insula Lero	1	2.00	−0.50
40	Antipolis	1	2.00	0.00
43	Albintimilium	1	2.00	0.00
49	Luna	1	2.10	−1.00
51	Populonia	1	2.00	−2.00
55	Cosa	1	2.20	−1.00
60	Alsium	1	2.00	0.00
62	Antium	1	1.90	−0.50
64	Astura	1	2.00	−0.60
65	Circeii	1	2.00	−1.00
66	Tarracina	1	1.80	0.00
68	Formiae	1	2.00	−0.50
71	Misenum	1	2.00	−3.20
72	Baiae	1	2.00	−5.50
72	Baiae	2	1.30	4.00
73	Puteoli	1	2.00	−4.00
73	Puteoli	2	1.30	5.80
74	Nesis	1	2.00	−4.00
75	Pausilypon	2	1.30	4.80
75	Pausilypon	1	2.00	−4.00
77	Herculaneum	1	2.00	−2.00
80	Surrentum	1	2.00	−2.00
84	Velea	1	2.50	0.00
93	Zankle	1	0.30	−0.50
96	Megara Hyblaea	1	2.30	−1.70
97	Thapsos	1	3.40	−0.90
98	Siracusa	1	2.50	0.00
104	Selinunte	1	2.50	0.00
105	Mazara	1	2.00	0.00
106	Lilybaeum	1	2.30	0.00
107	Motya	1	2.50	0.00
115	Leptis Magna	1	2.00	0.00
117	Sabrata	1	2.40	0.00
121	Thaenae	1	2.00	0.00
124	Alipota (Africa)	1	2.60	0.00
125	Thapsus	1	2.00	0.00
126	Leptis Minor	1	2.00	0.00
127	Ruspina	1	2.00	−0.30
128	Hadrumentum	1	2.00	0.00
129	Heraklea	1	2.00	−0.20
136	Carthago	1	2.30	−1.00
136	Carthago	2	2.00	−1.50
136	Carthago	3	1.30	−0.15

Archaeological data and sea-level change 255

1[a]	2[b]	3[c]	4[d]	5[e]
137	Utica	1	2.50	1.00
148	Rasguniae	1	2.00	−1.00
150	Tipasa	1	2.00	−0.50
151	Caesarea (Africa)	1	2.00	−1.00
160	Nora	1	2.40	−4.50
164	Neapolis	1	2.00	−1.00
168	Turris Libisonis	1	2.00	−1.00
174	Rosh Hanniqra	3	1.50	−0.50
175	Achzib	2	3.50	0.00
175	Achzib	4	2.00	0.00
176	Achzib, Islands	1	2.70	−0.40
176	Ros. Hanniqra Isle	1	2.50	−0.30
179	Shavei Zion (N)	1	2.00	0.00
180	Shavei Zion (S)	1	2.00	0.00
181	Yassif River	1	2.00	0.00
182	Acco	2	2.70	−0.50
182	Acco	6	0.80	−0.50
188	Tell Abu Hawam	1	3.30	0.00
190	Bat Galim	1	2.00	−0.40
191	Tell Shikmona	2	2.00	0.00
193	Atlit	1	10.00	−10.00
193	Atlit	2	6.00	−2.50
193	Atlit	3	3.80	0.00
193	Atlit	4	3.40	0.00
193	Atlit	5	2.70	0.00
193	Atlit	7	1.90	0.00
193	Atlit	8	0.80	0.00
194	Newe Yam	1	6.00	−2.00
195	Tell Nami	1	3.80	−0.50
195	Tell Nami	2	3.40	−0.50
197	Tell Nami (S)	1	0.80	−0.50
198	Dor (N)	1	2.00	0.50
199	Dor (Tantura)	1	3.20	−0.80
199	Dor	2	2.80	−0.90
199	Dor	3	2.60	−0.50
199	Dor	4	2.50	−0.40
199	Dor	5	2.40	−0.30
199	Dor	6	2.20	−0.10
199	Dor	7	1.90	0.20
199	Dor	8	1.50	0.80
199	Dor	9	1.30	0.50
199	Dor	10	0.80	−1.00
199	Dor	11	0.60	−1.00
199	Dor	12	0.20	−0.20
199	Dor	13	0.10	−0.10
200	Nahal Dalia	1	6.00	−4.00
201	Magan Mikhael	1	1.90	0.00
202	Crocodilopolis	1	2.20	−0.30
203	Caesarea (Israel)	2	2.00	−5.00
203	Caesarea (Israel)	3	0.80	−1.00
205	Mikhmoret	1	3.30	0.00
207	Apollonia (Israel)	1	1.90	0.00
207	Apollonia (Israel)	2	0.80	0.00
215	Palmachim	1	3.40	0.00
215	Palmachim	2	2.70	0.00
215	Palmachim	3	1.90	0.00
215	Palmachim	4	1.50	0.00
218	Tel Mor (Ashdod)	1	3.80	0.00

1a	2b	3c	4d	5e
218	Tel Mor (Ashdod)	2	3.40	0.00
220	Ashdod-Yam	1	1.40	0.00
221	Ashkelon	1	3.80	0.00
221	Ashkelon	2	0.80	0.00
223	Gaza Maiumas	1	1.50	0.00
224	Tell el Ajjul	1	3.80	0.00
225	Tell el Qatifa	2	2.80	−0.30
225	Tell el Qatifa	1	7.00	−1.00
226	Tell el Ridan	1	3.50	−0.50
229	Gytheum	1	1.50	−2.50
230	Trinasus	1	1.00	−1.00
231	Asopus	1	3.00	−3.00
231	Asopus	2	1.50	−2.00
232	Arkangelos	1	2.00	−0.20
233	Onugnathus	1	3.50	−4.00
234	Kythera	1	4.00	0.00
235	Antikythera	1	5.00	3.00
236	Boeae	1	1.50	−0.30
237	Minna	1	2.00	−1.00
238	Zarax	1	2.50	−3.00
239	Asine	1	3.00	−2.00
240	Halieis	1	2.40	−5.00
241	Lorenzon	1	1.00	−2.00
242	Epidaurus	1	2.00	−2.70
243	Cenchreae	1	2.50	−2.00
244	Lechaeum	1	2.00	−0.70
245	Phea	1	2.00	−1.00
246	Pylus Coryphasi.	1	2.00	−1.00
247	Methone	1	1.00	−1.50
248	Asine	1	0.60	−1.20
250	Cardamyle	1	1.00	−1.00
251	Leutra	1	2.00	−2.50
252	Pephnus	1	1.00	−1.00
253	Tigani	1	2.00	−0.50
254	Teuthrone	1	1.00	−1.00
255	Skoutari	1	1.50	−3.50
256	Acriae	1	3.00	−2.00
256	Elaea	1	2.30	0.00
258	Cyme	1	2.40	−1.00
259	Smyrna	1	4.00	−3.00
259	Smyrna	2	2.50	−1.00
260	Clazomenae	1	2.30	−1.00
261	Urla Beach	1	0.30	−0.50
262	Erithrae	1	2.40	−1.40
262	Erithrae	2	0.20	−0.50
262	Erithrae	3	0.10	−0.30
263	Yali	1	1.00	−0.50
264	Ilica	1	2.40	−0.50
265	Cesme	1	0.10	−0.20
266	Ciftlik	1	0.10	−0.10
267	Teos	1	2.30	−0.80
267	Teos	2	2.30	−0.80
267	Teos	3	0.10	−0.10
269	Notium (Claros)	1	2.60	−1.00
271	Miletus	1	3.20	−1.50
271	Miletus	2	2.00	−1.50
272	Heraclea Latmus	1	2.50	−1.50
273	Ghioucker I.	1	1.20	−1.50
274	Panormus	1	2.60	0.00

Archaeological data and sea-level change 257

1[a]	2[b]	3[c]	4[d]	5[e]
275	Iasus	1	2.00	−0.50
275	Iasus	1	1.00	0.00
276	Bargylia	1	2.00	−0.50
277	Caryanda	2	0.30	0.00
277	Caryanda	1	1.50	−0.30
279	Myndus	1	2.00	−1.20
280	Karatoprak	1	1.00	−0.50
281	Halicarnassus	1	2.40	−1.00
282	Cedreae	1	1.50	−0.30
283	Cnidus	1	2.30	0.00
284	(Old Cnidus)	1	2.50	0.00
285	Orhaniye	1	1.00	−1.00
286	Bozburun	1	1.00	−1.00
287	Loryma	1	1.50	−1.00
288	Saranda	1	1.00	−0.50
289	Telmessus	1	1.50	−1.20
291	Antiphellus	1	2.20	−2.20
291	Antiphellus	2	1.00	−1.00
292	Myra	1	1.00	−1.00
293	Chersonissos	1	1.90	−1.00
294	Malia	1	3.60	−1.00
295	Spinalonga	1	2.00	−1.00
296	Port Kolokithia	1	1.00	−1.00
297	Olous	1	2.00	−1.90
298	Minoa	1	2.20	−1.50
300	Psira	1	1.90	−0.75
301	Gemili	1	1.40	−2.00
302	Simena	1	2.50	−2.20
302	Simena	2	1.50	−1.20
303	Phaselis	1	2.50	−0.20
304	Lara	1	1.70	0.00
305	Side	1	2.20	0.00
306	Kara Burun	1	2.00	0.00
308	Coracesium	1	0.70	0.00
310	Syedra	1	1.50	0.00
311	Sellinus	1	1.80	0.00
312	Hamaxia	1	1.50	0.00
315	Anemuriam	1	1.50	0.00
316	Mamuriye	1	0.60	0.00
317	Arsinoe	1	1.50	0.00
319	Mellaxia	1	2.00	0.00
320	Celendris	1	0.50	0.00
321	Holmus	1	0.60	0.00
322	Agalimani	1	0.60	0.00
323	Agalimeni Fort	1	0.60	0.00
325	Persente	1	1.50	0.00
326	Corycus	1	1.50	0.00
328	Soli	1	2.00	0.00
329	Megarus	1	2.00	0.00
330	Aegeae	1	2.00	0.00
331	Baiae	1	0.50	0.00
332	Fort Bannel	1	2.00	0.00
333	Seleucia Pieria	1	1.80	0.00
334	Amnisos	1	3.60	−1.50
335	Nirou Khani	1	3.60	−1.75
337	Mokhlos	1	3.60	−1.75
337	Mokhlos	2	1.90	−0.25
338	Eteia	1	1.80	−0.80
339	Itanos	1	2.00	−2.20

1[a]	2[b]	3[c]	4[d]	5[e]
340	Capa Plaka	1	1.00	−0.20
341	Zakros	3	1.50	0.00
341	Zakros	2	2.00	−0.50
341	Zakros	1	3.00	−1.00
343	Akri Goudara	1	5.00	3.50
344	Hierpatra	1	1.80	−0.50
346	Moni Arvi	1	0.50	−0.25
347	Metallon	1	1.90	0.00
347	Metallon	2	1.10	−2.00
350	Plakios	1	1.90	2.00
351	Leuce	1	1.90	2.20
352	Lasea	1	1.80	0.00
353	Agia Marina	1	1.90	2.00
355	Sphakia	1	1.90	3.50
356	Phoenice	1	1.90	3.50
357	Kalamydes	1	1.90	7.50
360	Plakaki Krio	1	2.20	8.00
362	Musagores	1	2.20	8.50
363	Khrysoskalitissa	1	2.20	7.50
364	Sphinarion	1	2.20	7.00
365	Phalasarna	1	2.20	6.60
367	Agneion	1	2.20	6.00
368	Kisamos	1	2.20	5.20
369	Napia	1	2.20	5.00
370	Dhiktinaia	1	2.20	3.00
371	Burdroae	1	2.20	3.00
372	Chor Fakia	1	2.20	1.10
373	Marathi	1	2.20	1.50
375	Amphimalla	1	2.20	1.50
376	–	1	2.20	1.00
377	–	1	2.20	0.80
378	Rethymme	1	2.20	−0.20
379	Panormus	1	0.50	−0.50
380	Varignano	1	2.00	−0.50
382	Lapethus	1	2.00	0.00
384	Cerynia	1	2.30	0.00
385	Vrisi	1	3.00	0.00
386	Platymeis	1	2.00	0.00
388	Agios Thyrsos	1	2.00	−0.25
389	Carpasia	1	2.00	0.00
392	Salamis	1	1.70	−2.00
392	Salamis	2	1.00	−0.30
393	Ammochostos	1	2.00	−0.50
395	Dades	1	2.00	−0.20
396	Kitium	1	5.00	−7.00
396	Kitium	2	2.00	−0.50
399	Amathos	1	2.50	−0.75
399	Amathos	2	0.50	−0.20
400	Curias	1	0.50	−0.10
401	Dreamer's Bay	1	1.80	−2.00
404	Paphos	1	2.00	−0.30
405	Drepanum	1	2.00	−0.30
409	Agios Nikolaos	1	2.00	−0.75
410	Marium Polis	1	2.00	−0.75
413	Bema	1	2.00	0.00
414	Camiras	1	1.00	0.00
415	Rhodus	1	5.00	1.75
416	Zimbule	1	0.70	−0.25

Archaeological data and sea-level change 259

1[a]	2[b]	3[c]	4[d]	5[e]
417	Calithea	1	5.00	3.70
417	Calithea	2	2.40	2.70
417	Calithea	3	0.70	−0.25
418	Aphandos	1	5.00	5.24
419	C. Teodoco	1	5.00	3.40
420	Tsambika	1	5.00	3.25
421	Lindus	1	5.00	2.00
422	Merminga	1	5.00	1.00
423	C. Istros	1	5.00	0.33
425	Lephkos	1	2.00	−1.00
426	Vurgunda	1	2.00	−0.75
427	Tristoma	1	0.80	−0.25
428	Palatea	1	1.00	0.00
430	Vathi Potamus	1	2.00	−1.10
431	Makriyalo	1	2.00	−0.90
431	Makriyalo	2	1.00	0.00
433	St. Rocchino	1	2.60	−1.80
433	St. Rocchino	2	2.50	−1.40
438	St. Liberata	1	1.90	−0.60
439	Orbetello	1	2.30	−1.00
440	Giglio I.	1	1.90	−0.60
441	Pianossa I.	1	2.00	−0.90
444	Martanum 1	1	2.00	−0.50
447	Torre Valdaliga	1	2.00	−0.60
448	Mattonara	1	2.00	−0.60
449	Punta St. Paolo	1	2.00	−0.50
450	Punta de Vipera	1	2.00	−0.60
451	Fosso Guardiole	1	2.00	−0.70
453	Grottacce	1	1.90	−0.60
457	Porto di Claudio	1	1.90	−0.50
458	Il Palazzo	1	2.00	−0.50
459	Casarina	1	2.00	−0.50
460	Lago di Paolo	1	1.90	−0.50
461	Pisc. di Lucullo	1	1.90	−0.40
462	nr. Sperlonga	1	2.00	−0.50
464	Nave	1	2.00	−0.50
466	Scauri	1	2.00	−0.40
467	Ponza (Pontia)	1	2.00	−1.00
468	Ventotene	1	2.00	−0.90
471	C. Ognina	1	3.40	−1.00
519	Aquilaea	1	1.80	−1.50
519	Aquilaea	2	0.70	−4.00
567	Phalerura	1	4.00	−2.00
604	Troy	1	7.00	−2.00
604	Troy	2	4.50	2.00
604	Troy	3	3.00	0.00
604	Troy	4	2.00	0.00
638	Mytilene	1	2.20	−0.50
646	Phylakopi	1	3.20	0.00
710	Ravenna	1	2.50	−3.00
710	Ravenna	2	2.00	−2.00
710	Ravenna	3	1.50	−1.00
743	Naupactus	1	2.40	0.00
798	Anthedon	1	1.50	−0.20
899	Paros	1	7.00	−5.00
899	Paros	2	2.40	−3.00
923	Ras et Tarf	1	2.00	−0.30
931	Aradus	1	3.00	−1.00

1[a]	2[b]	3[c]	4[d]	5[e]
931	Aradus	2	2.80	1.00
939	Sidon	1	2.80	−1.00
939	Sidon	2	2.00	1.00
940	Sarepta	1	1.70	0.00
941	Tyre	1	2.40	0.00
949	Apollonia (Africa)	1	2.40	−2.10
949	Apollonia (Africa)	2	2.00	−2.70
950	Ptolemais	1	2.00	−1.50
952	Euhesperides	1	2.00	0.00
956	Phycus	1	1.50	−1.50
957	Cercina	1	2.20	−0.50
958	Governor's Beach, Gibraltar	1	12.00	−10.00
963	Saliagos	1	7.00	−4.50
964	Agios Petros	1	7.00	−10.00
965	Franchthi	1	6.50	−10.00
966	Hof Dado	1	6.00	−1.20
967	Kefar Samir	1	5.70	−5.40
969	Lagoon of Thau	2	5.20	−1.00
980	Vieille-Couronne	1	2.20	−0.50
981	Meryem	1	2.70	0.60
982	La Gaillarde	1	2.00	−0.40
985	Bouar	1	2.00	0.80
986	Machroud	1	3.00	−1.00
987	Akovitika	1	5.00	−11.00
987	Akovitika	2	2.80	−2.00
988	St. Cecile	1	2.50	−1.00
990	Son Real	1	4.00	0.00
990	Son Real	2	0.50	0.00
994	Port de Carro	1	2.20	−0.50
995	Anse de Verdon	1	2.20	−0.50
996	Beaumaderie	1	2.20	−0.50
997	St. Croix	1	2.20	−0.50
998	Lixus	1	2.00	0.00
999	Cotta	1	1.80	0.00
1000	Asilah	1	0.60	0.00
1001	Sidi Kacem	1	4.00	0.00
1002	Cave of Hercules	1	10.00	−10.00
1004	Malabata	1	0.60	0.00
1005	Ksar es Seghir	1	0.60	0.00
1006	Lattes	1	5.00	−3.00
1027	Sullectum	1	2.00	−1.00
1028	Bizerta	1	2.00	−0.50
1030	Koutsoundri	1	2.00	−0.20
1031	Tell Harez	1	6.00	−2.60
1039	Oikonomos I.	1	5.00	−3.00
1040	Aperlae	1	2.40	−2.50
1040	Aperlae	2	1.30	−2.50
1042	Naousa Bay	1	2.50	−3.00
1043	Venice	1	2.00	−0.90
1044	Tabbat el Hamman	1	2.90	0.00
1045	Phoukeri	1	1.60	−2.00
1048	Minturnae	1	2.20	0.00
1049	Can Picafort	1	2.00	0.00
1049	Can Picafort	2	0.50	0.00
1051	Caesarea B (Israel)	1	2.20	0.00
1051	Caesarea B (Israel)	2	2.00	0.00
1051	Caesarea B (Israel)	3	0.80	0.00

to install a tide gauge during the period of measurement, and very few observers have done this. I know of no case where this factor has been taken fully into account, although Flemming (1983, unpublished report) recorded fluctuations of daily mean sea-level of 0.4 m during a six-day study of Tel Qatif (225). I know of no case where the tidal/seasonal/barometric factor has been taken fully into account with long-term measurements of the state of the tide, or use of tide tables.

So far as is known to the author, the only sites at which observations were corrected by on-site short-term measurement of the state of the tide are the data reported by Schmiedt (1972), and Lambousa (382), Salamis, Cyprus (932), Kenchreai (244), Tel Qatif (225), Aghios Petros (964), Tel Nami (195), Dor (199), Caesarea (203/1051) and Pavlo Petri (970).

The most accurately observed data set assembled so far is that published by Schmiedt (1972). Schmiedt (1972) observed numerous fish tanks along the west coast of Italy and made hundreds of measurements at each tank, defining the height of the boundary walls, depth of floor, height of internal dividing walls, level of sills and sluice gates, level of channels, etc. He referred all measurements to mean sea-level, but did not state how mean sea-level was established locally. If it were the true annual mean sea-level, then correction factors should have been introduced to allow for the season with the highest or lowest monthly mean sea-level. Conversely, if the mean sea-level used as a reference were the mean level at the time of observation, then there is an uncorrected factor for barometric effects at that time, and the seasonal difference between that season and the highest and lowest monthly mean sea-levels. The need for these corrections was acknowledged and analysed by Schmiedt (1972), but the application of the corrections to the data was not shown.

The previous paragraph does not detract from the brilliant observational detail of Schmiedt's work. However, the table of measured sea-level changes (Schmiedt, 1972, p. 213) lists the estimates to the nearest 0.01 m, with no estimate of errors. From the present analysis it must be implicit that there are unresolved errors of the order of 0.2 m.

Since many structures were in use for two centuries or more, and since expected relative rates of sea-level change are of the order of 0.05–0.1 m per century, a time error bar of two centuries is equivalent to a vertical error bar of 0.1–0.2 m.

Given the seasonal, interannual, storm and barometric changes of sea-level that may endure for days or months or result in annual variations in mean sea level, uncertainty arises from our lack of knowledge about acceptable frequencies of flooding or drying out of structures. When a particularly useful type of building stone can be obtained conveniently by quarrying on the shore directly into a cliff, would the quarry-master find it acceptable to have the quarry floor flooded once a year? for the whole of

◀ Notes on table 8.2:
[a] = site number in MEDSITE database
[b] = site name, ancient name where known, otherwise modern name
[c] = sequence number of archaeological levels of different dates at the same site
[d] = age in thousands of years
[e] = vertical displacement relative to present sea-level, in metres, negative means site now submerged

one month during each year? or never? It is an observable fact that quarries were cut down in almost every case very close to sea-level: but were they cut to highest spring high tide, mean spring high tide, mean sea-level, or what? Even if we add a constant wave exposure factor, the uncertain range remains.

Good observation and instrumental methods result in an accuracy of the order of ±0.2–0.3 m in the best cases, but refinement beyond this point would require full tidal, storm, barometric and seasonal correction at every site, combined with subjective estimates of permissible frequency of flooding. No site has yet been observed in such detail.

In addition to those artifacts or structures that produce precise data, many other features can be classified as to whether they must have been dry (e.g. a road surface, mosaic floor, tomb, etc.); part in the water and part out (e.g. mole foundation, quay, slipway, etc.); or wholly submerged (e.g. floor of an entrance channel to a harbour, floor of a fish tank). If many structures exist at a single site, the assemblage may provide a very accurate bracket. Flemming and Webb (1986, Fig. 1) show an idealized cross-section of a site with structures that indicate sea-level. At some sites (e.g. Atlit, 193; Dor, 199) structures from different archaeological periods provide a relative sea-level time curve.

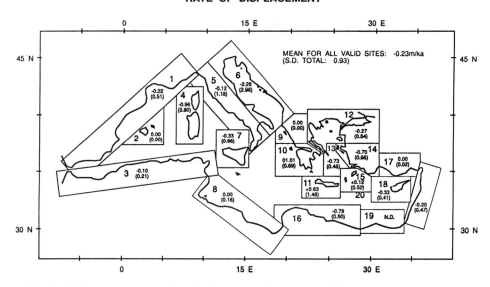

Fig. 8.1 Mean rate of vertical displacement of regions studied. The rectangles and polygonal boxes show the boundaries of the regions within which site data were analysed separately. The large numbers are the region identifiers, which are also used in Table 8.1 and Fig. 8.2. The mean rate of displacement for each region is shown in metres per thousand years (m/ka) with negative meaning land subsidence relative to the sea. The figures in brackets indicate standard deviation. The calculation is based on raw data without correction to remove eustatic sea-level change.

In reassessment of early data by the present author (especially Flemming 1968, 1969, 1972) derivations of sea-level and error bars have been revised, in some cases on the basis of subsequent experience, and further data on the site.

The accuracy of vertical displacement and time estimates may not be symmetrical. For example, a site may have been built at a known date or destroyed at a known date, but not both. The date error bar is therefore firmly limited on one side, but extends with diminished probability on the other. Similarly, a key structure, such as a mole foundation, may place a lowest possible level on sea-level, but there may be no data to limit the highest estimate. The method for allocating probability histograms to such data is given in Flemming (1972) and Flemming et al., (1973). In the present paper, since many of the sites have not been personally visited by the author, a more conventional estimate of errors is used, based on symmetrical error bars.

3 Interpretation of Archaeological Results

The present author has been most directly concerned with the analysis of archaeological data, and these will be described before the tide-gauge data. The principal results have been summarized by Flemming and Webb (1986).

The principal result, which can be deduced from the archaeological sea-level indicators after statistical analysis, is that the mean sea-level for each 200-year period has varied by less than 30 cm during the last 2000 years, and that each coastal region has a characteristic mean rate of vertical displacement, combined with very different local variability. The mean and standard deviation of vertical earth movement for each region in the study are shown in Fig 8.1.

The following discussion refers to Figs 11 and 12 taken from Flemming and Webb (1986) and reproduced here as Figs 8.2 and 8.3.

For the long narrow regions in Fig. 8.1 the predicted pattern of mean earth movements is shown in Fig. 8.3. There are no data south of the Bay of Naples in Italy, but much data from the Bay area itself. The result is that the data from this volcanic region severely distort the plot for western Italy. At this highly smoothed level, with a wavelength of hundreds of kilometres, the effects of local faulting are concealed, and these show up as high residuals. The most obvious effects are the broad subsidence of the Golfe du Lion-Rhône Delta region, uplift in Tunisia, and slight uplift in Syria. These phenomena are not altered, relatively speaking, by the application of small corrections from the eustatic curves (see Flemming and Webb, 1986 for details of the corrections).

The highest residuals, and/or rates of earth movement, are concentrated around Caesarea in Algeria, Naples, south-western Crete, the Cesme Peninsula of west Turkey, the Kekova-Kas region in south Turkey and Caesarea-Acco in Israel.

Regional patterns of earth movement have been plotted previously for the equiaxial Regions 10, 11, 14 and 15 in Fig. 1 (Flemming, 1978). The techniques used in the present study are similar to those used previously for these regions, although the computer software was different. Plots from

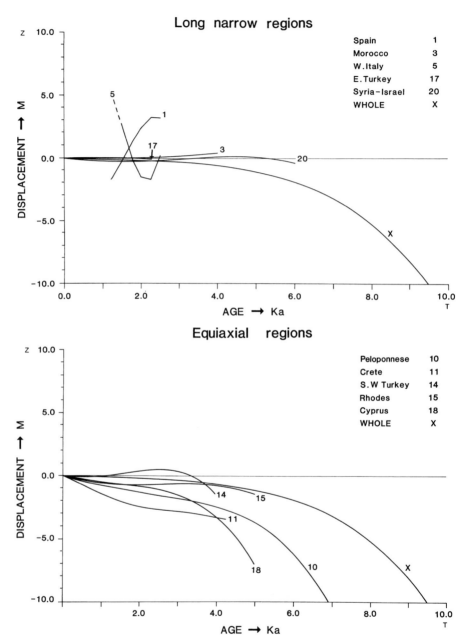

Fig. 8.2 Best-fit curves showing the relationship $Z = f(T^n)$, for site records for each region, after removal of geographically dependent components representing earth movements. Z = vertical displacement in metres; T = age in ka; A = curves for long narrow regions, uncorrected for transfer of linear component of rate of change between eustatic factor and earth movements; B = equi-axial regions, uncorrected for transfer of linear component of rate of change between eustatic factors and earth movement.

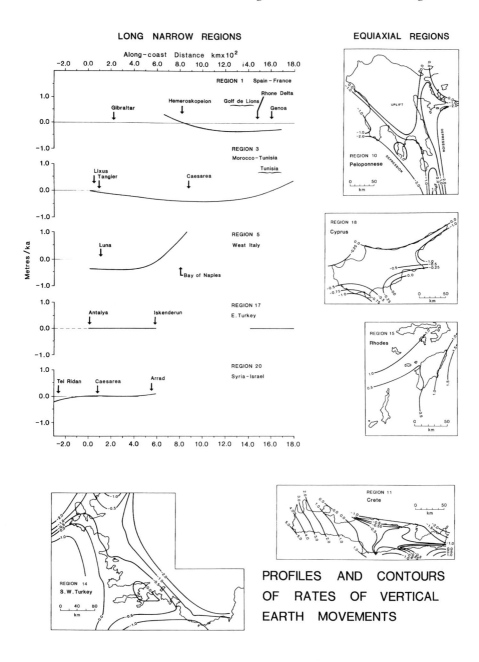

Fig. 8.3 Plots of regional earth movements, after removal of the eustatic component. For the long narrow regions the mean rate of displacement in m/ka is plotted along the coast as a function of distance in hundreds of kilometres. Distance is measured along the X-axis of the Region. For the equi-axial regions, plots are shown as contours of rates of vertical movements in m/ka.

the present study have been compared with the earlier results and are very consistent. Since the earlier work contained more detailed considerations than can be set out in the present work, the curves from Flemming (1978) are reproduced in Fig. 8.3.

The pattern of earth movements may be correlated with any of the following factors: plate boundaries in general; subduction zones; strike-slip faults; volcanism; continental shelf width/narrowness; continental marginal tilting; sedimentary isostatic loading; sediment compaction; aseismic warping. The factors listed above will be discussed briefly after the methods and results have been compared with those of other researchers.

3.1 Other Methodologies

Other field methods reveal sea-level data for the same time span as archaeological data. These include study of alluvial sediments (e.g. Kraft et al., 1975; Kraft et al., 1980; Gifford, 1978); study of solution notches (e.g. Hafemann, 1965; Flemming, 1978; Higgins, 1980; Guery et al., 1981; Pirazzoli et al., 1981); vermettus reefs and trottoir (Sanlaville, 1977; Paskoff et al., 1981; Paskoff and Sanlaville, 1983); and the study of organic materials dateable by C14 (Newman et al., 1980; Richards and Vita-Finzi, 1982; Marcus and Newman, 1983); and the study of low-level raised erosion terraces (e.g. Angelier, 1976; Le Pichon and Angelier, 1981; Peters, 1985).

Even if each of these methods is perfectable and produces internally consistent results, there is no reason why all the methodologies should calibrate identically with each other until proven. In the present study the only non-archaeological data used are when solution notches can be traced on cliffs between archaeological sites, and can be identified at some point intersecting archaeological structures with a before/after indication. All other types of indicator have been omitted. This may seem to be an arbitrary rejection of valid data, but in the present circumstances it is preferable to demonstrate that the restricted archaeological data set produces consistent results with a measurable error, and the merging of different data types can be carried out later when calibration is sure.

The different data types each have a bias towards certain sites of occurrence. Solution notches only occur on massive calcareous rocks; vermettus shelves only occur in the southeast Mediterranean; alluvial accumulations are strictly local forms. As an example of the relative bias which this can produce, one can cite the differences between most of the data reported in this chapter and that cited by Kraft and Aschenbrenner (1977). The data observed personally by the present author are restricted almost entirely to archaeological remains that can be seen with the naked eye, whether above or below present sea-level, and without excavation. This tends to exclude material buried in recent sediments, or even visible data separated from the present shoreline by prograding sediments, since there is seldom opportunity to level accurately across miles of marsh. The data are therefore biased slightly by the omission of such sites, probably amounting to a few per cent of the total number of sites (e.g. Oeniadae (541) and Ephesus (270), etc.).

Conversely, Kraft et al. (1975), Kraft and Aschenbrenner (1977) and Rapp and Kraft (1978) have gathered data almost exclusively from sites in alluvial and prograding areas. These zones in the Mediterranean are likely to occur in structural depressions which are themselves active grabens. Because the sediment may be liable to compaction, data from these areas would be expected to plot consistently below the data observed by (Flemming, 1968, 1978; Flemming et al., 1973), and this is indeed the case. With regard to their own data, Kraft and Aschenbrenner (1977, p. 39) state "Note the problem of the overwhelming variable tectonic effect". The purpose of this comparison is not to show that one or other method is incorrect, but to draw attention to systematic bias inherent in most samplir.g procedures that are dictated by methodology.

3.2 Correlation with Plate Boundaries

Figure 8.4 is a simplified diagram representing the location of plate boundaries in the Mediterranean, derived from the synthesis by Peters (1985) based on the work of several authors. The most extensive quiet zones revealed by the present study, Gibraltar to Genoa (Region 1) and Antalya to Iskenderun (Region 17), are both far from active plate boundaries. The Gulf of Sirte (Region 8) is probably also quiet, but there are insufficient data to be sure. Cyrenaica (Region 16) appears to be anomalous in view of its active depression, reversal of direction at Apollonia (949) and distance from the Hellenic subduction zone. However, it is probable that the compressive forces caused by the Mediterranean Ridge and the subduction process extend to the African foreland, causing tectonic activity (Biju-Duval, 1974).

Plate boundaries extend through northern Algeria, central Italy and the Yugoslav coast, all of which show some vertical activity. In general there is a broad correlation between proximity of a plate boundary and degree of vertical tectonic motion averaged over several thousand years.

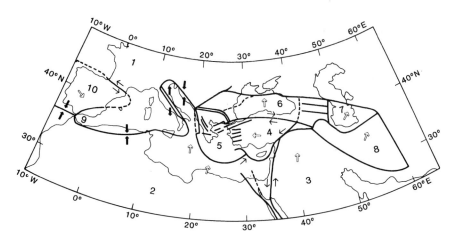

Fig. 8.4 Boundaries of plates and sub-plates simplified from Peters (1985), based on various authors. Arrows represent direction of movement. Numbers on plates as follows: 1 = Eurasian; 2 = African; 3 = Arabian; 4 = Turkish; 5 = Aegean; 6 = Black Sea; 7 = Caspian; 8 = Iranian; 9 = West Mediterranean; 10 = Iberian.

3.3 Correlation with Subduction Zones

There are two major arcuate subduction zones in the Mediterranean, the Calabrian Arc (Ghisetti and Vessani, 1982; Caputo, 1983) and the Hellenic Arc (Galanopoulos, 1973; Angelier, 1976, 1977, 1979; Dewey and Sengör, 1979; Pichon and Angelier, 1981; Peters, 1985). Neither arc is a simple subduction zone in the Pacific sense, since the radius of curvature is too sharp to allow for a coherent descending slab of lithosphere, and the proximity of approaching continents forces the back-arc area to develop anomalously (McKenzie, 1978; Dewey and Sengör, 1979). Notwithstanding these anomalies, the two areas do appear to be the principal zones of recent crustal consumption, and will be discussed as such.

The Hellenic Arc is associated with consistent evidence for rapid uplift on the outermost islands (Antikythera, Crete, Karpathos and Rhodes), the most rapid rates of movement observed anywhere in the Mediterranean (9.5 m in 2000 years, in south-western Crete). Pirazzoli et al. (1981) fitted a continuous contoured surface to the data for Antikythera and Crete to indicate a progressive distortion and tilt along the arc. At the suggestion of McKenzie (pers. comm.) Flemming tested computer-fitted regressions to the archaeological data for Antikythera and Crete, considering the data first as a single data set, and secondly as two data sets separated at the Antikythera Channel. The results are published by Flemming (1978). The residuals are much lower when the islands are considered separately, and it is therefore logical to consider them as de-coupled by the active faulting in the Antikythera Channel (see McKenzie, 1972, for evidence of recent seismicity). One concludes that the islands of the Hellenic Arc are broken into coherent slabs of typical dimension 50–100 km, which are tilting as monolithic de-coupled units in response to the underthrusting of the subduction zone. Peters (1985, p. 208) reports similar tilting on the Quaternary time-scale for the eastern region of Crete.

The evidence for the Calabrian Arc is less clear. Apart from the Messina earthquake of 1908, and the obvious volcanism of the Lipari Islands and the Bay of Naples, the majority of archaeological sites in the area fail to produce data. The Bay of Naples shows the greatest range of vertical movement, after south-western Crete, but this is associated with volcanism rather than the immediate effects of subduction (Günther, 1903a; Flemming, 1969).

The incidences of most rapid upward movement of the Mediterranean coast are thus from the arc areas, with rates of the order of 5 m/ka.

3.3.1 Correlation with strike-slip plate boundaries

The two major strike-slip plate boundaries are the Anatolian fault, running east–west across northern Turkey, and the Dead Sea fault, running north–south through the Gulf of Eilat/Aqaba and the Dead Sea. Neither fault is immediately adjacent to the Mediterranean coast, and the present survey has not included possibly helpful data from the Black Sea. The coast of Israel is moderately active, with vertical movements of 5 m at Caesarea (203/1051) and possibly 2 m at Dor (199) and 1 m at Acco (182). But these movements seem to be localized around the Carmel-Qishon Graben, rather than being direct effects of the plate boundary.

3.3.2 Correlation with local faulting

The southern and central Aegean probably should be regarded as an intensely faulted region, criss-crossed with normal extensional faults forming multiple horst-and-graben features (Flemming, 1978; Dewey and Sengör, 1979; Le Pichon and Angelier, 1981; Rotstein, 1985). The same model, in very broad terms, extends into the Gulf of Corinth (Brooks and Ferentinos, 1984; Vita-Finzi and King, 1985), where vertical movements at rates of 1–3 m per 1000 years are recorded for upward motion within 10 km of sites which have submerged at similar rates. Unfortunately, the present survey has revealed insufficient reliable data to construct a good model for the Aegean area. From such data one would expect to find upward and downward vertical displacement on adjacent islands, or on adjacent blocks separated by normal faults within an island. Flemming (unpublished report) noted a fault in the cliff at Vurgunda Karpathos (site 426) with a throw of 1 m, dislocating solution notches which were post-Roman. Similar but smaller faults can be seen in cliffs on Crete.

Other locations where vertical site displacement is associated with known faulting are Caesarea (203/1051), Dor (199) and Acco (182) near the Qishon Graben; probably Apollonia in Cyrenaica (949); Ruspina (127); Fethiye (289) to Kekova (302) south Turkey, Nora (160) southern Sardinia. King and Vita-Finzi (1981) analysed the fault structure associated with a modern seismic event near the coast of Algeria, just inland from Caesarea (151), which has the highest rate of subsidence in Region 3. In all cases the observed data indicate submergence, with the exception of one period at Dor. Vita-Finzi and King (1985) note that submergence is more easily detected than emergence even by contemporary observers of an earthquake event, and the same phenomenon is true of historical and archaeological data. Except for gross examples of uplifted harbour-works, uplift can go unreported. Nevertheless, the existing data do show a correlation of submergence with local faulting, although this may require further interpretation.

3.3.3 Correlation with modern and historical seismicity

Seismicity is associated with active faulting, plate boundaries and volcanism, so it is not independent of the other factors under discussion. The association is not exact in terms of the surface expression of these phenomena, and so direct evidence of seismicity should be reviewed.

The British Geological Survey provided a computer print-out map of modern seismicity of the Mediterranean showing earthquake events classified in terms of depth and magnitude. It is theoretically possible to compute the total number of large and small earthquakes expected over a prolonged period on the basis of recorded events over a few decades (Main and Burton, 1984), and hence to typify areas in terms of total energy release. Similar generalizations can be obtained from assessment of slip rates (North, 1974). For the present analysis, correlation was made subjectively with the BGS plots. It should be borne in mind that the seismicity refers only to the recent decades, and the distribution is not necessarily identical with the average over several thousand years.

Subjectively the correlation is evident. In the eastern Mediterranean, southern Turkey and northern Cyprus are aseismic and correlate with vertical stability. The Levant coast and Cyrenaica are moderate on both counts. The Hellenic Arc shows dramatically high rates of seismicity, with the greatest concentrations of large events in Rhodes and south-western Crete, where the highest rates of vertical movement are measured. We do not have sea-level data for the Ionian islands and Achaia, where similar seismicity is recorded. The vertical activity of the Gulf of Corinth correlates with that of high current seismicity.

As noted above, the data for the Calabrian Arc are not good enough to provide sure correlation, though this may be partly due to the occurrence of most epicentres offshore between the Lipari Islands and Naples. A belt of seismicity cuts across Tunisia and Algeria, intersecting the coast in the neighbourhood of the submerged sites of Sullectum (1027) and Caesarea (151). There is no recorded seismicity relating to the faults in southern Sardinia close to Nora (160), whilst the seismicity of south-east Spain is not correlated by data for vertical movement.

Although the present data set is not complete in some of the most interesting areas, observed relative vertical displacements apparently are correlated with local seismicity during recent years. Flemming (1978, p.444) showed that the data for the north-eastern Mediterranean correlate with the evidence from Ambraseys (1961, 1970) for historic seismicity. Historical seismicity for France (Vogt, 1979; Gagnepain-Beyneix et al., 1982, p. 274) and Italy (ENEL, 1977; Caputo, 1981) provide source material for further historical comparisons.

3.3.4 Correlation with continental shelf characteristics

The Mediterranean continental shelf is generally very narrow, with the exception of the Golfe du Lion, the northern Adriatic and the Gulf of Sirte. The narrowest shelf and deepest water are associated with the Hellenic Arc on the south side and, in this location, with high rates of vertical movements. In contrast, the very steep shelf off the French Côte d'Azur is associated with vertical stability.

The wide shelf of the Golfe du Lion is apparently subsiding; the Gulf of Sirte, at least in the region of Gabes, is associated with uplift; whilst the Adriatic is associated with submergence. Thus there is no general correlation.

If the coast is tilting about an axis parallel to the coast, high-sea-level indicators will be biased in one sense, and low-sea-level indicators in the opposite sense. Since the relief of the Mediterranean area is increasing, down-faulting or down-warping of the shelf relative to the hinterland is, on average, probable. On crenellated coasts, such as the west coast of Cyprus, the headlands will tend to show submergence (or at least lower rates of uplift) relative to the backs of the bays (Richards, 1984, field report). Contamination of the data due to coastal and shelf tilting is liable to introduce an average bias into sea-level studies, especially over periods longer than 5000 years. Many of the present data values may be determined by their location landward or seaward relative to a near-coastal axis of tilt.

The possible bias towards tilting of the continental margin is shown in some regions by the rapid subsidence of the outer end of the Cesme peninsula (sites 262 and 266); the subsidence of the headlands of western Cyprus (Richards, field report); subsidence and tilting of the shelf off Atlit

(193) Adler (1985); and the general doming pattern of the Peloponnese (Fig. 8.3).

3.3.5 Correlation with sedimentary isostatic loading

Sedimentary isostasy and sediment compaction would both tend to occur on large deltas and the large alluvial infilling of re-entrant bays and graben valleys. In locations where valid sites occur the correlation is confirmed, with depression below average eustatic curve in the Rhône Delta (sites 30, 31, 971, 978) and Po Delta (sites 588, 773, 1125, 1128). It is difficult to separate isostasy from compaction, but the data exist in some cases. For example, some Roman quarries in the Rhône area are on bedrock, where an isostatic factor might be detected independently of compaction. Sites in the Golfe du Lion (25, 969, 1006) are also submerged more than the adjacent coasts, which is compatible with long-term sediment accumulation and aseismic subsidence.

4 ANALYSIS OF TIDE-GAUGE DATA

Tide-gauge data have been used to estimate earth movements in Europe and the Mediterranean by Emery and Aubrey (1985), Pirazzoli (1987b), Emery et al. (1988) and Flemming and Woodworth (1988). Tide-gauge data provided by the Hellenic Navy (Flemming and Woodworth, 1988) show that the residual trend of earth movements, after removal of seasonal and interannual variability, is of the order of 5–20 mm/yr at 15 Greek ports, with movement of the coast as likely to be up as down. Pirazzoli (1987b, Table 5.1) provides data for 16 tide gauges in the Western Mediterranean and Adriatic, showing rates of sea-level change over 50–100 years averaging 0.3 to 2.3 mm/yr, all with relative subsidence of the coast with one exception. These data have not been corrected to separate earth movements from eustatic sea-level change.

Pirazzoli (1987b, p. 175) points out that the rate of rise indicated by the tide gauge at Marseille is much higher than the average over the last 2000 years, and that the rate of rise indicated by two closely dated archaeological features at Marseille approximately 2000 years ago also is higher than the average.

This effect, probably caused by aliasing, was studied explicitly by Flemming and Woodworth (1988). Decadel tide-gauge measurements of relative rate of change of sea-level are two to eight times the average rate over archaeological time intervals; archaeological estimates of average rates of vertical change exceed geological estimates averaged over 0.2–2.0 million years by a further order of magnitude. If a timeseries of relative sea-level change at a point on the coast consisted of a series of oscillations, with different frequencies the average rate of movement measured over different time-spans would tend to decrease with increased averaging interval. In theory the short-term rate could be in the opposite sense to the long-term average rate, and Flemming and Webb (1986) report sites showing reversal of direction. The great majority of sites with multi-purpose data show the same direction of movement on all time-scales. This suggests that the spectrum of vertical earth movements at a point is biased towards monotonic movement in one direction over time-scales of hundreds to tens of thousands of years. Reversals in direction are sufficiently short-lived not to show up often in tide-gauge records.

Flemming and Woodworth (1988) show that, in general, rates of vertical earth movement in the Aegean averaged over 2000 years are ⅙ x the rate of those measured at the same sites by tide gauges over 10–20 years. From this a general equation can be deduced for the region. Table 8.3. shows this correlation.

Emery and Aubrey (1985) deduce a rate of vertical movement from tide gauges on the Scottish coast that is only twice that based on Carbon-14 dating of beach deposits over several thousand years (Flemming, 1982). This suggests that the spectrum of variability of earth movements is much less noisy and has fewer high-frequency components for a region dominated by isostatic recovery. The process of isostatic post-glacial rebound is also probably slowing down.

The aliasing ratio for different regions gives a general indication of the reliability of trying to extrapolate from short records to longer periods.

Table 8.3

Assume equation $\dfrac{\text{Rate-}T}{\text{Rate-}0} = \left\{ \dfrac{T_0}{T} \right\}^{0.4}$

Where T = period of time over which rate of vertical movement is averaged, many tens to thousands of years; T_0 = a few years over which rate is measured by tide gauge; Rate $-T$ = rate of vertical movement averaged over time T; Rate -0 = rate of vertical movement averaged over a few years.

A. Ratios linking tide-gauge time-scales to archaeological time-scales

T_0 years	T years	Ratio = Rate-0/Rate-T
100	1000	2.51
100	2000	3.31
80	1000	2.73
80	2000	3.61
50	1000	3.31
50	2000	4.36
10	1000	6.3
10	2000	8.3

B. Ratios linking short tide-gauge time-scales to future multi-decade sea-level changes

T_0 years	T years	Ratio = Rate-0/Rate-T
10	40	1.74
10	50	1.90
10	60	2.04
10	70	2.18
10	80	2.30
10	90	2.41
10	100	2.51

The more high frequency components present, the higher the ratio. Large tectonic events separated by time-spans of the order of 500–1000 years would be very difficult to detect or predict.

Table 8.3a shows the ratios that would be expected by comparing tide-gauge records of several decades with archaeological periods of 1000 or 2000 years, if the equation for the Aegean were applied to all data. The ratios range from a minimum of 2.5 when 100-year tide-gauge data are compared to 1000-year archaeological data, through to 4.36 when 50-year data are compared with 2000-year data; and a maximum of 8.3 when 10-year data are compared with 2000-year data.

There is no region outside the Aegean where so many tide gauges can be compared accurately with co-located archaeological sea-level data. We therefore are reduced to comparing the average rate of displacement of regions over two time periods.

Region (1) (Flemming and Webb, 1986) contains seven tide gauges listed by Pirazzoli (1987b). The average length of tide-gauge records is 76 years, and the average relative displacement is 0.685 mm/yr relative rise of sea-level. The archaeological data are based on an average age of about 2000 years, and shows a relative subsidence of 0.23 mm/yr. The ratio is almost exactly 3.0. Table 8.3 shows that the aliasing ratio from 80 years to 2000 years would be 3.6 if the area were as active as the Aegean, which it is not.

Examination of the data from Pirazzoli (1987b) shows no other regions in which even this crude average comparison can be made with confidence. Region 5 contains only two reliable tide-gauge records, and Region 3 contains four. None of the available tide gauges co-locates exactly with an archaeological sea-level indicator site, and the distances of approximation become unacceptable.

Emery *et al.* (1988) list 29 tide gauges with records longer than 15 years, producing 18 records not included in Pirazzoli (1987b). Seven of these sites, in Israel, Turkey, and Italy, co-locate with archaeological data. The directions of movement correlate in every case, and the aliasing ratios vary from 2.1 to 11.0.

5 Predictions of Future Coastal Earth Movements

An irregular time series of earth movements is not attributable to a single physical cause, and therefore cannot be described by a simple spectrum relating amplitude or rate of movement to frequency. Nevertheless, rate of movement does correlate negatively with period. The constants in this empirical relationship would probably vary from region to region, with regions exhibiting different ratios between high-frequency and low-frequency events. This model suggests a way of predicting the probable rate of earth movements in each region during the next 50–100 years, even in the absence of long-term tide-gauge data.

The method of prediction, and the certainty, depends upon the type of data available locally and the known characteristics of the spectrum of earth movements in time and space. The different conditions of data availability are as follows:

1) tide-gauge data with 50–100 years of accurate well-calibrated data at the study site;
2) tide-gauge data with 50–100 years of accurate well-calibrated data available at a site within 10–20 km of the study site;
3) tide-gauge data with 10–20 years of well-calibrated data at the study site;
4) tide-gauge data with 10–20 years of modern well-calibrated data within 10–20 km of the study site;
5) archaeological estimate of relative sea-level change and deduced earth movements over 1000–2000 years available at the study site;
6) archaeological estimate of relative sea-level change and deduced earth movements over 1000–2000 years within 10–20 km of the study site;
7) no tide-gauge or archaeological data within 10–20 km of the study site.

Each condition above will be considered in turn:

1) A typical tide-gauge record of annual mean sea-level shows fluctuations that may be due to inter-annual variations in absolute mean sea-level, or faster and slower changes of land level. In order to determine the earth movement component, the global mean sea-level change over the same time-scale should be subtracted. The best estimate of this is subject to some debate (Gornitz et al., 1982; Barnet, 1984; Douglas, 1991) and I do not intend to recommend a best estimate. The residual earth movement then can be extrapolated for the same period into the future with reasonable confidence.

Various cross-checks should be applied. If the area is tectonically active and if archaeological data suggest that the long-term rate of change is faster than would be suggested by the general $1/T^{0.4}$ aliasing relationship, then some allowance should be made for the probability that rapid rates of movement can occur from time to time, which have not manifested themselves in the last 100 years but may occur in the next 100.

In all cases of tectonic movement it is possible for the direction of movement to reverse for periods of decades to centuries. Flemming and Woodworth (1988) show that for co-occurring pairs of tide gauges and archaeological data sites the direction of movement is the same in every case. The probability of reversal would be increased if the archaeological record provides clear evidence of reversals in the past (e.g. Matala (347), Caesarea (203/1051), Zimbule (416)), or if the average rate of movement is significantly less over the archaeological time-scale than suggested by applying the aliasing factor to the tide-gauge data.

If the cause of earth movement at a site is sediment compaction or isostatic subsidence under sediment loading, this probably will proceed at a more even and steady rate than tectonic subsidence. On the other hand, deltaic sediments may actually slump on rare occasions.

2) If there is no long tide-gauge record at a site, but one exists within 10–20 km, then the prediction can be extended laterally with reasonable confidence but some caveats. The lack of coherence between nearby sites of course is more marked in areas of extreme tectonism than in areas which are aseismic. The data provided by Flemming

(1969), Pirazzoli (1976b), Flemming (1978) and Flemming and Webb (1986) show that there are regions of the Mediterranean coast that are stable and laterally coherent (e.g. southern Turkey from Antalya to Iskenderun); regions that are moderately stable and coherent (e.g. Spain and southern France); regions that are inherently noisy and laterally incoherent (e.g. much of Greece and the Aegean coast); and finally regions that are unstable, but tilting or warping in coherent blocks, that can be contoured for trends in rate of vertical movement (e.g. Crete).

The factors mentioned above, together with knowledge of local geology, should be used to see if it is justifiable to extrapolate the predicted rate of earth movements from the site of a known tide gauge for a distance of 10, 20 or 30 km along the coast. Knowledge of local faulting, presence of alluvium, settlement of sediments in deltas, etc. should be taken into account on the basis of immediately local geological data.

3) A tide-gauge record of less than 20 years is relevant, but prediction from such a record for 50–100 years into the future is statistically less certain than with a longer record. The short record may exaggerate the rate of movement through aliasing, or fail to reveal movement if it is periodic with a period longer than the record. If a consistent rate of movement of the earth is detected over a period of the order of 10 years, it is likely to be an overestimate of the rate of movement to be expected over 50 years by a factor of 2, and an overestimate of the rate to be expected over 100 years by a factor of 2.5 (see Table 8.3B). These ratios apply to tectonically active areas, such as the Aegean, and probably the coast of Yugoslavia and Albania, and parts of Italy and Algeria. In areas of subsidence and sediment loading, the overestimate from short tide-gauge records is probably less exaggerated.

A short tide-gauge record should be cross-checked against co-located or nearby archaeological data if possible (see below).

4) If neither short-term nor long-term tide-gauge records are available on site, but a short-term tide-gauge record exists within 10–20 km, then further approximations must be adopted. Archaeological and geological data permit the region to be characterized in terms of local lateral coherence as in (2) above. The short-term tide-gauge estimate of the vertical rate of movement of the coast should be corrected for the extrapolation in time on site, and then for the extrapolation laterally along the coast. Archaeological data should be used wherever possible to cross-check the tide-gauge prediction, using appropriate correction factors from Table 8.3.

5) If there are archaeological data on site, then a fairly accurate assessment should be available of the vertical rate of earth movement averaged over 1000–2000 years. This average figure, in some cases, will be accurate to better than 0.2 mm per year. The problem is that the average can severely underestimate possible short-term movements, and therefore the archaeological indicated rate should be increased by a conversion factor (Table 8.3). The conversion factor is derived from the Aegean, and therefore should be conservatively safe for other areas.

6) If there are no archaeological data available at the site, but such data exist within 10–20 km, then the regional pattern should be studied, as in (2) above. Flemming (1972, 1978) and Flemming and Webb (1986) show pronounced regional patterns in crustal tilting, so that the rates of vertical movement can be interpolated with some confidence between sites in some areas. The interpolated values then should be corrected for aliasing, as in (5) above.
7) If there are no tide-gauge data and no archaeological indicators of sea-level within 10–20 km of a site, the prediction becomes much more subjective and less certain. This is, however, the most general case. Examination of regional geological data, seismicity, faults, alluvium, etc. provides a general view of the expected level of vertical movement in the region within say 100–200 km of the site. Archaeological sea-level data within the region, cross-checked with tide-gauge data where possible, will confirm the general picture. Publication by Vita-Finzi and King (1985), Le Pichon and Angelier (1981), Flemming and Webb (1986), Sanlaville (1977), Paskoff and Sanlaville (1981) and others provide general information on the regional rates of crustal deformation, tilt, faulting, etc.

6 CONCLUSIONS AND RECOMMENDATIONS

1) Tide-gauge data are the most accurate source of information on vertical coastal land movements, but a record of at least 10–20 years is needed before a reasonable prediction on the 50–100 year time-scale can be made; the record should preferably be of the order of 50–100 years. There are only 16 published long records, and 15 short records in the Mediterranean Action Plan area, with adequate quality control.
2) Because of the lateral variability in coastal earth movements, data at one site can only be interpolated or extrapolated laterally for a distance of the order of 10–20 km. If a large number of data sites is available for a region, then deduction of trends, tilts, general stability, etc. enables more certain prediction between data points, but even then local anomalies cannot be ruled out.
3) There are 335 good estimates of sea-level change based on archaeological data over the 1000–2000 year time-scale, with an average spacing for the whole Mediterranean Action Plan area of 134 km. In practice the regions with good archaeological data are much more densely sampled, whereas some regions are almost totally without data. If the coasts of Jugoslavia, Albania, Egypt and the north and central Aegean are excluded, the published archaeological data provide a sampling density of one site per 75 km of coast. This is quite close to an adequate sample.
4) If a predictive study is urgently needed to cover the whole coast of the Mediterranean Action Plan area, the following steps are recommended:
   a) Obtain further tide-gauge data from ports where records exist but have not been published, e.g. Ashdod, Haifa, and others.
   b) Calculate the ratios of rates of movement for co-located tide-gauge

and archaeological data for other regions than Greece, so as to find the correction ratios to be applied in different regions.

c) Obtain archaeological-historical data for the poorly documented regions. A regional study of a national coastline several hundred kilometres long can be conducted in one or two years, and is not expensive. Although the data are not as accurate as tide-gauge data, they are available almost immediately, whereas a tide-gauge installed now will have to be maintained for at least 10 years before it provides valid data on coastal earth movements.

5) The methodology suggested in this chapter has been worked out rather quickly and could be considerably refined. The data sets available for each region should be carefully assessed, and the best regional principles derived for prediction of the vertical rate of movement predicted over the next 50–100 years, with confidence limits.

6) The precise gradient of the shore at each point should be recorded in a database so that the extent of land flooded can be calculated for each additional 10 cm rise of relative sea-level, summed for eustatic and earth movement causes.

7 GENERAL NOTE

The present author has worked personally on over 300 coastal archaeological sites during the last 30 years. Several sites have been revisited at intervals during that period. The rate of destruction of coastal and submerged archaeological sites from both natural and human causes is considerable. Whilst some sites are probably well protected for natural reasons, and have remained unchanged for centuries, others have been seriously damaged in 1–10 years. The submerged classical city of Apollonia in Libya was severely damaged by winter storms between 1958 and 1959; the Bronze Age walls and Classical Theatre at Plitra-Asopos in Greece were destroyed by wave action between 1968 and 1979. Neolithic remains near Atlit (Israel) are exposed by natural sand movements each year on the sea bed, and organic materials are damaged when exposed. In other areas, classical slipways, harbour works and buildings in shallow water have been damaged or destroyed by the construction of coastal roads, yacht marinas and land reclamation.

If the unique record of the origins of marine technology in the Mediterranean is to be preserved for humanity, it is recommended that most threatened sites should be identified and a decision taken as to whether they should be protected, or whether detailed surveys and records should be made.

8 REFERENCES

Adler, E., 1985. The submerged Kurkar ridges off the northern Carmel coast. MA Thesis, University of Haifa, Israel, p. 106 + maps.

Ambraseys, N.N., 1961. The seismic history of Cyprus. *Revue pour l'étude des Calamitées*, Geneva, 1–26.

Ambraseys, N.N., 1970. Value of historical records of earthquakes. *Nature*, London, **232**, 375–379.

Ambraseys, N.N., 1975. Studies in historical seismicity and tectonics: Near and Middle East. In: Brice, W. (ed.), *Historical Geography of the Middle East*, Academic Press, London.
Angelier, J., 1976. La néotectonique cassante et sa place dans un arc insulaire: l'arc Egéen meridional. *Bull. Soc. Geol. Fr.* (7), **18**, 1257–1265.
Angelier, J., 1977. Sur les movements égéens depuis le Miocène supérieur: l'évolution récente de la courbure sud-hellénique (Grèce). *C.r. hebd. Seanc. Acad. Sci.* Paris (D) **284**, 1037–1040.
Angelier, J., 1979. Néotectonique de l'arc égéen. *Soc. Géol. Nord.*, **3**, p. 418.
Barnett, T.P., 1984. The estimation of "global" sea-level change: A problem of uniqueness. *J. Geophys. Res.*, **89**, 7980–7988.
Biju-Duval, B., 1974. Tertiaire du domaine Mediterranéene, 1:25 million (map). Paris: Institut Franais du Petrole, CNEXO.
Blackman, D.J., 1973. Evidence of sea-level change in ancient harbours and coastal installations. *Colston Pap.* **23**, *Marine Archaeol.*, 114–137.
Blackman, D.J., 1982. Ancient harbours in the Mediterranean. Part 1. *Int. J. Naut. Arch.*, **11**, 79–104.
Blackman, D.J., 1982. Ancient harbours in the Mediterranean. Part 2. *Int. Jour. Naut. Arch.*, **11**, 185–212.
Brooks, M. and Ferrentinos, G., 1984. Tectonics and sedimentation in the Gulf of Corinth and the Zakynthos and Keffalinia channels, western Greece. *Tectonophysics*, **101**, 25–54.
Caputo, M., 1981. Study of the ENEL catalogue of Italian earthquakes from 1000 through 1975. *Rass. Lav. Publ.*, **2**, 3–16.
Caputo, M., 1983. The occurrence of large earthquakes in south Italy. *Tectonophysics*, **99**, 73–83.
Dewey, J.F., Pitman, W.C. and Ryan, W.B.F., 1973. Plate tectonics and the evolution of the Alpine system. *Geol. Soc. Am. Bull.*, **84**, 3137–3180.
Dewey, J.F. and Sengör, A.M., 1979. Aegean and surrounding regions: complex multiplate and continuum tectonics in a convergent zone. *Geol. Soc. Am. Bull.*, **90**, 84–92.
Douglas, B.C., 1991. Global sea level rise. *J. Geophys. Res.*, **96**, 6981–6992.
Emery, K.O., and Aubrey, D.G., 1985. Glacial rebound and relative sea-levels in Europe from tide-gauge records. *Tectonophysics*, **120**, 239–255.
Emery, K.O., Aubrey, D.G. and Goldsmith, V., 1988. Coastal neo-tectonics of the Mediterranean from tide-gauge records. *Marine Geol.*, **81**, 41–52.
ENEL, 1977. Catalogue of Italian earthquakes from the year 1000 through 1975, in tape form. (cited by Caputo, 1981, q.v.).
Fairbridge, R.W., 1961. Eustatic changes in sea-level. In: *Physics and Chemistry of the Earth*, Pergamon Press, London, **4**, 99–185.
Flemming, N.C., 1968. Holocene earth movements and eustatic sea-level in the Peloponnese. *Nature*, London, **217**, 1031–1032.
Flemming, N.C., 1969. Archaeological evidence for eustatic changes of sea-level and earth movements in the Western Mediterranean in the last 2000 years. *Spec. Pap. Geol. Soc. Am.*, **109**, 1–125.
Flemming, N.C., 1972. Eustatic and tectonic factors in the relative vertical displacement of the Aegean coast. In: Stanley, D.J. (ed.), *The Mediterranean Sea*, Stroudsberg, Dowden, Hutchinson and Ross, 189–201.
Flemming, N.C., Czartoryska, N.M.G. and Hunter, P.M., 1973. Archaeological evidence for eustatic and tectonic components of relative sea-level change in the South Aegean. *Colston Pap.* **23**, *Marine Archaeol.*, 1–63.
Flemming, N.C., 1978. Holocene eustatic changes and coastal tectonics in the northeast Mediterranean: Implications for Models of Crustal Consumption. *Phil. Trans. Roy. Soc. London*, A. **289**, 405–458.

Flemming, N.C., Raban, A. and Goetschel, C., 1978. Tectonic and eustatic changes on the Mediterranean coast of Israel in the last 9000 years. *Progress in Underwater Sciences*, Pentech Press, London, **3**, 33–93.

Flemming, N.C., 1979. Archaeological indicators of sea-level. *Oceanis, Journal of the Institut Océanographique*, Paris, **5**, 149–166.

Flemming, N.C., 1982. Multiple regression analysis of earth movements and eustatic sea-level change in Britain in the last 9000 years. IGCP Report Volume, IGCP-Project–61, M.J. Tooley (ed.), Proc. Geol. Assoc. **93**, Vol 1, 113–125.

Flemming, N.C., 1983. Survival of submerged lithic and Bronze Age artifact sites: A Review of Case Histories. In: Masters, P.M. and Flemming, N.C. (eds.), *Quaternary Coastlines and Marine Archaeology*, 133–173, Academic Press, London, p. 641.

Flemming, N.C. and Webb, C.O., 1986. Tectonic and Eustatic coastal changes during the last 10,000 years derived from archaeological data. *Zeitschrift. Geomorf. N.F. Supl. Bd.* **62**, 1–29.

Flemming, N.C. and Woodworth, P.L., 1988. Monthly mean sea-levels in Greece 1969–1983 compared to relative vertical land movements measured over different timescales. *Tectonophysics*, **148**, 59–72.

Frost, H., 1972. Ancient harbours and anchorages in the eastern Mediterranean. In: *Underwater Archaeology, A Nascent Discipline*, (Unesco), 95–114, Published by the United Nations, Unesco, Paris, p. 306.

Galanopoulos, A.G., 1973. Plate tectonics in the area of Greece as reflected in the deep focus seismicity. *Ann. Geofis.*, **26**, 84–105.

Gagnepain-Beyneix, J., Haessler, H. and Modiano, T., 1982. The Pyrenean earthquake of February 29, 1980: An Example of Complex Faulting. *Tectonophysics*, **85**, 273–290.

Ghisetti, F. and Vessani, L., 1982. Different styles of deformation in the Calabrian Arc (southern Italy): Implications for Seismotectonic zoning. *Tectonophysics*, **85**, 149–165.

Gifford, J.A., 1978. Paleogeography of archaeological sites of the Larnaca lowlands, southeastern Cyprus. PhD thesis dissertation, University of Minnesota, p. 192.

Gnirs, A., 1908. Beobachtung über den Fortschritt einer säkularen Niveauschwankgung des Meeres wahrend der letzten zwei Jahrtausend. *Geogr. Gesell. Wien Mitt.*, **51**, 1–56.

Gornitz, V., Lebedeff, S. and Hansen, J., 1982. Global sea-level trend in the past century. *Science*, **215**, 1611–1614.

Graaf, J., 1980. An investigation of the frequency distributions of annual sea level maxima at ports around Great Britain. *Estuarine Coastal and Shelf Science*, **12**, 389–450.

Guery, R., Pirazzoli, P.A. and Trousset, P., 1981. Les variations du niveau de la mer depuis l'antiquité à Marseille et à la Couronne. *Histoire et Archéologie*, **50**, 8–27.

Günther, R.T., 1903a. Earth movements in the Bay of Naples. *Geogrl. J.*, **22**, 121–269.

Günther, R.T., 1903b. The submerged Greek and Roman foreshore near Naples. *Archaeologia*, **58**, p. 62.

Gutenberg, B. and Richter, C.F., 1954. *Seismicity of the Earth*. Princeton University Press, Princeton, NJ, p. 310.

Hafemann, D., 1960. Anstieg des Meerespiegels in Gesichtlicher Zeit. *Umschau*, **60**, 193–196.

Hafemann, D., 1965. Die Niveauveränderungen an den Küsten Kretas seit dem Altertum. *Abh. math.-naturw. Kl. Akad. Wiss. Mainz.*, **12**, 608–688.

Higgins, C.G., 1980. Nips, notches and the solution of coastal limestone: An Overview of the Problem, with Examples from Greece. *Estuarine and Coastal Marine Science*, **10**, 15–30.

King, G.C.P. and Vita-Finzi, C., 1981. Active folding in the Algerian earthquake of 10 October 1980. *Nature*, **292**, 22–26.
Kraft, J.C. and Aschenbrenner, S.E., 1977. Paleogeographic reconstruction in the Methoni enbayment in Greece. *J. of Field Archaeology*, **4**, 19–44.
Kraft, J.C., Kayan, I. and Eroc, O., 1980. Geomorphic reconstruction in the environs of ancient Troy. *Science*, **209**, 776–782.
Kraft, J.C., Rapp, G. and Aschenbrenner, S.E., 1975. Late Holocene Paleogeography of the coastal plain of the Gulf of Messenia, Greece, and its relationship to archaeological settings and coastal change. *Geol. Soc. Am. Bull.*, **86**, 1191–1208.
Le Gall, J. (ed.), 1981. Ports et villes engloutis. *Histoire et Archéologie*, **50**, 5–87.
Lehman-Hartleben, K., 1923. Die antiken Hafenanlagen des Mittlemeers. *Klio. Monograph*, **14**.
Le Pichon, X. and Angelier, J., 1981. The Aegean Sea. *Phil. Trans. R. Soc. London*, **300**, 357–372.
McKenzie, D.P., 1972. Active tectonics of the Mediterranean region. *Geophys. J.R. Astr. Soc.*, **30** (2), 109–185.
McKenzie, D.P., 1978. Active tectonics of the Alpine-Himalayan belt: The Aegean Sea and Surrounding Areas. *Geophys. J.R. Astr. Soc.*, **55**, 217–254.
Main, I.G. and Burton, P.W., 1984. Information theory and the earthquake frequency-magnitude distribution. *Bull. Seism. Soc. Am.*, **74**, 1409–1426.
Marcus, L.F. and Newman, W.S., 1983. Hominid migrations and the eustatic sea level paradigm: A Critique. In: Masters, P.M. and Flemming, N.C. (eds), *Quaternary Coastlines and Marine Archaeology*, 63–85, Academic Press, London, p. 641.
Masters, P.M. and Flemming, N.C. (eds), 1983. *Quaternary Coastlines and Marine Archaeology*. Academic Press, London, p. 641.
Mörner, N.A., (ed.), 1980. *Earth Rheology, Isostasy and Eustasy*. John Wiley, New York.
Negris, P., 1904. Vestiges antiques submergées. *Athenischer Mitt.*, **29**, 230–363.
Newman, W.S., Marcus, F., Pardi, R.R., Paccione, J.A. and Tomacek, S.M., 1980. Eustasy and deformation of the geoid: 1,000–6,000 radiocarbon years BP. In: Mörner, N.A. (ed.), *Earth Rheology, Isostasy and Eustasy*, Wiley, New York, 555–567.
North, R.G., 1974. Seismic slip rates in the Mediterranean and Middle East. *Nature*, **252**, 560–563.
Paskoff, R., Trousset, P. and Dalongeville, R., 1981. Sur les côtes de Tunisie, la montée des eaux depuis 2000 ans. *Histoire et Archéologie*, **50**, 52–59.
Paskoff, R. and Sanlaville, P., 1983. Les Côtes de la Tunisie, variations du niveau marin depuis le Tyrrhénien. *Maison de l'Orient Meditérranéen*, **14**, Lyon, 192 pp.
Peters, J.M., 1985. Neogene and Quaternary vertical tectonics in the south Hellenic arc and their effect on concurrent sedimentation processes. University of Amsterdam, GUA Papers of Geology, Series 1, No. 23. Drukkerij Elinkwijk BV, Utrecht, 247 pp.
Pirazzoli, P.A., 1976a. Sea-level variation in the Northwest Mediterranean during Roman times. *Science*, **194**, 519–521.
Pirazzoli, P.A., 1976b. Les variations du niveau marin depuis 2000 ans. *Memoire de Laboratoire de Géomorphologie de l'Ecole Pratiques des Hautes Etudes*, No. 30, 1–421, Dinard.
Pirazzoli, P.A., 1987a. Recent sea-level changes and related engineering problems in the Lagoon of Venice (Italy). *Proceedings in Oceanography*, **18**, 323–346.
Pirazzoli, P.A., 1987b. Sea-level changes in the Mediterranean. In: *Sea-level Changes*. Basil Blackwell, Oxford, 152–181.
Pirazzoli, P.A., Thommeret, Y., Laborel, J. and Montaggioni, L.F., 1981. Les rivages emergés d'Antikythera (Cerigotto): corrélations avec la Crète occidentale et

implications cinématiques et géodynamiques. CNRSUniversité de Paris, Institut de Géographie, 49–65.
PSMSL, 1976. Monthly and annual mean heights of sea-level. Permanent Service for Mean Sea Level, UNESCO-FAGS. Printed by Institute of Oceanographic Sciences, Bidston, England, computer printout, about p. 1000.
Pugh, D.T., 1987. Tides, surges and mean sea-level: a handbook for engineers and scientists. John Wiley & Sons, Chichester, p. 472.
Raban, A., 1981. Recent maritime archaeological research in Israel. *Int. Jour. Naut. Arch.*, **10**, 287–308.
Raban, A., 1983. Submerged prehistoric sites off the Mediterranean coast of Israel. In: Masters, P.M. and Flemming, N.C. (eds.), *Quaternary Coastlines and Marine Archaeology*, 215–232, Academic Press, London, p. 641.
Rapp, G. and Kraft, J.C., 1978. Aegean sea-level changes in the Bronze Age. In: Doumas, C. (ed.), *Thera and the Aegean World*, 183–194, Thera and the Aegean World, London, p. 813.
Richards, G.W. and Vita-Finzi, C., 1982. Marine deposits 35,000 to 25,000 years old in the Chott el Djerid, southern Tunisia. *Nature*, **295**, 54–55.
Rickards, L.J., 1985. Report on sea-level data collected during the MEDALPEX experiment from 1st September 1981 to 30th September 1982. Intergovernmental Oceanographic Commission, UNESCO, Paris, p. 170.
Rossiter, J.R., 1967. An analysis of annual sea-level variations in European waters. *Geophys. J. Roy. Astr. Soc.*, **12**, 259–299.
Rotstein, Y., 1985. Tectonics of the Aegean block: rotation, side arc collision and crustal extension. *Tectonophysics*, **117**, 117–137.
Sanlaville, P., 1977. Etude géomorphologique de la region littorale du Liban. Public Univ. Libanaise, Sect. Et. Geogr., Beyrouth, p. 859.
Schmiedt, G., 1972. Il livello antico del mar Tirreno. Leo Olschki, Firenze, p. 323.
Spratt, T.A.B., 1865. Travels and Researches in Crete. Vols 1 and 2, J. van Vorst, London.
Striem, H.L., 1974. Storm surges and unusual sea-levels on Israel's Mediterranean coast. *Int. Hydrogr. Rev.*, **51**, 59–70.
Walcott, R.I., 1972. Past sea-levels, eustasy and deformation of the Earth. *Quaternary Res.*, **2**, 1–14.
Warwick, R. and Oerlemans, J., 1990. Sea Level Rise, pp. 260–261. In: Houghton, J.T., Jenkins, G.J. and Ephraums, J.J. (eds.) *Climate Change, The IPCC Scientific Assessment*, Cambridge University Press, p. 365.
Van de Plaasche, O., 1986. (ed.). Sea-level research: a manual for the collection and evaluation of data. (A contribution to projects 61 & 200 of IGEP), Geo-books, Norwich, p. 618.
Vita-Finzi, C. and King, G.C.P., 1985. The seismicity, geomorphology and structural evolution of the Corinth area of Greece. *Phil. Trans. R. Soc. Lond., A.*, **314**, 379–407.
Vogt, J., 1979. Les tremblements de terre en France. *Mem. Bur. Rech. Geol. Min.*, **96**.
Vrsalović, D., 1974. Istrazivanja i zastita podmorskih arheoloskih spomenka u S.R. Hrvatskoj. Republicki Zavod za Zastitu Spomenika Kultutr, Zagreb, p. 245.
Yorke, R.A. and Dallas, M.F., 1968. Cambridge Illyricum Expedition. Expedition report, p. 29.
Yorke, R.A. and Davidson, D.P., 1969. Roman harbours of Algeria. Maghreb Project 1968. Expedition report, p. 42.
Yorke, R.A., Davidson, D.P. and Little, J.H., 1973. Pentapolis Project 1972: A Survey of Ancient Harbours in Cyrenaica, Libya. Expedition report, p. 12.

9

Implications of a Future Rise in Sea-Level on the Coastal Lowlands of the Mediterranean

S. Jelgersma
(Geological Survey of the Netherlands, Haarlem, The Netherlands)

G. Sestini
(Applied Earth Science Consultant, London, UK

Abstract

The physical characteristics, state of population and land use of the Mediterranean low-lying coastal areas are reviewed in the context of the impacts of sea-level rise. Many coasts, especially near deltas, are presently subject to erosion and retreat, largely because of reduced sediment supply (caused by dams, and by river-bed and beach-sand mining) and the manipulation of beaches by fixed constructions. The formerly extensive wetlands have been much reduced and continue to be threatened by human activities.

Sea-level rise prediction in the Mediterranean region is marred by the controversial evidence for rise in the last decades, because of insufficient tidal gauge measurements, and because of tectonic activity (subsidence versus uplifted parts). Nevertheless, a possible future relative rise of sea-level (locally) of up to 100 cm would effect important physical changes to shorelines, sand dunes, marshes, lakes and lagoons. It would increase the cost of river and tidal flood protection, as well as adjusting harbour and coastal defence structures. Saline intrusion would seriously affect lagoons and wetlands, and all recently reclaimed lands. Sea-level rise impacts could be minimized, however, by adequate land-use planning. Multiple scenarios of possible changes should be considered in the current design of all coastal infrastructures.

A set of urgent research options include the collection of data on sea-levels, on coastal dynamics, delineation of local geological setting and the identification of risk levels. Particularly stressed are coastal zone planning, of controlling land reclamation, and of suspending all dumping of toxic substances near the shores.

1 Introduction

This contribution is a general review of the physical changes that could affect the shorelines, sand dunes, wetlands, lakes and lagoons at the margins of the Mediterranean Sea, and of the ensuing problems, in consequence of a significant future rise of sea-level. The coastal regions mentioned are shown in Fig. 9.1. More detailed and specific information can be found in the case study reports in this volume dealing with the deltas of the Ebro, Rhône, Po, Axios and Nile rivers.

An increasing body of evidence suggests that in the coming decades a global warming due to the greenhouse effect will lead to a substantial rise in sea-level. Estimates for the next 100 years range from 0.5 to 3.5 m. This

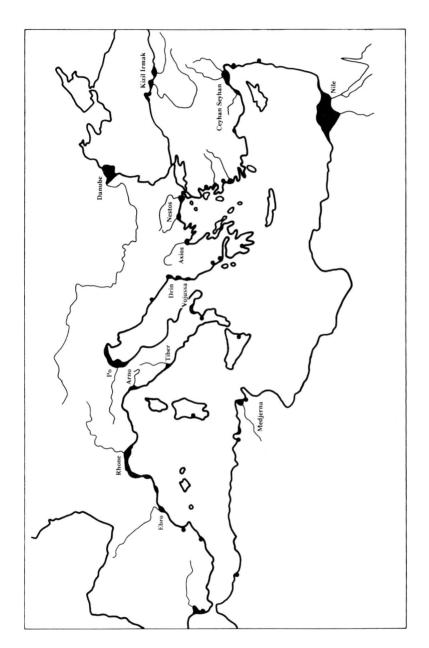

Fig. 9.1 Map of the Mediterranean, indicating in black the most important coastal lowlands.

range is due to the fact that the input to the models has many uncertainties. G. de Q. Robin (1985) summarized the problem as follows: "With our lack of knowledge of certain aspects of the hydrological cycle and the dynamics of the oceans and the polar ice sheets, forecasting of global changes of sea-level involves considerable extrapolation and speculation."

Because a considerable part of the population of maritime countries lives in low-lying areas near the sea, a rise between 0.5 and 3.5 metres would have an important impact on society. The tourist industry, so extensively developed on the Mediterranean coasts, has interfered with and caused the destruction of natural shoreline defences, such as sand dunes, for the provision of hotel accommodation and other facilities.

Most lowland coastal areas, especially deltas, are subject to a degree of slow tectonic subsidence that will accentuate the predicted sea-level rise. In some deltaic areas with dense population and heavy industrial development, additional human-induced land subsidence due to the overdrawing of ground water has become a serious problem (i.e. Venice, Ravenna, Tokyo, Bangkok, Mexico City etc.; case studies in Poland, 1984). Subsidence also has been caused by the draining of land and the making of polders.

During the past few years, five conferences have been devoted, specifically or in part, to the entity and the impacts of a future sea-level rise. The first was the international conference on the assessment of the role of carbon dioxide and of other greenhouse gases in climate variation and associated impacts in Villach, Austria, in October 1985 (UNEP, ISCSU and WMO) (Bolin *et al.*, 1986). The second was an international conference on the Health and Environmental effects of changes in stratospheric ozone and global climate, organized by the US Environmental Protection Agency and UNEP in June 1986 at Crystal City, USA, which included a workshop on sea-level rise (Titus, 1986). In August 1986 the Delft Hydraulics Laboratory organized an international workshop in The Netherlands on the impact of sea-level rise on society (Wind, 1987). The fourth conference was a workshop on climatic change, sea-level, severe tropical storms and associated impacts held at the Climatic Research Unit of the University of East Anglia, in Norwich, UK, in September 1987. The fifth, a session on the impact of a future rise in sea-level on the European Coastal Lowlands during the European Workshop on Interrelated Bioclimatic and Land-use Changes in The Netherlands, Noordwijkerhout, in October 1987 (Jelgersma, 1987).

The general conclusions on the causes and the effects of a future sea-level rise that were reached at the UNEP–EPA workshop are given below:

1) *Causes*
a) *The projected global warming would accelerate the current rate of sea-level rise* by expanding ocean water, melting alpine glaciers, and eventually causing polar ice sheets to melt or slide into the oceans.
b) *Global average sea-level has risen 10 to 15 cm over the last century.* Ocean and glacial studies suggest that the rise is consistent with what models would project, given the 0.4°C warming of the past century. However, no cause and effect relationship has been conclusively demonstrated.
c) *Projected global warming could cause global average sea-level to rise 10 to 20 cm*

by 2025 and 50 to 200 cm by 2100. Thermal expansion could cause a rise of 25 to 80 cm by 2100; Greenland and alpine glaciers could each contribute 10 to 30 cm through 2100. The contribution of Antarctic deglaciation could be between 0 and 100 cm; however, the possibility cannot be ruled out that (i) increased snowfall could increase the size of the Antarctic ice sheet, thereby offsetting part of the sea-level rise from other sources; or (ii) meltwater and enhanced calving of the ice sheet could increase the contribution from Antarctica as much as 2 m.

d) *The disintegration of the West Antarctic ice sheet could raise sea-level an additional 6 m over the next few centuries.* Glaciologists generally believe that such a disintegration would take at least 300 years, and probably as long as 500 years. However, a global warming might result in sufficient thinning of the Ross and Filcher-Ronne ice shelves in the next century to make the process irreversible.

e) *Local trends of land subsidence or emergence must be added to or subtracted from the estimates of the rise of sea-level at particular locations.*

2) *Effects*

a) *A substantial rise in sea-level would eventually invade wetlands and lowlands, accelerate coastal erosion, aggravate coastal flooding, and increase the salinity of estuaries and aquifers.*

b) *Bangladesh, Egypt, China and the estuaries of north-west Europe appear to be among the regions most vulnerable to the rise of sea-level in the next century.* Up to 20% of the land in Bangladesh could be flooded by a 2 m rise in sea-level. Although less than 1% of Egypt's land would be threatened, over 20% of the Nile Delta, which contains about half of the nation's people, would be involved.

c) *A large fraction of the world's coastal wetlands could be lost, with negative effects on fisheries.* A rise in sea-level of 1–2 m by 2100 could destroy 50–80% of the present-day US coastal wetlands. Although no study has been made to estimate the worldwide impact, this figure is probably representative.

d) *Erosion caused by sea-level rise could threaten recreational beaches throughout the world.* Case studies have concluded that a 30 cm rise in sea-level would result in beaches retreating 20 to 60 metres or more. Because the first row of houses or hotels is often generally less than 20 metres from the shore at high tide, the recreational beaches throughout the world would be seriously threatened by a 30 cm rise unless major beach preservation efforts are undertaken or existing defences improved.

e) *Sea-level rise would increase the costs of flooding, flood protection and flood insurance in coastal areas.* Flood damage would increase because higher water levels would provide a higher base for storm surges; erosion would increase the vulnerability to storm waves; and decreased natural and artificial drainage would increase flooding during rainstorms.

f) *Future sea-level rise should be included in the planning and design of coastal drainage and flood-protection structures.*

g) *Some adverse impacts of sea-level rise could be ameliorated through anticipatory land-use planning and structural design changes.*

h) *Increased salinity from sea-level rise may convert marshes and swamps to open water and threaten drinking water supplies.*

i) *Other consequences of global warming might offset or aggravate the impacts of sea-level rise.* Increased droughts might amplify the salinity impacts of sea-level rise. Increased hurricanes and storm surges, and rainfall could amplify flooding in coastal areas. Warmer temperatures might enable mangrove swamps – which can accrete vertically more rapidly than salt marshes – to advance into higher latitudes, perhaps offsetting wetland loss caused by sea-level rise.

As mentioned above, the estimates of the amount of a future rise in sea-level during the next 100 years vary greatly due to our lack of knowledge of various physical aspects. However, the main conclusion of both the Villach and Norwich conferences were that the maximum amount of sea-level rise during the coming 100 years could be in the order of 1 m. Further modelling of the involved parameters that global average sea-level might be 20 cm higher by 2025, 65 cm higher by the late 2000s (Warwick an Oerlemans, 1990).

In this study a general evaluation will be made of the likely impact of a projected sea-level rise of 1 m on the coastal lowlands that occur around the Mediterranean Sea. It can be expected that further modelling of the involved parameters will change this amount of expected sea-level rise.

2 Environment, Land Use and Shoreline Behaviour

2.1 Environment of the Coastal Lowland

Coastal lowlands are influenced directly by waves, tides and currents, as well as indirectly by the interaction of salt and fresh water. Coastal lowland sediments include peats, clays, silts and sands, generally lying in the intertidal and subtidal zone. Inland coastal wetlands show a zonation from salt and brackish-water environments, marshes and lagoons to fresh-water peats.

The salt-water environment of the supratidal zone is strongly influenced by climate, particularly rainfall. In general, the Mediterranean climate is characterized by long, dry warm summers and by rainy winters. In the southern and eastern Mediterranean the climate is warmer and precipitation less than in the coastal zones of the northern Mediterranean. Accordingly while the latter areas have brackish marshes and lagoons, the former have evaporites and hyper-saline lagoons with desiccated mud flats and algal mats (sebkhas).

In many areas the coastal zone is bordered by low dunes. Due to climatic circumstances and to the presence of calcite, they can be cemented; eolianites, are found in Israel and West of the Nile Delta.

The Mediterranean shorelines include important river deltas such as the Ebro, Rhône, Po, Seyhan and Nile. They show great morphological variations due to the controlling processes of waves, currents and fluvial sediment input.

The complex of sub-environments of the shore zone (intertidal flats, marshes, channels, lagoons, fresh swamps and sebkhas) has shifted landwards due to the Holocene rise in sea-level caused by the melting of the Late Pleistocene continental ice caps. It should be kept in mind that relative sea-level changes derive also from tectonic movements and compaction of sediments due to the effect of loading. Deltaic areas like the Po and the Nile delta, are situated in slowly subsiding basins (several centimetres per century).

Most of the Mediterranean region, however, is situated in a tectonically active zone, characterized by the occurrence of earthquakes and volcanic eruptions. Areas of Greece and Italy are extremely active tectonically (Flemming, 1992). An example is the earthquake that occurred in 1908 at Messina, Italy which increased the relative sea-level by 57 cm. The active tectonics is due to the position of the Mediterranean between the African-Arabian and the Eurasian continents. A series of deep sedimentary basins follows the junction between the continents. Pirazzoli (1987) has presented geomorphological evidence for a major tectonic uplift in the eastern Mediterranean between 300 and 550 AD.

The geoid, the relief of the sea surface topography, is related to the variations in density of the earth crust. Variations in the mean sea-surface topography of as much as 50 m (Fig. 9.2), probably reflect the changes in density of the underlying earth crust (Barlier et al., 1983). In such a tectonic area these densities are likely to shift, affecting the shape of the geoid, and consequently causing changes of sea-level.

Shoreline processes in the Mediterranean are characterized by low tides, winter storms with directions south-west, north-west to north-east, by pronounced longshore currents, and by rivers with reduced discharges, mainly because of human interference.

2.2 Land Use in the Coastal Zone

Since prehistoric times most coastal lowlands have been sites of human settlement and thereby strongly influenced by human activities. As compared to the south margins of the North Sea, the extent of coastal low-lying plains and deltas on the Mediterranean shores is fairly limited (Fig. 9.1). The demographic, economic and ecological importance of these areas is nevertheless very considerable on local, national and international scales. Besides settlements involving millions of people, vital agriculture and fishing resources, as well as industrial, commercial and communication centres, and the increased recreational use of beaches, most areas still contain patches of natural ecosystems of irreplaceable value. It also should be mentioned that the discovery and exploitation of the giant oil and gas fields of North Africa (e.g. the Polignac basin in Algeria and the Syrte basin in Libya) have resulted in the construction of oil harbours, pipelines and oil refineries all around the Mediterranean.

A brief description is given below of the coastal lowlands around the Mediterranean.

Spain: Low-lying coastal stretches are mainly associated with Neogene tectonic depressions, in the form of large sandy bays, coastal lagoons, coastal dunes and marshes. The most important lagoons are La Albufera de Valencia and El Mar Menor in the south and the coastal lowlands of the Bay of Rosas in the north. In addition to fisheries and agriculture these areas have acquired considerable economic value owing to the numerous tourist resorts. Further to the north there are the deltas of the rivers Llobregat and Ebro (Fig. 9.3).

The Ebro Delta is small (285 km^2) but is an important rice-producing and fishing area (mainly shellfish). Part of the delta is still a wetland, of great

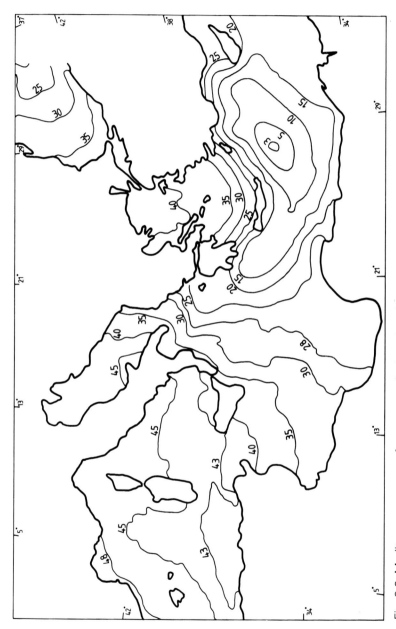

Fig. 9.2 Mediterranean sea-surface topography based on SEASAT altimeter data. Contour interval: 5 m (redrawn after Barlier et al., 1983).

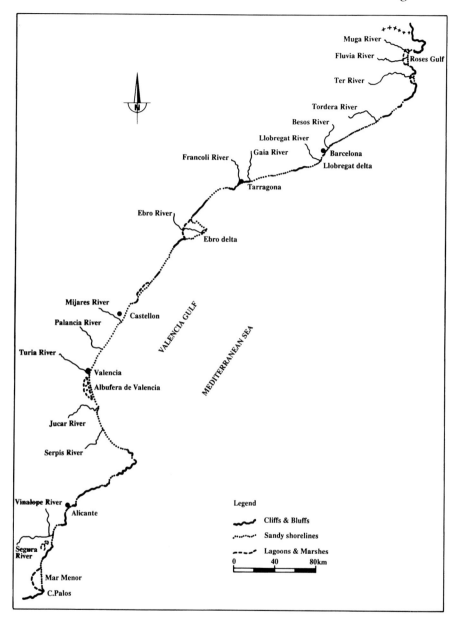

Fig. 9.3 Coastal lowlands of Mediterranean coast of Spain.

value for migrating birds. At present the delta is subject to erosion due to the construction of dams and reservoirs in the upstream part of the Ebro basin. About 96% of the river sediment is trapped in the reservoirs (Marino, 1992).

The other sandy beaches along the Mediterranean shoreline are also subject to erosion, which at many places is counteracted by groins.

France: The coastal lowland segment of southern France is situated between the mountainous shores of the Pyrenees and those of Provence. The coastal plain of Roussillon and Languedoc is a barrier coast with lagoons and low plains. Economic activities include agriculture, such as rice, vegetables, vineyards and cattle breeding, saltpans and industry. During the last 20 years more than 40% of the shoreline has been developed into beach resorts. In parts the shoreline is receding, in parts advancing (L'Homer, 1987; Corre, 1992).

The Rhône Delta is situated at the eastern end of the Golfe du Lion. It covers about 170 km^2 and is characterized by the shifting course of the delta distributaries. The morphology and shape of the delta is the result of the combined effects of waves and fluvial sediment output (Van Andel et al., 1960). In recent times, however, the Rhône river catchment area has been harnessed by many large dams that have reduced the sediment discharge to the sea from 40 million tons a year at the end of the 19th century to 12 million tons in 1956 and to 4 million tons in 1970 (Guilcher, 1985). Accordingly, erosion prevails nowadays along the delta shoreline. Part of the delta is of great importance as a natural reserve for migrating and nesting birds. This, and the natural environment character of the Camargue are attracting an increasing number of tourists in search of open space recreation (Corre, 1992).

At the eastern end of the Rhône lowlands is the new industrial port of Fos de Mer (oil refineries and chemical industries), the Etang de Berre (important for fishing) and the commercial port and city of Marseille (population 1 million).

Italy: The Italian coastline extends for 7500 km with an alternation of rocky and low, sandy shorelines. The coastal lowlands by the Tyrrhenian Sea are related to the deltas of the rivers Arno, Ombrone, Tevere and Volturno. On the Tyrrhenian are located some of the most important harbours, strategic industries and major cities of Italy: Genoa, Livorno, Rome and Naples. In the south and in the islands, valuable lowlands are situated near the cities of Cagliari, Catania and Taranto.

On the Adriatic Sea, the most extensive lowlands are found at the edge of the Po River basin, the most important agricultural area of Italy. This subsiding basin has accumulated a large amount of sediments brought down by rivers from the Alps and the Apennines mountains. The coastal lowlands of Veneto and Romagna are bordered by low sandy beaches enclosing marshes and lagoons. Well known are the ancient cities of Venice and Ravenna, the Venice and Grado lagoons, and the Po D elta (wetlands of international importance) and the string of beach resorts that yearly attract millions of tourists. About 2 million people live in the coastal plains, engaged in agricultural, commercial and industrial activities. The industries of this area, oil refineries, ship building and chemical plants are of great importance for the north of Italy. Sestini (1992a) presents an extensive discussion of the Po Delta.

Most of the coastal lowlands in Italy are subject to erosion largely caused or enhanced by human activities, like a reduction of the sediment load of rivers due to the construction of dams and reservoirs. For more information refer to Zunica (1985).

Yugoslavia: During the post-glacial rise of sea-level, the sea invaded a coastal area with a topography of parallel limestone ridges separated by wide valleys, creating a coast with many bays and a number of long islands stretching parallel to the shore. The littoral plains were reduced to small patches overlooked by strips of bluffs beyond which are peaks as high as 1900 m. As a result the Adriatic coast has few coastal plains. Nevertheless, the regions of Rijeka, Zadar and Split have important harbours and vital industries.

Albania: In the south-eastern Adriatic the 385 km long coast of Albania includes 190 km of low sandy beaches, marshes, lagoons and deltas related to the rivers Drin, Buna, Matia, Skumbini, Semeni and Vojussa (Sestini, 1940; Paskoff, 1985a). The lowlands have a population around 700,000 and constitute a large proportion of prime agricultural land in a country that is mostly mountainous and rugged. Vlore and Durres are the country's ports and main industrial centres. Due to the high rate of sediment discharge by the rivers, many parts of the Albanian low-lying coast have advanced or have been stable during the past hundred years. Those that are eroded, in spite of a low-energy wave regime, are abandoned delta lobes. The Drin and Mat rivers still contribute a substantial sand load, although they have been dammed for hydropower generation.

Greece: Greece has a very long shoreline in comparison to its area. The entire coast and country are strongly effected by tectonic activity. This has created an extremely rugged topography along the entire coastline. In relation to this tectonic activity a large number of horsts and grabens extend in a north-west-south-east direction; the coastal lowlands of Greece correspond to the graben. Compared to the whole shoreline coastal lowlands are few but are of high economic value as they afford the only sites for harbours and related strategic industries. The most significant deltaic lowlands of primary, agricultural and industrial importance is at the mouth of the Axios River (Thermaikos Gulf) near the city of Thessaloniki, the second largest in Greece (Georgas and Perissoratis, 1992).

Figure 9.4 gives an indication of the topographically low areas of Greece and the western part of Turkey. In the latter region coastal lowland also are related to the occurrence of grabens (Tziavos and Kraft, 1985).

Turkey: The Mediterranean coastline of Turkey has a length of more than 5000 km and is structurally controlled by tectonic movements (graben and horst). It is a shore of rocky cliff coasts alternating with deltaic lowlands and sandy shorelines. Small, but economically important deltaic plains occur on the Aegean coast: the Gediz Delta at Izmir, the Havran at Edrenut and the Kuçuk and Buyuk Menderes deltas.

On the Mediterranean there are the Aksu deltaic plain near Antalaya and the large Cukurova plain near Adana. The latter is the product of two large rivers, the Seyhan to the west and the Ceyhan to the east. South-easterly waves have formed broad beach accretion features bordering the shoreline. Much of the plain was formerly swampy or marsh but is now a large well-drained agricultural area of great economic importance. In and around this plain live about 2 million people, and the cities of Adana and Mersin are important industrial centres. Shoreline erosion is not a problem here as river sediment input is still sufficient to keep the coasts in equilibrium with sea-level rise and tectonic subsidence.

About 600 km² of the Cukurova plain is still a wetland; lagoons and mudflat bordered seawards by coastal dunes. These coastal dunes, with a length of 100 km, are the most important in Turkey. The wetland itself is of great importance for migrating and wintering birds. Sea turtles are present by the coast. Figs 9.4 and 9.5 illustrate the most important coastal lowlands of Turkey (Erol, 1985).

Syria and Lebanon: The coastline of Syria and Lebanon is about 400 km long and is structurally controlled; it runs parallel to the north-south Jubal Alaouite anticline in Syria, and to the north-north-east–south-east-west Lebanese mountains the western slope which falls more or less steeply to the sea. In spite of the vicinity of mountains, the shoreline is generally low, though often rocky. Coastal plains are few, the most important is by the Bay of Akkar (Fig. 9.6).

Banyas and Tripoli are important oil export harbour, connected by pipeline to the oilfields in Syria.

Israel: The straight coastline of Israel is also structurally controlled. The coast has a low relief with only two mountains by the shore: Mount Carmel near Haifa and the Rosh Hanigra Ridge. Elsewhere the foothills are separated from the sea by a fairly wide coastal plain in the south (up to 40 km) that narrows towards the north, reaching only a few hundred metres

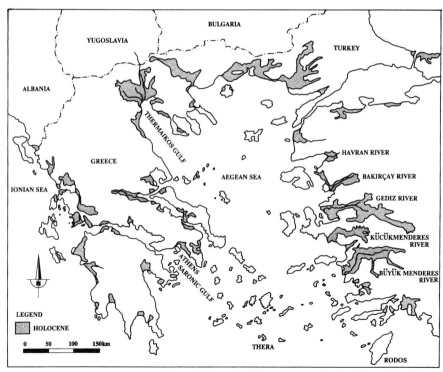

Fig. 9.4 Coastal plains of Greece and western Turkey.

Coastal lowlands and sea-level change 293

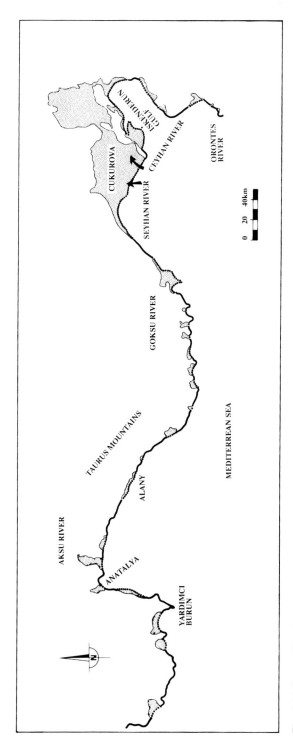

Fig. 9.5 Coastal lowlands of southern Turkey.

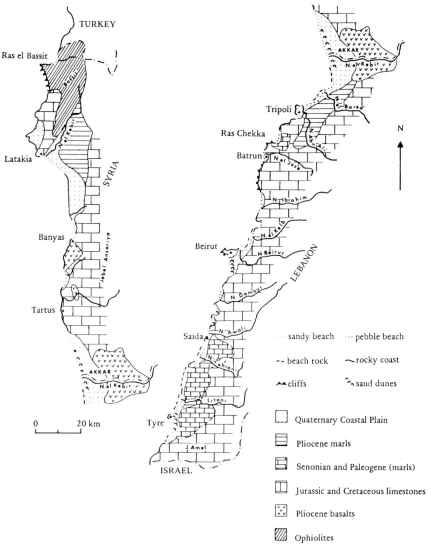

Fig. 9.6 Coastal morphology of Syria and Lebanon (after Sanlaville, 1985).

in the Carmel Plain north of Haifa. Most of the coastline is characterized by cliffs composed of carbonate cemented sands and loams.

About 13 ephemeral rivers (wadis) drain to the coastline from the mountains. Because these rivers carry only a limited sediment load during the wet season, deltaic plains are absent in Israel. Beach erosion is a problem, thought to be caused by human activities along the beaches (constructions and sand mining) and by the reduction of the amount of sand transported by littoral currents from the Nile Delta (Nir, 1985).

Egypt: In Egypt the deltaic plain of the Nile River is the most important coastal lowland of the Mediterranean shoreline. Before the Aswan High

Dam sediment cutoff, the delta could be classified as wave and river dominated. The coast of the delta consists of a series of lagoons separated by sandy barriers from the sea. Most of the lagoons have low salinities, but in the Bardawil lagoon and in a cut-off portion of Lake Mariut the water can reach high salinities, and at times, in parts, the lagoons become salt pans. Of the several distributaries of the Nile that once existed, only the Rosetta and Damietta branches are left. The present deltaic plain overlies a Neogene sedimentary basin containing 2–3000 m of sediment, due to input of the River Nile and to tectonic subsidence (Said, 1981; El-Ashry, 1985; Sestini, 1989).

The Nile Delta region is of great economic importance for Egypt, not only because it contains 50% of the country's population but also because of its main commercial outlets and industrial and recreational centres of Alexandria and Port Said. Agriculture and fishing account respectively for 45 and 60% of national production. One-third of the migrating and wintering birds of the whole Mediterranean area are estimated to use the deltaic lagoons.

The shoreline of the Nile Delta has undergone conspicuous changes during the last 100 years (Sestini 1992b). In the last few decades erosion has occurred at several important locations: El Gamil, Ras el Bar, Baltim and El Burg, Rosetta, Abu Qir and Alexandria (Fig. 9.7). At the same time shoaling is occurring near the entrance of the Suez Canal in front of Port Said breakwater, causing navigational problems and requiring expensive dredging. Severe shoreline erosion is mainly due to damming of the Nile River, which has dramatically decreased the sediment load supplied to the nearshore zone by its two branches. The closure of the Aswan High Dam in 1964 brought the hydrology of the river under full control, reducing its flow into the sea to minor quantities.

Libya: The shoreline of Libya is about 1900 km long. A coastal lowland stretches along the Gulf of Sirte, a barrier lagoon coast locally with dunes and backed by extensive sabkhas (salt pans). Agriculture is locally possible. In this area the oil trans-shipment facilities from the inland Sirte oil and gasfields (El Zueitina, El Brega, El Sider) are of high economic value for the country.

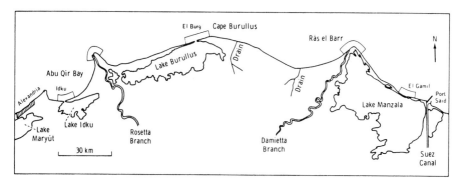

Fig. 9.7 The Nile Delta (after El-Ashry, 1979). Boxed: erosion areas.

Tunisia: The coastline of Tunisia is about 1700 km long, with coastal landforms as indicated on Fig. 9.8. One of the most important lowlands is the deltaic plain of the River Medjerda on the Gulf of Tunis. The Medjerda Delta had been steadily growing in the past centuries, due to deforestation and soil erosion, and later to flood control works, until a channel deviation, in the plain, and major river dams were constructed upstream in the 1970s. Near the lake of Bizerte, Lake Ichkeul is an important ecological wetland and a bird sanctuary (Hollis, 1992). Other coastal lowlands are found along the Gulf of Hammamet and the Gulf of Gabes.

The lowlands of Tunisia are of great economic value in view of agriculture, urbanization, harbours and important industry and of the extensive tourist resorts development. The latter has become a major priority of the authorities since 1960, becoming the second source of foreign exchange for the country. However, the building that has accompanied the tourist boom has caused unwise human intervention in the shoreline, and it is thought to be responsible for the recent serious shoreline recession. The erosion and aggradation stretches of the Tunisian coast are indicated in Fig. 9.8 (Paskoff, 1985b).

Algeria: The Algerian coastline, about 1100 km long, is almost completely bordered by chains of mountains that plunge abruptly into deep water, with intervening small stretches of low sandy coasts. The latter are related to various short rivers that reach the coast. On the whole, the extent of important lowlands is limited and a significant tourist industry has not yet been established (Mahrovur and Dagorne, 1985). The oil transshipment facilities in the harbour of Arzen, Algiers, Beyaia and Skikda built to export the products of the Polignac oil and gas basin are of high economic value for the country.

Morocco: The Mediterranean coast of Morocco is strongly influenced by the Rif Mountains: rocky cliffs correspond to the folds and low shores to the synclines. The occurrence of coastal lowlands is very limited, south-east of Melilla the Nador Lagoon is separated from the sea by a thin sand barrier and the River Moulowya has built a small delta (Weisrock, 1985). Fishing is carried out in the lagoon and the low-lying coastal plain has a good sandy beach near Al Hoceima, now a tourist resort.

2.3 Shoreline Behaviour

From the brief review of the Mediterranean low-lying coasts, it must be concluded that a good many part are at the present time subject to erosion (Bird and Paskoff, 1979; Bird and Schwartz, 1985).

The Po and Ebro deltas are very young. They have grown mainly after the deforestation of their drainage basins since the Middle Ages. Accordingly, a high amount of eroded sediment reached the coast. During the last decades, however, the Ebro and Po river basins have been subject to damming and reservoir construction, and sediment supplied to the beaches and deltas has been greatly reduced. This decrease has resulted in shoreline erosion. A further reduction of sediment discharge has been caused by the industrial dredging of sand from river beds.

In the case of the Nile Delta, the consequences extend to the coast of Israel. The littoral and drift currents in the south-eastern corner of

Coastal lowlands and sea-level change 297

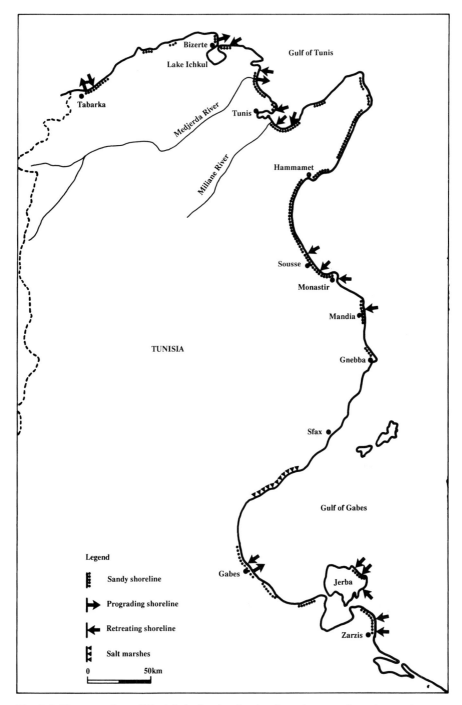

Fig. 9.8 The coastline of Tunisia indicating lowlands and areas of erosion and aggradation (redrawn after Paskoff, 1985b).

the Mediterranean now feed the beaches of Israel with sand derived from the erosion of the Nile Delta coast. The strong reduction of the sediment load is thought to be one of the causes of shoreline erosion in Israel. In notable contrast, relatively little coastal retreat has been noted at the deltas of Albania and of Mediterranean Turkey, in spite of recent dam construction.

Outside the delta areas the following data can be mentioned. In Israel human activities along the beaches have had mostly destructive results, either due to quarrying large amounts of sand or to the construction of groins, harbours and breakwaters which accumulate sands, thus removing them from the littoral drift system (Nir, 1985). In Tunisia, Paskoff (1985b) has indicated that recent shoreline recession has increased dramatically by unwise actions related to the tourist boom. Besides sand mining on the beach in many places the coastal dunes have been completely destroyed by hotels built close to the water edge. The removal of the dead leaves of *Posidonia*, which accumulate on the shore and inhibit sunbathing, has increased wave action; the latter has been enhanced by degradation of the infralittoral *Posidonia* meadows damaged by sewage pollution.

The same facts mentioned by Nir and Paskoff are valid for most of the Mediterranean beaches. The conclusion is that many beaches are subject to erosion, not only because of the recent slow rise of sea-level but mostly because of man's economic and social activities. Particular interference with coastal processes has occurred at places where the building of stone jetties, and concrete or wooden groins, was meant to stop the sand drift. This method, however, is disputable, and it seems that as a reaction to these artificial works, erosion of the unprotected shoreline is increasing. Beach renourishment with sand supplied from the sea bottom seems to be a more successful method. A general description of the critical balance of the Mediterranean shoreline is found in Fabbri and Bird (1983) and Bird and Schwartz (1985).

Due to human activities, not many natural coastal wetlands, like marshes and lagoons, remain in the Mediterranean. The most important areas left are found in the deltas of the Rhône, Ebro, Po and the Cukurova plain. In Tunisia, Lake Ichkeul is a reserve. For the European part of the Mediterranean an inventory of marshes is given in a map edited by the Council of Europe. It should be recommended to extend it to the rest of the Mediterranean coast.

In summary, man's intervention on the coastal zone has resulted in the following:
- Drained and reclaimed wetlands for agriculture, commercial, industrial and housing have accelerated subsidence due to compaction.
- The creation of reservoirs and the dredging sand from rivers have strongly reduced sediment supply to the coast, leading to erosion.
- Construction of breakwaters and harbours, as well as excavation in the wetlands, produces an interruption of longshore transport, hence coastal erosion.
- Removal of sand from the beach also contributes to coastal erosion.
- Ground water, gas and oil extraction have caused local land subsidence.

- Erection of dikes along creeks, channels and lagoonal inlets have reduced sediment supply for the wetlands.

3 The Impact of a Future Sea-Level Rise

To predict the impact of a sea-level rise of 1 m in the coming 100 years on the coastal lowlands of the Mediterranean, we need to analyse the following points:
1) The relative movements of sea-level derived from tide-gauge observations.
2) Data about erosion and/or aggradation of the various shorelines.
3) The frequency of storm surges.

A premise is that: on a given shoreline in the Mediterranean, tide gauges indicate a relative rise of sea-level of 10 cm/century. Shoreline observations and measurements have indicated shoreline retreats of 50 m/century.

If it is accepted that the amount of historical retreat (e.g. Flemming, 1990) can be directly correlated with the observed rate of sea-level rise, then a future increase in the rise of sea-level will cause increased shoreline erosion. By this, a certain prediction is possible and the decision must be either to defend the shoreline or choose for a planned retreat.

As regards tide-gauge measurements, during the last 100 years along parts of the coasts of Europe there has been a rise in sea-level between 10 and 15 cm. This rate of rise is greater than in the recent geological past in Europe where from Roman times to present a total subsidence of 1 m, indicates only 5 cm/century (Flemming, 1990).

The recent relative rise in sea-level is thought to be partly caused by the greenhouse effect. In the Mediterranean the number of tide gauges with recordings dating back to the last 100 years is very limited and unevenly distributed: more in the western than in the eastern part. Pirazzoli (1987) analysed the available stations and concluded that no appreciable increase in sea-level has occurred in recent years. Various tectonic movements strongly influence the local sea-level rise. Also, for tide-gauge observations, especially those monitored for less than 30 years, the derived signal could either be representative or due to noise.

More data are known from the Nile Delta basin and the Adriatic coast. These basins show a continued tectonic subsidence superimposed on the sedimentary compaction of the clays and peats. Tide-gauge measurements in the eastern part of the Mediterranean indicate a sinking rate of 4.8 mm/yr, in the Port Said region of Egypt and an uplift of 2.8 mm/yr at Haifa, Israel. In the Po Delta and in the Venice area a subsidence of 1.2 mm/yr has been observed (Emery et al., 1988). Locally during the period 1950–1970 this subsidence increased dramatically (to 10 cm/year) due to excessive ground-water extraction.

4 Conclusions and Proposed Research

Regarding shoreline behaviour, the following can be concluded and recommended: during the Inqua Congress in Ottawa (1987) the Inqua Shoreline Commission advised its members to start working during the inter-congress period on the impact on shorelines of a future sea-level rise. The same item was stressed at the European Workshop on Interrelated Bioclimatic and Land Use Changes in Nordwijikerhout, Holland (Tooley et al. 1987). During a special session of this workshop 12 contributions were presented on the impact of a future rise in sea-level on the European coastal lowlands, four dealing with the coastal lowlands of the European part of the Mediterranean, one with the Nile Delta. An outline of the recommendations and research proposals of that session, which generally apply to the coastal lowlands of the Mediterranean are given below.

1) Monitoring the movement of sea-level by employing the tide-gauge data maintained and updated by the Permanent Service for Mean Sea-Level, IOS, Bidston, UK, to determine the increased rate of sea-level rise and local variations.
2) Monitoring the impacts of sea-level rise, of neotectonic movements (earthquakes) and of man's economic and social activities on shorelines and coastal lowlands using satellite imagery analysis.
3) Investigations on the coastal dynamics of shorelines to determine changes in the sediment budget as a result of sea-level rise and changing output of sediments by rivers.
4) Investigations on the recent geological history of coastal lowlands and the determination for each lowland of the rates of sea-level change and the responses in terms of sediment type and distribution, of land forms and of paleogeography. These investigations can serve as analogues for future sea-level changes.
5) The development and application of objective criteria to be applied to land-use planning in the coastal lowlands and the production of maps showing high-, intermediate- and low-risk zones in relation to the EPA sea-level scenarios to AD 2100.
6) The updating of the Council of Europe's map of salt marshes of Europe, and an extended inventory to include coastal wetlands and natural areas. The impact of sea-level rise, using different sea-level scenarios, should be undertaken for selected areas of international significance for the world conservation strategy as identified by the UN and IUCN and within the framework of the EC, UNEP, INQUA and the Man and Biosphere programme.

5 Recommendations

5.1 Present Actions:

- Control coastal development to minimize risks to human life from sea-level rise.
- Control land reclamation to reduce the area of coastal lowlands susceptible to inundation.
- Control ground-water exploitation to reduce subsidence and salt-water intrusion.
- Zone lowlands into high-, medium- and low-risk categories. Strategic industries, such as electricity generation, should be located away from high-risk lowlands.
- Suspend all dumping/storage of toxic and radioactive wastes in high-risk zones in coastal lowlands susceptible to long-term inundation sea-level rise.

5.2 Action Over the Next Five Years
- Sponsor research programmes to generate new data from coastal lowlands.
- New and existing environmental data and social and economic data will be integrated in Geographical Information Systems (GIS), developed on personal computers to permit interchange of information.
- Models of sea-level change will be tested rigorously against empirical data collected from the coastal lowlands at different spatial scales from the local scale (individual estuary) to the regional scale (Mediterranean basin).
- An evaluation of the cost-effectiveness of shoreline protection of coastal lowlands will be undertaken and an inventory of European and North African coasts made to determine which segments of coasts need protection and which segments can be sacrificed.

An inventory of the occurrence and frequency of storms along the various coastal segments of the Mediterranean is also an important tool to understand historical shoreline behaviour. Also the damages to the shoreline due to the different storms should be analysed.

In general, it can be stated that a future rise in sea-level will affect the Mediterranean low-lying shorelines by:
- increasing inundating reclaimed lowlands;
- accelerating coastal erosion;
- increasing the risk of flood disasters for cities and industries located on coastal floodplains;
- creating problems with respect to drainage systems;
- increasing salt-water intrusion into ground water, rivers, bays and farmland;
- damaging port facilities and coastal protection structures;
- destroying quality farmland;
- disrupting fisheries and bird habitats, and disrupting ecological balance in coastal systems;
- resulting in the loss of recreational beaches;
- shifting sedimentation in rivers upstream, and hampering shipping.

Another important factor that can influence the coastal areas is the climate change due to the predicted rise of temperature. If the weather pattern and climate change will the runoff from rivers increase in winter and decrease in summer? Drier summers would have serious consequences for the Mediterranean.

Increased runoff can give problems for the embankments, whereas decreased runoff could accentuate salt-water intrusion upstream in rivers.

Another aspect of a change in climate can be an increase in storminess along certain parts of the coasts. The most important damages on coastlines occurs during storm surges at high tide. If this increase in storminess should occur along the coasts of the Mediterranean, the results for the coastal lowlands, with a high concentration of population and industry, would be disastrous. Also this would seriously damage coastal areas where the important tourist industry is concentrated.

Unless remedial actions are taken now, as have been recommended in the European Workshop Report, the social and economic impacts will be profound and widespread.

6 REFERENCES

Barlier, F., Bernard, J., Bouiri, O. and Exertier, P., 1983. The geoid of the Mediterranean deduced from Seasat data. *Proc. 2nd Int. Symp. Geoid in Europe and the Mediterranean* (Rome, 1982), Inst. Geogr. Militare, Firenze, 14–35.

Barth, M.E. and Titus, J. (eds), 1984. *Greenhouse Effect and Sea-level Rise: A Challenge for this Generation*. Van Nostrand Reinhold, New York.

Bird, E.C.F. and Paskoff, P., 1979. Relationships between vertical changes of land and sea-level and the advance and retreat of coastline. In: Suguio, K. (ed.), *Coastal Evolution in the Quaternary*, Univ. of São Paolo, Brazil, 29–40.

Bird, E.C. and Schwartz, M.L. (eds), 1985. *The World's Coastline*. Van Nostrand Reinhold, New York.

Bolin, B., Doos, B.B., Jager, J. and Warrick, R.A. (eds), 1986. *The Greenhouse Effect, Climatic Change and Ecosystems*. John Wiley, New York.

Charlier, H.C. and Piety, J.W., 1985. Greece. In: Bird, E. and Schwartz, M.L. (eds), *The World's Coastline*, Van Nostrand Reinhold, New York, 439–442.

Corre, J.J., 1992. Implications des changements climatiques dans le Golfe de Lion. Chapter 11, this volume.

El-Ashry, M.T., 1985. Egypt. In: Bird, E. and Schwartz, M.L. (eds), *The World's Coastline*, Van Nostrand Reinhold, New York, 513–517.

Emery, K.O., Aubrey, D.G. and Goldsmith, V., 1988. Coastal neo-tectonics of the Mediterranean from tide-gauge records. *Marine Geology*, **81**, 41–52.

Erol, O., 1985. Turkey and Cyprus. In: Bird, E. and Schwartz, M.L. (eds), *The World's Coastline*, Van Nostrand Reinhold, New York, 491–500.

Fabbri, P. and Bird, C.F. (eds), 1983. Coastal problems in the Mediterranean area. IGU Commiss. Co. Environments, Bologna, Italy.

Flemming, N.C. and Webb, C.O., 1986. Tectonic and eustatic coastal changes during the last 10 000 years derived from archaeological data. *Z. Geomorph. Suppl.*, Bd **62**, 1–29.

Flemming, N.C., 1992. Prediction of relative coastal sea-level change in the Mediterranean, based on archaeological, historical and tide-gauge data. Chapter 8, this volume.

Georgas, D. and Perissoratis, C., 1992. Implications of future climatic changes on the inner Thermaikos Gulf. Chapter 13, this volume.

Guilcher, A., 1985. France. In Bird, E. and Schwartz, M.L. (eds), *The World's Coastline*, Van Nostrand Reinhold, New York, 385–396.

Hollis, G.E., 1992. Implications of climatic change in the Mediterranean Basin. Chapter 15, this volume.

Jelgersma, S., 1987. The impacts of a future rise in sea-level on the European lowlands. *European Workshop on Interrelated Bioclimatic and Land Use Changes*, Noordwijkerhout, The Netherlands, October 1987.

L'Homer, A., 1987. The impact on Rhône coastal lowlands of the projected sea-level rise of 1 m. In *Proc. of European Workshop on Bioclimatic Changes and Land Use*, Noordwijkerhout, The Netherlands.

Mahrovur, M. and Dagorne, A., 1985. Algeria. In: Bird, E. and Schwartz, M.L. (eds), *The World's Coastline*, Van Nostrand Reinhold, New York, 531–536.

Mariño, M.G., 1992. Carbon dioxide buildup impact on western Mediterranean. The Ebro Delta Case. Chapter 10, this volume.

Marques, M.A. and Julia, R., 1985. Spain. In: Bird, E. and Schwartz, M.L. (eds), *The World's Coastline*, Van Nostrand Reinhold, New York, 397–410.

Nir, Y., 1985. Israel. In: Bird, E. and Schwartz M.L. (eds), *The World's Coastline*, Van Nostrand Reinhold, New York, 505–511.

Paskoff, R., 1985a. Les côtes d'Albanie, Aspects géomorphologiques. *Bull. Ass. Geogr. Franc.*, Paris, **2**, 77–83.

Paskoff, R., 1985b. Tunisia. In: Bird, E. and Schwartz, M.L. (eds), *The World's Coastline*, Van Nostrand Reinhold, New York, 523–530.

Pirazzoli, P.A., 1986. The early Byzantine tectonic paroxysm. *Z. Geomorph. N.F. Suppl.*, Bd. **62**, 31–49.

Poland, J.F. (ed.), 1984. Guidebook to studies of land subsidence due to groundwater withdrawal. UNESCO (Studies and Reports in Hydrology, No. 40), Paris, 305.

Robin, G. de Q., 1986. Changing the sea-level. In: Bolin, B., Doos, B.R., Jager, J. and Warrick, R.A. (eds), *The Greenhouse Effect, Climatic Change and Ecosystems*. John Wiley, New York, 323–359.

Said, R., 1981. *The River Nile*. Springer-Verlag, Berlin, 151.

Sanlaville, P., 1985. Syria and Lebanon. In Bird, E. and Schwartz, M.L. (eds), *The World's Coastline*, Van Nostrand Reinhold, New York, 501–504.

Schwartz, M.L., 1985. Libya. In: Bird, E. and Schwartz, M.L. (eds), *The World's Coastline*, Van Nostrand Reinhold, New York, 519–521.

Sestini, A., 1940. Le pianure costiere dell'Albania. *Boll. Reale Soc. Geogr. Ital.*, sr. **7**(5), 513–527.

Sestini, G., 1987. The impact of climatic changes and of sea-level rise on two deltaic lowlands of the Mediterranean Sea. *European Workshop on Interrelated Bioclimatic and Land Use Changes*, Noordwijkerhout, The Netherlands, October 1987.

Sestini, G., 1992a. Implications of climatic changes for the Nile Delta. Chapter 14, this volume.

Sestini, G., 1992b. Nile Delta depositional environments and geological history. In: Whately, K.G. and Pickering, K.T. (eds), *Deltas, Sites and Traps for Fossil Fuels*, Blackwell Scientific, Oxford.

Shuisky, Y.D., 1985. Albania. In Bird, E. and Schwartz, M.L. (eds), *The World's Coastline*, Van Nostrand Reinhold, New York, 443–444.

Titus, J.G. (ed.), 1986. Proceedings of the International Conference on Health and Environmental Effects of ozone modification and climatic change. Crystal City, Virginia, USA.

Tooley, M.J. and Jelgersma, S., 1987. European Workshop on interrelated bioclimatic and land use change Noordwijkerhout, The Netherlands. Conclusion and recommendations of parallel session 1a and 2a. *Impact of a future in Sea level on the European Coastal Lowlands*.

Tziavos, Ch. and Kraft, J.C., 1985. Greece. In: Bird, E. and Schwartz, M.L. (eds), *The World's Coastline*, Van Nostrand Reinhold, New York, 445–453.

Van Andel, Tj.H. and Curry, J.R., 1960. Regional aspects of modern sedimentation in northern Gulf of Mexico and similar basins and palaeogeographic significance. In Shephard, F.P., Phleger, F.B. and Van Andel, Tj.H. (eds), *Recent Sediments Northwest Gulf of Mexico*, Amer. Assoc. Petrol. Geol., Tulsa, 345–364.

Warwick, R.A. and Ooerlemans, J., 1990. Sea level rise. In: Houghton, J.T., Jenkins, J.G. and Ephraums, J.J. (eds), *Climate Change*. The IPCC Scientific Assessment, Dress Syndicate of the Univ. of Cambridge, 261–285.

Weisrock, A.L.E., 1985. Marocco. In: Bird, E. and Schwartz, M.L. (eds), *The World's Coastline*, Van Nostrand Reinhold, New York, 537–544.

Wind, H.G. (ed.), 1987. Impact of sea-level rise on Society. Report of a project-planning session, Delft, 27–29 August 1986. A.A. Balkema, Rotterdam.

Zunica, M., 1985. Italy. In: Bird, E. and Schwartz, M.L. (eds), *The World's Coastline*, Van Nostrand Reinhold, New York, 419–429.

10

Implications of Climatic Change on the Ebro Delta

M.G. Mariño
(*Instituto de Salud Carlos III
Majadahonda, Madrid, Spain*)

Abstract

The Ebro Delta, the fourth largest delta in the Mediterranean Sea, is a triangular-lobate delta, with a surface area of about 285 km², located on the coast of north-east Spain. The population is currently about 19,000. Primary economical activities in the Delta are agriculture and fisheries. About 20% of Spanish rice production and up to 25% of the fish and 40% of the molluscs obtained in Catalonia come from this area. The natural values of the Delta are also of great importance as the low-lying parts and the lagoons that form the "Natural Park" represent essential nesting and resting areas for large numbers of migratory aquatic birds.

The Delta shoreline is being reshaped by the sea because of the reduction of the sediments and water river output, as up to 96% of the river sediment is retained by upstream dams. The Ebro mouth is receding and the Trabucador isthmus is being eroded, cutting occasionally the connection with the Alfaques peninsula and exposing Alfaques Bay, where the main aquaculture facilities are based, to the sea.

The principal consequences of the predicted climatic change and sea-level rise will be the aggravation of human-induced erosion and reshaping of the Delta coastline, mainly on the front and southern lobes, affecting the wetlands and the natural areas which may eventually disappear. Although subsidence is not an important factor today, it may become so in the future if water management is not planned wisely and water extraction from the aquifers increases significantly.

Little consequence is expected over the short term with respect to economic activities or infrastructures in urbanized areas, as the tourist development of the Delta is moderate and the main infrastructures are located inland. The main land-based activities that could be impacted are the aquaculture plants in the Alfaques bay and the salt-production facilities in the Alfaques peninsula. The possible change in the nutrients cycle due to the reshaping of the bays may also affect the productivity of the adjacent sea and reduce the fisheries in the area.

The Ebro Delta is a good example of the pre-eminent role played by human activities in the problems attributed to the climatic change. It is clear that the main problems will arise from human-induced modifications; climatic change will only aggravate them.

1 Introduction

Scientific interest on the Ebro Delta in recent years has produced many important research programmes on such topics as pollution control, coastal dynamics, marine productivity and agricultural development.

Both the Spanish State Government and the Catalonian Autonomous Generalitat have promoted these studies as a response to economic development and the modifications of Ebro River discharge. This report is based on the available results of these programmes, specially those related to coastal erosion, the development of agriculture, aquaculture, tourism and the sanitary and economic infrastructure of the delta.

2 Mediterranean Setting

In the Mediterranean region, erosion and the loss of agricultural land are already common problems and forest fires are not infrequent during the summer season, especially in forests containing introduced tree species. Adequate water supply, both quantity and quality, is scarce, and salinization together with sea-water intrusion are normal problems in many coastal areas, where agricultural, residential and industrial development tends to concentrate. Superimposed on these activities, summer tourists exert additional demands that exceed available resources in many of the new coastal resorts.

Gradual and episodic sea-level rise and associated coastal dynamics and erosión will be of special significance on the Mediterranean coast due to five main factors:

1) In many Mediterranean countries the littoral strip of land holds most, if not all of the main economic activities, lacking the inland development common in other European countries. Also, some of the most important areas of natural interest are located on the coast, as is the case of the wetlands where many European migratory bird species find sanctuary.

2) There is a lesser tradition of coastal defence in the Mediterranean countries than in northern Europe, and the economic capacity to undertake them is generally insufficient.

3) The most productive areas are located in low-lying lands like deltas and lagoons, where a rise in the sea-level would have definite effects, either by the flooding of these areas or by the alteration of the quality of the water used for irrigation.

4) Apart from their agricultural value, the lowlands and the shorelines of the Mediterranean have acquired an increasing value for recreational activities based on the development of beaches, dunes and coastal lagoons, with huge investments in the form of residential complexes, on the assumption of a continuing stability of the coast.

5) Although social and technical transformations of the next decades will have a far greater impact on the Mediterranean than the climatic change and the sea-level rise, these may aggravate social problems and increase the occurrence of catastrophic peak events.

Recent history of the Mediterranean offers many examples of changes in the coastline that have caused the decline of areas dedicated to agriculture or commerce (i.e. Rosas, on the Catalonian coast). People have adapted to these changes by moving their activities inland or to other more advantageous places; natural areas have been substituted by new ones. This flexible response is not easy today: pressure on the coast has increased dramatically, reducing the available land and raising property values. This, together with the lack of fertile areas inland, leads to a population which is unwilling to move and puts the emphasis on coastal defence, now technically more feasible.

As a result of this attitude, any increase in temperature and sea-level will especially affect the natural reserves, as the defence infrastructures will be concentrated in the areas of greatest economic value, not allowing for the transformation of present lowlands into wetlands, while the existing wetlands disappear as the sea rises and the changing dynamics alters the configuration of the coastline.

On the other hand, it is expected that the main ports and engineering works in the Mediterranean will be adapted as the sea rises gradually, with little disruption in their operations. Although the total investment will be high in the long run, climatic changes will occur gradually, and thus their effects will be treated more as natural events, rather than as a human-induced modification.

3 THE EBRO DELTA

3.1 General Aspects

The Ebro Delta, located on the Mediterranean coast of north-east Spain, is a triangular-lobate delta with a surface area of about 285 km^2. It is the fourth largest delta in the Mediterranean, the whole deltaic system covering about 40,000 ha (Fig. 10.1).

Together with the delta of the Guadalquivir River, the Ebro Delta is one of the most valuable coastal ecosystems in Spain, with wetlands that are essential to a great number of migratory birds. As in the case of the Guadalquivir, the Ebro Delta also sustains a very important agricultural activity, which is based on the quality of the soils and the amount of available regulated waters for irrigation. It also houses a very productive fishing fleet and aquaculture development; up to 20% of the Spanish rice production and up to 25% of the fish and 40% of the molluscs obtained in Catalonia come from this area.

The main contemporary problems of the Ebro Delta are associated with the reduced flow of the river water and sediment. The decreased sediment load is considered to be one of the main reasons for the progressive erosion of the frontal parts of the delta. The need for a safe supply of irrigation water of adequate quality also makes the delta vulnerable to future reductions in water resources. Although the levels of mercury and PCB's in some sediments indicate the existence of industrial discharges into the river, there presently are no serious pollution problems in the area (Mariño, 1983), and only high water temperatures in the bays during summer cause some occasional high death rates among the mussel farms located in the area.

Most of the points mentioned for the Mediterranean situation with regards to the predicted climatic change also can be felt in the Ebro Delta. As a result, the main important impacts of this change will surely be the flooding of the low-lying wetlands, an increased need for fresh water and a worsening in the existing coastal dynamics, both because of a further decrease of river output and because the sea-level rise will modify the present equilibrium profile of the shoreline.

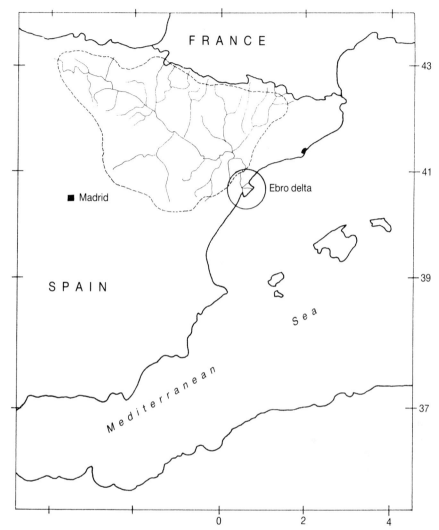

Fig. 10.1 Index map of Ebro Delta with Ebro drainage basin.

3.2 Present Situation

3.2.1 Socio-economic setting

The population of the municipalities comprising the delta is 70,000, while about 19,000 people live in the delta itself; 10,187 in Deltebre, 3446 in Sant Jaume, 3390 in Aldea and 2914 in Camarles. The major economic activity is agriculture, with rice, wheat and vegetables as the main products; 77% of all delta surface is cultivated under irrigation, while the rest is unproductive. Rice, with around 17,000 ha (69% of the delta surface), represents 18% of the whole Spanish production; orchards cover 2400 ha, and cereals, mainly wheat, account for another 4700 ha (Seró and Maymó, 1972; Anuario del Ministerio de Agricultura, 1986).

Fisheries around the delta and in the lagoons are also very productive. In terms of the value of the catches, the port of Sant Carles de la Rápita is the fourth in importance among the Catalonian fishing ports, and the port of l'Ametlla is sixth. L'Ampolla, Deltebre and Las Casas d'Alcanar also contribute to the total fish production of the area. Crustaceans, mainly shrimps, are one of the most important products of the Ebro delta marine waters. In the bays protected by the delta lobes there is also an increasing number of mussel and oyster farms, especially in "Puerto de los Alfaques" in the south. Both sea water from the bays and ground water from the saline aquifer are utilized in these new farms although surface water has problems due to temperature oscillations and pollution from the overflow of the cultivated areas.

Total fish production in the delta ports during 1986 (Anuari de la Comisió de Ports de Catalunya, 1986) reached 10,218 metric tons (molluscs 1570, crustacea 990, fish 7650), with a commercial value of 2860 million pesetas (US$260 million). This production represents 28% of that of Catalonia; for molluscs, crustacea and fish the delta contributes respectively 43%, 71% and 24% of the catches. Fisheries in the lagoons are also very important, producing 84 tons of highly valuable species.

Industry is limited to two saltworks producing sea salt through evaporation of sea water in shallow ponds; they are located in the Alfaques bay. The largest (Salinas de la Trinidad) has a yearly output of 25,000 tons, which are mainly exported to northern Europe (*Gran Enciclopedia Catalana*, 1979). Smaller saltworks have been abandoned and are now being transformed into fish farms.

Two tourist resorts have recently been opened in the delta, one on the northern shore (Ruimar), the other in the south (Els Eucaliptus), just in front of the white sandy beaches that surround the whole area. Although they account now for only a small per cent of the whole delta population, there are plans for the construction of 1000 new apartments and houses in the Els Eucaliptus area (Loaso, pers. comm.). Three port facilities also have been installed for recreational boats in Tortosa, Amposta and Deltebre, and a navigational channel between Tortosa and the sea is under study by the Generalitat (Canals, pers. comm.).

Even if these projects become a reality, it is expected that the main economic activities in the delta will presumably continue to be agriculture and fisheries. Tourist development will remain limited to the two areas already urbanized, especially that on the coast above the Trabucador isthmus (Els Eucaliptus). The only probable change in the economy of the delta for the future is the further expansion of the shellfish cultivation in the bays behind the northern and southern lobes, where high productivity and natural defences against the sea are most conducive. These areas will thus become amongst the most productive and valuable in the delta and in the whole Spanish Mediterranean coast.

3.2.2 Climate
Rainfall ranges from a minimum of 18 mm in July to maximum of 78 mm in September and 69 mm in October; annual precipitation averages 550 mm. Average daily temperatures range from a maximum of 25°C in July and August to a minimum of 9.3°C in January and February; the

average annual temperature is 16.6°C. Humidity remains rather stable at around 60% through the year, while evapotranspiration varies from 90–170 mm/month, with an annual loss of about 1400 mm. Annual sunshine hours are over 2700 (Seró and Maymó, 1972; Observatorio de Roquetes data cited by Riba *et al.*, 1976; Canals, pers. comm.; MOPU, 1978).

Winds blow rather strongly in the Ebro Delta area. During September and October the "mistral" winds, coming from the north, north-west and west, reach their highest velocities (wind speeds over 200 km/h are frequent). "Vents marins", mainly from the east, predominate during spring, pushing sea water into the river mouth, elevating the sea surface and obstructing drainage operations from cultivated areas. Storms from east and north-east during winter and spring months are also frequent ("llevantades"), causing surges that flood the Trabucador isthmus for a number of days. During the summer months the predominant wind comes from the south-west ("garbi") with lower velocities (Font, 1983).

3.2.3 Oceanography

The oceanographic features of the Ebro Delta margin are dominated by the river discharge, coastal currents and the dynamics of the bays. The river discharge controls most of the balance of nutrients and productivity. Alvarez and Masso (1983) studied the basic oceanographic parameters of the area, where stratification during the summer season is common (see Fig. 10.2), with temperatures as high as 25°C outside the delta (it reaches 32°C inside the bays), (Loaso, pers. comm.).

Nutrient concentrations are high due to the river discharge. Phosphorous concentration reaches 179 μg PO_4-P/L and ammonia 137 μg NH_3-N/L. Secchi disc depths range from 18 m in front of the river mouth to 10 m at the 20 m depth contour. Chlorophyl has its peak in February with 50 μg/L. The dissolved oxygen content in the water tends to remain over 40 mg/L, with minima during summer periods in the bays.

General water circulation in the inner shelf in front of the Ebro Delta is dominated by the flow imposed from offshore in the Catalan sea, which in turn is influenced by the flow coming from the Balearic sea. Winds perturb the overall currents, modulating the larger externally forced flow, which can be felt mainly in the bottom, with velocities in the range of 12 cm/s towards the south-south-east.

The average current in summer is in the range of 7 cm/s, although flows as high as 25 cm/s running southwards for 70 days have been measured (Han *et al.*, 1983). Summer stratification produces an upper layer that moves under wind fluctuations, while the bottom layer flows independently southwards (Font, 1983). Wind effects also can be felt in winter, when the surface and bottom currents seem to flow northwards in response to strong north-west winds, with an average velocity of 2.5–3.0 cm/s. This can be due to the blocking of the strong mistral winds by the coastal mountain ranges, which produces a concentrated jet centred in the Ebro Delta. This local wind maximum creates an Eckman suction through the curl of the wind stress, and this torque is compensated by a pair of centro-rotating gyres in the flow (Han *et al.*, 1983). This situation has been modelled by Mariñas and Tejedor (1983).

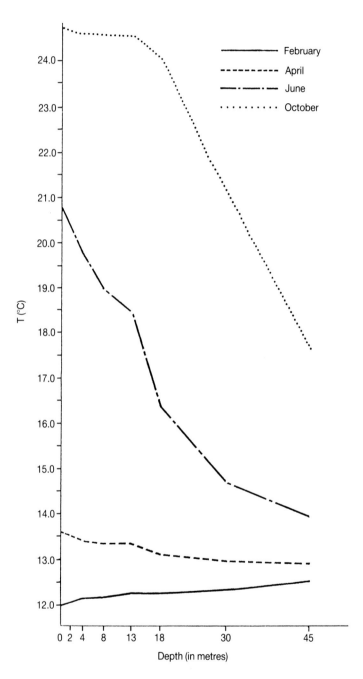

Fig. 10.2 Temperature profiles.

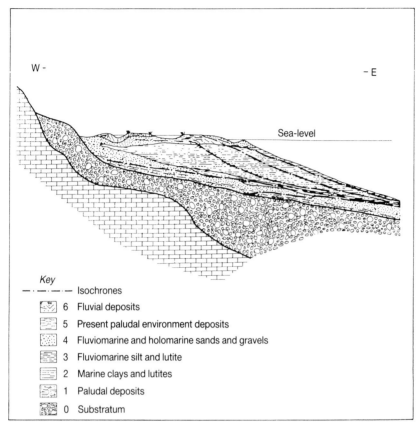

Fig. 10.3 Schematic profile W-E through the axis of the delta complex (from Maldonado, 1975).

The oceanography of the bays has been studied extensively by Camp and Delgado (1987). They describe them as salt-wedge estuaries in which the sea water penetrates as a salt wedge along the bottom while the low-salinity water tends to flow seaward over it. The northern bay (Fangar) has a surface of about 12 km² and an average depth of 2 m, with a connecting channel of 1 km in width. The figures for Alfacs, in the south, are 50 km², 4 m and 3 km, respectively. The total volume of Fangar is in the order of 16 hm³, while Alfaques reaches 200 hm³.

Fresh water enters the bays laterally from the lagoons through a series of channels and from the aquifer, causing a preferential low-salinity distribution in the landward margin. Stratification tends to predominate in the bays until wind induces the mixing or fresh-water input and produces instability of the water masses, which are renewed by sea water in periodic fluctuations after 1–2 days in Fangar and 10–20 days in Alfaques, as residence time in these bays is controlled by the ratio between the fresh-water inflow and the total volume of the bay (Camp and Delgado, 1987). As these fluctuations persist in winter, when the irrigation channels are dry, Camp and Delgado suggest the possible existence of a large ground-water inflow into the bays and its importance for the renewal frequency and stability.

Also related to the fresh-water inflow are the nutrient and phytoplancton distribution in the bays, that tend to show higher values around the channel discharge points (Delgado, 1987; Delgado and Camp, 1987). However, as nutrient inflow through the irrigation channels explains only 25% of the high primary productivity, Delgado and Camp (1987) mention organic mineralization processes in the sediments as a possible source of phosphorus to the water column. This high productivity, coupled with regular renewal in the bays, results in a periodic carbon outflow towards the sea on the order of 100 tons/week (Camp, pers. comm.).

Nutrients and productivity also are influenced by seagrasses, whose spatial distribution and biomass in the bays were studied by Pérez and Camp (1986). Seagrasses cover in Fangar reaches 65% of the total bay surface and 100% of shallow areas, while they are only present in 26% of the Alfaques bay shallow parts, representing a total biomass of 3600 tons.

3.2.4 Coastal physiography

The development of the Ebro Delta began at the end of the last glacial stage (Maldonado, 1975; Díez, pers. comm.). At the maximum of the Würm glaciation, about 20,000 years ago, the sea stood 85–90 m below present sea-level. Because the shoreline was located near the present edge of the shelf, most of the sediments borne by the river were carried by turbidity currents and deposited on the deep submarine Ebro fan (Fig. 10.4). During the subsequent sea-level transgression thick overlapping fluvial gravel deposits were laid down in the fluvial valley. Later, transgressive sequences were deposited, extending across the shelf over the gravel and Pleistocene deposits.

With the temporary stabilization of the sea-level at -10 m, the prograding delta actively built an extensive deltaic plain. Later, with the continuing rise in sea-level, this former deltaic plain was covered by the sea and shallow marine sediments were formed. The small extent of these marine deposits points to a sharp decrease in the rate of sea-level rise and indicates that the fluvial supply of sediments was sufficient to prevent an extensive marine drowning of the deltaic plain.

Most of the deposits of the present deltaic complex have been laid down during the last 8000 years. There has been therefore a sharp increase in the effective rate of sedimentation in this final stage of the delta building, concurrent with the slowing down of sea-level rise.

The present Ebro deltaic plain consists of three pronounced delta lobes extending 26 km seaward. The development of these lobes has notably increased the delta plain during the past four centuries until a major flood in 1937 produced a fourth lobe, as the river opened a new mouth (Gola de Sorrapa), abandoning the older lobe in less than 20 years (Maldonado, 1977) (see Fig. 10.5). Older mouths of the river have also been abandoned or closed artificially; Gola de Midgorn in 1915 and Gola de Tamuntana in 1957 during the 20th century, la Platjola, el Galeró and Sol de Riu in the 19th century (Loaso, 1988, pers. comm.).

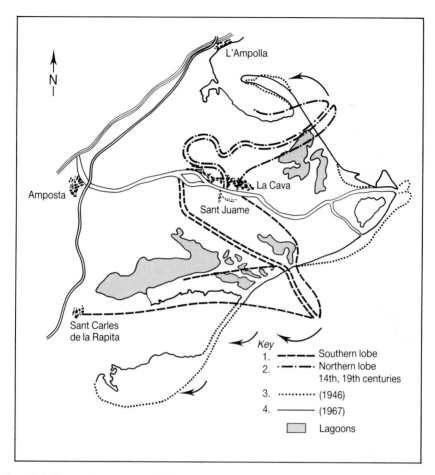

Fig. 10.4 Formation process of lobes in the deltaic plane (from Maldonado, 1983).

In historic times the Ebro Delta underwent remarkable modifications, growing from an estuary which reached Tortosa in Roman times, to the modern delta of today (Fig. 10.5). The southern delta lobe is the oldest, dating from the early 16th century. The northern delta lobe mainly developed through the 17th and 18th centuries extending into the early 19th century. The active period of the central delta lobe coincided with that of the northern delta lobe during its final stage of development. Initially the central delta lobe had a northern distributary active through the early 18th century. The eastern distributary reached its maximum development in about 1946.

Since 1970 the growth of the delta has almost stopped, while the front and south lobes have been subjected to accelerated erosion due to the damming of the Ebro and to modifications to the discharge mouth and the hydrographic system. Because of this process, Maldonado (1983) calculates the rate of loss of sediments in the coastal zone to be in the range of 19,500 t/yr during the second half of this century.

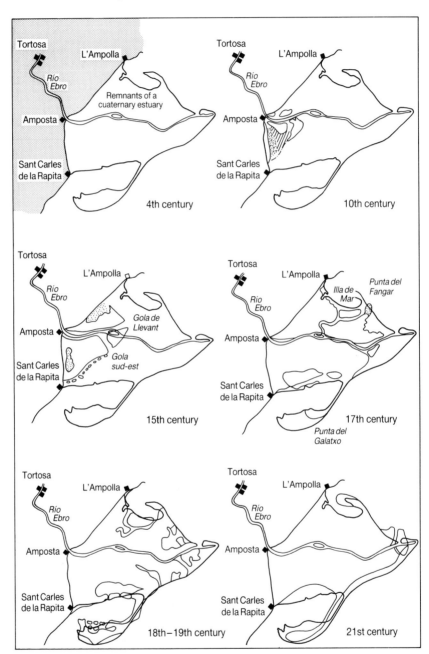

Fig. 10.5 Evolution since Roman times (from MOPU, 1978).

The river damming has resulted in the reduction of the sediment discharge, from about 4,000,000 t/yr before 1965 (Bayerri, 1935, and Catalán, 1969, as cited by Maldonado, 1983), to less than 400,000 t/yr at present (Varela *et al.*, 1983), although the study of the bathymetry of the Ribarroja and Mequinenza dams shows that they have accumulated 8,800,000 t/yr of sediments during the last 20 years that would otherwise have reached the delta (Varela *et al.*, 1983). The discharge point also has shifted from the Gola Midgjorn, in the east, to the Gola Nord and then to its present location, pointing northwards (the latter was closed artificially and is presently occupied by an eel nursery).

As a result of these modifications there is now an active reshaping of the delta shoreline by the sea, particularly in the Isle of Buda and the former Migjorn mouth, along the southern and northern shores (Maldonado, 1983). As shown in Fig. 10.6, the loss of sediment is concentrated at those points where the dynamic equilibrium is more precarious, as in Cabo Tortosa. The Alfaques lobe, in the south, is also under erosion as is the Trabucador isthmus, now in danger of being breached by the sea. On the other hand, the northen lobe (El Fangar) is growing, although at a much slower rate.

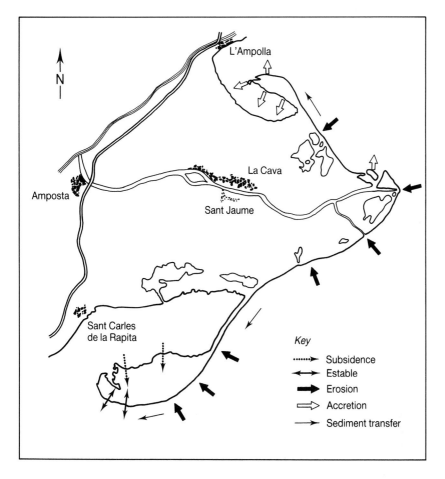

Fig. 10.6 Main coastal dynamic processes (from Maldonado, 1983).

316 *Climatic Change in the Mediterranean*

3.2.5 Surface and ground water

Average monthly River Ebro discharges at Tortosa (basin surface of 85,500 km²) ranged from 120 m³/s in August to 800 in January–February for the period 1951–1965, with a yearly average discharge of 600 m³/s (*Gran Enciclopedia Catalana*, 1979). These figures represent a decrease of almost 20% with respect to the 1912–1935 period, before the dams along the lower Ebro were constructed.

There is an irrigation channel system of more than 1000 km that covers most of the delta and an important drainage network with several pumping stations which discharge through the lagoons to the bays. Less than 10% of the irrigation flow goes back to the river or directly to the sea (Camp and Delgado, 1987). Two main irrigation channels follow both river banks, starting from the Xerta dam, 27 km upstream from the delta; they transport 21 and 17 m³/s each. Total guaranteed regulated flow in the Ebro is in the order of 60–70 m³/s.

The ground water in the Ebro Delta area flows within a multilayer aquifer, in which gravel and clays contain water with very different salinities (Loaso and Herrán, 1988) (Fig. 10.7). The system has a very low hydraulic

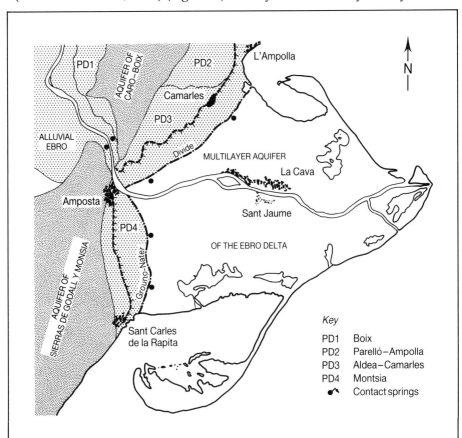

Fig. 10.7 Aquifers (from Loaso and Herrán, 1988).

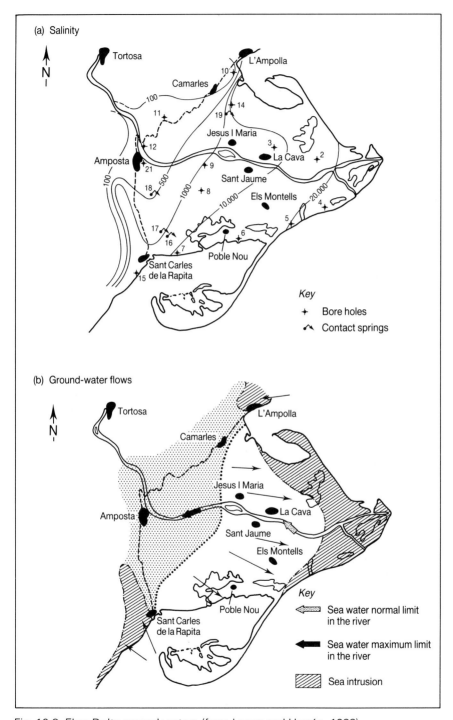

Fig. 10.8 Ebro Delta ground waters (from Loaso and Herrán, 1988).

gradient that produces a slow flushing of the original saline waters. It is only partially connected with the alluvial aquifer located upstream (salinity under 2000 μS/cm) and with the "piedemonte" (foothill) aquifers of Montsiá and Aldea-Camarles which emerge in the "ullals", small ponds located along the contact line between these aquifers and the peats and clays of the delta (Fig. 10.8).

Apart from the extraction for aquaculture, ground-water use in the delta is very limited, as irrigation depends almost exclusively on surface water; only a little was pumped until recently during the winter to water the orchards, when the irrigation channels were drained for repairs. Nowadays, ground water is produced at a small rate out of the Aldea-Camarles aquifer (8 hm^3/yr) and out of the alluvial aquifer (15 hm^3/yr), which represents an important resource for the future (Loaso and Herrán, 1988).

3.3 Ecological Importance

Two of the lagoons in the southern part of the delta (Tancada and Encañizada) have been designated protected areas because of their value for many migratory birds which either stop there on route or come to winter (see Fig. 10.9). Isla de Buda, which is located on the eastern front of the delta, is another important area for aquatic birds.

Natural vegetation is found mainly in the lagoons and along the coastal sand dunes, while most of the delta is presently dedicated to rice, wheat and orchards. Ammophiletea, Phragmitetea and Arthrocnemetea communities have been described in the "Landscape vegetation map of Ebro Delta" (Camarasa et al., 1976), which is summarized in Fig. 10.10.

Forty-nine km^2 of the northern part of the delta have been declared "Parque Natural". Five sectors of the Parque Natural are qualified as "Areas de Especial Interés Natural" (Isles of Buda and San Antonio, Canal Vell and Olles lagoons and Fangar peninsula), comprising 25 km^2 (56% of the Parque Natural), (Fig. 10.11). According to the Parc Natural proclamation, these legal designations aim at "the efficient conservation of the natural ecosystems, making them compatible with the maintenance of the traditional activities and the promotion of the contacts between man and nature".

4 Evaluation of Climatic Change Impacts

4.1 Physical Environments

4.1.1 Climate and oceanography

The first step in evaluating of the possible impact of the climatic change on the Ebro delta is evaluating the climatic change itself. Even a preliminary assessment of possible impacts requires using models on a regional or subregional scale and that include not only the Mediterranean but also the Cantabrian and Pyrenees mountain chains from where most of the Ebro River water originates.

The necessary information needed to complete such model is not entirely available yet. Therefore, the analysis has to be made on the basis of general climatic models that have a resolution in the order of 5–10°, or

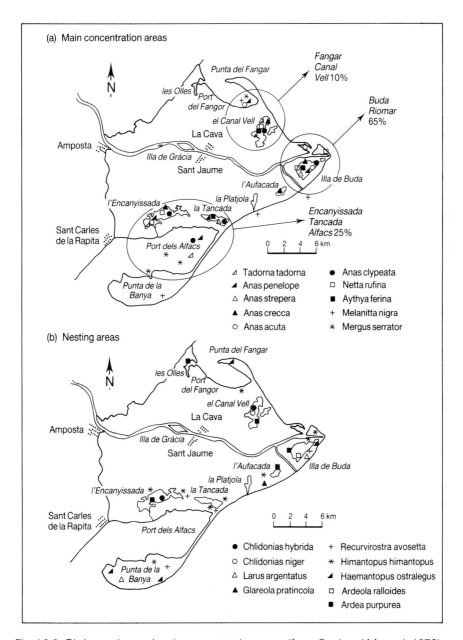

Fig. 10.9 Birds nesting and main concentration areas (from Seró and Maymó, 1972).

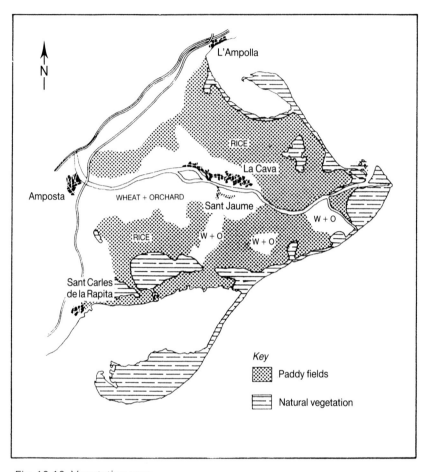

Fig. 10.10 Vegetation map.

of extrapolations of paleoclimatic and recent climatic data. Most assumptions on climatic change made in this chapter use the results of the GISS model and those derived from instrumental data (Pitovranov, 1987). From available models (see Wigley, Chapter 1) the following climatic scenarios for the Ebro Delta and for the Cantabrian and Pyrenees mountains have been assumed:

- In spring, cold and wet weather conditions will be more frequent.
- Summers will be longer, hotter and dryer.
- Rainfall will increase 0.2–0.4 mm/day during spring and between 0 and 0.2 mm/day in winter, it will remain similar to today's rate for autumn and will decrease about 0.2 mm/day for the summer period.
- Precipitation variability will decrease slightly, especially in winter.
- Evapotranspiration will increase.
- It also can be assumed that the temperature change will mean an increase in the frequency of episodic events associated with storms and floods.

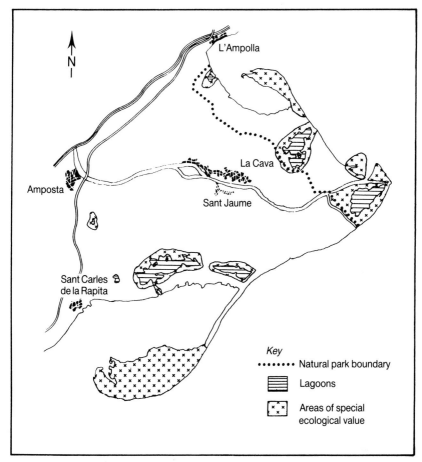

Fig. 10.11 Parc Natural of the Ebro Delta.

Thus, while total rainfall in the catchment area will be similar to today, or even higher, the pattern will be different. This agrees with paleoclimatic data for the first part of the Holocene, which show that during the climatic optimum, around 6500 BP, when temperatures were 2–3°C higher than today, humid conditions predominated in the present-day subtropical zone, with the existence of large lakes in the Sahara (Janssen and Kuhn, 1987).

The general sea circulation in the western Mediterranean is not expected to be affected by this climatic change, as it mainly dominated by currents and only partially modulated by the strong winds that flow through the Ebro river opening in the coastal mountains (Mariñas and Tejedor, 1983; Han et al., 1983). Summer stratification, on the other hand, may be enhanced, raising the water temperature in the bays and affecting water renovation.

Hydrodynamics by the coast will become more energetic, especially during storms and sea surges associated with high easterly winds. Wave action on the shore is also expected to increase, accelerating the reshaping of the delta, which is now wave-dominated as a result of the decreased sediment input.

4.1.2 Surface and ground water

As most surface water in the delta comes from the Ebro through the irrigation system, the quality of surface and, to a large extent, ground water will be governed by the quantity and characteristics of the river water. With longer dry summer periods it can be assumed that the water balance in the delta will become increasingly negative, although this could be offset in the short term by the predicted precipitation patterns in the Ebro catchment basin, which can be regulated by the existing network of dams and irrigation system. Consequences of this water disbalance will be felt mainly in the areas surrounding the delta, leading to more water extraction from the river upstream and to more severe forest fires in the upland areas. Water extraction from the aquifers will surely increase, requiring careful management to avoid salinity intrusion.

In the long term, water management will need to be reassessed as a result of the predictable filling up of most of the dams of the basin and the loss of their regulating capacity. This infilling could be accelerated if soil erosion increases with the expected increase in evapotranspiration and floods occurrence (Kwadijk, 1987) and remedial actions are not undertaken to control it.

Water quality, in terms of salinity, will decrease if present trends do not change. However, as this is mainly due to a mismanagement of irrigation, water salinization might be controlled or even minimized well before it presents a problem for the delta.

Other difficulties could arise from a rise in sea-level, both gradual or due to episodic events. Drainage of cultivated land will be more difficult, needing more frequent use of pumps. Periodic flooding of low lands during storms could affect the quality of ground water and soils in those areas. The salt wedge in the estuary could move landward also, from its actual normal position between Isle of Gracia and Amposta, to Tortosa; this would affect lateral aquifers (the alluvial and the Aldea-Camarles) with serious consequences for the continued exploitation and overall future use of this most important aquifer of Catalonia (Loaso, pers. comm.).

A sea-level rise also would affect the flow of ground water in the delta, which is governed by a very low hydraulic gradient (Loaso and Herrán, 1988), reducing the flushing of saline water and favouring salt intrusions.

4.1.3 Erosion and coastal stability

Although coastal dynamic studies of the delta are far from complete, a preliminary assessment of the possible problems can be delineated from the description of the present-day situation. It seems clear that the main impact of the expected sea-level rise will come through the increase in coastal dynamic processes. Sea-level may rise by as much as 1–4 cm per year (about 1 m in 50 years); this rate is similar to the fastest documented Holocene rise of the last 10,000 years (Aloisi, 1986). More reasonable estimates, however, suggest that the rise will be more on the order of 10–20 cm (see Milliman, Chapter 2).

Even this more gradual rise in sea-level will lead to the retreat, erosion and more regular drowning of the Trabucador isthmus, which is only 50–100 m wide and 0.5 m above the average sea-level. Should this happen the erosion processes now acting on the Alfaques peninsula to the south will be accelerated, exposing the Alfaques Bay and the southern lowlands of the delta to the sea. In the long run it is not pessimistic to assume that important parts of the delta could be eroded, returning it to the shape it had in the 16th and 17th centuries.

In the shorter term it is likely that low-lying areas, such as the peninsulas, Isla de Buda, Balsa del Pall in the north and Lagunas de la Tancada and la Encañisada in the south, will be regularly invaded by the sea, changing their fresh-water characteristics. Peninsula del Trabucador, Isla de Buda and especially Cabo Tortosa will recede in the face of the sea action, moving the delta front landwards. Apart from an increase in wave action, the rise of sea-level also may reduce the bedload carried by Ebro River, decreasing even more the sediment output at the mouth (Canals, pers. comm.).

Díaz Guerrero (pers. comm.) suggests that due to the retreat of the coastline, Alfaques Bay in the south and Fangar Bay in the north would gradually become closed, resulting in restricted lagoons whose communication with the sea would be reduced to small channels, mainly controlled by storm surges. This closing of the bays would take place because of the displacement of the peninsulas, which would be faster than that of the main body of the delta.

Díaz Guerrero also points out that, although the present state of the Trabucador isthmus does not favour its migration, the progressive destruction of the Isle of Buda in the north, which will be accelerated by sea-level rise, could initiate an important drift of eroded materials that would mitigate its retreat. Nevertheless, flooding would certainly become more common during storms and became fully emergent only during long periods of good weather.

4.2 Ecosystems

Effects of climatic changes on the Ebro Delta ecosystems can arise from modifications of marine and bay oceanographic conditions, alteration of the configuration of lagoons and marshes, and changes in the distribution of species because of the temperature increase.

Oceanographic conditions are expected to remain as they are now, except for the bays, where increased stratification and higher water temperatures are likely. As a consequence of these changes oxygen depletion in these bays may become more frequent and some loss of productivity can be envisaged. As ground- and surface-water characteristics may change due to greater extraction and better water management techniques, the renewal mechanisms of bay waters also could be altered.

Apart from direct impact on the cultivation of molluscs, it is difficult to assess possible repercussions of these changes on the indiginent species or general productivity of the whole area. Nevertheless, it seems possible that chlorophyl loads from these bays will be reduced as will be the conditions for mussel and oyster cultivation.

Present erosion processes already are reshaping the delta coastline, with a tendency to retreat at Cabo Tortosa and to break down Isla Buda behind it. If sea-level rises, most of these ecologically valuable areas of the delta will be covered by sea water. Conservation of these natural areas will depend greatly on defence actions that could be taken and on the stability of the Trabucador isthmus, but its future undoubtly will be compromised as priorities almost surely will be given to the agricultural lands. Water-management schemes and shoreline protection will certainly be needed to preserve these important ecological lagoons and marshes.

Finally, the impact of expected changes in temperature and rainfall patterns on species distribution is, at present, unpredictable. Speculations can vary from changes in the flora of the main wetlands to reintroduction or proliferation of pests and parasites (such as mosquitoes) that in the past proved a barrier for delta development. Nevertheless, available data do not suggest significant or rapid changes in the ecology of the delta.

4.3 Social and Economic Structures

The main impacts of the climatic changes on the social and economic structures of the delta can be grouped into the following areas:

Residential. Most of the towns in the delta are located in its central part which also has the highest elevation (3 to 4 m above sea-level). Therefore the main effect of the expected sea-level rise will be suffered by the tourist developments on the coast which represent only a small proportion of those existing on the Mediterranean coast of Spain and only a marginal value for the economy of the delta.

Agriculture and Fisheries. The main activity in the delta is agriculture. In the short run the effect on agriculture will consist of the loss of shorefront low lands and in the increase in salinization (which perhaps can be offset by better irrigation practices that can optimize the use of the Ebro water, if its quality remains and the present increased salinization is stopped or reversed). Crop changes are not thought to be necessary.

Although the Ebro discharge, which is the main source of nutrients, probably will remain unchanged, fisheries are expected to decrease because of the loss of nursery grounds in the northern and, especially, in the southern bays. Also, aquaculture development in the southern bay will be in jeopardy if it is opened to the sea by breaching of the Trabucador isthmus and exposed to wave action, or as a consequence of higher summer water temperatures and increased stratification associated with a reduction of the renewal rate because of the progressive closing of the bay mouths.

Infrastructure. There are no main infrastructures in the delta at the moment, and the only important engineering work is the San Carlos de la Rapita port in the south. Because the sea rise will be gradual and because this port is located just off the delta, no major structural problems are expected there, as defences will surely be adapted in time. The problems for this port could come from the increased flux of the sediments eroded from the Alfaques peninsula; depending on this factor the port could be silted up, or require additional dredging to keep it open.

There are no main roads in the delta and those which cross it are rather inland, thus minimizing the impact of sea-level rise and erosion.

The saltworks are the only important activity that will certainly be affected by climatic change. Nevertheless, with adequate protection measures for shore regeneration or stabilization, these works could be saved, adding an extra protection to the southern bay in the short term. However, in the long term it is doubtful that their preservation is feasible in the face of a higher sea-level and with the expected increase in storms and associated lowest and high tides.

5 CONCLUSIONS AND RECOMMENDATIONS

The delta's shoreline already is being reshaped because of decreased sediment input from the Ebro River. A climatic change and sea-level rise in the predicted range would accelerate the present trend. The southern part of the delta is the most threatened, mainly because of possible rupture of the Trabucador isthmus and from the erosion and reshaping of the peninsula.

In the short term, climatic changes probably will be felt by an increased occurrence of episodic peak events, such as storm surges and floods. More frequent stratification in the bays and oxygen deficits in these areas during the summer season also may occur as a consequence of these changes and modification of the general hydrological system. Water quality and quantity in the delta may not change if the increased evaporation is offset by better management of the irrigation system and by the adequate regulation of the Ebro basin. Over the longer term, however, it may be necessary to redesign the present water management approach.

In order to evaluate and control the major negative effects of the expected climatic and sea-level changes, the following activities should be undertaken:

1) Basic research on the delta should be continued, with regular and continuous data collection, especially with regards to the dependence of ecological areas on water characteristics and the impact on fisheries of the possible changes in the northern and southern bays.

2) Detailed mapping should be done on the delta, both in the emerged area and in the adjacent shallow waters. Adequate high-precision geodetic mapping should be used to evaluate relative sea-level rise and possible local subsidence.

3) Coastal reshaping and sediment transport should be studied and control measures adopted, taking into account sea-level rise and the consequent flooding of lowlands during storm surges.

4) Long-term continuous data series are essential to documenting sea-level, nutrient and phytoplankton concentrations, the dynamics of the bays, river output and shoreline change; data needed to evaluate and correct the possible negative effects of the expected climatic change should be collected as soon as possible.

5) Based on the predicted temperature increase, insect population changes should be studied. Possible health and agricultural consequences need to be analysed, together with the side-effects of necessary control actions (i.e. increase in the use of pesticides and herbicides and the runoff of these products into the lagoons and the bays).

6) The Ebro basin and dam system management needs to be redefined to react to predicted changes in soil erosion, sediment transport and water demands. Ground-water quality and management and resource allocation in the delta also should be reassessed, taking into account possible future conditions.

7) Planned and expected economic development on the delta should be re-evaluated. This applies especially to agriculture, aquaculture, transport (mainly port facilities), salt production and tourism infrastructure.

6 Acknowledgments

Dr Belén Alonso has been of great help in obtaining the information used in this chapter. The written contribution received from Carlos Loaso, Ignacio Díez and M. Canals and the comments from Jordi Camp and Jose Luis Monsó, who reviewed the manuscript, have been greatly appreciated. I also have to thank my friend Dr Giuliano Sestíni, who corrected the spelling and gave many important suggestions about the text.

7 References

Aloisi, J.C., 1986. Sur un modele de sedimentation deltaique. Contribution a la cannaisance des marges passives. These de 3me cycle. Univ. de Perpignan. Perpignan, p. 162.

Alvarez de Meneses, A. and Masso, C., 1983. Características Oceanográficas del Delta del Ebro. In: M. G. Mariño (ed.), *Sistema Integrado del Ebro*. Madrid, 1986. 61–81.

Camarasa, J.M., Folch i Gillén, R., Masalles, R.M. and Velasco, E., 1976. Carta del Paisatge Vegetal del Delta de L'Ebre. Ed. Institució Catalana d'Historia Natural, Barcelona, 1976.

Camp, J. and Delgado, M., 1987. Hidrografía de las bahias del delta del Ebro. *Investigaciones Pesqueras*, **51**(3), 351–369.

Centro de Estudios Hidrográficos (C.E.H.) 1975. Datos de Salinidad del Bajo Ebro. CEDEX-MOPU, 5p + Annexes.

Delgado, M., 1987. Fitoplancton de las bahías del delta del Ebro. *Investigaciones Pesqueras*, **51**(4), 517–548.

Delgado, M. and Camp, J., 1987. Abundancia y distribución de nutrientes inorgánicos disueltos en las bahías del delta del Ebro. *Investigaciones Pesqueras*, **51**(3), 427–441.

Font, J., 1983. Corrientes permanentes en el borde de la plataforma continental frente al Delta del Ebro. In: Castellví, P. (ed.), *Estudio Oceanográfico de la Plataforma Continental*. Barcelona, 1983, 149–161.

Gran Enciclopedia Catalana, 1979. **6**, p. 443.

Han, G., Ballester, A. and Kohler, K., 1983. Circulation on the Spanish mediterranean continental shelf near río Ebro. In: Castellví, P. (ed.), *Estudio Oceanográfico de la Plataforma Continental*. Barcelona, 1983, 137–147.

Hekstra, G., 1986. Will Climate Changes Flood the Netherlands? Effects on Agriculture, Land Use and Well-Being. *Ambio*, **15**, 316–326

Janssen, C. R. and Kuhn, M., 1987. Indications for climatic changes from past analogues. *European Workshop on Interrelated Bioclimatic and Land Use Changes*. Vol. B: Paleoclimatic data.

Kwadijk, J., 1987. Identification of climate sensitive processes and landforms. In: *European workshop on interrelated bioclimatic and land use changes*, Vol. C: Impact of

Climatic change and CO_2- enrichment on exogenic processes and biosphere: General, 1–30, Noordwijkerhout, Netherlands.
Loaso, C. and Herrán, F.J., 1988. Aportaciones al conocimiento de la Hidroquímica e Hidrología del acuifero del Delta del Ebro. Su importancia en relación con el desarrollo de la Acuicultura. *IV Simposio de Hidrogeología*.
Maldonado, A., 1975. Sedimentation, Stratigraphy and Development of the Ebro Delta, Spain. In: Broussard, M.L. (ed.), *Delta Models for Exploration*, Houston Geological Society. Houston (Texas), 311–338.
Maldonado, A., 1977. Introduccion Geologica al Delta del Ebro. In: *Els Sistemes Naturals del Delta del Ebre*. Institucio Catalana d'Historia Natural (ed.), Barcelona, 7–45.
Maldonado, A., 1983. dynamica Sedimentaria y Evolucion Litoral reciente del Delta del Ebro. In: Mariño, M. (ed.), *Sistema Integrado del Ebro*. Madrid 1983, 33–60.
Mariñas, J. and Tejedor, L., 1983. Modelo Numérico de Simulación Hidrodinámica del Delta del Ebro. In: Mariño. M. (ed.), *Sistema Integrado del Ebro*. Madrid 1983, 157–173.
Mariño, M.G. (ed.), 1983. Sistema Integrado del Ebro. Resultados del Estudio Cooperativo Hispano Americano 793028. Madrid.
MOPU, 1978. Plan Indicativo de Usos del Dominio Público Litoral de la Provincia de Tarragona. Ed. Servicio de Publicaciones del Mopu. Madrid.
Pérez, M. and Camp, J., 1986. Distribución espacial y biomasa de las fanerógamas marinas de las bahías del delta del Ebro. *Investigaciones Pesqueras*, **50**(4), 519–530.
Pitovranov, S. E., 1987. The assessment of impacts of possible climatic changes on the results of the IIASA rains sulfur deposition model in Europe. *European Workshop on Interrelated Bioclimatic and Land Use Changes*. 17–21 October 1987.
Seró, R. and Maymó, J., 1972. Les Transformacions Economiques al Delta de l'Ebre. Ed. Banca Catalana. Servei d'Estudis. Barcelona.
Varela, J., Gallardo, A. and Lopez de Velasco, A., 1983. Retencion de Solidos por los Embalses de Mequinenza y Ribarroja. Efectos sobre los Aportes al Delta del Ebro. In: Mariño, M. (ed.). *Sistema Integrado del Ebro*. Madrid, 1983. 203–219.

11

Implications des Changements Climatiques Etude de Cas: le golfe du Lion

Implications of Climatic Changes on the Golfe du Lion

J.-J. Corre
(*Consultant, Institut de Botanique,
Montpellier, France*)

Avec de concours de
A. Berger, J.-P. Béthoux, S. Castanier, A. Crivelli, Ph. Daget, M. Desse,
P. Duncan, F. Gagnier, P. Grillas, O. Guelorget, H. Hafner,
P. Heurteaux, A. Johnson, Mme A.-Y. Le Dain, A. L'Homer,
J.-M. Miossec, J.-P. Perthuisot, B. Picon, J.-P. Quignard, F. Rueda,
A. Tamisier, F. Teral, J. Trayssac et J.Y. Vourgères

English Abstract

The international conference of the World Climate Programme, held in Villach (Austria) in 1985, forecast that a doubling of the percentage of CO_2 would cause a warming of the atmosphere by 1.5 to 4.5°C, the lower value being the most likely. The changes hypothetically predicted by the GISS general circulation model for the French Mediterranean coast are a warming of 4°C in January, between 3 and 4°C in July east of Montpellier, and 4°C to the west, with an annual mean increase of 4°C. Precipitation would increase by 0.5 mm/day (+15.5 mm/month) in January; there would be practically no change in July (a slight increase east of the Camargue, a slight decrease to the west), and the annual mean precipitation would remain the same as today. Evaporation would increase by 0.25 mm/day, that is 7.5 mm/month, in January; it would increase slightly in July to the east of the Rhône and decrease slightly to the west.

In contrast to these predictions, Wigley (1987) and Pitrovanov (1987) have concluded that the consequences of a general warming in the Mediterranean coastal area could be minor winter cooling (0.5°C), but with greater interannual variations. Spring and summer temperatures would remain unchanged or increase gradually, with a warmer autumn. Rainfall should be more important in winter, less in summer. Cyclonic circulation could become more active with a global rise of temperature, as noted in the Atlantic during the last period of warming (Rognon, 1981), and extreme situations, such as a succession of cold winters, dry summers, etc., could lead to greater interannual variability in climate.

The Golfe du Lion forms an arc about 270 km long between the crystalline massif of the Albères in the south-west and the Estaque calcareous chain in the north-east (Fig. 11.1). Most of the littoral zone is flanked by a low shore, frequently lagoonal, interrupted by few rocky capes (Cap Leucate, the La Clape Mountains, Cap d'Agde, Mont Saint Clair). The Rhône Delta, situated at the eastern end of the Golfe, covers an area of 173,640 ha. As much as 40% of the Golfe shoreline has been formed by the alluvial deposits of the Rhône River.

The shoreline of the Golfe du Lion since the end of the last glaciation and during past historical periods has been subject to much greater changes than those that are foreseen for the next century. In the absence of human occupation and activities, its evolution would be of no great consequence, but because of the present degree of permanent occupation and economic investment, the predicted changes are worrisome, particularly as they probably cannot be entirely controlled.

GEOLOGY AND GEOMORPHOLOGY

The area from the north of the Albères to the Canet is characterized by subsidence filled with Neogene sediments. The Flandrian transgression led to the formation of a coastal strip accompanied by deltaic and lagoonal sedimentation. The present-day lakes are the remnants of lagoons filled by river sedimentation. Further north, under the Leucate sand bar, a regressive marine facies is underlain by sands.

The dune system has a width of only a few hundred metres and at the back of the beach it is mostly limited to a simple ridge.

The lake of Thau (48.2° lat., 1.4° long.) is an old gulf that occupies a basin closed by a narrow, flat sand bar. From Sète to the Grande Motte, the coast does not seem to have developed very much during the Flandrian period. The system of dunes is relatively wide and low in altitude (<10 m), greatly altered by the local winds.

East of the Grande Motte, the sand bar is divided into several fan-like sandy strips with interspersed lagoons. This system extends throughout the Petite Camargue and gives a coastal formation several kilometres wide. It is abruptly cut off near the Petit Rhône by a fault that lies perpendicular to the coast. Each one of the bars appears to mark various stages of considerable progradation of the coast, beginning in the Flandrian transgression (Bazille, 1974). They continued to develop until modern times as witnessed by the formation of Espiguette Cape. The system of dunes is very diversified, but altitudes, although highest on the entire coast, do not exceed 12 m.

East of Saintes-Maries-de-la-Mer, the landscape consists of a network of ponds and earth banks, remnants of dunes partially filled by alluvial clay deposits. This landscape is the result of the combined action of fluvio-swampy deposits, predominant in the north, and lagoonal-marine deposits in the south (Fig. 11.20).

On the basis of surface deposits we can distinguish three ecological sectors from north to south: the Haute Camargue, with altitudes between 4.5 and 1 m NGF[1] and with low-lying marshes; the Moyenne Camargue, with a lower general altitude and marshes locally below sea-level, including the

Vaccarès Pond; and the Basse Camargue, characterized by more or less distinct rows of dunes separated by ponds; altitudes vary between +7 m NGF for certain dunes and -0.3 m NGF in the ponds. The Haute and Moyenne Camargue were formed from a network of distributory channels encircling the marshes, the Basse Camargue through lagunal-marine sedimentation (sands and clayey silts).

Today the coast of the delta is advancing seaward through two large blunted capes that are migrating westward: the Beauduc headland, at right angles to the Camargue, reflects the discharge of the Grand Rhône; the Espiguette headland, normal to the Petite Camargue, corresponds to that of the Petit Rhône.

THE CLIMATE TODAY

The climate of the Golfe du Lion is Mediterranean subhumid, with cool to temperate winters. It also has been called "transitional" because some years it displays oceanic characteristics and more rarely continental climate characteristics. Annual rainfall ranges between 400 and 750 mm (540 mm in the Camargue), with 50 to 95 rainy days a year. Mean annual temperatures range between 14 and 15°C.

In the Golfe du Lion region the prevailing winds have a north-west–south-east orientation with a strong preponderance in the north-west sector. Winds are frequent and at times violent; they blow 208 days per $year^{-1}$ at more than 38 km/h^{-1} and for 11 days per $year^{-1}$ at more than 74 km/h^{-1}; peak velocities reach 135–165 km/h^{-1}.

The rise of the Azores anticyclone in the summer generally protects the region from perturbations from the Atlantic, but some years the anticyclone allows the penetration of low pressures from the west. In winter, the relief of the hinterland and the continental thermal high pressures displace the perturbated currents northwards; however, the weakening of these high pressures may allow western circulation perturbations to pass through. In any event, two to three of the perturbations affecting the region are local in origin, caused by meridian circulations of cold polar air reaching the Mediterranean. In particular, they give rise to south easterly winds that can be violent and bring torrential rain.

The average minimum temperature is especially important for vegetation, as it affects the distribution threshold of various species; it decreases to the north-east (Fig. 11.32) to +4°C on the Albères coast (Cap Béar); the decrease diminishes to +0.9°C in Montpellier and it increases to +1.7°C in the Camargue and +8°C in Monaco, near the Italian border. The median part of the Gulf is thus relatively cold in the winter and this causes a biogeographical gap.

Climatic variability has been studied in detail on the basis of continuous observations from 1945 and 1987 at the biology research station of La Tour-du-Valat and the National Scientific Research Centre (CNRS) in the Rhône Delta. During that period, precipitation and temperature have been quite variable. At first sight there seems to be a considerable yearly variation: a wet or cold year often followed by a dry or hot year. This important variability, however, masks a very complex general evolutionary trend, though relatively uneventful before 1942–1945. No statistical trends can be detected and therefore the climatic variations of the last 44 years give no indication on the behaviour of climate in the next decades.

THE SEA

The tides are weak, the amplitude not exceeding 30 cm at average springtide, but variations due to oscillation of atmospheric pressure and winds are important. Sea-level increases with south-easterly winds and falls with north-westerly winds. Greslou (1984) reports record values of +1.80 m east of the Camargue (toward the Gulf of Fos) and -0.5 m NGF at Port-La-Nouvelle (south of the Golfe du Lion). According to French National Geographic Institute data there has been a net sea-level rise of 10 cm from 1885 to 1979 with an acceleration between 1944–1955 and a decrease afterwards.

In the northern part of the Golfe (Sète-Camargue), the prevailing orientation is from the south-west or the south-east; in the south, the main directions are east, south-east or north-north-east. Maximum 100-year wave height is calculated at 5–10 m depending on the site; the annual probability is from 3 to 6 m.

The general oceanic circulation affects coastline dynamics, but currents caused by waves that are oblique to the beach also play an important role in the longshore transport of coastal sediments.

THE SOCIO-ECONOMIC SETTING

The Past

The coastal zone of the Golfe du Lion had long been considered marginal because it lies outside the principal economic activities of the region. Thus dominant activities included fishing and shellfish growing in the ponds, exploitation of salt, hunting, agriculture, extensive breeding of horses (*manades*), etc. The harbours of Sète and Port-la-Nouvelle remained the principal centres of commercial and industrial activities on the coast, and summer tourism was active.

The environment had reached a certain level of balance. When the agricultural or fishing activities required, the land owners took various measures of protection against the sea. They aimed primarily at preserving the continuity of the dune strip and at managing the channels (*graus*) between the sea and the lakes. In 1867, Régy, a governmental civil engineer, proposed a management plan for the Hérault coastline as well as a plan for the protection of the coast (planting of protective vegetation, creation of artificial dunes, etc.) all of which gives a clear picture of the concern by the authorities to minimize coastal erosion and innundation by the sea.

In the Camargue, the first attempts to control the environment go back to the 12th century, but the true concern for the protection of the coastline did not arise until the 19th century when salt production allowed stretches of the Basse Camargue to revert to a natural condition after unsuccessful attempts to cultivate them. In 1859, a sea wall was completed. In 1869, the arms of the Rhône were dammed permanently; this put an end to the shifting of the river and to the risks of flooding.

In 1929, most of the Basse Camargue was made into a reserve. This was done because of the conflict between salters and farmers over the management of water resources and because of the importance of the area as a biotope. The status of the reserve for the delta has fostered both international tourism and large-scale scientific activities (Biology Station of the Tour du Valat, CNRS laboratory – until 1986, Camargue National Reserve, Regional Park, etc.).

The Present Situation
The present interest of the coast of the Golfe du Lion and the financial stakes concerning it have developed greatly under several large-scale development programmes (Figs. 11.1a-c).
1) *Agriculture*: development of irrigation and drainage networks in the Camargue for rice growing and management of marshes for water fowl. In the Languedoc, the construction of the Canal du Bas-Rhône and large-scale infrastructure projects of irrigation and drainage for local diversification of crops.
2) *Tourism*: for the entire coastline (except in the Camargue) mosquito control has made the beaches more attractive. A concerted construction plan was launched (new towns, extension of existing urban areas, etc.) which made the entire Golfe du Lion one of the new summer resorts of Europe. Currently more than 40% of the coastline has been developed.
3) *Industry*: mainly the creation of a new harbour, metal works and a oil refining complex at Fos-sur-Mer (Annexe III).
4) *Environment*: creation of the Regional Nature Park of the Camargue and designating the Camargue as Reserve of the Biosphere. These actions underscore the importance of this region both for its role in the preservation of the European avifauna and for the uniqueness of its biotopes (for instance, the Riège forest of *Juniperus phoenicea*), as well as for the richness and diversity of its wetlands (506 natural stretches of water of 0.5 ha or more area, covering 40% of the Camargue and ranging from fresh and lightly saline waters to highly saline waters (Britton and Podlejski, 1981).

CONTEMPORARY COASTAL CHANGES

A systematic study of the development of the coast based on aerial photographs and bibliographical data for the years 1942–1946 to 1970–1980 shows the following trends:
1) Eastern Pyrénées and Aude (from Albères to the Vendres lagoon): 90 km of coast, 27 km of which are bordered by dunes. Generally the coast is advancing, with some recession near the mouths of coastal rivers (Tech and Tet) and around rocky headlands (Albères, Port Leucate): 53% of the coastline is prograding by 0.5 to 1.5 m/yr^{-1}, locally more than 3 m/yr^{-1}; 27% of the coastline recedes by 0.5 to 1.2 m/yr^{-1}, in places more than 3 m/yr^{-1}; 20% of the coastline is stable (variations <0.3 m/yr^{-1}) taking into account the margin of error in measurement.
2) Hérault (from the Vendres lagoon to the Grande Motte): 84 km of coastline, 62 km of which are bordered by dunes. The coast shows

a similar recession near river mouths and rocky headlands. This is, however, less systematic, because of existing coastal protection. In summary, 9.5% of the coastline advances, 24.4% of the coastline recedes, 66.1% of the coastline is stable.
3) Gard and Bouches-du-Rhône (from the Grande Motte to the gulf of Fos): 103 km of coastline; of the 33 km of the Gard, 17.6 km are coastal dunes. Coastal dynamics are linked with the gradual westward movement of the sedimentary deposits of the two mouths of the Rhône. Espiguette Point advances by 18 m/yr^{-1}; construction of dikes at its tip has changed the direction of this forward movement and in spots advances of 20–25 m/yr^{-1} has been observed. Beauduc Point advances by 11 m/yr^{-1}. Near the Petit Rhône mouth coastal regression is considerable, more than 4 m/yr^{-1}. This last figure can be correlated with the decrease in the solid discharge of the river, which has decreased by 90% since the end of the last century. This section of coast can be summarized as follows: 28% of the coastline advances, 62% of the coastline recedes, 10% remains stable.

Based on observations from the Hérault coastline (Le Dain et al., 1987), accretion and erosion occurs in cycles as has been observed on other coasts (Bird, 1986). Major storms have particularly formidable consequences: a three-day storm in 1982 resulted in local coastal recession of several dozen metres and produced breaches in the coastal dunes.

AN EVALUATION OF THE IMPACTS OF CLIMATIC CHANGE

Physical Oceanography
The dynamic circulation of the Mediterranean Sea depends upon water temperature and salinity differences between the Atlantic and the Mediterranean and other factors such as sea-level, winter temperatures, fresh-water inputs and winds. Despite any change in temperature due to climate change, the Mediterranean will continue to have a continental climate compared to the Atlantic. The western basin will continue to be a basin in which evaporation exceeds fresh-water input, and a sea-level change of a few to several decimeters cannot modify appreciably the flow across the sills of Gibraltar and Sicily. On a seasonal or annual scale, modification of rainfall (with greater intensity in winter) and colder winters will tend to magnify the Ligurian current, although generally within ranges already observed.

Marine Fauna
The physico-chemical features of the Golfe du Lion, salinity and temperature in particular, explain a large part of the faunal distributions as well as their recent changes and modifications. As the Golfe constitutes a biogeographic boundary for a number of species, any modification of these factors could have far-reaching consequences.

Physical Effects on the Coast
The predicted coastal changes in 2025 AD (Fig. 11.22a and b) only shows the general trends. Taking into account the capacity of this deltaic coast to reconstruct itself after major storms, and to rise gradually in phase with the

average rise of sea-level, no catastrophic changes are envisioned during the next 40 years. Nevertheless, those stretches of the shore that are already unstable or threatened will be even more so in 2025. These are, from east to west:
- the spit of La Gracieuse, which may become detached;
- the Courbe-à-la-Mer and the inlet of La Dent;
- the mouth of the Petit Rhône and beaches situated on both sides of the promontory that protects Saintes-Maries-de-la-Mer;
- the shore of the Petite Camargue;
- the embayment between the point of Espiguette and the Grau-du-Roi, which risks shoaling and silting up.

There are, however, still several outstanding questions:
- How will the offshore coastal defences behave, such as the breakwaters of the Courbe-à-la-Mer in the Petite Camargue?
- Will the mouth of Grand Rhône shift eastwards? Or, more likely, will a new mouth develop towards the south–south-west?
- Will subsidence continue at Beauduc, threatening inundation of the lagoons in that sector of the delta?

The coast presents different situations in its response to changes:
a) In dunal areas not bordered (landward) by roads, tourist resorts or areas of intensive agriculture, and that are in a good condition (e.g. continuous and high dune belt, with homogeneous grain size), the whole system will retreat (Chapter 4). It is possible that in the shorter term the retreat may be slower or faster.
b) In the dune areas that are damaged or have been intensively modified/occupied by man, the dunes have variable elevations, breaks and anomalous gradients. In these cases degradation will accelerate, both by seaward removal of sand (i.e. littoral drift) and landwards by overwash. There will probably be a thinning of beaches and the progressive formation of inland dune belts, fragmented and oriented parallel to the dominant onshore wind (the Mistral or Tramontane). The littoral coastal landscape will gradually become a succession of flat surfaces, with or without wetlands, and of inland dunes, with a general north–south orientation, not east–west as today. The attainment of such morphology will probably take half a century.
c) In the beach-dune barriers that separate the sea from the lakes, sea-level rise and especially storm surges could lead to the formation of new inlets.
d) In the strongly urbanized areas, there has been in general a decrease of beach-slope gradients due to the smothering or removal of dunes for the construction of sea-front roads, of buildings, camp sites, and a reduction of beach width because of the close spacing of defence structures.

These two aspects of coastal degradation emphasize the risk of washovers by storm waves and the deposition of sand outside the beach zone, especially where beaches are narrow. Once moved onshore, the sands are lost to littoral drift. Thus, the damping effect of the beach to bigger waves will decrease; beach profiles will steepen and retreat; they become less functional, even in the stretches where coastal erosion does not seem important (for example, the sea front of Valras Beach after the storm of 1982, and the western shore of Palavas after the October 1987 storm). In these strongly transformed sectors, it is likely that a significant rise of sea-level will cause much greater damage to facilities because of the increased frequency of storm washovers.

In the event of a clearly identified rise of sea-level, it is probable that coastal protection will be focused on the urbanized stretches and on the protection of the coastal lakes and lagoons, the areas of major economic importance in the Languedoc-Roussillon. Proper identification of the areas of probable degradation will permit faster response for necessary protective measures. Given added time, costs may be optimized.

Changes in the Lagoonal Zone

Assuming that the littoral barriers will weaken (without human intervention), inlets to the coastal lakes should widen and increase, followed by the gradual return of a certain number of lagoonal basins to a marine environment. In the absence of general control, one can expect a widespread shifting of the paralic systems. In several cases this should result in primary productivity (and therefore shellfish culture) in the lagoonal basins and a consequent impoverishment of the adjacent marine belt (Frisoni, 1984; Guelorget and Perthuisot, 1983; Guelorget, 1985).

In the strongly confined lakes, and particularly in most of the lakes of the Camargue, two alternatives can be considered: with or without intervention to the hydraulic system.

Without intervention

The lakes of the Camargue are separated from the sea by a dike; communication is through a series of sluices. About 60 million m^3 were discharged during 1986 and 1987, representing an outflow of 1.5 million tons of dissolved solids.

Outflow is enhanced by the north–north-west mistral wind, which pushes waters southward. In practice, outflow presently occurs only at the edge of the lower lakes, at elevations of >0.10 m NGF. Clearly, the rise of sea level will increasingly decrease this outflow. Normally, a rise of the level of the lakes should accompany that of the sea, but because the upper parts of the Vaccarès system are conditioned by commercial activities, they hamper the drainage of the delta; the result could be marine flooding of the lowlands.

It is foreseen by P. Heurteaux that without intervention, the Vaccarès and the lower lakes of the Camargue will be transformed into an hypersaline system (50–70 g/l^{-1}). This will result in dramatic biological changes; for instance, the salinity will be too high for fish colonization. The permanent salt marshes located around the system Vaccarès of the lower lakes will not be affected, due to a controlled (and large) inflow of fresh water. The salinity of the temporary, uncontrolled salt marshes will probably increase especially in the winter, due to an increasingly brackish water table. The dry period could be shortened by decreased drainage capacity caused by a rise of sea-level. Changes will be more notable, as the salt marshes are close to the water table. Change in the macroflora of the marshes should be minimal, but species sensitive to salt (like the rare *Damasonium stellatum*) could be seriously affected.

In general, one can expect a first phase of de-oxygenation in the coastal basins with localized distrophic tendencies (the development of sulphate-reducing bacteria and methanogenesis favoured by the anoxic environment and by temperatures above 21°C). This first phase will be quite significant, especially in the basins dedicated to intensive fish and shellfish culture. The second phase would be a re-equilibration of the systems by the bacterial activity together with a decrease of primary production which will be damaged by the reduced flushing of the inland areas.

With intervention
It is rather unlikely that people will remain passive in the face of a drastic change in salinity and drainage conditions in the Camargue. Intervention will be directed to counteract the increased salinities in the Vaccarès and the lower lakes, as well as to ensure the flushing of rain and waste waters. According to the measures taken, the salinity of the Vaccarès could be reduced considerably, down to its present values or lower. The lake would then be a favourable environment for the development of an important macroflora, though one cannot foresee what kind. Fish culture could continue, provided there is access to the sea between October and March (when fish migrate). The other lakes, which communicate with the sea, should bear the same fish populations as today.

The managed salt marshes around the Vaccarès and the lower lakes would not be modified. Human intervention could, however, improve the drainage of the Camargue, and by that slightly shorten the duration of the submergence of the temporary natural marshes.

In some parts, the increased temperature could increase phytoplankton development and the growth of benthos. But it also will systematically favour *in situ* bacterial activity, as well as the turn-over of bacterial populations. This increased bacterial activity would take place at the expense of the organic matter in each basin. With the decrease of primary productivity, however, the reserve of organic matter eventually could be exhausted, thus limiting bacterial blooms and the recycling of nutrients.

Changes in Flora and Plant Cover
As sea-level rises, plant cover will change and dunes probably will become more unstable. A drier environment, colder winters or major storms could cause part or total destruction of plant populations (*Thymelaea hirsuta*, *Otanthus maritimus*, etc.) along large parts of seashore. As salinity change will depend on specific location, extensive pasturelands jammed between ponds may be particularly affected.

Changes in the Bird Fauna
As birds are at the end of the food chain, the impacts of climatic changes on them are difficult to foresee. The expected deterioration of the ichtyofauna in the central part of the Camargue will certainly affect wintering fish-eating species, especially the cormorants, grebes and herons. As regards the nesting birds, the wellbeing of the Ardeids (e.g. the grey heron) could depend on the persistence of permanent fresh water, or slightly brackish marshes, and of their fish and amphibian fauna.

The flamingo, more than any other species, will benefit from the more saline waters. The Laro-Limicoles will not be affected, as they are not fish-eating. The survival of the sterns will depend on the persistence of fresh-water marshes.

The effect on migrating birds will depend on their habitats. Species like the *Chevalier cocorli* will benefit from more saline waters, while visitors that live in fresh-water marshes (e.g. the *Chevalier sylvain*) will suffer from their decreased area.

SOCIO-ECONOMIC IMPACTS

The Future of the Camargue

It is not certain whether the present-day harmony between human beings and nature has prevailed over the former approach of subjugation or outright domination of nature. However, there is a general consensus about the kind of response that coastal societies must have to the threat of sea-level rise.

In 1965 André Malraux, Minister for Culture, indicated that for the Camargue it would be better to control human activities and to oppose intended modifications to the natural environment. That present coastal activities are immutable is a questionable approach. There is nothing to indicate, for instance, that farmers will be prepared to increase water pumping to save activities that are already threatened, like rice growing. If increased to an area of 10,000 ha, through the financial assistance of the Ministry of Agriculture, would rice culture survive new EEC restrictions and the freeze on acreage expansion? After the failure in 1975 to grow reeds for paper-making (Picon, 1978) and the more recent (1980) failure of aquaculture (Picon, 1985), the future of new agricultural investments in increasingly uncertain natural conditions is rather doubtful.

There is nothing to suggest that the Henin Bank will be ready to invest the needed capital to elevate the salt marshes. Instead, it is quite possible that activities will move towards natural recreation. A survey conducted among visitors to the Camargue in 1984 indicates a strong demand for natural space (Picon, 1987): 82.5% of the visitors were attracted by nature. Furthermore, the persistence of a very low population density since Roman times (Table 11.1), a sign of demographic adaptation to the limitations of the environment, could be an advantage in response to renewed shoreline instability.

Therefore, rather than impose the need for new projects of coastal management and defence, rising sea level may provide the opportunity for the further development of "green" tourism. This type of recreation, increasingly popular with the urbanized middle classes of western Europe, does not necessitate infrastructures like marinas or yachting harbours. After 1981, when camping was forbidden on the beach and dunes of Sainte-Maries-de-la-Mer, summer campers (65% of whom were workers, clerks and craftsmen from the region) were replaced by hikers of upper-middle class origin from European cities (Picon, 1987). If social demand, natural conditions and economic profit are thus combined, it is possible to envisage that the physically evolving coast will produce a situation in which development projects will minimize the economic loss of changing natural conditions. Recent investigations (Picon *et al.*, 1987) have shown that in the lower Camargue using land as natural spaces can be as profitable for the landowner as using it for agricultural or industrial production. This awareness on the part of landowners appears to derive from a progressive modification of the long-held belief that the land should be used for agricultural economy. Initially hostile in principle to the creation of nature reserves and tourism, they appear to have modified their position considerably. Those who conceived the saline and lacustrine environments of the lower Camargue as a hindrance to agricultural development have gradually begun to realize that in the context of a post-industrial society it can be considered a resource.

The future of the Languedoc-Roussillon coast

Climatic changes. An increase in temperature could be a positive element in regard to tourism and to associated activities. An extension of the warm season would improve the regional potential, allowing quicker returns in investment, thanks to more intensive utilization of facilities over a longer period. It also could increase the duration of seasonal activities that depend on the flow of mass tourism.

On the other hand, longer or more frequent periods of drought could jeopardize the availability of water supplies needed for a large concentration of population.

Sea-level rise. Sea-level rise, together with longer or more storm surges, could cause a deterioration of beaches and infrastructures resulting in escalating costs of protection and repairs, and from increasing technical and financial problems of coastal defense.

Such new physical parameters will require the intervention by public authorities to solve the complex problems regarding financing as well as the designation of the responsibilities of local communities, departments or regional authorities and the State. Beyond the expected physical effects, climatic changes could induce modifications in the behaviour of society and the business community in a more general way.

Because business involves both elements of time and risk, investments for the maintenance of existing installations and for tourist uses will require increased risk. In economic terms then, man's continued use of the coast for tourism may be questioned, as the shore becomes increasingly unstable and therefore less hospitable.

Responses to climatic changes could include a relocation of tourist activities, for instance a greater concentration on the more stable rocky coasts, although this prospect should be considered as an extreme scenario. The Languedoc coast then would return to conditions that prevailed at the end of the 19th and early 20th century.

CONCLUSIONS

It is fortunate that the insalubrity of the coast in previous centuries caused most of the large urban centres of the Golfe du Lion (Perpignan, Narbonne, Béziers, Montpellier, Nîmes, Arles) to be built sufficiently back from the beach, so that they are protected from the serious impact of a foreseeable rise of sea-level (Fig. 11.1). On the other hand, the coast, with its tourist and harbour installations, is particularly vulnerable because of the generally narrow strip of sand for most of its length, the low altitude of the dunes and coastal ridges, their levelling for construction purposes, and their decayed condition due to the period of abandonment between the farming and the tourist development phases.

Presently, the major risk would appear to be from the increased frequency and severity of the storms linked to cyclonic circulation. The great storm of 1982 acted as an alarm for public authorities who subsequently launched a number of actions and control measures, either experimental or definitive, to reinforce existing defence policies needed to respond to the most critical situations. The aim of these actions and measures is three-fold: to control tangential transits (rockfill), to help to bring back and hold the sand on the beach and the dunes (by a network of semi-permeable barriers, nets, planting schemes, etc.), and to create an obstacle to storms while also maintaining a wide enough strip of usable beach (e.g. artificial dunes, dikes for the bottom of the beach, etc.).

RECOMMENDATIONS

1) Complete documentation of the recent climatic evolution for the western Mediterranean region.
2) Establish seashore fluctuations, with special attention to seasonal and yearly changes to ascertain cycles of erosion/accretion.
3) Establish a detailed classification of plant or geomorphological indicators in relation to the vulnerability linked with sea and (or) wind erosion.
4) Establish (to a scale of about 1/5000) provisional charts showing the vulnerability to the short-, medium- or long-term (year, decade, century).
5) Increase, by arranging and diversifying experimental designs for shore protection.
6) Complete evolution schemes for spontaneous or introduced plant cover in relation to sea and wind erosion, and in relation to conservation of seashore; in order firstly to understand, then to select best mechanisms for dune defence.
7) Precise qualitative and quantitative schemes for evolution of lagoonal populations in both "confining" and "non-confining" environments.
8) Develop studies on the role of environmental diversity for the maintenance of animal communities.
9) To document land use in the littoral and lagoonal environments to identify changes in socio-economic pressures and competition for use.

---

1 INTRODUCTION

1.1 Hypothèses sur les variations du climat

Avec un réchauffement général du globe on devrait s'attendre pour le littoral méditerranéen à un refroidissement hivernal peu important de l'ordre de $<= 0,5°C$ mais présentant une plus grande variabilité interannuelle, un printemps et un été sans beaucoup de changements avec peut-être une tendance à la hausse, un automne plus chaud. Les pluies devraient être plus importantes en hiver, moindres en été.

A cela s'ajoute des circulations cycloniques qui peuvent devenir plus actives avec l'élévation de température du globe comme cela a été observé dans l'Atlantique au cours de la dernière période de réchauffement (Rognon, 1981).

Même si les changements sont peu importants, ils ne doivent pas faire perdre de vue qu'ils auront des incidences sur les situations extrêmes (Wigley, 1985): successions d'hivers froids, d'étés secs, etc. ayant une très grande importance biologique. Ces évènements extrêmes doivent être envisagés avec d'autant plus d'attention que l'on risque d'être confronté à une plus grande variabilité interrannuelle.

1.2 Le Golfe du Lion

Le Golfe du Lion qui sera l'objet de notre étude dessine un vaste arc de cercle d'environ 270 km, ouvert sur le Sud-Est, situé entre 42. 5° et 43. 6° de latitude Nord, 3° et 5° de longitude Est. Les limites sont, au Sud-Ouest le massif cristallin des Albères et, au Nord-Est la chaîne calcaire de l'Estaque. La majeure partie du littoral est bordée d'une côte basse souvent lagunaire.

1.3 Les problèmes abordés

L'intérêt d'une telle étude se mesure par l'importance des changements prévisibles sur le milieu et les communautés qui y vivent, mais aussi par la nature des activités socio-économiques qui y seront effectuées.

Afin de permettre aux décideurs de juger des enjeux (risques ou avantages) nous analyserons dans différents domaines les tendances à envisager en s'appuyant sur les mécanismes et les pressions qui ont conduit à la structure actuelle (physique, biologique, sociale, économique) de la région.

S'agissant d'une analyse interdisciplinaire qui se veut aussi concrète que possible l'exposé a été construit en équipe, chaque membre assumant la responsabilité d'un chapitre.
– Enjeux économiques:
  Pour la Camargue
    B. Picon, Centre de recherche en écologie sociale
    (C.N.R.S.).[1]
  Pour le Languedoc-Roussillon
    A. Berger et F. Teral, Centre régional de productivité et des études économiques (Université de Montpellier I/C.N.R.S.).
  Occupation de l'espace
    J. Trayssac, F. Gagnier, J. M. Miossec, M. Desse et J. Y. Vourgères, Laboratoire d'aménagement des littoraux et organisation de l'espace (Université Montpellier III).
– Le climat:
    Ph. Daget, Institut de botanique (Université Montpellier II, C.N.R.S.).
– Les composantes du milieu:
1) le milieu marin
   océanographie physique

J.- P. Béthoux, Laboratoire de physique et chimie marines (Université P. et M. Curie),
relations climat-faune ichthyique du Golfe du Lion
J. P. Quignard, Laboratoire d'ichthyologie et de parasitologie générale (Université de Montpellier II),
2) le milieu terrestre, géologie, géomorphologie, sédimentologie A. L'Homer, Bureau de recherches géologiques et minières; hydrologie lagunaire J. -P. Perthuisot, O. Guelorget et S. Castanier, Groupe d'études du domaine paralique (Universités de Nantes et de Montpellier II). Dans ce chapitre, P. Heurteaux (C.N.R.S.) y expose les problèmes relatifs aux relations entre eaux souterraines et plans d'eau; communautés animales et végétales, relations avec le milieu: pour la Camargue, A. Crivelli, P. Duncan, P. Grillas, H. Hafner et A. Johnson, Station biologique de la Tour du Valat, P. Heurteaux; pour la Camargue et le Languedoc: J. J. Corre, Institut de botanique (Université de Montpellier II), et Tamisier (C.N.R.S.).
3) Protection du littoral
Madame A. Y. Le Dain, Institut des aménagements régionaux et de l'environnement; F. Rueda, Service maritime et de navigation du Languedoc Roussillon.

2 LES ENJEUX SOCIO-ECONOMIQUES

Le golfe du Lion a longtemps été marginalisé sur le plan économique. Les divagations du Rhône à son embouchure, l'absence de port naturel susceptible d'accueillir des navires de haute mer, la présence d'importantes lagunes entretenant une certaine insalubrité rendaient la côte et le proche arrière-pays peu attractifs. Seules quelques activités, vont se localiser sur le littoral. La pêche, l'exploitation salinière, l'agriculture créeront en certains lieux privilégiés des zones de peuplement et d'activité, laissant cependant à l'état naturel la plus grande partie de ce territoire.

L'évolution socio-économique sera fortement dépendante des particularités régionales. Le delta du Rhône (incluant la Camargue) restera essentiellement rural alors que le reste du littoral va subir une forte poussée urbaine sous la pression du tourisme. Par delà des contraintes similaires du milieu physique, les enjeux socio-économiques vont sensiblement différer ce qui peut hypothéquer les réactions futures.

2.1 Delta du Rhône: La Camargue

2.1.1 Histoire de la relation entre l'homme et la nature

Les particularités sociales en Camargue se caractérisent par une coupure passant grossièrement vers la latitude nord du Vaccarès et permettant de distinguer la haute et la basse Camargue. De formation fluviatile, la haute Camargue se compose essentiellement de terres alluviales riches, vestiges des anciens cours du Rhône entre lesquelles se dessinent des bassins dépressionnaires composés de marais et d'étangs (Picon, 1978).

De formation laguno-marine, la basse Camargue est constituée de dépôts marins et éoliens très largement stérilisés par la présence de sel. Cette stérilité du sol associée à des conditions de vie très précaires puisque soumises aux caprices du fleuve et de la mer a entrainé à la fois la dispersion de l'habitat et une exploitation extensive du sol.

Dans le cadre d'une économie rurale, ce type d'exploitation du sol a consisté à se protéger du fleuve et de la mer, à accroître ensuite la productivité agricole. Cette politique a abouti à la création des digues du Rhône en 1859. Les crues hivernales qui dessalaient naturellement la terre cessant du fait de l'endiguement des deux bras du fleuve, il fallut pallier ce manque par des moyens artificiels et creuser tout un réseau complexe d'irrigation et de drainage.

Les frais considérables que nécessitèrent de tels investissements supposèrent l'apport de capitaux extérieurs au delta et ainsi une ouverture très forte sur le monde économique environnant.

L'absence de communautés rurales traditionnelles provoquée par ce mode de faire-valoir supprima un des freins habituels aux innovations économiques dans le monde rural, ce qui se traduisit par une soumission très forte du delta aux fluctuations économiques nationales.

2.1.2 Caractères du paysage socio-économique camarguais
Les grandes propriétés (60 propriétés de 200 à 2000 ha s'étendent sur les 2/3 du delta), l'industrie salinière sur 13000 ha d'étangs jouxtant la Réserve Nationale de 15000 ha au bord de la mer, l'élevage taurin et chevalin, l'avifaune aquatique protégée par la Réserve ont contribué à donner à la Camargue une exceptionnelle image de nature.

On estime à 1 million par an le nombre de visiteurs attirés sur le delta par les 30 km de plage restées vierges.

Une autre catégorie, moins nombreuse, vient bénéficier de l'extraordinaire rente de situation dont jouissent les grands domaines périphériques de la Réserve, sur lesquels des marais artificiellement entretenus deviennent de très giboyeux terrains de chasse à la sauvagine. La gratuité de l'accès au domaine public maritime opposée au prix élevé de l'action de chasse président à une stratification sociale très caractéristique du tourisme et des loisirs en Camargue: tourisme de classe sur les grands domaines agricoles, tourisme de masse sur le domaine public maritime.

2.1.3 L'avenir de la Camargue
Il n'est pas sûr qu'aujourd'hui, où le discours de l'harmonie Homme-Nature tient plus de place que celui de la soumission ou du rapport de force, le consensus soit total dans la réponse que peuvent faire les sociétés littorales en face de la menace de la montée des eaux.

L'hypothèse selon laquelle les activités côtières d'aujourd'hui sont intangibles est discutable. Rien ne permet de dire que les agriculteurs ou riziculteurs soient prêts à effectuer des pompages supplémentaires pour sauver une activité déjà menacée et subventionnée (le riz).

Après l'échec de la tentative de la culture de la canne de Provence à usage papetier (1975: Picon, 1978), et l'échec plus récent (1980) de la reconversion à la pisciculture d'eau douce (Picon, 1985), on peut s'interroger sur l'avenir de nouveaux investissements agricoles dans des conditions naturelles plus contraignantes qu'auparavant.

Rien ne permet d'affirmer que la Banque la Hénin soit prête à investir des sommes importantes pour surélever ses marais salants. Il n'est pas absurde d'imaginer qu'au nom d'une certaine forme de prise en considération de la nature, de ses caprices compris, la basse Camargue agricole et salinière ne bascule un peu plus dans la sphère des loisirs "naturels". Après tout, les sondages, fait auprès des visiteurs de Camargue (Picon, 1987), font apparaître une très forte demande de consommation "d'espaces naturels"

Une enquête effectuée en 1984 auprès d'un échantillon de visiteurs de Camargue met en évidence que 82,5 % d'entre eux viennent pour voir de la nature (paysages, oiseaux, etc.).

Les activités taurines et équestres qui y sont liées sont à la recherche de pâturages extensifs.

De plus, la permanence d'une très faible densité de population dans ce delta depuis l'époque romaine (Table 11.1), si elle est signe d'une adaptation démographique aux contraintes du milieu constitue peut-être dans cette optique, une ressource pour cette région confrontée à une nouvelle instabilité du littoral.

Tableau 11.1 Population de la Camargue

Epoque romaine	entre 4000 et 9000		
XVIe siècle Quiqueran de Beaujeu	10000		
1968 (INSEE)[2]	8100		
1984 (INSEE)	9645	Arles, Camargue	4700
		Stes-Maries-de-la-Mer	2045
		Salin-de-Giraud	2900

En bref, ne peut-on poser l'hypothèse qu'une nouvelle instabilité de la zone littorale de Camargue, loin d'induire en certains points de nouveaux aménagements de protection et de défense, ne serait pas plutôt le prétexte, sous-tendu par les idéologies d'un nouveau rapport à la nature, pour développer les activités nouvelles de tourisme et de loisirs "verts"? Ceux-ci sont de plus en plus demandés par les couches moyennes urbanisées de l'Europe entière et ne nécessitent pratiquement pas d'infrastructures de type marinas ou ports de plaisance.

Si demande sociale, conditions naturelles et rentabilité économique se combinent ainsi, il n'est pas impossible de penser que les caractéristiques prévisibles d'évolution du littoral ne débouchent non sur une nouvelle artificialisation de celui-ci mais sur un nouveau laisser faire qui bloquerait les projets d'aménagements au nom des risques naturels et permettrait ainsi de mettre à la disposition du public de vastes espaces de nature.

Des recherches récentes (Picon et al., 1987), montrent qu'en Basse Camargue, la production d'espaces naturels peut être aujourd'hui aussi rentable pour son propriétaire que la production agricole ou industrielle.

Cette prise de conscience des propriétaires du sol semble bien s'assortir d'une modification progressive des représentations qu'ils se faisaient de leur propre environnement.

Eux qui ne percevaient les milieux salés et lacustres de Basse Camargue que comme des contraintes au développement agricole, commencent insensiblement à se rendre compte que, dans un contexte de société post-industrielle, la contrainte peut dorénavant devenir ressource.

2.2 Le Languedoc-Roussillon

2.2.1 *Espace littoral et activités sensibles*

Deux domaines d'inégale importance économique s'opposent le long de ce littoral. On a d'une part un linéaire côtier changeant qui combine, sur une bande étroite une relative instabilité morphologique, un fort développement urbain et des créneaux d'espaces protégés. Ce mode très contrasté est en gros conforme au plan d'aménagement touristique lancé dans les années 1960. D'autre part on a le "domaine littoral interne", plus ou moins fractionné par les étangs, qui est plus nuancé et nettement moins dynamique. Son occupation repose sur une ségrégation entre secteurs cultivés, étendus jusqu'aux rives des étangs et friches palustres, de contours assez mouvants en raison de la progression du colmatage. Son orientation essentiellement agricole détermine une occupation à la fois plus continue et bien moins dense. Les espaces protégés, plus compacts,

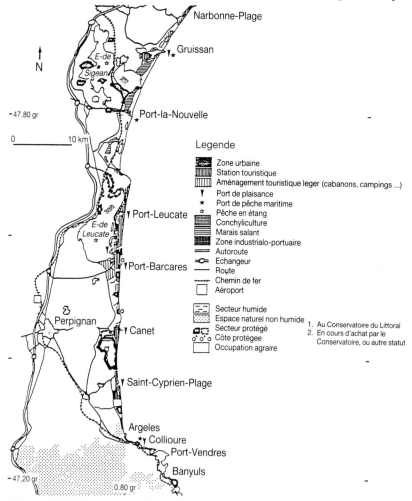

Fig. 11.1a Carte d'occupation. Roussillon-Languedoc Sud.

s'articulent surtout sur les étangs et les marais, sans créer de ruptures comparables à celles qui marquent le linéaire côtier. Les modifications artificielles de surface et de tracés de rivages se cantonnent à des secteurs réduits, limitrophes des agglomérations (ex.: emprises de Fréjorgues et de Palavas, au Sud de Montpellier). L'extension de certains plans d'eau a globalement compensé ces pertes (étang du Ponant près du Grau du Roi, étang de Luno près d'Agde). La moindre valeur foncière, l'espace disponible rendent possible une politique de recul face à une expansion des nappes d'eau libre (Fig. 11.1a,b,c).

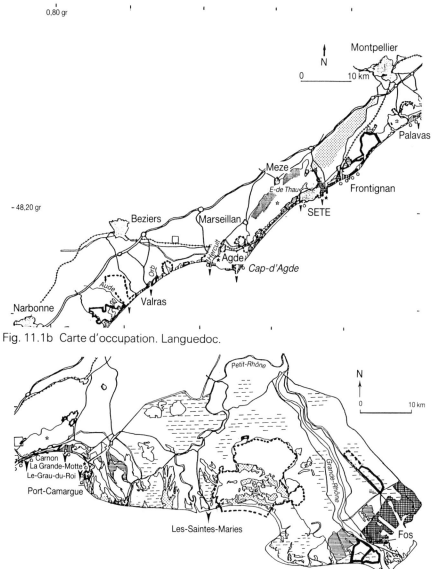

Fig. 11.1b Carte d'occupation. Languedoc.

Fig. 11.1c Carte d'occupation. Petite et Grande Camargue.

Plusieurs types d'activités économiques sont directement concernées par les événements climatiques prévus:
- les activités touristiques, à travers la diversité des investissements et des aménagements réalisés, en cours ou en projet, ainsi qu'à travers la consommation touristique;
- les activités agricoles aussi bien en limite avec la frange littorale que plus à l'intérieur des terres;
- les activités aquacoles en bord de mer et dans les lagunes.

Parmi toutes ces activités, le tourisme est la plus préoccupante, en raison de son importance actuelle et de ses développements prévisibles. Son évolution est soumise au jeu d'un certain nombre de paramètres (qu'ils soient naturels ou bien créés par l'homme à travers l'aménagement).

Si au niveau du consommateur (c'est-à-dire le touriste), la pérennité n'a que peu de valeur, elle est par contre une contrainte essentielle pour l'investisseur, qu'il soit public ou privé.

Dans le cas des touristes, on vérifie le caractère fluctuant de la consommation lors des conditions climatiques défavorables: par exemple (pluie en juillet entraînant une chute brutale des fréquentations), ou bien du fait d'évènements politiques ou sociaux (évènements brutaux au pays Basque ou grève par exemple).

Pour les investisseurs, la durée est une donnée majeure dans le calcul des amortissements comme de la rentabilité. Une remise en cause du cadre naturel, la nécessité d'entreprendre des investissements de protection, des dépenses de réparation et d'entretien modifient les perspectives de rentabilisation de l'aménagement touristique et peuvent déterminer une remise en cause totale ou partielle de ce dernier.

2.2.2 Mise en place des structures touristiques littorales
Initialement le tourisme d'été, fortement limité par les nuisances du milieu (moustiques en particulier), était le fait exclusif des populations de la plaine viticole et des agglomérations les plus proches.

A partir des années cinquante, le tourisme devient une activité, de plus en plus ouverte sur la masse. Il s'ensuit, sous la pression de la demande, une phase d'expansion rapide des stations, expansion plus souvent anarchique qu'organisée.

Le tourisme balnéaire reste cependant dans la région encore largement contrôlé par les villes languedociennes. Les grands flux touristiques qui font déferler les populations de la France et de l'Europe septentrionale vers le Sud, ne s'arrêtent guère sur le littoral languedocien, sauf de manière inorganisée et même sauvage (exemple sur le lido entre Sète et Agde). C'est la volonté de modifier cette situation et d'essayer de profiter de la marée touristique qui conduira les Pouvoirs Publics à entreprendre dans les années soixante une vaste opération d'aménagement du territoire, destinée à donner à un milieu, encore naturel et sauvage pour une large part mais au pouvoir attractif indéniable, un potentiel d'accueil à l'échelon européen (Racine, 1980).

Précédé d'un aménagement hydraulique, dans la partie orientale de la région à vocation première agricole, qui permet une maîtrise de l'eau et sa distribution sur une partie du littoral, l'aménagement touristique languedocien portera essentiellement sur la bordure maritime.

L'assainissement du milieu, avec la démoustication, la construction d'infrastructures d'accès et d'accueil, vont déclencher un bouleversement considérable sur toute la frange côtière du Languedoc-Roussillon, modifiant totalement son usage socio-économique.

2.2.3 Le flux touristique sur le littoral

Durant les années soixante-dix, au fur et à mesure de l'avance de l'aménagement du littoral, des enquêtes ont permis de suivre l'évolution des flux touristiques durant la saison estivale (Table 11.2). La croissance des flux est allée de pair avec la création des structures d'accueil.

Tableau 11.2 Evolution des flux touristiques pour l'ensemble du littoral Languedoc-Roussillon entre 1971 et 1976

Année	Nombre de de séjours	Durée moyenne des séjours	Nombre de nuits (en milliers)
1971	1002	21.6	21255
1972	1071	21.8	23436
1973	1210	21.9	26552
1974	1402	20.6	28908
1975	1690	20.6	34832
1976	1682	20.4	34292

Sources: CRPEE-INSEE[3]

La fin des enquêtes de mesure des flux ne permet plus d'avoir une information très fiable quant à l'évolution du nombre des touristes et leur condition de séjour. Cependant, des estimations donnent une idée de cette évolution et de son dynamisme. Ainsi, le nombre des séjours, qui atteignait 1 680 000 durant les quatre mois d'été en 1976, serait passé à 2 500 000 en 1979 et à 4 000 000 en 1985.

Dans le même temps, les capacités d'accueil ont continué à croître, même si le rythme s'est ralenti. De 25 000 logements en 1963, on est passé à 120 000 en 1986, dont plus de la moitié sont mis en location. Il faut ajouter à cela, pour 1986, 9000 lits d'hôtel, 9600 places en résidence locative, 200 000 places en camping et 13 000 places en villages-vacances.

2.2.4 Les conséquences de la politique touristique et du développement régional

Si le tourisme a été un facteur prédominant pour l'évolution économique du littoral languedocien, cette dernière est soumise aux pressions d'un espace bien plus vaste dont seule la résultante est mesurable.

Après avoir abrité des activités relativement réduites et le plus souvent traditionnelles, l'espace littoral devient un lieu de concentration d'hommes et d'activités nouvelles, favorisant l'émergence d'espaces récréatifs, touristiques et même urbains lorsque la proximité d'une ville en expansion le permet.

Le littoral languedocien: espace de peuplement

Deux facteurs vont jouer conjointement dans le sens de la dynamique démographique (Fig. 11.2).

L'essor touristique, encouragé et accompagné par l'action d'aménagement du littoral, avec la construction de nouvelles stations touristiques, le réveil et le développement des anciennes stations balnéaires va favoriser sur le pourtour méditerranéen du Languedoc la création d'un parc immobilier important. Destiné en premier lieu à accueillir en été des centaines de milliers de touristes, il permettra de manière secondaire l'établissement d'une population permanente.

Parallèlement à cette impulsion, la région Languedoc-Roussillon connait au cours des vingt-cinq dernières années une croissance démographique spectaculaire fondée essentiellement sur des courants migratoires et qui la place dans le peloton de tête des régions françaises. Les villes de la plaine, et en particulier celle de l'Ouest de la région, vont être les principales bénéficiaires de ces afflux permanents de population.

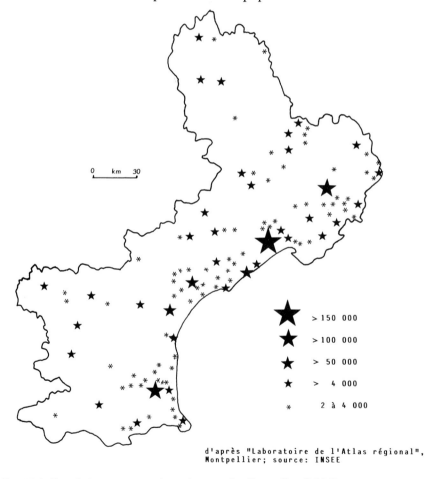

Fig. 11.2 Population communale en Languedoc-Roussillon (1982).

La proximité du littoral et des zones urbaines en croissance, favorisera alors l'urbanisation de certaines zones de la côte. Le phénomène sera particulièrement sensible dans sa partie héraultaise, ainsi que sur la côte catalane.

L'expansion du domaine bâti
L'accroissement du parc immobilier trouve donc son explication dans la conjugaison du renouveau démographique du Bas-Languedoc depuis les années soixante, et de la mise en oeuvre de l'aménagement.

Globalement, entre 1962 et 1982, le nombre de logements sur le littoral est multiplié par quatre. La croissance du domaine immobilier est plus particulièrement spectaculaire aux deux extrémités du Golfe du Lion, dans la zone qui s'étend de Port-Camargue jusqu'à Sète et le long de la côte sableuse catalane.

Cette conquête du littoral par le béton, est le fait à la fois des résidences principales et des résidences secondaires (Fig. 11.3). Ces dernières sont à

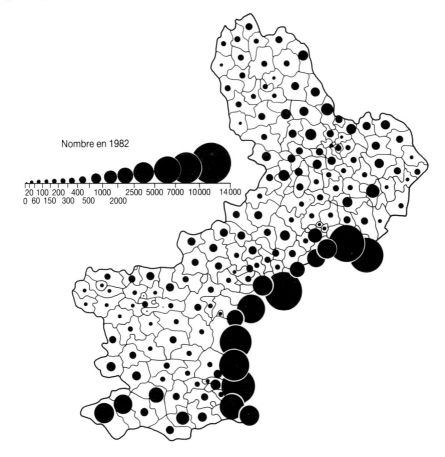

Sources: J.L. Escudier, op. cit.

Fig. 11.3 Languedoc-Roussillon. Les résidences secondaires en 1982.

l'origine de l'essor résidentiel le plus spectaculaire, elles sont cependant responsables dans leur dynamique de fluctuations conjoncturelles.

Un renouvellement des activités économiques sur le littoral
L'intensité des mouvements de population, saisonniers avec le tourisme, permanents avec la dynamique propre de la région languedocienne, est en relation étroite avec les mutations qui affectent l'économie littorale.

La régression des activités traditionnelles, relevant le plus souvent du secteur primaire (agriculture et pêche), une augmentation sensible du secteur secondaire (en particulier le bâtiment) et une progression spectaculaire du tertiaire sont les principales caractéristiques de l'évolution des activités.

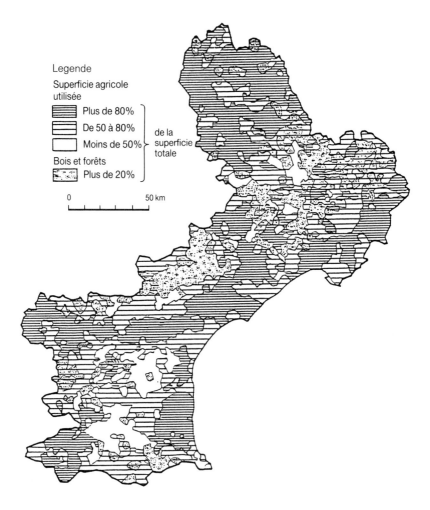

Fig. 11.4 L'occupation du sol (1980) en Languedoc-Roussillon.

(i) *La régression des activités traditionnelles*

Les activités agricoles sont de moins en moins nombreuses dans les communes du littoral. D'une manière global le nombre des exploitants a diminué de 27 % entre 1970 et 1979[4], la surface agricole utilisée de 7%, les surfaces plantées en vigne, principale spéculation, régressent de 22 % pour la même période (Fig. 11.4).

Le recul de l'agriculture est le plus net dans les parties du littoral où la pression foncière d'origine non agricole est la plus forte (Hérault et Pyrénées-Orientales).

Comme l'agriculture, les activités liées à la pêche connaissent depuis quelques années un processus de repli, dû essentiellement à l'industrialisation et à la concentration de l'effort de pêche.

L'aquaculture, dans ses formes modernes, est souvent présentée dans la région comme l'activité d'avenir, assurant la relève d'une exploitation "archaïque" du milieu marin naturel. Les expérimentations en cours depuis de nombreuses années, si elles ont donné des résultats scientifiques indéniables en matière d'écloserie et de grossissement de certaines espèces (crevettes, dorades, loups...), n'ont pas jusqu'à présent confirmé sur le plan économique les espoirs répandus par le discours scientifique et technologique. Sans pour autant préjuger de l'avenir, aucune des "entreprises" aquacoles situées sur le littoral ne donne encore les preuves de la viabilité économique de leurs activités.

On peut considérer que le processus de régression des activités primaires va se poursuivre dans les années qui viennent, débouchant dans les zones les plus urbanisées du littoral sur une disparition totale de l'agriculture comme de la pêche.

(ii) *Une industrialisation limitée*

A l'écart de la révolution industrielle du XIXème siècle, le Languedoc a longtemps porté comme une tare son faible niveau d'industrialisation. Seuls sur le littoral quelques lieux privilégiés ont vu s'implanter et se maintenir des entreprises industrielles, Sète, Frontignan, Narbonne. Aujourd'hui, la "crise" contribue à une redistribution spatiale des unités de production de type industriel (Fig. 11.5), en même temps qu'elle remet en cause l'existence d'entreprises depuis longtemps présentes (exemple: la raffinerie de pétrole de Frontignan).

Seul le secteur du batiment et ses fournisseurs ont pendant une vingtaine d'années profité largement de l'aménagement touristique et des marchés importants qui s'ouvraient alors. Aujourd'hui, l'ensemble du secteur connait un tassement avec l'achèvement des grands travaux d'infrastructure et le ralentissement de la construction sur la côte.

(iii) *L'explosion du secteur tertiaire*

Le recensement des activités présentes dans les communes du littoral permet de mettre en évidence la spécificité affirmée du milieu.

L'importance du secteur tertiaire, et en particulier des activités de commerce et des services privés, la place qu'occupent la restauration et l'hôtellerie donnent à l'ensemble du littoral une structure économique presque homogène, destinée en priorité à répondre à une demande touristique à caractère saisonnier.

Si la taille moyenne des commerces se situe autour de trois ou quatre emplois, on observe une dispersion assez grande selon les branches

Fig. 11.5 Tendance d'évolution de l'emploi industriel entre 1954 et 1982.

commerciales et surtout leur localisation. C'est ainsi que la proximité de la zone urbaine dynamique Sète-Montpellier-Nîmes influence profondément la nature et la taille des commerces et des services privés implantés sur le littoral.

Les activités tertiaires dans ces zones prennent un caractère plus souvent permanent, répondant à la fois à la demande urbaine régionale et à la demande touristique estivale.

2.2.5 L'avenir du littoral Languedoc-Roussillon
Vis-à-vis de la situation matérielle
– Les changements climatiques: dans l'éventualité d'un réchauffement des températures, c'est un élément positif vis-à-vis du tourisme et des activités qui lui sont liées. Un allongement de la saison chaude renforcerait les potentialités régionales par rapport au littoral espagnol, italien ou d'Afrique du Nord, en permettant un amortissement plus rapide des investissements grâce à une utilisation plus intense sur une période plus

longue. Elle pourrait également augmenter la durée de fonctionnement d'activités saisonnières jusqu'à présent strictement dépendantes des flux liés au tourisme de masse.

Par contre des périodes de sécheresse plus longues ou plus fréquentes risquent de rendre plus difficile et plus onéreuse la satisfaction des besoins en eau face à de fortes concentrations de population.

– Les coups de mer plus fréquents avec une hausse du niveau général marin peuvent provoquer d'année en année des dégradations aux infrastructures et aux plages, occasionnant des coûts d'intervention de plus en plus importants et posant des problèmes techniques et financiers de protection non encore solutionnés dans le contexte présent.

De telles contraintes physiques nouvelles vont nécessiter la mise en place d'actions publiques dont l'organisation, la cohésion à l'échelle de l'ensemble du littoral, le financement comme la désignation des responsables et des maîtres-d'œuvre des actions (particuliers, collectivités locales, départementales, régionale, Etat, etc.) poseront de multiples problèmes à résoudre.

Au-delà des effets directs des processus physiques qui sont envisagés, les mutations climatiques peuvent entraîner des modifications dans le comportement des hommes et des agents économiques d'une manière plus générale.

Tout comportement économique, intégrant dans sa logique le temps, intègre également la notion de risque. L'investissement nouveau, l'entretien des équipements déjà réalisés, la consommation touristique vont se trouver sensibilisés par une augmentation des risques.

Il se posera à plus longue échéance le problème de l'usage touristique par les hommes, d'un espace littoral de plus en plus fragile et par là inhospitalier.

Les modifications accompagnant les changements climatiques pourront entraîner un déplacement de la localisation des activités touristiques (et une concentration plus grande de ces dernières vers les littoraux rocheux, moins sensibles et donc moins fragilisés). Cette dernière perspective doit être considérée comme un scénario extrême.

Le littoral languedocien du golfe du Lion pourrait alors retrouver, dans cette hypothèse extrêmement pessimiste, la situation qui était la sienne, avant la colonisation par l'homme à la fin du dix-neuvième siècle et durant le vingtième.

Situation juridique
Si l'on se place sur un plan institutionnel et juridique, les perspectives de réchauffement climatique, la montée prévisible du niveau des eaux appellent plusieurs observations.

– Les problèmes juridiques liés à la prévention des risques d'immersion.

La fragilisation physique du littoral en raison de l'accentuation des risques cycloniques pose d'abord sur le plan réglementaire l'hypothèse de la mise en oeuvre d'une "servitude d'exposition aux risques" faisant référence à l'avancée des eaux qui devrait être aménagée par rapport à la réglementation actuelle et qui serait mise en oeuvre par une autorité suffisamment affranchie des pressions locales.

– Les problèmes juridiques liés à la protection des infrastructures menacées par l'immersion et l'érosion du littoral.

En droit, c'est au propriétaire du bien qu'il appartient, à ses frais et sous sa responsabilité, de mettre en oeuvre les mesures de protection. En réalité, la puissance publique (plus généralement les collectivités) se trouve incluse rapidement dans cette problématique juridique.

On assiste alors sans cadre juridique précis, à une formule de "cogestion du risque". L'organisation de la défense contre la mer étant prise en charge pour sa plus grosse partie par la collectivité, (commune, région, département, Etat) signe de sa "responsabilité inavouée" et de la défense d'un patrimoine collectif.

– L'avancée de la domanialité publique naturelle:

Les limites du domaine public maritime correspondent à la délimitation du rivage de la mer, fixée en Méditerranée depuis le XVIIIème siècle ". . . au plus grand flot de mars, mais à l'exclusion des tempêtes exceptionnelles".

L'élévation du niveau de la mer et l'accentuation des régimes cycloniques devraient faire avancer ces limites au détriment des droits des riverains.

La rupture du cordon littoral et la création naturelle de graus feront basculer nombre d'étangs privés, appartenant à des particuliers ou des collectivités dans le domaine public maritime si ces communications "directes et naturelles" acquièrent une certaine pérennité.

Dès lors, les limites du domaine public pourraient être reportées vers l'intérieur des terres de plusieurs kilomètres. Les conséquences juridiques, outre les droits des propriétaires qui se trouveraient suspendus ne sont pas minces.

L'ensemble des actes négociés sur la base des titres initiaux ne seraient pas opposables à la puissance publique et surtout aux usagers du domaine public. C'est dire que la pêche et la navigation seraient libres sur ces étangs, que leur gestion relèverait "de plano" de la puissance publique sans que cette dernière puisse se voir opposer des actes de gestion ou de disposition antérieurs.

3 Premiere Partie: Tendances Climatiques au Nord de la Mediterranee de 1944 á 1987

3.1 Introduction

Le devenir du climat au cours des 45 dernières années sur les rivages du Nord de la Méditerranée sera examiné à partir de la station de La-Tour-du-Valat. Cette station d'observations biologiques est située à l'Est du Rhône dans le delta de la Camargue; ses coordonnées sont 43°30'N et 4°40'E, avec une altitude de un mètre (Fig. 11.32).

Des observations météorologiques sont faites régulièrement depuis son ouverture en 1944.

Dans les discussions qui vont suivre, nous diviserons la période d'observations en quatre parties de onze années que nous appellerons par convention des "décennies", et que nous numéroterons I, II, III, IV (Table 11.3).

Tableau 11.3 Années extrêmes

Décennies	Années très froides	très chaudes
I	0	6
II	2	2
III	4	0
IV	3	0

La proportion des années très chaudes par décennie décroît de façon significative ($P = 0.007$).

3.2 Températures

3.2.1 *Moyennes annuelles*

La température moyenne annuelle varie de 13,2°C l'année la plus froide, en 1956, jusqu'à 15,2°C en 1950 et 1961, les années les plus chaudes. Elle ne présente donc qu'une faible amplitude de variation de 2°C. La température moyenne annuelle sur l'ensemble de la période d'étude est de 14,2°C. La variation interannuelle est représentée sur la Figure 11.6 avec la moyenne courante sur 5 ans.

Sur 22 années qui ont une température moyenne supérieure à la moyenne générale, 11 sont situées avant 1956, c'est dire que les premières années sont sensiblement plus chaudes que les suivantes. En première approximation, la température décroit régulièrement depuis 1944 de façon très significative ($P < 0,001$). En seconde approximation, il apparait, là aussi de façon très significative, deux phases dans l'évolution thermique:
– décroissance nette dans les décennies I et II;
– relative stabilité en III et IV, s'achevant par une faible remontée.

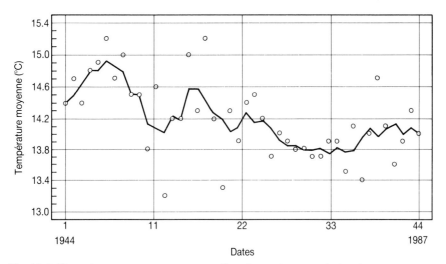

Fig. 11.6 Températures moyennes annuelles observées et variation de la moyenne courante sur 5 ans.

Celle-ci étant tout à fait insuffisante pour rattraper la première phase ne paraît pas sur la comparaison des moyennes décennales: 14,63 – 14,22 – 13,96 – 13,95 qui sont significativement différentes (P = 0,002).

En ce qui concerne les extrêmes, les 9 années les plus froides et les 8 années les plus chaudes se répartissent d'une manière opposée de façon hautement significative.

3.2.2 Moyenne des minimums du mois le plus froid

La moyenne des minimums du mois le plus froid présente une moyenne de 1,2°C sur la période d'observation et varie de –4,7°C en 1956 à 4,2°C en 1955 soit une amplitude de variation de 8,9°C. La variation interannuelle est représentée sur la Figure 11.7 avec la moyenne courante sur 5 ans.

Il faut d'abord souligner que le minimum de –4,7°C observé en 1956 reste une valeur anormalement basse qui, statistiquement, peut être considérée comme aberrante. Sans elle, l'amplitude de variation est de 6,6°C et les minimums s'étendent entre –2,4°C en 1945 et 4,2; cette valeur doit tout de même être considérée comme le rappel que les catastrophes sont toujours possibles. Pour suivre le détail des aléas climatiques, et de ce qui peut influer sur les êtres vivants, il ne faut pas perdre de vue que les minimums extrêmes sont bien plus faibles que leur moyenne; le plus faible atteint –15°C le 10 février 1956.

La variation des minimums ne montre aucune tendance significative à une évolution dans un sens ou dans l'autre. Il est donc possible de considérer qu'aux variations aléatoires près, la moyenne des minimums du mois le plus froid reste stable pendant toute la période considérée.

3.2.3 Moyenne des maximums du mois le plus chaud

Les moyennes sont très classiquement calculées par

$$t = \frac{M + m}{2}$$

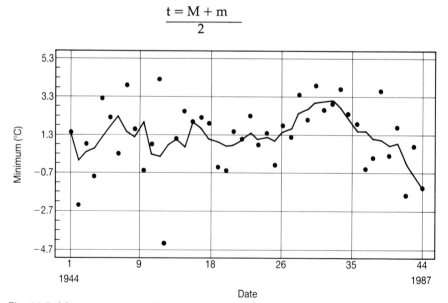

Fig. 11.7 Moyennes des minimums du mois le plus froid et variation de la moyenne courante sur 5 ans.

d'où :
M = 2 t − m

mais, on vient de voir que t est très variable et m pratiquement stable, il en résulte que les variations observées seront plus nettes pour les maximums que pour les températures moyennes.

Les moyennes des maximums du mois le plus chaud varient donc de 25,8°C en 1977 à 31,8°C en 1949 avec une valeur moyenne de 28,7°C. La variation interannuelle est représentée sur la Fig. 11.8 avec la moyenne courante sur 5 années. Malgré une variation interannuelle très importante, un été chaud succédant très souvent à un été frais, en première approximation, la moyenne des maximums du mois le plus chaud diminue régulièrement de façon significative pendant toute la période d'observation ; les maximums présentent des phases évolutives rappelant celles des moyennes, mais en plus accentué :
– diminution importante jusqu'en 1970 ; suivie d'une remontée nette, mais insuffisante pour retrouver les valeurs d'origine.

Les moyennes décennales suivent le même rythme, et constituent un ensemble de valeurs dont les différences sont significatives :

$$29,9 - 28,3 - 28,0 - 28,6.$$

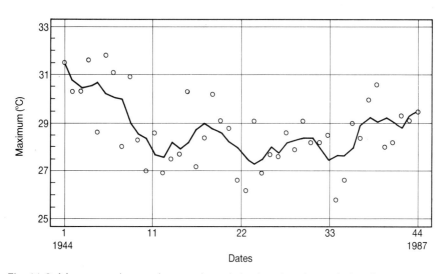

Fig. 11.8 Moyennes des maximums du mois le plus chaud et variation de la moyenne courante sur 5 ans.

3.2.4 Amplitudes annuelles

L'amplitude thermique peut être définie comme la différence M-m. Au cours de la période d'observation, les amplitudes (Fig. 11.9) varient entre 32,7°C en 1945 et 22,1°C en 1977 avec une moyenne de 27,5°C. Il y a une succession hautement significative d'une année dont l'amplitude est élevée suivant une année dont l'amplitude est nettement plus faible. Les amplitudes ne présentent pas de variation progressive régulière, au contraire, un ajustement quadratique très significatif ($P < 0,001$) montre

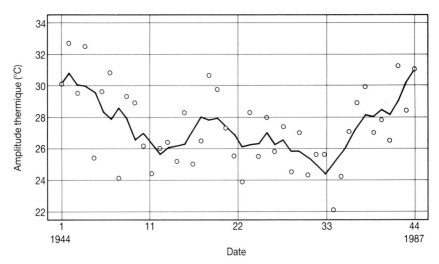

Fig. 11.9 Variation de l'amplitude thermique moyenne annuelle et de sa moyenne courante sur 5 ans.

une diminution jusqu'en 1966 suivie d'une augmentation insuffisante pour rattraper la différence. Les résidus par rapport à cette tendance générale montrent une variation sinusoïdale de période 20 ans avec un maximum en 1945, 1963 et 1982, qui, combinée avec la variation tendantielle, donne le modèle général de la Fig. 11.10.

Il semble que l'on puisse rattacher cette valeur assez forte des amplitudes thermiques à l'influence du Mistral qui a pour effet de faire baisser les minimums thermiques.

3.2.5 Extension spatiale
Pour la ville d'Agde, située à une altitude de 8 m et à 120 km à l'Ouest de la station examinée ici (Fig. 11.32), il existe une série d'observations de 1956 à 1984. Elles permettent certaines comparaisons.

Les températures moyennes annuelles varient de 14,4 à 15,6°C avec une moyenne de 14,2; les minimums varient de –5,6 à 4,2 avec une moyenne de 1,1°C, tandis que les maximums passent de 26,7 à 30,1 avec une moyenne de 28,3°C. L'amplitude annuelle moyenne est de 27,6°C et varie entre 21,4 et 35°C.

On note donc qu'à Agde les moyennes thermiques sont pratiquement les mêmes qu'en Camargue, mais avec des amplitudes de variation sensiblement plus faibles, ce qui semble pouvoir être relié à l'absence de Mistral.

3.2.6 Conclusions sur les températures
Pendant les 44 années d'observations, les variations de températures observées sur la côte nord-méditerranéenne sont dues à des différences estivales, les hivers restant homogènes. Trois phases sont enregistrées:
- rafraîchissement rapide des étés pendant la première décennie, dont la moyenne des maximums passe de 31,5°C à 27,5°C;
- deux décennies d'étés frais avec un M moyen de 28,2°C;
- léger réchauffement jusqu'à 29,5°C pendant la dernière décennie.

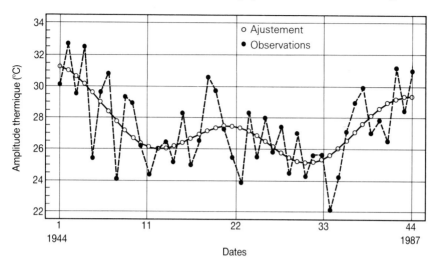

Fig. 11.10 Modèle général de la variation de l'amplitude thermique entre 1944 et 1987.

3.3 Précipitations

3.3.1 *Module annuel*

Au cours des 44 années d'observations (Fig. 11.11), le module pluviométrique annuel oscille entre 320,4 mm en 1945 et 955,4 mm en 1969, avec une moyenne de 595,2 mm. La figure fait apparaître une variation interannuelle très importante qui se traduit par le fait qu'à une année très arrosée succède, en général, une année très peu arrosée; cette alternance est significative (P = 0,03).

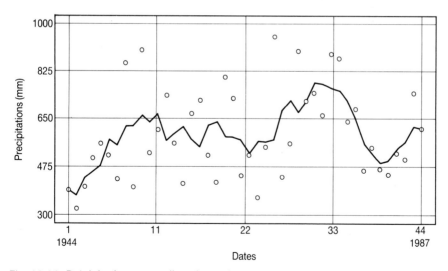

Fig. 11.11 Précipitations annuelles observées et variation de la moyenne courante sur 5 ans.

Sur 19 années dont les précipitations sont supérieures à la moyenne, 2 ont lieu pendant la première décennie puis 5 et 6 pendant les trois suivantes; plus particulièrement 8 surviennent entre 1972 et 1979. On démontre toutefois que cette tendance agrégative n'est pas significative.

Le Tableau 11.4 montre le nombre d'années exceptionnellement humides et sèches par décennie.

Tableau 11.4 Années extrêmes

Décennies	Années Humides	Sèches
I	2	5
II	1	2
III	3	2
IV	2	0

En ce qui concerne les premières, on y retrouve la séquence précédente. L'examen de la longueur de la période de retour entre deux années humides montre qu'il n'y a aucune tendance au rapprochement ou à l'éloignement de ce type d'années. Par contre, il existe une tendance non significative à l'éloignement progressif des années sèches due à l'absence d'années très sèches depuis 17 ans, ce qui est significativement anormal.

Au cours de la période d'observation, il y a une légère tendance non significative à l'augmentation du module pluviométrique annuel (P = 0,14); par contre, il passe presque significativement (P = 0,051) par un maximum au cours de la décennie III. Les résidus, après cette variation tendantielle, suivent une variation sinusoïdale de période 20 ans.

Cette remarque faite, il subsiste une variation des résidus présentant de façon significative (P = 0,002) une période de 1,5 ans et une de 20 ans telle que les maximums se produisent en 1954 et en 1974, les minimums en 1944, 1964 et 1984. La Fig. 11.12 représente le modèle d'ensemble de la variation des pluies sur 44 ans; il est très significatif (P = 0,008), mais ne rend pas compte de l'alternance interannuelle déjà signalée.

3.3.2 *Module estival*
En ce qui concerne l'été, c'est-à-dire les mois de Juin, Juillet et Août, le module oscille (Fig. 11.13) entre 6,8 mm en 1949 et 174,7 mm en 1977 avec une moyenne de 70,8 mm. Il n'y a pas de tendance évolutive régulière, mais présence significative d'étés plus arrosés au cours des décennies II et III (P = 0,03).

3.3.3 *Module du mois le plus arrosé*
Le mois le plus arrosé (Fig. 11.14) reçoit entre 64,2 mm en 1945 et 377 mm en 1953 (c'est-à-dire plus que toute l'année 1945) avec une moyenne de 70,8 mm.

3.3.4 *Elargissement dans le passé*
La station toute proche de Salin-de-Giraud a enregistré les précipitations de 1884 à 1962. Pendant 14 des 18 années communes avec La-Tour-du-Valat, les observations des deux stations donnent des résultats très voisins

Gulf of Lion and climatic change 361

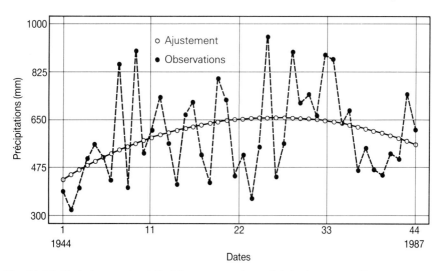

Fig. 11.12 Variation tendantielle à long terme des précipitations annuelles.

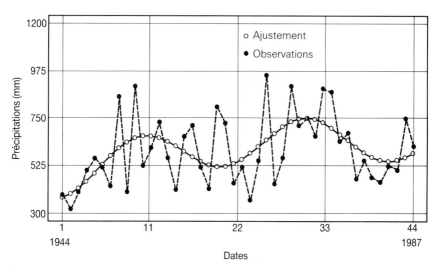

Fig. 11.13 Modèle d'ensemble de la variation des précipitations en Camargue de 1947 à 1987.

ou même identiques. Il n'est donc pas excessivement aventuré de rabouter les deux séries, ce qui permet une estimation de la marche des pluies sur plus d'un siècle.

Le minimum annuel descend à 300,4 mm en 1896, mais le maximum reste celui de 1969; un second maximum en est très proche avec 929,3 mm en 1907. La moyenne centenaire est 562,8. La Fig. 11.15 représente la succession des 104 valeurs avec la moyenne courante quinquennale, tandis que la Figure 11.16 donne un ajustement périodique (P = 0,07) qui

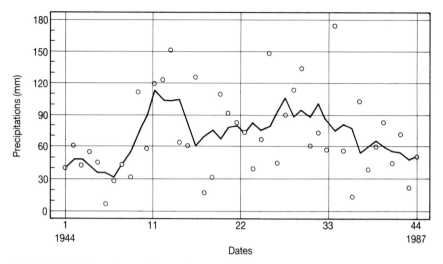

Fig. 11.14 Variation du module estival des précipitations et de sa moyenne courante sur 5 ans.

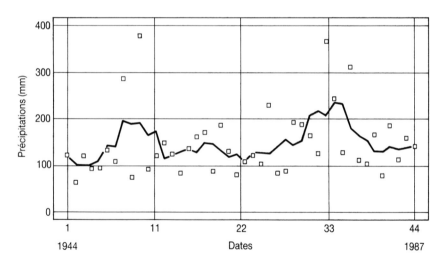

Fig. 11.15 Variation du mois le plus arrosé et de sa moyenne courante sur 5 ans.

souligne, en éliminant le bruit des variations interannuelles, les grandes lignes de la marche des précipitations en Camargue. Cette courbe laisse la gamme des possibilités d'évolution largement ouverte pour les 20 années à venir.

3.3.5 Extension spatiale

Pour Agde, il existe une série d'observations pluviométriques commençant en 1874 et malheureusement interrompue en 1984 (Fig. 11.18). L'examen de ces 109 années montre que la moyenne centenaire est, avec 570,3 mm du même ordre qu'en Camargue, mais par contre la variabilité est

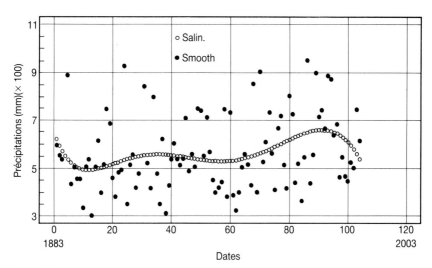

Fig. 11.16 Variation tendantielle à long terme des précipitations annuelles en Camargue de 1884 à 1987.

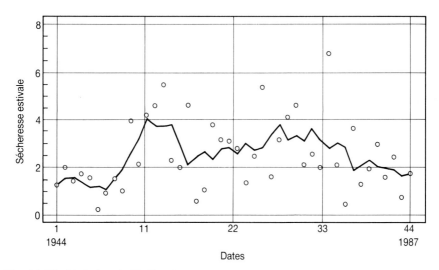

Fig. 11.17: Variation de l'indice de sécheresse estivale et moyenne courante sur 5 ans.

nettement plus forte (le coefficient de variation est de 33% contre 28% en Camargue). Cela se traduit par un plus grand intervalle de variation, depuis un minimum de 237 mm en 1947 jusqu'à 1149 mm en 1920. Comme en Camargue, en première approximation, l'analyse ne dégage aucune variation tendantielle à long terme, mais une approche plus fine permet de mettre en évidence (Fig. 11.18) une évolution tendantielle bimodale rappelant celle de Camargue, mais plus prononcée, avec un dernier maximum plus précoce et une diminution des dernières années plus

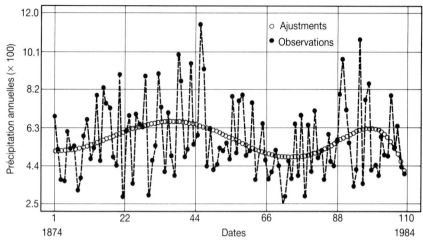

Fig. 11.18 Variation tendantielle à long terme des précipitations annuelles à Agde de 1874 à 1984.

franche. Comme en Camargue, il n'apparaît pas de modification très sensible dans la répartition des années très sèches et très humides.

Le mois le plus arrosé reçoit, en moyenne, 159,5 mm avec une variation de 54,5 à 439 mm, presque le double du module annuel de l'année la plus sèche! Le module estival est, à Agde, de 73,4 mm en moyenne; il varie de 0 à 269 mm.

Dans le détail, il apparaît des différences assez considérables entre les deux séries de données. Ainsi, la moitié seulement des années très sèches (ou très humides) en Camargue le sont aussi à Agde; et des inversions complètes sont possibles, comme en 1921 qui est en Camargue une des années les plus sèches du siècle tandis qu'à Agde c'est une des plus humides. Le coefficient de concordance entre les deux séries est pratiquement nul (tau = 0,04), ce qui traduit une indépendance presque complète des chutes de pluie annuelles entre ces deux emplacements.

3.3.6 Conclusion sur les précipitations

On sait que, pour faciliter les comparaisons de moyennes, une norme a été fixée: les moyennes sont calculées sur les 30 années de la période 1931–1960. Les précipitations normales sur la côte nord de la Méditerranée (en Camargue) peuvent donc être calculées à Salin de giraud et à Agde.

Salin de giraud
- Précipitations annuelles: 546,7 mm
- Agde module du mois le plus arrosé: 80,5 mm en octobre
- Module estival: 64,9 mm
- Précipitations annuelles: 504,5 mm
- Module du mois le plus arrosé: 65,9 mm en décembre
- Module estival: 71,3 mm

Avant 1943, les précipitations étaient plus faibles que le module normal; ensuite, elles ont augmenté, progressivement et de façon complexe, jusque vers 1970 et légèrement baissé depuis sans atteindre la normale. Cette complexité fait que l'évolution dans les vingt années à venir est très problématique.

3.4 Diagnostic général des climats annuels

3.4.1 Sécheresse estivale

Les précipitations estivales varient au cours de la période d'observation d'une manière complexe (Fig. 11.14), tout autant que les températures estivales (Fig. 11.8), mais de manière différente. Or, la sécheresse d'un climat ou d'une période s'exprime par le bilan entre l'apport d'eau météorique et le pouvoir évaporant de l'air pendant la même période; ce dernier est lié à la la température de l'air. La sécheresse est un phénomène physique difficile à quantifier directement, mais il existe des mesures indirectes et l'une des plus commode est l'indice de sécheresse estivale d'Emberger-Giaccobe; $S = PE/M$ (où PE est le module estival et M la moyenne des maximums du mois le plus chaud, paramètres qui ont été discutés dans les paragraphes précédents). La variation en Camargue de cet indice est représentée sur la Fig. 11.17. Elle met en évidence que, sur 44 ans, seules 2 années, 1969 et 1977 présentent une valeur de S supérieure à 5 et ne peuvent donc être considérées que comme sub-méditerranéennes; à l'opposé, c'est en 1949 que l'été fut le plus sec avec $S = 0,21$).

3.4.2 Diagnostic général

Les climats méditerranéens sont classés dans le système d'Emberger sur la base de trois critères:
- Le contraste thermique
- L'humidité annuelle
- La rigueur hivernale

Le premier est mesuré par le coefficient de Gorczinski modifié, le second par le quotient pluviothermique annuel et le troisième par la moyenne des minimums du mois le plus froid. Les deux derniers servent à distinguer les étages bioclimatiques et leurs variantes. La reconnaissance se fait, après le calcul des indices cités, par référence à un climagramme, ou de façon automatique.

Dans le cas présent, cette approche confirme que les années 1969 et 1977, dont la nature sub-méditerranéenne a déjà été soulignée, sont proches du type océanique; pour les autres, les dénombrements effectués donnent:
- Pour le contraste thermique
  - 10 années semi-continentales atténuées
  - 31 années littorales
  - 2 années sub-méditerranéennes à tendance océanique
- Pour l'humidité annuelle
  - 1 année semi-aride inférieure (1949)
  - 15 années semi-arides supérieures
  - 22 années sub-humides*
  - 6 années humides
- Pour la rigueur hivernale
  - 1 année à hiver très froid (1956)
  - 9 années à hiver froid*
  - 27 années à hiver frais
  - 7 années à hiver tempéré

(* y compris une année sub-méditerranéenne à tendance océanique.)

Aucune des distributions des trois critères précédents ne s'écarte de façon significative d'une distribution uniforme; c'est-à-dire que d'une décennie à l'autre, il n'apparait aucune tendance évolutive dans le climat global de la rive méditerranéenne nord.

Lorsque les 44 années sont reportées sur un climagramme d'Emberger, on note que les années de la première décennie semblent un peu plus souvent semi-arides à hiver frais, alors qu'au cours des trois suivantes, elles sont plus souvent subhumides à hiver frais; cependant, les quatre distributions ne sont pas significativement différentes; toutefois, les années 1945, qui fut semi-aride inférieure à hiver froid, et 1956, qui fut sub-humide à hiver très froid s'écartent sensiblement de l'ensemble des autres.

3.5 Conclusion générale

Les précipitations comme les températures de la côte nord de la Méditerranée sont très variables. A un premier niveau, il apparait une très grande variation d'une année à la suivante: à une année très arrosée, ou très froide, succède, en général, une année sèche ou chaude. Cette très importante variation masque une tendance évolutive générale très complexe, mais relativement calme avant 1942–45. Elle se décompose en termes périodiques, les uns très apparents, les autres masqués, mais dont le jeu ne permet pas de tirer de prévisions à un terme supérieur à quelques années. Deux années, depuis 1944, se distinguent nettement: 1945, très sèche, et 1956, très froide; mais, une fois encore, il n'est pas possible d'envisager une période de reproductibilité.

Deuxième Partie: Les vents

Le long de la côte du golfe du Lion, les vents soufflent selon une direction NW-SW nettement dominante mais selon des sens opposés, ce qui détermine des vents de terre et des vents de mer (Fig. 11.19).

Les vents de terre sont de secteur nettement NW dans le Sud du golfe, le secteur NE prenant de l'importance dans le Nord. Ils sont secs et froids. Leur rôle est important dans la mobilisation des sables émergés et les transferts d'eau des lagunes vers la mer. Ces vents sont fréquents et peuvent être violents avec des vitesses de pointe dépassant 140 km/h. Une statistique des fréquences en fonction de la vitesse (Clique *et al.*, 1984) donne les résultats suivants:

37 à 47 km/h	....	103 j/an
50 à 58 km/h	....	64 j/an
59 à 72 km/h	....	30 j/an
> 72 km/h	....	11 j/an

Les vents de mer sont de secteur SE dominant. Ils sont chauds et humides. Ils sont moins fréquents mais peuvent être violents favorisant les entrées d'eau de mer dans les lagunes et déterminant l'agressivité de la mer le long de la côte.

Sauf cas particulier lié à la forme du rivage les fortes houles (supérieures à 1,25 m) sont liées à cette orientation du vent ainsi qu'en témoigne la statistique suivante:

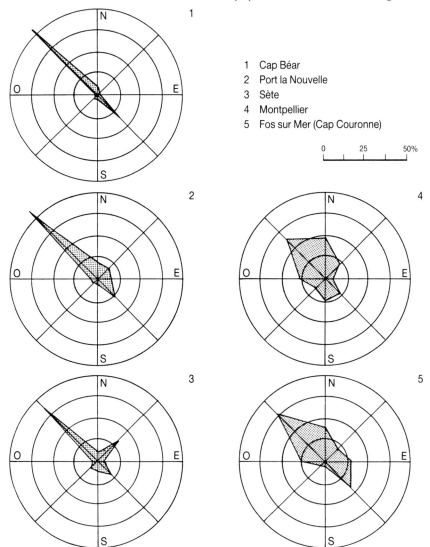

Fig. 11.19 Rose des vents le long du golfe du Lion (d'après Crique et al., 1984, Greslou, 1984).

Cap Béar (sud du Racou), par ordre de fréquence décroissante:
                          N, NE, E, SE.

Saint Cyprien   (houles de 0,5 à 2,5 m), par ordre de fréquence décroissante:
                          SE.

Port la Nouvelle, par ordre de fréquence décroissante:
                          E, SE.

Cap d'Agde,   par ordre de fréquence décroissante:
                          SE, S, W.

Sète, par ordre de fréquence décroissante:
SE, E, ESE.
Cap Couronne sans indication de hauteur, par ordre d'énergie décroissante:
SSE à E et SW à WSW.

4 LE DOMAINE MARIN

4.1 Océanographie physique

4.1.1 Structure hydrologique et circulation des masses d'eau

La dynamique du bassin Nord-Occidental dépend directement de celle de la Méditerranée dans son ensemble. Les structures hydrologiques et notamment les circuits cycloniques du Golfe du Lion, de mer Ligure et de mer Catalane, et leurs importants mouvements horizontaux et verticaux relèvent, à priori, du fonctionnement de la Méditerranée, caractéristique d'un bassin de concentration. Sur ces structures, le climat local et le vent surimposent des cycles saisonniers et des effets transitoires. Par ailleurs, la côte Nord-occidentale alimente de ses apports fluviaux la partie septentrionale de la circulation, de l'île d'Elbe jusqu'en mer Catalane, avec notamment les apports de l'Arno, du Var, du Rhône et de l'Ebre. L'étude du milieu marin dans le bassin Nord-occidental nécessite donc la prise en compte de ces différentes dynamiques.

À grande échelle de temps et d'espace, la Méditerranée est un bassin de concentration. La quantité d'eau perdue par évaporation y est supérieure aux apports d'eau douce par les fleuves et précipitations. Les eaux atlantiques, peu salées, sont progressivement transformées en eaux méditerranéennes, plus salées, qui ressortent vers l'Atlantique en s'écoulant, en profondeur, sur le seuil de Gibraltar. L'entrée d'eau atlantique superficielle est initialement motivée par le déficit en eau (de l'ordre de 1 m par an). Cependant, l'équilibre simultané des bilans en eau, en sel et en énergies thermique et mécanique provoque une amplification d'un facteur 20 environ des flux. Au lieu d'une entrée équivalente à une lame d'eau méditerranéenne de 1 m, c'est l'équivalent d'une épaisseur d'eau de 21 m qui entre annuellement par Gibraltar, alors que s'échappe en profondeur une quantité équivalente à une lame d'eau de 20 m. Cette amplification des flux entrant et sortant fait que dans toute la Méditerranée, et notamment dans le bassin occidental, existent des circulations horizontales et verticales intenses, les principaux flux étant de l'ordre de 10^6 $m^3.s^{-1}$.

La prise en compte des différents bilans a permis de proposer un modèle simplifié de fonctionnement climatique de la Méditerranée, à échelle de temps annuelle ou pluri-annuelle (Béthoux, 1979a). Les caractéristiques des eaux atlantiques étant des données extérieures au système proposé, le modèle relie les flux sur le seuil de Gibraltar (ou de Sicile), les conditions climatiques (évaporation, précipitation) et les caractéristiques des eaux profondes. En supposant des conditions climatiques différentes des actuelles (optimum climatique ou maximum glaciaire), a été envisagé le fonctionnement de la Méditerranée au cours du passé (Béthoux, 1979 b, 1984). Les paramètres physiques déterminants sont: la forme du détroit (plus précisément la section minimale offerte à l'écoulement), la profondeur de l'interface entre flux entrant et flux sortant, et l'aire de la surface de la mer, ces trois paramètres variant avec le niveau de la mer.

Au large du golfe du Lion, le courant Ligure constitue le trait dominant de la dynamique marine. Ce courant longe la Riviera italienne depuis le golfe de Gênes, puis la Côte d'Azur française et le littoral de l'Estérel puis les Maures. Il est relativement moins bien connu au large du Golfe du Lion, mais il trouve son prolongement en mer Catalane. Trait d'union entre le bassin Algéro-provenal, la mer Tyrrhénienne, le golfe du Lion et la mer Catalane, avec un flux moyen de 1,8 10^6 m^3.s^{-1} (Béthoux *et al.*, 1982; Béthoux et Prieur, 1984), il joue un rôle important dans la circulation en Méditerranée occidentale.

Ce courant intéresse essentiellement la couche superficielle, 0–250 m, mais sa dynamique peut également influer sur la position et la circulation de la couche d'eau "intermédiaire" située entre 200 et 600 m de profondeur. Le long des côtes Nord-occidentales, il constitue la partie septentrionale d'un ou de plusieurs circuits cycloniques de la couche superficielle. Un flux d'origine "atlantique" remonte dans le bassin occidental et se dirige vers les côtes occidentales de la Corse. Au Nord du cap Corse, sa jonction avec un flux superficiel provenant de la mer Tyrrhénienne, constitue le courant Ligure. Après avoir longé, à seulement quelques milles du littoral, le golfe de Gênes et la mer Ligure, à l'aplomb de l'isobathe 1000 à 1500 m, la topographie et le plateau continental étendu du golfe du Lion l'obligent à s'éloigner de la côte, jusqu'en mer Catalane où il se rapproche à nouveau du littoral. Une partie du courant ligure boucle le circuit cyclonique en mer Catalane et se joint au flux "atlantique" en direction du cap Corse, tandis qu'une partie s'échappe de la mer Catalane par le détroit d'Inice, et rejoint la circulation d'origine atlantique à la sortie de la mer d'Alboran.

4.1.2 *Formation hivernale d'eau profonde*

La Méditerranée Nord-Occidentale, à savoir les zones hauturières du golfe du Lion, de la mer Ligure et de la mer Catalane, est le siège du processus hivernal de formation d'eaux denses. Sous l'action du climat, les effets conjugués du déficit en eau et des transferts thermiques vers l'atmosphère se traduisent par une augmentation de salinité et une diminution de température des eaux superficielles, d'où une augmentation de densité entrainant des mélanges verticaux, aidés par les coups de vent. De tels mélanges peuvent, en février, atteindre toute la colonne d'eau et conduire à la formation d'eau profonde. Une modification du climat devrait modifier l'intensité du processus de formation d'eau profonde, et donc également, en retour, l'intensité d'une partie de la circulation cyclonique.

4.1.3 *Le forçage du vent*

Il a été montré (Béthoux *et al.*, 1982), que le vent ne pouvait être le forçage principal du courant ligure au large. Dans la partie la plus ventée, au débouché du golfe du Lion, le vent le plus fréquent, le Nord, Nord-Ouest, tend à renforcer le courant ligure passant à l'aplomb des isobathes 1000m. Cependant, sur le plateau continental, la circulation liée à la composante côtière du courant ligure peut, selon les saisons et le vent, être totalement modifiée en direction et intensité. D'après Millot (1979; 1983) et Millot et Wald (1980), en été, par vent faible, la composante côtière du courant ligure progresse vers l'Ouest avec des vitesses de l'ordre de 0,25 m.s–1. Elle est cependant souvent contrariée par les vents de Nord et Nord-Ouest qui induisent alors une circulation complexe. Il peut y avoir dans ce cas, développement rapide "d'upwelling" le long des côtes du Languedoc et de Provence, et de "downwelling" le long des côtes du Roussillon. A plus petite échelle, des courants induits par le vent s'organisent en cellules autour des zones d'upwelling, pouvant atteindre des vitesses de 0,5 à 1 noeud. Mais, toujours d'après Millot (1983), la fréquence des vents de Nord-Ouest dans le golfe du Lion ne doit pas masquer l'effet, en automne et hiver, des forts vents de Sud-Est. Ces coups de vent, qui durent plusieurs jours, lèvent une mer forte et entraînent une circulation importante le long de la côte, avec probablement des courants vers le Sud-Ouest supérieurs à 0,50 m.s–1. De tels coups de vent, en automne et hiver, sont directement en relation avec le déplacement de perturbations atmosphériques.

En conclusion, la circulation dans la partie septentrionale de la Méditerranée Nord-Occidentale, à savoir en mer Ligure, dans le golfe du Lion et en mer Catalane, est sous l'influence prépondérante du courant ligure. A différentes échelles de temps ce courant dépend du climat:

– à l'échelle décennale et séculaire, il dépend du fonctionnement climatique de la Méditerranée, lui-même lié au caractère continental de son environnement, et aux conditions limites, températures et salinités, des eaux atlantiques, ainsi qu'au niveau de la mer. Ce dernier facteur joue sur la dynamique des détroits (voir par exemple: Béthoux, 1979 b, 1984; Armi et Farmer, 1985; Bryden et Stommel, 1984).

Il dépend également des caractéristiques des hivers: certains hivers "froids" ayant été reconnus comme conduisant à un processus de formation d'eau profonde, apparemment sur une plus grande zone marine et pendant un temps plus long, donc devant, à priori, conduire à un volume d'eau formé plus grand.

– à l'échelle saisonnière, l'intensité du courant ligure (volume transporté et vitesses) dépend du cycle des apports d'eau douce, avec une circulation nettement plus intense en hiver qu'en été. Son impact sur le littoral du golfe du Lion dépend également de l'occurence et de la direction de vents forts, donc du déplacement des dépressions sur l'Europe et le bassin méditerranéen.

4.1.4 Impacts et implications des changements climatiques

Le fonctionnement dynamique de la Méditerranée, dans la gamme de variation de température proposée, ne dépend pas de la valeur absolue de la température, mais de la variation de température entre la Méditerranée et l'Atlantique. La variation de température de l'air envisagée ne peut être que zonale, et dans ce cas, la Méditerranée aura toujours un climat continental par rapport à l'Atlantique. Le fonctionnement du bassin occidental ne peut donc être que celui d'un bassin de concentration identique à l'actuel (pertes d'eau par évaporation supérieures aux apports d'eaux douces). Ce n'est pas la variation du niveau de la mer de quelques décimètres qui peut modifier sensiblement les flux sur les seuils de Gibraltar et de Sicile. Un tel fonctionnement en bassin de concentration a déjà été supposé tant lors d'optimums climatiques que de maximums glaciaires.

À l'échelle saisonnière ou interannuelle, une modification des pluies (intensité accrue en hiver), de même que des hivers plus froids, tendront à intensifier le courant ligure, mais dans des gammes de variations déjà rencontrées actuellement, lors d'hivers froids et/ou de fortes pluviosités automnales ou hivernales.

En conclusion, les conditions limites hauturières sont fixées par un courant ligure plus ou moins renforcé, et la circulation littorale résiduelle, sur le plateau du golfe du Lion, est soumise à l'effet de coups de vent.

4.2 Faune ichthyique du Golfe du Lion

Dans le cadre méditerranéen, le golfe du Lion présente une certaine originalité biogéographique. Cette originalité tient évidemment à sa position par rapport au détroit de Gibraltar, à ses caractéristiques hydroclimatiques actuelles (eaux relativement peu salées et fraîches) et également à la présence d'une frange lagunaire importante d'eaux peu profondes oligo à euhalines. Enfin, pour comprendre le peuplement ichthyique du golfe du Lion il faut tenir compte des conditions climatiques lointaines et principalement de l'alternance des périodes glaciaires et interglaciaires du quaternaire.

4.2.1 Aspect actuel du peuplement du Golfe du Lion

Il résulte des faits évoqués précédemment que le golfe du Lion est le "refuge-frontière" le plus méridional pour des espèces originaires de l'Atlantique septentrional. Ces espèces sont de véritables relictes des peuplements glaciaires. Les plus connues sont: *Sprattus sprattus, Platichthys flesus, Lampetra fluviatilis, Myxine glutinosa, Taurulus bubalis*, etc. Le biotope golfe du Lion permet le plein développement de ces animaux par ailleurs assez euryvalents.

Par contre, on note la pauvreté spécifique en espèces subtropicales. De nombreuses espèces présentes sur les côtes d'Afrique du Nord et sur celles du Sud de l'Espagne sont absentes du golfe du Lion. La famille des Sparidés qui compte 23 représentants dans ces secteurs n'en a que 14 dans le golfe du Lion.

Il faut souligner ici le plein épanouissement d'espèces typiquement laguno-estuariennes dont le cycle de vie est fortement lié à la possibilité de séjour en eau dessalée et parfois fraîche. Ces espèces à affinités septentrionales sont: *Pomatoschistus microps, P. minutus* qui ne se trouvent que dans le golfe du Lion. Mais aussi l'abondance d'espèces plus répandues est à noter: *Gasterosteus aculeatus, Atherina boyeri, Syngnathus abaster, Symphodus cinereus staitii*.

Enfin, certains poissons peuvent être (encore) considérés comme endémiques: *Ophidion rochei, Alosa fallax rhodanensis, Gobius ater, Gammogobius steinitzi*.

4.2.2 Modifications faunistiques récentes à court terme

Il est difficile d'apprécier les changements faunistiques dans le golfe du Lion d'après les apports des pêches commerciales. En effet, dans ce secteur en perpétuelle mouvance technique, l'apparition d'une espèce ou l'augmentation des quantités débarquées d'une autre espèce sont difficiles à interpréter.

Malgré cela, on doit constater que durant la dernière décennie certains poissons ayant des affinités pour les secteurs tempérés chauds et même subtropicaux sont de plus en plus fréquents et abondants dans les pêches débarquées au Grau-du-Roi et à Sète. Citons par exemple: *Lichia amia, Lithognathus mormyrus, Puntazzo puntazzo, Sarpa salpa, Oblada melanura, Seriola dumerili, Trichiurus lepturus, Lepidopus caudatus, Solea senegalensis, Solea aegyptacia*, etc. Par contre, *Sprattus sprattus* et *Platichthys flesus*, animaux ayant des affinités septentrionales, sont nettement plus rares que vers les années 1970-1975 et avant!

Cette progression de la "faune chaude" et régression de la "faune froide" sont, peut-être, à mettre en relation avec les étés particulièrement cléments et le déficit hydrique de ces dernières années.

4.2.3 Perturbations ponctuelles des populations

Les hivers particulièrement froids de 1984-1985 et de 1986-1987 ont eu des effets immédiats, décelables sur certains éléments de la faune lagunolittorale. Par exemple, dans les lagunes peu profondes du Languedoc, la mortalité des petits gobies sédentaires, *Pomatoschistus microps* a été importante particulièrement au niveau des individus les plus âgés. De plus, on a noté que la croissance ultérieure des survivants fut mauvaise malgré le rétablissement de conditions favorables. Par contre, un autre gobie lagunaire *Pomatoschistus minutus* qui migre en mer à l'entrée de l'hiver pour pondre n'a pas été affecté par le froid et a même connu une certaine prospérité. Cette prospérité est à mettre en relation avec une diminution de la compétition intralagunaire entre ces deux gobies.

4.2.4 Conclusion

Le golfe du Lion présente une originalité faunistique certaine par le fait qu'il est le point le plus méridional de distribution de certaines espèces boréales et préboréales et le plus septentrional d'espèces "subtropicales" mais ces dernières sont moins abondantes que les précédentes. De plus, le système lagunaire qui le borde permet le développement de peuplements à caractéristiques estuarines.

Les qualités physico-chimiques des eaux du golfe du Lion, surtout température et salinité expliquent en grande partie la structure générale du peuplement, ainsi que les modifications ou perturbations récentes. Le golfe du Lion étant une frontière biogéographique pour un certain nombre d'espèces, toute modification de ces facteurs revêt une importance incontestable.

5 LE MILIEU TERRESTRE LITTORAL

5.1 Définition

La délimitation du domaine littoral comprendra tous les territoires terrestres et aquatiques influencés par la mer, c'est-à-dire: le système plage/dune, les étangs ou les lagunes et une frange terrestre périlagunaire influencée soit par les submersions marines ou saumâtres, soit par un aquifère saumâtre.

Il inclut le "domaine paralique" au sens où l'entendent Guelorget et Perthuisot (1983): "Le qualificatif de paralique (du grec *para*: à côté et *halos*: sel et, par extension, la mer) appliqué à une aire, un bassin ou un milieu signifie simplement qu'ils possèdent une –certaine– relation avec la mer". La dénomination est dans la pratique limitée au domaine aquatique situé en lisière des domaines marins et continentaux.

Pour comprendre les mécanismes de mise en place et d'évolution de ces milieux, les composantes physiques et biologiques doivent être associées. Dans le compartiment biologique du système, tous les peuplements animaux ou végétaux n'ont pas la même valeur soit comme indicateur de structure, soit par leur importance économique, soit par l'attrait qu'ils nous procurent. Ne pouvant être exhaustif, des choix ont été fait. C'est ainsi que l'on associera:
- le système lagunaire et ses communautés benthiques et pélagiques;
- le système dunaire et sa végétation;
- les zones périlagunaires et leur végétation.

Dans ce classement, l'avifaune doit être mise à part. Les relations très complexes qui s'établissent entre elle et les différents types de milieu qui viennent d'être définis rendraient sa compréhension difficile si on l'incorporait à ces subdivisions, aussi un chapitre particulier lui sera consacré.

Nous adopterons donc ce plan après avoir donné un aperçu détaillé de la structure et de l'évolution géologique récente de la côte.

5.2 Géologie, processus sédimentaires et évolution de la côte

5.2.1 Formation de la plate-forme du Golfe du Lion
Située à l'extrémité Nord-Ouest du Bassin méditerranéen où elle forme le golfe du Lion, la marge rhodano-languedocienne constitue l'exemple type d'une marge progressive caractérisée par une plateforme continentale qui peut atteindre 70 km au droit du delta du Rhône avec une pente moyenne de 0,5 % (Aloisi, 1986).

Cette morphologie de plateforme et la présence d'exutoires fluviatiles importants sont les facteurs déterminants de l'existence quasi continue d'une côte basse sableuse sur tout le pourtour du golfe du Lion. Cette plateforme est encadrée à l'Est par la marge provençale et à l'Ouest par la marge pyrénéo-catalane plus étroite, toutes deux à pentes plus fortes.

En certaines zones, notamment sur les bordures du delta du Rhône, des mouvements tectoniques affectent les terrasses alluviales édifiées au cours du Plio-Quaternaire par le Rhône (Costières du Gard) et par la Durance (plaine de la Crau).

Ces manifestations de la "néotectonique" sont actives jusqu'au Quaternaire moyen (fin du Riss). Seuls subsistent ensuite quelques effets de subsidence.

5.2.2 Le remblaiement holocène
La succession des dépôts durant la dernière remontée du niveau marin nous est bien connue grâce aux études des carottes des sondages effectués par la Compagnie SHELL (Kruit, 1951; Lagaaij et Kopstein, 1964; Oomkens, 1970), et par la Compagnie nationale pour l'aménagement de la région du Bas-Rhône et du Languedoc (Astier *et al.*, 1970).

La trace des premiers stationnements marins post-Wurm, depuis la côte –100 a été reconnue sur le plateau continental sous forme de cordons plus ou moins bien conservés (Aloisi et Duboul-Razavet, 1974).

La synthèse des données recueillies sur le plateau continental, au sein des dépôts des deltas du Var et du Rhône, ainsi que le long du littoral languedocien (Aloisi *et al.*, 1984), fait apparaître plusieurs stades du stationnement lors de la remontée marine depuis le début du Tardiglaciaire wurmien.

Vers 7200 BP le niveau marin se rapproche de l'actuel, pour presque l'atteindre vers 6500 BP (Bazille, 1975).

A partir de cette période le niveau marin tend à se stabiliser suivant une série d'oscillations dont l'amplitude s'amortit progressivement.

Nous pénétrons dans le domaine des constructions visibles en surface (série terminale) qui appartiennent à l'histoire récente du delta et dont l'interprétation peut être faite par exploitation des photos aériennes ou des images satellitaires (L'Homer *et al.*, 1981).

5.2.3 *Processus hydrodynamiques*
Domaine continental

Le golfe du Lion est le réceptacle d'un réseau fluviatile diversifié. A l'Ouest et au Nord, il est alimenté par des rivières pyrénéo-languedociennes (Tech, Têt, Aude, Orb, Hérault). Pour tous les cours d'eaux du secteur pyrénéo-languedocien, la combinaison du climat méditerranéen et de reliefs élevés entraîne un régime torrentiel. Il se caractérise par de brusques apports massifs liés aux fortes crues périodiques.

A l'extrémité orientale, le Rhône joue un rôle particulièrement important tant du point de vue hydrologique que sédimentaire, même si son influence ne concerne pas la totalité du golfe. Ses principales caractéristiques sont résumées dans le Tableau 11.5.

Actuellement son cours et son régime sont très fortement influencés par l'homme (rectification de tracé, endiguement, barrages, etc.). En 1973 (Ritter, 1973) les principaux aménagements hydroélectriques s'élevaient à 115 unités en service et 7 en projet pour l'ensemble du bassin versant.

Les alluvions qu'il rejette en mer sont essentiellement constituées de suspensions argilo-silteuses, tandis que les charriages de sable sur le fond ne s'observent plus qu'en période de crue (Vernier, 1972).

La réduction des volumes de sédiments amenés à la côte serait une des causes de destabilisation du littoral du delta. En particulier, la réduction de la charge solide du Petit Rhône, enregistrée depuis le début des années 1950, s'est traduite par une accélération des processus d'attaque du littoral de part et d'autre de son embouchure. Celle-ci ne progresse plus de nos jours que très lentement vers le large.

On constate également une réduction importante de la charge solide du Petit Rhône qui se traduit par des processus d'érosion active aux abords de son embouchure.

Domaines littoral et circalittoral

Dans des conditions de calme météorologique, le courant général géostrophique porte vers l'Ouest, ce qui explique la répartition des sables tout au long de la côte depuis les embouchures du Rhône.

Tableau 11.5 Principales caractéristiques hydrologiques du Rhône

Bassin versant 95500–97000 km² selon les évaluations
Longueur 812 km
Débit liquide
    Evaluation publiée en 1924 (Parde, 1924, in Brun, 1961)
        module: 1650–1700 m³ . s⁻¹
    Evaluation publiée en 1973 (C.N.R., 1973)
        étiage: 560 m³ . s⁻¹
        module: 1400 m³ . s⁻¹
        hautes eaux 4250 m³ . s⁻¹
        crue décennale 8400 m³ . s⁻¹
        crue centennaire 11200 m³ . s⁻¹
        crue millénaire 14000 m³ . s⁻¹
    Evaluation publiée en 1984–85 (Greslou, 1984; Anonyme, 1985)
        module: 1760–1790 m³ . s⁻¹
    Extrêmes observés
        minimum: 360 m³ . s⁻¹
        maximum: 13000 m³ . s⁻¹ en novembre 1840
    Régime
        type:
            nivo-glaciaire en amont de Génissiat
            atlantique-méditerranéen en aval de Lyon
        valeurs moyennes (publiées en 1924)
        1ᵉʳ maxi., mai 1950 m³ . s⁻¹
        1ᵉʳ mini., septembre 1220 m³ . s⁻¹
        2ᵉᵐᵉ maxi., novembre 1810 m³ . s⁻¹
        2ᵉᵐᵉ mini., janvier 1590 m³ . s⁻¹
    9/10 du débit s'écoule par le Grand Rhône (est)
    1/10 du débit s'écoule par le Petit Rhône (ouest)
Débit solide
    D'après Surrel (1847): 21.10⁶ m³ . an⁻¹ de limons (noter que les sables ne sont pas inclus)
    D'après Van Straaten (1957):
        5,5 . 10⁶ t . an⁻¹
        = environ 4.5 . 10⁶ m³ . an⁻¹
    D'après Blanc (1977):
        2,2 . 10⁶ t . an⁻¹
        = environ 1,8 . 10⁶ m³ . an⁻¹
    Cette dernière valeur est à mettre en relation avec une évaluation de 3 . 10⁶ t . an⁻¹ (environ 2,5 . 10⁶ m³ . an⁻¹) pour l'ensemble des apports en suspension dans le golfe du Lion (in Fernandez, 1984).

En compensation du courant liguro-provençal, peut se former localement le long de la côte un contre-courant languedocien dirigé d'Ouest en Est. Le voisinage même de la côte est caractérisé par un régime complexe de courants qui dépendent essentiellement des vents et des trains de houle de mer dominants (NW, W-NW, E, E-SE, . . .). Ces courants combinés avec les effets des vagues déferlantes sont responsables des processus érosifs et des transits littoraux des sédiments.

L'action des houles de mer ou des vents est limitée au domaine littoral (c'est-à-dire jusqu'à 40 m de profondeur). Aloisi (1986) considère que les houles les plus fortes, peu fréquentes (coups de vent d'Est à Sud Est)

doivent jouer un rôle assez faible dans la remobilisation des sédiments des fonds de la zone infralittorale.

Les côtes débordant le plus vers le Sud (secteur Faraman-Vieux Rhône et littoral de la Petite Camargue aux Saintes-Maries) sont les plus exposées aux houles dominantes de Sud-Sud-Est et de Sud-Ouest.

(i) *Processus et aires de sédimentation au large*
D'après Gadel et Pauc (1973), aux environs de l'embouchure, la lutte d'influence entre les apports fluviatiles et les courants marins contrôle la distribution des dépôts. Les facteurs majeurs qui impriment leurs effets sont au nombre de trois: distance à l'embouchure, effets de courants, phénomènes de floculation.

Les sables littoraux passent graduellement, en direction du large à des sables fins vaseux puis à des vases. Ces vases terrigènes, par leur teneur importante en carbonates (pouvant aller jusqu'à 40 %), sont de véritables "marnes en cours de formation" (Roux).

Van Straaten (1957) indique un taux de sédimentation pouvant atteindre 35 cm.an^{-1} à la sortie du fleuve, impliquant une accrétion de 30 m.an^{-1} de l'embouchure.

Les vitesses de sédimentation, évaluées grâce aux rejets radioactifs des usines installées sur le bassin versant du Rhône rendent compte d'un dépôt très rapide, dans les deux premiers milles marins en suivant le trajet des eaux fluviatiles. Les fonds se remblaient à une vitesse supérieure à 2 cm.an^{-1} (Got et Pauc, 1970). Cette zone voit se déposer environ la moitié de la charge turbide des eaux (Pauc, 1970).

La Figure 11.20 représente schématiquement la nature des fonds. Les pourcentages exprimés sont ceux de la teneur en lutites (fraction < 63 μ) du sédiment. Les dépôts sableux se répartissent suivant une bande étroite, parallèle à la ligne de côte. Les médianes granulométriques y oscillent entre 150 et 250μ (Kruit, 1951). Les replats sableux en face des Saintes-Maries, puis de Faraman et de la Pointe de la Gracieuse ont été interprétés par Van Straaten comme des "erosional environments". Il s'agit là de plateformes résiduelles situées aux emplacements d'anciens débouchés deltaïques majeurs.

La couverture de sédiments meubles est localement interrompue par des bancs rocheux en bandes plus ou moins parallèles au rivage actuel (au large des Saintes-Maries-de-la-Mer et des plages depuis la Grande Motte, Carnon, Palavas, et jusqu'à Frontignan). Ces bancs rocheux amortissent quelque peu les effets des houles du large.

(ii) *Processus littoraux d'érosion et de sédimentation*
Grosso modo on peut classer le littoral en trois types de contextes:
– côtes en recul où les apports ne compensent pas les départs par érosion (côtes en régression);
– côtes qui progressent régulièrement en empiétant sur le domaine maritime (côtes progradantes);
– côtes à peu près stables; elles correspondent aux zones d'équilibre relatif situées à la jonction des deux types précédents.

a) Côtes en voie de régression.
C'est le cas le plus fréquent. Ce processus est dû fondamentalement au fait que toute ancienne embouchure se trouve derechef en porte à faux avec le nouvel équilibre hydrodynamique: elle est donc fortement

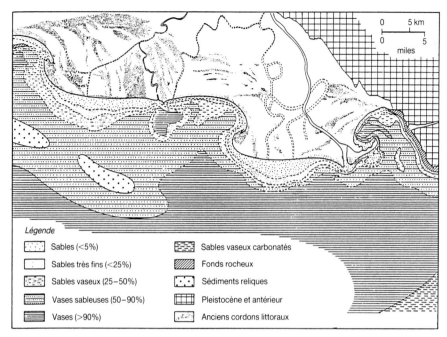

Fig. 11.20 Répartition des sédiments au large du delta.

attaquée. Cette attaque ne se ralentira que lorsque la côte aura retrouvé un tracé sensiblement rectiligne.

A ce phénomène s'en ajoute un autre: pour ne pas céder de terrain sous le coup de l'attaque des vagues lors des fortes houles, une côte sableuse doit recevoir des apports compensateurs -or ceux-ci ont fortement diminués-, comme on l'a vu précédemment, en liaison avec l'importante réduction des charges solides déversées aux embouchures du Rhône. Par voie de conséquence, force est de considérer qu'aujourd'hui une bonne partie des sables qui transitent le long des côtes sont remaniés soit à partir des côtes en voie de recul soit depuis les fonds sableux des anciens promontoires deltaïques en cours d'abrasion.

Par ailleurs si certaines défenses du littoral ont le mérite de protéger la zone du littoral où elles ont été implantées, elles peuvent provoquer une accentuation des processus érosifs de part et d'autre:
- b) Côtes progradantes
    Les sables ne gagnent sur le domaine marin que suivant trois types d'appareils sédimentaires privilégiés.
- b1) Barres d'embouchure

A l'embouchure du Grand Rhône, là où les apports sont excédentaires sur les départs, une partie importante des sables et débris organiques accumulés à la sortie de l'embouchure sous forme de bancs sont remaniés par les fortes houles et rejetés à la côte pour y constituer des barres d'embouchure ("mouth bars").
- b2) Flèches longitudinales (ou flèches de lido) orientées W. E.

En s'allongeant, ces flèches peuvent provoquer l'isolement de lagunes littorales. La flèche de la Gracieuse, aujourd'hui menacée, est de ce type.
- b3) Les flèches courbes convexes emboitées, à orientation grosso modo S.N.

Elles constituent une forme de stockage latéral des sables arrachés aux côtes en voie de recul. Lorsque les courants assurant le transit côtier des sables d'Est en Ouest débouchent dans des contextes de golfe (Beauduc, Aigues-Mortes), leur vitesse s'amortit: il se produit une perte de compétence avec dépôts de sables. Ceux-ci constituent tout d'abord des rides prélittorales qui s'exhaussent ensuite en faisant progresser vers l'Ouest la ligne de rivage. On a relevé des vitesses annuelles de 11 m pour Beauduc et de 18 m pour la pointe de l'Espiguette, au droit du camping.

5.2.4 Processus éoliens

Le vent dominant est le Mistral, de secteur Nord, vent froid et sec qui peut souffler plus de 210 jours par an avec des vitesses allant de 40 à 75 km.h^{-1} et des pointes extrêmes pouvant dépasser 135 km.h^{-1} et même atteindre exceptionnellement 150 km.h^{-1}. Les vents marins de secteur SE à SW, pour moins fréquents qu'ils soient, ont une grande importance dans les processus érosifs des plages et le transit littoral des sables.

Effets du vent sur les plans d'eau (mer, lagunes)
L' action se manifeste de cinq façons différentes:
- en générant des trains de houles de mer (vents violents du grand large),
- en générant des houles dites "houles de vent" (vent à la côte),
- en créant des courants: action sur les eaux de surface avec possibilité de courants de fond compensateurs,
- en faisant varier le niveau des eaux près des rivages aussi bien en bord de mer qu'en bordure des grandes lagunes,
- en provoquant des transports éoliens avec sédimentation dans les lagunes proches ou dans la mer.

Effets du vent sur la barrière littorale
(i) *Edification d'appareils dunaires*
Dans les secteurs où les sables n'ont pas été fixés (constructions, aménagements, plantations) les vents ont une action importante qui peut aboutir à la formation de dunes.

Cette action du vent est particulièrement spectaculaire là où la côte prograde rapidement. Les études des cartes anciennes montrent qu'une bonne partie des dunes proches du bord de mer des pointes de Beauduc et de l'Espiguette se sont constituées entre 1850 et 1920. Ce processus qui se poursuit aujourd'hui, mais de façon plus restrictive, est développé ci-dessous.

Pour le reste du littoral rhodanien, notamment le long de la plage des Saintes-Maries les dunes sont relatively stables dans la mesure où elles ne sont pas entaillées par l'érosion littorale comme cela été longtemps le cas le long du littoral de la Petite Camargue.

Entre l'Espiguette et Carnon, le vent modèle les cordons proches de la mer dont le sable a la granulométrie la plus fine. Elles prennent une allure en rateau, alors que le cordon interne à tracé rectiligne et constitué en majorité de matériau grossier, se prête plus difficilement à des remobilisations dunaires (Corre, 1987).

Notons également qu'il se crée régulièrement de nouvelles dunes de part et d'autres de l'embouchure du Rhône. Un cordon dunaire limité se reconstitue ou se complète au fur et à mesure du recul du rivage de la plage Napoléon, immédiatement à l'Ouest de la flèche de la Gracieuse.

(ii) *Cas de la Pointe de l'Espiguette*

La plupart des grandes accumulations dunaires (y compris celles aplanies pour exploitations agricoles) situées entre les Baronnets et Port Camargue se sont toujours édifiées aux abords de la zone où se trouvait alors la bande de sable la plus large du littoral non fixé.

Un tel contexte se trouve aujourd'hui théoriquement juste au Sud de Port-Camargue. Toutefois les divers aménagements récents (digue destinée à bloquer la progression des sables en mer, implantations de brise-vents, plantations . . .) concourent à modifier la dynamique des sables et à restreindre la création de nouvelles dunes.

Plus au Sud, en direction du phare de l'Espiguette des bourrelets de sable se sont constitués sensiblement sur l'alignement du rivage correspondant à 1920. Sur la plage et l'arrière-plage, les sables s'accumulent en grandes rides plates orientées grosso modo NE-SW. Si on ne remarque pas de création de nouvelles dunes importantes depuis 1937 (premières photos aériennes exploitables), par contre les bouleversements de structure y sont importants.

(iii) *Cas de la Pointe de Beauduc*

Le système progradant de la flèche de Beauduc est peu affecté par des aménagements, à la différence de la Pointe de l'Espiguette. Il garde par conséquent une certaine capacité d'édification ou d'accroissement d'accumulations sableuses dunaires, principalement dans le secteur proche du grau de Galabert et des Cabanes du Sablon. Certaines dunes sont indiquées avec des côtes de 5 à 6 m (chiffres qui mériteraient d'être contrôlés aujourd'hui).

On remarque, par contre, dans le secteur compris entre la pointe des Sablons et l'étang de Beauduc, que les dunes se sont fortement amoindries sous les effets de l'érosion par les eaux des lagunes. Ce processus a peut-être été facilité par un léger effet actuel de subsidence (voir ci-après).

(iv) *Données complémentaires relatives aux secteurs du Languedoc-Roussillon*

En se déplaçant de Carnon vers le Sud, en direction des Albères, le cordon littoral a tendance à se simplifier dans sa morphologie. Il constitue un lido unique séparant de la mer une série de lagunes à différents stades de comblement, ou forme un bourrelet accolé au socle continental. Les systèmes dunaires successifs d'âges différents se resserrent et la nature des matériaux change. En particulier la granulométrie tend à devenir plus grossière (Clique *et al.*, 1984; Greslou, 1984). Au delà de Carnon le système littoral tend à se résoudre en un seul cordon dunaire voir un simple bourrelet modelé plus par la mer et l'action de l'homme que par le vent.

5.2.5 Facteurs de variation du niveau marin
Si on exclut les variations d'ordre barométrique qui ont leur importance, on trouve les variations dues aux marées et aux vents.

Effets des marées
Ils sont considérés comme négligeables puisque le marnage de 30 cm, en vives eaux moyennes, n'est pas susceptible de créer des courants de marée perceptibles (Greslou, 1984).

Effets des vents
(i) *Sur le littoral*
Le niveau de la mer monte généralement par vent de Sud-Est pour descendre par vent de secteur Nord-Nord-Ouest.

Lors des tempêtes de Sud-Est, l'élévation peut atteindre et même dépasser nettement la cote +1 m NGF; ce qui se traduit localement par une submersion du cordon et un envahissement des terres basses et des étangs.

Lors des fortes tempêtes, le cordon du They de la Gracieuse peut être submergé. Tel fut le cas le 28 janvier 1978 où la variation de niveau aurait atteint +1,80 m dans le fond du golfe de Fos.

Enfin rappelons le "raz de marée" du 6 août 1985, qui submergea subitement la côte entre l'embouchure du Grand Rhône et les Saintes-Maries. Ce phénomène fut attribué à une onde de tempête.

Remarque importante:
Si les derniers faits signalés présentaient un caractère exceptionnel et une ampleur telle que le cordon sableux ne fut ni assez élevé, ni assez large en certains points pour résister aux assauts de la mer, par contre le système côtier est naturellement construit pour encaisser les élévations courantes du niveau marin atteignant 50 cm voire même 1 m. Il importe donc de souligner que les parties de la côte sableuse qui sont capables de supporter naturellement des variations brusques de plusieurs décimètres du niveau marin sans céder, supporteront aisément a fortiori une élévation lente et progressive de 20 cm du niveau marin étalée sur 40 ans. Le problème majeur résidera dans la capacité du système à résister aux tempêtes violentes et à se cicatriser pendant les périodes d'accalmie.

(ii) *Sur les rives des étangs*
En période de hautes eaux les vagues produites par les vents de tempête sappent les bordures des dunes et des berges élevées abruptes (anciennes levées entaillées). C'est le cas des rives situées au Nord et à l'Est du Vaccarès où on enregistre des reculs allant de 100 à 150 m depuis 1837.

Les bordures des dunes situées au milieu des étangs de la Petite Camargue sont régulièrement attaquées, mais plus encore les dunes au sein des étangs englobés dans la Pointe de Beauduc.

On remarque, compte tenu de la faible tranche d'eau existant dans ces étangs du littoral, que toute élévation même faible du niveau moyen des eaux en relation avec l'élévation du niveau marin y entraînera de façon sensible une accentuation de ces phénomènes d'érosion des "rives internes" du delta.

Ce processus "d'approfondissement" des étangs, consécutif à une élévation du niveau marin, sera d'autant plus sensible que depuis l'achèvement de l'endiguement des cours du Rhône au milieu du siècle dernier, il n'existe plus les apports compensatoires des limons déposés jadis lors des inondations.

Effet de subsidence
C'est un aspect particulièrement important, et cela pour deux raisons:

La première est d'ordre fondamental. Si nous mettons en évidence une région affectée par une subsidence de 20 cm (ou plus), depuis 40 ans, nous possédons alors grâce à cet artifice "une région-test" pour étudier l'impact d'une élévation générale du niveau des eaux à la côte et dans les lagunes côtières.

La seconde raison se situe au niveau de l'appréciation du risque. Il est clair que les risques de détérioration de l'environnement littoral seront fortement accrus dans un secteur où l'effet d'élévation du niveau général des eaux se cumulera avec l'effet d'affaissement des terrains généré par la subsidence.

(i) *Mise en évidence de la subsidence à l'échelle historique*
L'examen des images sattelitaires fait apparaître que les traits morphologiques liés aux anciens cordons sont bien marqués dans la partie occidentale du delta alors qu'ils sont effacés dans la moitié orientale. Ceci est dû au fait qu'à l'Ouest d'une ligne correspondant grosso modo au cours de l'ancien Rhône de Saint-Ferreol, le delta a été affecté par la subsidence (L'Homer, 1987). Celle-ci avait déjà été pressentie par Oldham (1930).

Le taux d'affaissement semble s'accroître du Vaccarès (environ 1,50 m) en direction du Sud-Est (4 à 6 m pour les restes submergés du port romain de Fos-sur-Mer).

En compensation, l'aile occidentale du delta s'est très légèrement soulevée. Ceci explique pourquoi les cordons édifiés lors des stades de haut niveau marin depuis 6500 BP sont restés si apparents à l'Ouest (L'Homer, *et al.*, 1981).

Une accentuation de cette subsidence entre le VIème et le IXème siècle, combinée avec la transgression flandrienne, serait à l'origine du déplacement du cours principal du Rhône vers l'Est avec la création de la branche du Grand Rhône (L'Homer, 1987).

(ii) *Hypothèse d'une poursuite actuelle de la subsidence*
La comparaison entre les morphologies représentées sur les éditions de la carte IGN à 1/50000 de 1951 (levés 1947) et de 1983 (révision 1980) des Saintes-Maries-de-la-Mer pour le secteur de la pointe de Beauduc semble révélatrice à cet égard. On constate en effet une large extension systématique des surfaces submergées par les eaux allant de pair avec un grignotage et une réduction généralisée spectaculaire de tous les édifices dunaires figurés en 1947. Ces modifications sont pour nous caractéristiques d'un affaissement des terrains par rapport au niveau moyen de la mer.

Il est certes délicat de diagnostiquer une subsidence dans ce secteur par simple comparaison entre deux topographies à 33 ans de distance, car des transformations dues aux mises en eau des lagunes par la Compagnie des salins ont dû jouer un rôle. Il faut cependant souligner que ces modifications spectaculaires ne trouvent leur équivalent ni dans la région

382 *Climatic Change in the Mediterranean*

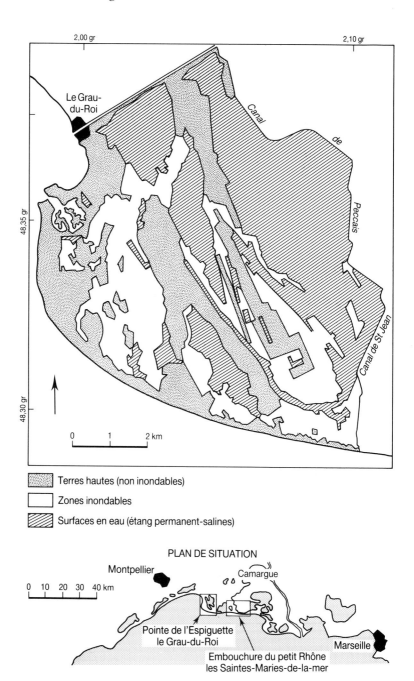

Fig. 11.21a Pointe de l'Espiguette – Le Grau du Roi.

Gulf of Lion and climatic change 383

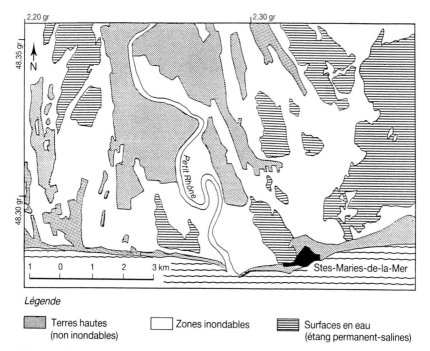

Fig. 11.21b Embouchure du Petit Rhône – Les Saintes-Maries-de-la-Mer

Légende des figures 11.21a et b: – Les terres hautes non inondables appartiennent à des cordons dunaires actuels ou anciens. Il faut remarquer l'extrême morcellement de l'espace en relation avec les conditions de submersion.

comprise entre le Vaccarès et la plage des Saintes-Maries, ni dans les étangs de Petite Camargue.

(iii) *Conséquences d'une subsidence encore active*

Si l'hypothèse d'une subsidence dans le secteur Beauduc-Fangassier se trouvait confirmée, on conçoit que la région devrait faire l'objet d'une surveillance régulière.

L'élévation prévue du niveau marin y serait particulièrement ressentie avec l'accentuation des phénomènes déjà observables aujourd'hui. On peut prévoir ainsi:

– À la côte:
- risque accru d'attaque du littoral (augmentation de la profondeur et de la pente en bas de plage),
- sape active de la digue qui longe la mer,
- risque de déferlement des vagues par-dessus cette digue avec possibilité de rupture de celle-ci,

– Dans les lagunes côtières:
- extension des surfaces occupées par les lagunes et approfondissement,
- enfoncement encore plus marqué des édifices dunaires, forte érosion et disparition de la majeure partie de ceux-ci,

- perturbations dans le système (circuit) de circulation des eaux pour les salins avec nécessité de réajustements de celui-ci,
- augmentation des infiltrations d'eaux marines.

5.2.6 Evolution du rivage depuis 100 ans
Cas général
Pour A. Y. Le Dain *et al.* (1987), si l'on s'en tient aux prévisions concernant l'horizon 2025, horizon maximum envisageable pour des considérations ayant un caractère économique et budgétaire, les reculs du trait de côte en 2025 seraient approximativement les suivants:

Ces chiffres applicables aux côtes rectilignes du Languedoc-Roussillon doivent être nuancés pour le delta du Rhône compte tenu des divers processus dynamiques qui l'affectent (Table 11.6).

Delta du Rhône
Pour la Camargue, l'évolution a été traîtée à partir de nombreux documents tous ramenés à l'échelle du 1/50000.

L'étude de l'évolution récente a été focalisée sur la période de 1947 à 1987 couverte par les photos aériennes, seuls documents apportant une précision suffisante. Le recul ainsi obtenu sur les 40 dernières années apparaissait théoriquement idéal pour éclairer l'évolution de la côte pour les prochaines 40 années à venir (horizon 2025). Là encore, les modifications intervenues dans les conditions hydrodynamiques fluviales et littorales ont eu un impact tel que seule la séquence des 10 dernières années peut être considérée comme significative pour l'avenir (moyennant quelques correctifs). Pour cette raison la carte de l'évolution récente ne comporte que des valeurs applicables aux seules dix dernières années (Fig. 11.22a).

Commentaires de la projection du littoral camarguais en 2025
L'échelle adoptée pour la carte donnant une projection du littoral à l'horizon 2025 (Fig. 11.22b) ne permet de figurer que les grandes tendances de l'évolution prévisible.

Compte tenu de la capacité de ce littoral deltaïque, construit par les houles, à se reconstruire après les tempêtes et à s'exhauser progressivement en phase avec la variation du niveau moyen des mers, il n'y a pas d'évolution catastrophique à redouter pour les 40 prochaines années.

Cependant plusieurs secteurs déja sensibles ou menacés le seront encore plus à l'horizon 2025. Ce sont, d'Est en Ouest:
- la flèche de la Gracieuse (qui risque de se scinder),

Tableau 11.6 Ordre de grandeur en mètres du recul du trait de côte

Remontée du niveau marin	Sables fins à faible pente	Sables grossiers à forte pente
13 cm	3–5	1–3
60 cm	17–25	4–7

Gulf of Lion and climatic change 385

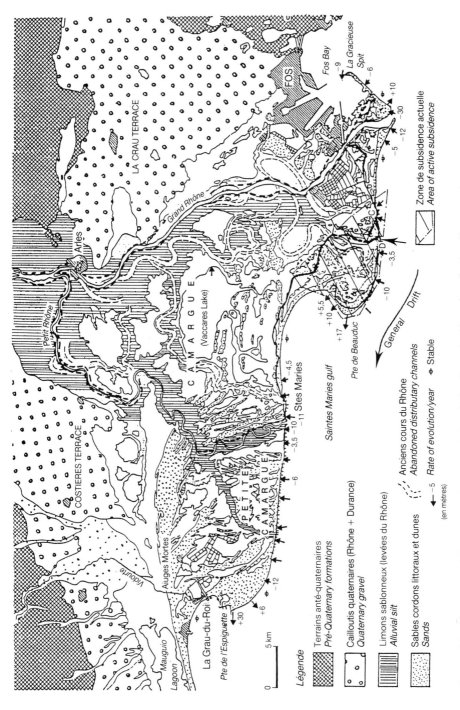

Fig. 11.22a Taux annuel de recul ou de progradation de la côte calculé sur les dix dernières années.

386 *Climatic Change in the Mediterranean*

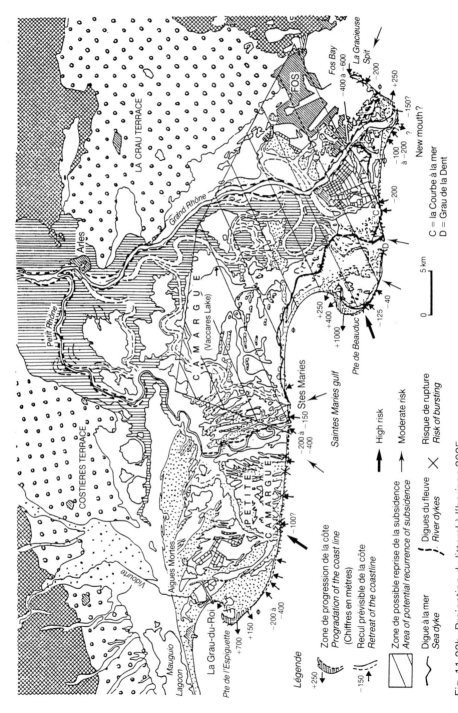

Fig. 11.22b Projection du littoral à l'horizon 2025.

- la "Courbe à la mer" et le grau de la Dent,
- l'embouchure du Petit Rhône et les plages situées de part et d'autre du promontoire de défense des Saintes-Maries-de-la-Mer,
- le littoral de la Petite Camargue,
- l'anse comprise entre la pointe de l'Espiguette et le Grau-du-Roi (risques à terme d'ensablement et d'envasement).

Plusieurs points d'interrogation subsistent toutefois:
- Comment résisteront les défenses à la mer (épis de Petite Camargue, Courbe à la mer . . .)?
- L'embouchure du Grand Rhône se dérivera t-elle vers l'Est? ou ce qui est probable, se formera-t-il un nouveau débouché en direction du Sud – Sud-Ouest?
- la subsidence se poursuivra-t-elle à Beauduc en créant un risque réel de submersion pour les étangs de ce secteur du delta?

5.3 Le domaine paralique (étangs et lagunes)

Sur l'ensemble du golfe du Lion les étangs et lagunes occupent une surface d'environ 68800 ha soit 38500 ha en Languedoc-Roussillon y compris la Petite Camargue (8700 ha) et 30300 ha en Camargue.

Ces plans d'eau appartiennent à plusieurs types géomorphologiques:
- type lagunaire (s. str.); il représente une portion du domaine marin plus ou moins séparée de celui-ci par un ou plusieurs cordons littoraux

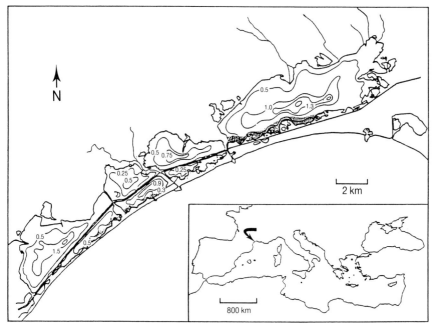

Légende

Profondeurs en mètres;
En hachures: sables et sables vaseux;
En blanc: vases.

Fig. 11.23 Carte bathymétrique simplifiée des étangs palavasiens (Hérault, France).

récents. Ceux-ci se formant plus ou moins parallèlement à la côte initiale, il en résulte que, dans bon nombre de cas, les lagunes présentent une forme allongée dans la même direction. Ce type est le plus répandu (Fig. 11.23);
- type bahira-lagune (Guelorget et Perthuisot, 1983) il s'agit d'anciennes cuvettes continentales envahies par la mer lors de la remontée eustatique holocène, ayant évolué en lagune par formation d'un cordon littoral. L'exemple le plus typique en est l'étang de Thau près de Sète (7500 ha);
- type estuaire-lagune; c'est un estuaire qui à la suite d'un changement de régime hydraulique (déviation du cours du tributaire, changement climatique, etc.) présente des courants de fonds insuffisants pour déblayer l'embouchure qui progressivement sera barrée par un seuil, puis un cordon. Ce type (Fig. 11.24) est assez rare mais se rencontre tant en Languedoc (La Grande Maire) près de l'actuelle embouchure de l'Orb qu'en Camargue (Vieux Rhône . . .);
- bassins artificiels; ils correspondent essentiellement aux salines d'Aigues-Mortes à l'Ouest du Petit Rhône et de Salin-de-Giraud à l'Ouest du Grand Rhône.

Ces plans d'eau sont peu profonds, certains se dessèchent partiellement ou entièrement pendant l'été. Les profondeurs moyennes se situent entre 0,3 et 1,5 m; certains comme l'étang de Thau sont plus profonds: 4,5 m avec un maximum de 10 m. Les fonds sont meubles, plus ou moins sableux près des cordons littoraux. Leurs rives sont

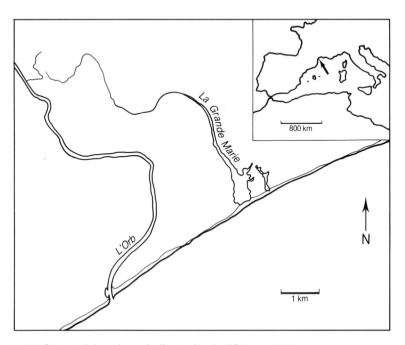

Fig. 11.24 Carte schématique de l'estuaire de l'Orb et de l'estuaire-lagune de la Grande Marie (France).

souvent marquées d'un bourrelet constitué par des rejets de matériaux biogéniques (coquilles, débris végétaux) qui en fixe le contour.

5.3.1 Fonctionnement physico-chimique du milieu paralique;

L'ensemble des étangs à l'exception des salines alimentées seulement en eau de mer reçoivent des apports d'eau douce et marins. L'eau douce provient de cours d'eau la plupart du temps de faible importance, de canaux, de sources. L'alimentation en eau salée provient des passes (graus) avec la mer et pour une part de la nappe aquifère.

Les échanges avec la nappe aquifère

Au niveau des côtes basses sableuses, donc poreuses, la mer fait sentir son effet à l'intérieur des terres par intrusion dans l'aquifère d'un biseau d'eau marine. En Camargue, le coin marin doit pénétrer à l'intérieur des terres d'une quinzaine de km. Il est encore nettement perceptible à La-Tour-du-Valat à 10 km à vol d'oiseau du golfe de Beauduc.

La nappe aquifère sous les étangs est partout salée et le plus souvent plus salée que les eaux de surface.

Si pour beaucoup d'étangs le fond argileux peut être considéré comme étanche à l'échelle d'un bilan annuel (Vaccarès, par exemple), d'autres comme les étangs au Sud de la Camargue (étangs inférieurs) reposent sur des sédiments relativement perméables (k de l'ordre de 10 cm.sec^{-1} soit 0,1 m.jour^{-1}) et il existe une solution de continuité entre les eaux de surface et les eaux souterraines sous-jacentes, si bien que les différences de pressions régnant dans l'aquifère jouent un rôle capital dans l'hydrodynamique des étangs. Les eaux de surface et les eaux souterraines sont, par différence de densité, séparées par une interface qui se déplace verticalement en fonction des pressions régnant dans l'une ou dans l'autre. Cette interface (Fig. 11.25) peut donc se trouver soit au niveau du fond (1), soit au-dessus (3), soit au-dessous (2).

Au point Z dans la nappe, situé à la profondeur H sous le fond, on a l'équilibre

$$g' \,(he.\,de + H.\,dn) = g'.\,hn.\,dn$$

où g', accélération de la pesanteur, intervient dans les deux membres de l'égalité, et donc s'annule; he est la hauteur d'eau de type étang; de est la densité de l'eau de l'étang dn est la densité de l'eau de la nappe et hn est la hauteur piézométrique de la nappe à partir de Z.

Dans les cas (1) et (2), l'interface est réelle sans être nette. Il y a un gradient de salinité croissant de haut en bas du fait du phénomène de diffusion qui assure un transfert de sels de la nappe vers l'étang. Actuellement, à l'échelle de l'ensemble des "étangs inférieurs" (4600 ha), ce phénomène apporte quelques dizaines de milliers de tonnes de sels par an aux eaux de surface. Dans le cas (3), l'interface est virtuelle dans la nature. Il y a un apport d'eau et de sels aux étangs et mélange par brassage (éventuellement éolien). C'est ce que nous appellerons l'effet de source diffuse. Le débit de la source, donc l'apport d'eau et de sels aux étangs varie avec la position théorique de l'interface au-dessus du fond. Actuellement, au cours des années les plus favorables (niveaux bas des étangs (ex.: 1967, 68, 71) on évalue l'apport maximal à 200 000 tonnes (0,2Mt) par an aux "étangs inférieurs" qui, pour une salinité moyenne

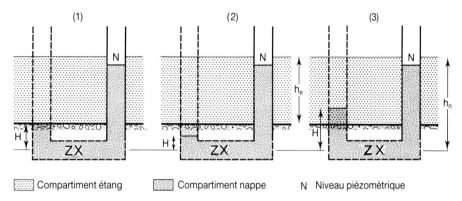

Fig. 11.25 Schéma des relations piézométriques entre eaux de surface et nappe aquifère saumâtre (étang).

de la nappe de 60 g.l^{-1} représente quelque 3 Mm^3 soit une hauteur d'eau de quelque 7 cm sur les étangs inférieurs et 3 cm sur l'ensemble du système Vaccarès. Ces échanges nappe-surface expliquent que les étangs temporaires s'assèchent sans se transformer en salines et se remettent en eau avec une salinité non nulle même en l'absence de relations avec la mer.

Les circulations dans les bassins paraliques
(i) *Les bassins de type estuarien*
En général, dans les bassins de type estuarien, la circulation des eaux est active et les échanges avec la mer ouverte importants.
(ii)*Les bassins de type lagunaire (lagunes et bahiras)*
Dans tous les cas, deux types de mouvement se combinent:
– Des courants de marée qui se font sentir au moins dans la ou les passes et à leur voisinage. Ceux-ci assurent les échanges avec le large.
– Une circulation générale dans le bassin engendrée par les dérives littorales sous la dépendance essentielle du régime des vents. L'importance relative de ces deux types de circulation varie suivant les conditions climatiques et météorologiques. C'est ce type de circulations périlittorales qui domine pour les lagunes méditerranéennes à faible effet de marée.

A ces deux types de mouvements, surtout horizontaux, s'ajoutent éventuellement les circulations dans le plan vertical (upwelling, stratification, etc.).

Lorsque la bathymétrie est différenciée avec des seuils et des dépressions, ces dernières sont souvent occupées par des corps d'eau stagnante souvent anoxique séparés de la circulation générale du bassin par des clines thermiques et/ou halines.

Enfin, dans un grand nombre de cas, notamment en ce qui concerne les lagunes méditerranéennes, les eaux, à l'issue de leur parcours dans le bassin, aboutissent dans une zone où elles stationnent avant d'être progressivement évacuées par les courants de marée. Ces zones constituent les "ombilics hydrauliques" (Guelorget *et al.*, 1984) qui ont des caractéristiques biogéologiques particulières (absence de macrophytes, diminution des suspensivores au profit des détritivores, abondance de la phase organique dans le sédiment).

Hydrochimie des eaux de surface
D'une façon générale, les milieux paraliques se caractérisent par des salinités globales différentes de la mer et, sauf exception, par les variations à la fois spatiales et temporelles de celle-ci dans chaque bassin. Ces variations de la salinité s'accompagnent de variations dans la composition ionique des eaux.

Elles sont commandées par différents paramètres externes au bassin considéré, notamment climatiques, météorologiques et hydrogéologiques qui déterminent la balance en eau douce de chaque bassin et par des paramètres internes d'ordre essentiellement hydrologiques (mais aussi biologiques) qui déterminent l'intensité des échanges avec la mer et à l'intérieur du bassin.

(i) *La balance globale en eau douce et en sels*
Le taux d'échange du bassin avec la mer peut être appelé le "confinement global" du bassin; cette grandeur correspond au temps de renouvellement global des eaux du bassin considéré.

Plus le confinement est petit plus la salinité globale du bassin est voisine de celle de la mer; il en va de même plus le déficit hydrique global est petit c'est-à-dire plus la balance hydrique est équilibrée, et vice-versa.

(ii) *Les gradients de salinité*
Dans la grande majorité des cas, les bassins paraliques présentent des champs de salinité plus ou moins stables. L'écart de salinité avec la mer d'un point donné est fonction schématiquement de la balance hydrique globale du bassin et du temps que mettent les éléments d'origine marine (donc l'eau) à atteindre le point considéré. Cette dernière grandeur, assimilable au temps de renouvellement des éléments d'origine marine en un point donné doit encore être appelée "confinement".

A confinement égal deux points, d'un même bassin ou de bassins différents dans une même ambiance climatique, ont des salinités voisines. Par contre sous des climats très différents, deux points d'égal confinement ont des salinités très différentes. Dans tous les cas, lorsque le confinement augmente, la différence de salinité (ou de concentration) avec la mer augmente, sauf si le bilan hydrique est voisin de zéro: dans ce cas, la salinité reste voisine de celle de la mer, quel que soit le confinement. Il est bien évident que, dans la nature, plus un milieu est confiné, donc enclavé dans le domaine continental, moins il a de chance de se situer dans un environnement dont le bilan hydrique soit équilibré, sauf s'il s'agit d'un confinement bathymétrique: dans le cas des bassins stratifiés, le corps d'eau inférieur se caractérise par un déficit hydrique nul. Enfin, rappelons-le, si le confinement est nul, le point considéré appartient au domaine marin (Fig. 11.26).

(iii) *Les variations de la composition ionique*
On ne peut entrer dans le détail des variations ioniques des eaux d'origine marine au cours de leur transit sur les aires paraliques. Les phénomènes biotiques et abiotiques dans l'eau elle-même, les échanges avec l'eau des sédiments, les apports continentaux, etc. agissent dans des sens trop divers. On peut cependant dire que la composition ionique d'une eau paralique se modifie en fonction croissante du confinement.

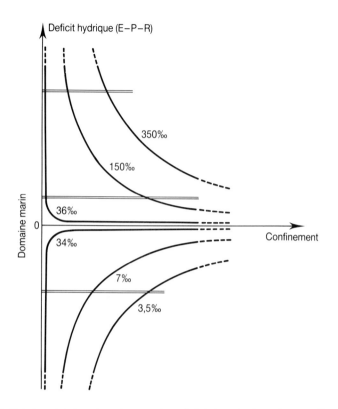

Fig. 11.26 Schématisation des champs de salinité dans le domaine paralique en fonction du déficit hydrique et du confinement.

Les ions majeurs les plus sensibles sont ceux qui interviennent dans les processus biosédimentaires: (HCO_3^-, Ca^{++}, Mg^{++}, SO_4^{--}), ainsi évidemment que les nutriments azotés, phosphatés, siliciques.

Seuls Cl^- et Na^+ interviennent peu dans les processus et peuvent être utilisés comme référence pour les variations relatives des autres ions, du moins tant que les apports continentaux ne sont pas trop chargés en sels.

Enfin, lorsque le confinement et le climat conduisent à des dépôts évaporitiques, tous les ions sont affectés par ces processus.

Organisation sédimentologique
(i) *Généralités*
La sédimentation dans les systèmes paraliques dépend évidemment d'un très grand nombre de facteurs internes et externes, qui par ailleurs agissent dans des conditions changeantes (Fig. 11.27). Chaque cas est donc original mais dans tous les bassins paraliques, on retrouve trois composantes principales:
- l'influence de la mer,
- l'influence du continent,
- l'activité du bassin lui-même.

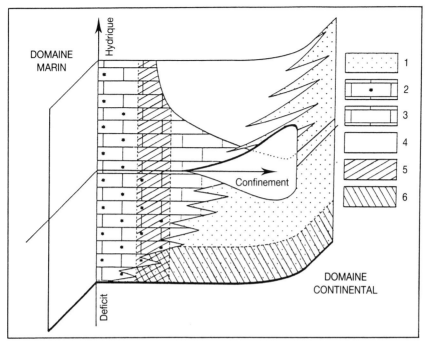

Légende

1. Détritiques terrigéniques;
2. Carbonates biogéniques, surtout biodétritiques;
3. Carbonates biogéniques (microbiens) et évaporitiques;
4. Évaporites (gypse–sel);
5. Matière organique autochtone (pétroles ?);
6. Matière organique allochtone (charbons ?).

Fig. 11.27 Schéma de l'organisation sédimentologique du domaine paralique en fonction du déficit hydrique et du confinement.

(ii) *Les apports détritiques*

Ils viennent pour l'essentiel du continent soit par les cours d'eau, soit par les vents. Les apports fluviatiles sont évidemment plus importants lorsque le bassin reçoit des tributaires. D'une façon générale, les sédiments de bassins paraliques dessalés comportent une importante phase terrigénique, notamment dans leurs zones les plus continentales.

Cependant, la dérive périlittorale en mer ouverte est susceptible d'apporter au bassin des quantités importantes de matériaux détritiques et biodétritiques soit directement par la passe avec les courants de marée montante ou les tempêtes, soit par transfert éolien et/ou fluviatile au-dessus du lido (Guelorget *et al.*, 1984). Dans le premier cas il existe généralement, de part et d'autre de la passe, des "deltas tidaux" formant des hauts fonds bombés plus ou moins encombrés de flèches et d'îlots.

Ces apports détritiques sont ensuite répartis dans le bassin selon sa dynamique propre. Pour les bassins de type lagunaire, le dispositif apparaît souvent plus ou moins concentrique à cause du vannage des marges.

Les éléments fins ont tendance à se concentrer dans les zones centrales profondes, les sédiments des marges étant plus grossiers.

Les formations biologiques (herbiers, récifs) peuvent encore compliquer le dispositif en freinant ou immobilisant les particules en transit.

(iii) *Les carbonates*

La sédimentation de la phase carbonatée primaire (généralement calcique) résulte pour l'essentiel de l'activité biologique du bassin lui-même qui produit des particules biodétritiques, issues de la fragmentation plus ou moins intense des tests et coquilles d'organismes paraliques, des particules biogéniques directement produites par les êtres vivants (Bactéries, Cyanophycées), ainsi que des particules carbonatées précipitant sous l'effet de la photosynthèse.

Par ailleurs, une partie de la phase carbonatée peut provenir de la sursaturation évaporitique du milieu ou encore du mélange d'eau à des degrés différents en concentration saline.

Les carbonates diagénétiques proviennent de l'activité soit de bactéries calcifiantes soit de bactéries sulfo-réductrices. On peut alors obtenir des phases magnésiennes: calcite très magnésienne, dolomite, magnésite, huntite (Perthuisot, 1971, 1975).

L'essentiel de la phase carbonatée des sédiments paraliques est donc biogénique.

(iv) *Les évaporites*

Les évaporites paraliques n'existent évidemment que dans les aires paraliques à fort déficit hydrique soit sous forme épigénétique dans les sédiments de platiers intertidaux: il s'agit alors souvent de lentilles de gypse et parfois d'anhydrite nodulaire (Perthuisot, 1980), soit sous forme syngénétique. Elles correspondent alors à des sédiments déposés sous une certaine tranche d'eau (Busson et Perthuisot, 1986). C'est ce dernier processus qui est mis en oeuvre dans les marais salants (Perthuisot, 1983).

(v) *La matière organique*

Elle a une triple origine possible: une petite partie de la matière organique sédimentée peut provenir de la mer: celle-ci amène au bassin des organismes sténobiontes qui à leur arrivée dans le bassin meurent et se sédimentent. Une autre partie peut provenir du continent, amenée par les tributaires, il s'agit en général de débris de végétaux supérieurs. Cet apport peut donner lieu à de véritables dépôts sédimentaires images actuelles déformées des bassins houillers. Mais la source fondamentale de la matière organique sédimentaire des aires paraliques est l'activité biologique (production primaire) des bassins eux-mêmes, l'essentiel étant le fait du phytoplancton et du microphytobenthos (tapis algaires).

Dans les bassins où la marée ne peut évacuer les particules fines, la matière organique s'accumule, notamment dans les zones d'ombilic hydraulique où elle peut constituer une fraction notable du sédiment (jusqu'à 25 ou 30 % en poids sec).

Ces zones sont souvent néfastes pour les activités aquacoles car elles constituent des foyers possibles de dystrophie (développement à partir de 23°C de l'activité des bactéries sulforéductrices). Elles font cependant partie de l'évolution biologique normale des écosystèmes paraliques très confinés, qu'en quelque sorte, elles régénèrent.

5.3.2 Biologie des milieux paraliques
Composition spécifique
Le milieu paralique présente une unité biologique caractérisée par des espèces qui lui sont inféodées. Parmi les poissons on peut citer:

 Atherina boyeri   *Syngnathus abaster*
 Blenius pavo   *Gobius niger*
 Pomatoschistus microps   *Symphodus cinereus*
 Parmi les espèces du benthos:
 Hydrobia acuta   *Pirenella conica*
 Cerastoderma glaucum   *Nereis diversicolor*

etc.
Chez les végétaux:
 Ulva lactuca   *Ruppia cirrhosa*
 Enteromorpha gr. intestinalis

En raison des communications avec la mer d'une part, des apports d'eau douce d'autre part, la faune se diversifie. Parmi les espèces de poissons les plus fréquentes migrant en mer, on relèvera:

 Mugil cephalus   *Belone belone*
 Liza aurata   *Diplodus annularis*
 Anguilla anguilla   *Platychtys flexus*
 Syngnathus acus   *Sparus aurata*
 Dicentrarchus labrax   *Solea vulgaris*

La faune d'eau douce ou légèrement saumâtre est moins fréquente, on notera cependant:

 Cyprinus carpio   *Gasterosteus aculeatus*
 Gambusia affinis

Distribution des peuplements lagunaires
L'observation montre que la structure des peuplements (richesse, densité) est très dépendante des échanges avec la mer d'où l'idée de prendre en compte la vitesse du renouvellement des eaux avec celles de la mer aux différents points de la lagune. Cette vitesse mesurera le degré de "confinement" vis à vis de la mer. Elle intervient sur la salinité (cf ci-dessus) mais aussi sur le cycle de substances qualitativement importantes pour les êtres vivants (oligo-éléments). Cette notion est ici particulièrement importante en raison des variations prévisibles des échanges mer-lagune si le niveau marin change.

Dans l'état actuel des techniques le "confinement" n'est pas quantifiable par des mesures physiques, mais il est possible de l'apprécier et de définir une échelle de référence en s'appuyant sur le benthos. Avec des échanges mer-lagune de moins en moins important, on aura:
- Zone I: macrofaune thalassique
- Zone II: macrofaune représentée par: *Mactra corallina, Audouina tentaculata, Portumnus latipes, Paracentrotus lividus*, etc. et macroflore représentée par: *Cymodocea nodosa, Zostera noltii, Caulerpa prolifera*; disparition de *Posidonia oceanica*.
- Zone III: disparition des échinodermes; présence de *Scrobicularia plana, Corbula gibba* . . .
- Zone IV: présence d'espèces exclusivement paraliques: *Abra ovata, Nereis diversicolor* . . .

- Zone V: les diatomées et les cyanophycées sont dominantes; présence de crustacés détritivores (*Sphaeroma hookeri* . . .) de gastéropodes brouteurs (*Hydrobia acuta* . . .) de Chironomides . . .
- Zone VI: tapis ou édifices stromatolithiques de Cyanobactéries.

Au confins de cette zone s'arrête le "proche paralique" avec la disparition des Foraminifères et débute le "paralique lointain" dans lequel le rôle du confinement vis-à-vis de la mer semble s'atténuer au profit d'autres paramètres.

Les degrés de confinement sont corrélés avec les éléments mobiles de la communauté tel le phytoplancton ou la faune de poissons. Toutefois dans ce dernier cas, il semble que l'organisation du peuplement soit surtout en relation avec l'importance de l'aire occupée par chaque zone d'un type donné de confinement.

Dans le cas particulier de la Camargue où tous les échanges sont régis au travers d'un réseau de passes entre les étangs, la nature des peuplements en un lieu donné est fortement dépendante de la possibilité physique du recrutement (ex.: ouverture ou fermeture de la digue à la mer, ou des vannes entre étangs) et la période pendant laquelle se font ces ouvertures et fermetures dans l'année (la colonisation se faisant à des périodes bien précises selon les espèces).

5.3.3 Pronostic sur l'évolution du domaine paralique
Conséquence de l'agressivité de la mer et de la remontée du niveau
En considérant l'hypothèse selon laquelle on aurait affaiblissement du cordon littoral, ceci devrait se traduire, pour les étangs de première ligne, sans intervention humaine, dans un premier temps par l'élargissement des graus, par la multiplication de ceux-ci, puis par le retour progressif au domaine marin d'un certain nombre de bassins lagunaires. En l'absence d'aménagement, on peut s'attendre d'une manière générale au déconfinement global des systèmes paraliques. Dans de nombreux cas, ceci devrait entraîner une chute de la productivité phytoplanctonique (et donc de la production conchylicole) dans les bassins lagunaires et un appauvrissement conséquent des zones marines adjacentes (Frisoni, 1984; Guelorget et Perthuisot, 1983; Guelorget, 1985).

Dans les étangs à fort confinement et en particulier la plupart des étangs de Camargue, on retiendra deux cas de figures; avec ou sans interventions sur le système hydraulique.

(i) *Sans intervention*

L'ensemble des étangs camarguais est isolé de la mer par une digue. Les communications se font au moyen de systèmes de vannes. Actuellement un seul est efficace. Il a permis un écoulement gravitaire vers la mer de quelque 60 millions de m^3 pour l'ensemble des années 1986 et 1987, ce qui représente un départ de sels de 1,5 millions de tonnes (Mt).

Les sorties gravitaires à la mer sont d'autant plus importantes que le niveau des étangs est surcoté par rapport à la mer. A cet égard, le mistral, vent de secteur Nord-Nord-Ouest qui repousse les eaux au Sud des étangs joue un rôle important. Pratiquement, dans l'état actuel des choses, les sorties à la mer n'interviennent que pour les cotes des étangs inférieurs >0,10 m NGF). A l'évidence, la remontée progressive du niveau marin entravera cet écoulement gravitaire qui pour être quantitativement appréciable exigera des niveaux d'eau toujours plus hauts dans les étangs. L'élévation du niveau des étangs doit normalement accompagner l'élévation du niveau marin, mais les hauts niveaux du système Vaccarès sont préjudiciables aux activités économiques autres que la pêche, car ils entravent le drainage gravitaire du delta dans les bassins culturaux même poldérisés et font courir le risque d'inondations généralisées des basses terres.

Selon les prévisions de P. Heurteaux, sans intervention le Vaccarès et les étangs inférieurs de Camargue se tranformeraient en un système hypersalin (50 à 70 g.l^{-1}). Ils perdraient alors beaucoup de leur intérêt biologique, et toute macroflore. Cette salinité serait trop élevée pour permettre l'établissement d'un peuplement de poissons. Les marais permanents situés autour du système Vaccarès/étangs inférieurs ne seraient pratiquement pas affectés tant ils sont artificialisés (salinité et régime hydrique) par des apports massifs d'eau douce, essentiellement pour des raisons cynégétiques, et donc économiques. Les marais naturels non aménagés (temporaires) verraient probablement augmenter leur salinité moyenne et surtout hivernale par des remontées salines de la nappe plus importante. Leur période d'assec pourrait être raccourcie par les pertes de capacité de drainage qu'occasionnerait une remontée du niveau de la mer. Les modifications seraient d'autant plus marquées que les marais se situent à une altitude faible (proche de la nappe). Cela ne devrait modifier que légèrement (sinon très localement) la macroflore des marais. Toutefois les espèces les plus sensibles au sel, mais qui sont aussi parmi les plus rares en Camargue (*Damasonium stellatum* par exemple) pourraient être sérieusement affectées.

Les poissons devraient subir peu de modifications. Bien qu'un appauvrissement du nombre des espèces présentes soit prévisible en raison de l'absence des espèces du milieu saumâtre (ex. anguille, muge) qui ne pourront plus coloniser la Camargue, si ce n'est qu'accidentellement par les unités de pompage le long du Rhône.

(ii) *Avec interventions*

Il est assez peu probable que l'homme reste sans réaction devant une modification radicale des conditions de salinité et de drainage en Camargue. Une intervention viserait alors à contrecarrer l'augmentation de salinité du Vaccarès et des étangs inférieurs qui pourraient être saumâtres ou polysaumâtres, ainsi qu'à assurer l'évacuation des eaux de pluie ou de colatures. Selon l'ampleur de l'intervention, la salinité du Vaccarès pourrait être considérablement réduite, jusqu'aux valeurs actuelles ou en-dessous. L'étang serait alors favorable au développement d'une macroflore importante sans que l'on puisse prédire de quelle nature. Un peuplement de poissons du type de celui qu'on rencontre actuellement pourrait s'établir, à la condition impérative qu'il existe des ouvertures à la mer entre les mois d'octobre et mars. Si de telles ouvertures étaient inexistantes, il y aurait peu de chances d'avoir des poissons dans ces étangs. Les autres étangs, dans le cas d'ouverture, devraient avoir des peuplements identiques à ceux d'aujourd'hui.

Les marais aménagés situés autour du système Vaccarès + étangs inférieurs ne seraient pas modifiés. Une intervention humaine pourrait par contre contribuer à améliorer le drainage en Camargue et par là raccourcir légèrement la durée de submersion des marais temporaires naturels, sans empêcher la légère augmentation de salinité décrite précédemment. Ces marais temporaires naturels resteraient peu affectés dans la plupart des cas.

Conséquence de l'augmentation de température
Dans certains cas, l'augmentation de température peut favoriser les développements phytoplanctoniques et la croissance des êtres benthiques. Mais, elle favorisera systématiquement l'activité bactérienne in situ ainsi que le "turn-over" des populations bactériennes. Cette activité bactérienne accrue se fait au dépens de la réserve en matière organique de chaque bassin. Du fait de l'abaissement de la productivité primaire, il est possible qu'à terme cette réserve s'épuise limitant ainsi les proliférations bactériennes et le recyclage des nutriments.

On peut s'attendre en général à une première phase de désoxygénation des bassins paraliques avec des tendances dystrophiques localisées (le développement des bactéries sulfato-réductrices et des méthanogènes étant favorisé par l'anoxie du milieu et des températures supérieures à 21°C). Cette première phase sera surtout sensible dans les bassins voués à la conchyliculture et à l'aquaculture intensive.

La deuxième phase prévisible est une remise en équilibre des systèmes d'une part par l'activité bactérienne elle-même, et à travers elle, des grands cycles de la matière notamment celui de l'azote et d'autre part par la diminution de la production primaire laquelle sera défavorisée par le moindre lessivage des aires continentales.

5.4 Dunes et terrains salés

5.4.1 Le rôle des paramètres édaphiques
Le système dunaire
(i) *Les milieux*
L'instabilité du milieu est déterminée par la mer et par le vent, ce qui conduit à distinguer deux unités écologiques où l'action de l'une ou de l'autre est prédominante: le système plage/haut de plage et le système dunaire (Fig. 11.28; Corre, 1987).

- Le système "plage/haut de plage" est soumis à l'influence prédominante de la mer qui mobilise les sédiments. Leur transport se fait tangentiellement et perpendiculairement à la rive. Il est important surtout lors des tempêtes.

Outre la plus ou moins grande instabilité du substrat, les plages peuvent se distinguer par la plus ou moins grande humidité ou salinité du substrat selon leur profil topographique. Si elles acquièrent un profil de pente concave, l'affleurement de la frange capillaire issue de la nappe aquifère leur donnera un caractère humide et plus ou moins salé.

- Les dunes sont régies par l'action prédominante du vent, mais leur évolution est également influencée par l'histoire des apports sédimentaires (origine, conditions de mise en place . . .). C'est ainsi que le long du "lido" de Carnon au Grau-du-Roi, on observe (Fig. 11.29) la juxtaposition de cordons dunaires se comportant différement sous l'action du vent et se différenciant par leur couvert végétal.

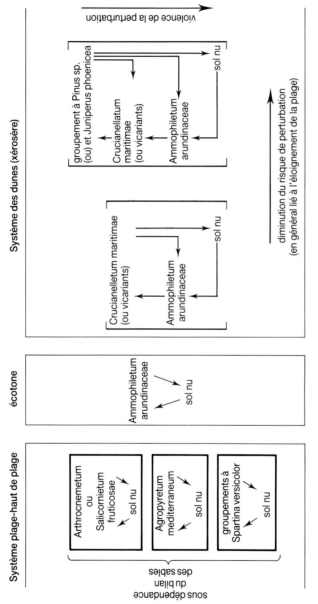

Fig. 11.28 Schéma de la structure et de la dynamique des systèmes littoraux dans les secteurs dunaires du Golfe du Lion.

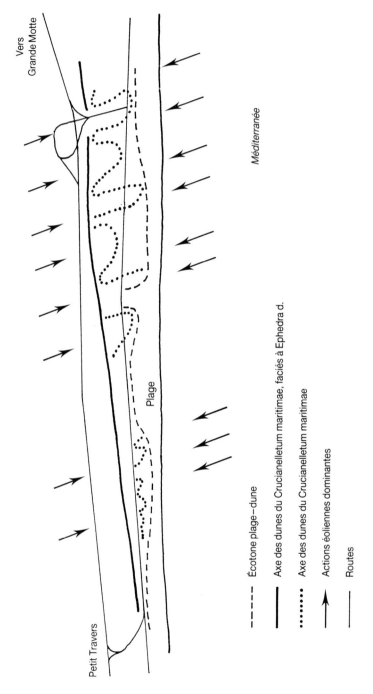

Fig. 11.29 Schéma de structure des cordons dunaires entre Carnon et la Grande Motte.

Schématiquement on distinguera en front de mer une zone "cicatricielle" qui peut être considérée comme un écotone entre le haut de plage et la zone dunaire proprement dite. Il porte autant la marque des fortes tempêtes que de l'action du vent. En arrière lui fait suite une zone en cours de stabilisation par suite d'une action éolienne moindre sur l'accrétion ou la déflation du sable.

Le système dunaire est le siège d'accumulations souterraines d'eaux douces d'origine pluviale, douces ou peu salées. Elles ressurgissent à la base des massifs selon un débit fluctuant saisonnièrement (Corre, 1976).

(ii) *La distribution des peuplements végétaux*
En haut de plage le cycle saisonnier particulièrement contraignant des tempêtes ne permet en général qu'une végétation d'espèces annuelles (*Cakile maritima, Salsola kali*). Une modération des transports de sable peut permettre une certaine pérennité des peuplements. Ils se caractériseront par l'apparition de vivaces tolérantes à la salinité. Selon les particularités édaphiques les espèces dominantes seront: *Elymus farctus, Arthrocnemum glaucum, Spartina versicolor* (Fig. 11.28).

Dans le système dunaire (Fig. 11.30), la zone cicatricielle formant écotone est occupée par un peuplement à *Ammophila arenaria*. La zone en voie de stabilisation est très diversifiée avec une différenciation régionale des peuplements: par exemple, *Artemisio-Teucrietum maritimi* en Camargue,

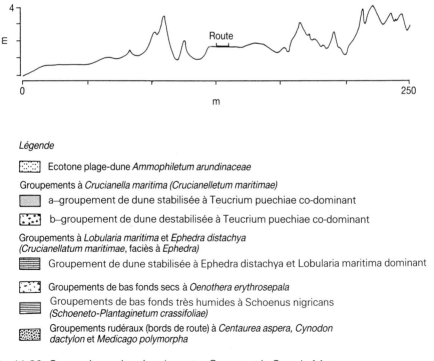

Fig. 11.30 Coupe du cordon dunaire entre Carnon et la Grande Motte.

Crucianelletum maritimae en Languedoc, *Diantho-Corynephoretum* et *Cladonio-Ephedretum* en Roussillon (Baudière et Simonneau, 1974; Molinier et Tallon, 1965), ce qui est certainement à la fois en relation avec la diversité de nature du substrat et probablement aussi celle des conditions climatiques.

Dans les bas fonds les ceintures de végétation se répartissent en fonction des gradients d'humidité et de submersion et selon la dynamique saisonnière des apports d'eau douce et des émergences d'eau salée.

La distribution des peuplements constituant la zonation est aussi sensible à des facteurs régionaux (climat, nature du substrat: Corre, 1987). En Petite Camargue et en Languedoc on pourra avoir du plus sec au plus humide, du plus doux au plus salé.

Holoschoenetum romani
  → *Schoeno-Plantaginetum crassifoliae*
    → *Artemisio-Staticetum virgatae*
      → *Salicornietum fruticosae*

Les terrains salés périlagunaires
(i) *Les mécanismes structurant le milieu*
Schématiquement le pourtour de la lagune peut être subdivisé en trois niveaux (Fig. 11.31):
– Le niveau proximal: il est justifié par les conditions de submersion. La salinité qui a un rôle important dans la différenciation des peuplements est fortement dépendante de celle des eaux de surface avoisinnantes. Du point de vue écologique, il a un statut d'écotone entre lagune et milieu terrestre (Corre *et al.*, 1982). Comme le milieu terrestre, il est influencé

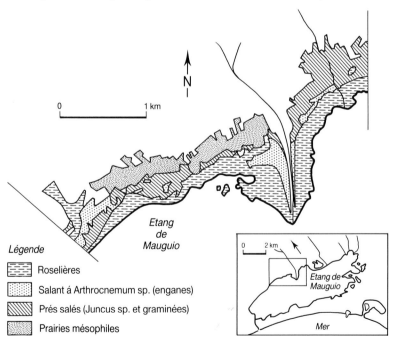

Fig. 11.31 Répartition de la végétation au bord d'une lagune (étang de Mauguio).

par les nappes aquifères. De la lagune il reçoit des apports de matières organiques et sa dynamique sédimentaire en subit l'influence. Certains peuplements, en particulier d'invertébrés lui sont caractéristiques.

– Le niveau intermédiaire: le facteur submersion est secondaire; l'altitude étant plus élevée, ce niveau est émergé pendant la majeure partie de l'année. La salinité joue un rôle prépondérant. Elle est sous la dépendance de la nappe aquifère lagunaire. Les phénomènes de salant peuvent y être intenses en été.

– Le niveau distal: la salinité et les conditions de drainage régissent la distribution des milieux. Avec l'éloignement de la rive, les nappes aquifères fluviales interviennent de façon plus marquée sur le bilan hydrique pouvant créer un milieu doux et humide.

Ce schéma de structure fonctionne selon un cycle hydrique et salin saisonnier, ce qui le différencie nettement du système slikke/schorre des mers à marée et le rapproche des systèmes de lacs salés continentaux. Il est rendu dans les faits plus complexe par:

– l'hétérogénéité des apports sédimentaires qui interfèrent sur les bilans salins et hydriques;

– la formation de bourrelets de laisses, en bordure d'étang qui créent une eutrophisation ponctuelle et un compartimentage du milieu à l'égard des circulations d'eau de surface;

– l'action de l'homme: rehaussement des terres, réseau de drainage, piétinement des troupeaux.

(ii) *La distribution des peuplements végétaux*

a) Niveau proximal

Selon l'importance de la salinité des eaux de surface, la végétation sera dominée par des chénopodiacées halophiles annuelles ou vivaces: *Suaeda sp., Salsola sp., Salicornia europaea, Arthrocnemum sp.* (salure élevée) ou des roseaux et des scirpes: *Phragmites australis, Scirpus maritimus* (salure modérée). Des variations de salure pouvant se manifester selon des cycles pluriannuels provoquent une alternance de l'un ou l'autre de ces types de végétation.

b) Niveau intermédiaire

Les peuplements végétaux sont dominés par des chénopodiacées vivaces (*Arthrocnemum sp.*) seules ou en mosaïque dans les "prés salés"; le processus de salinisation peut conduire localement à la disparition de toute végétation.

c) Niveau distal

Il est occupé par une végétation de "prés salés" caractérisée par des graminées et des joncs (*Aeluropus littoralis, Juncus gerardi, Elymus repens x Elymus farctus*, etc.).

La transition avec le milieu non salé se fait par des prairies mésophiles à *Brachypodium phoenicoides* et/ou des boisements avec *Populus alba, Ulmus campestris, Fraxinus oxyphylla* souvent traités en bocage.

Pronostic

(i) *Evolution de la végétation littorale: des faits habituels*

Des changements majeurs dans la végétation littorale, liés à des modifications de substrat, se sont déjà produits au cours de l'Holocène. Ils ont été le corollaire de la remontée flandrienne de plus de 90 m du niveau marin.

En Camargue, vers 5 350 ± 150 BP (Pons *et al.*, 1979), une végétation de graminées, cypéracées, *Sparganium*, composées, puis plus tard de chénopodiacées vient remplacer une végétation de type arboré. Elle est la conséquence d'un changement dans la nature de l'alluvionnement; elle fait suite à une pause de la remontée flandrienne et à la constitution d'un puissant lido isolant un système lagunaire.

Le long du rivage languedocien une séquence de niveaux saumâtres dans le Quaternaire récent, avec présence de pollen de chénopodiacées s'entrecroise dans les sondages avec des niveaux marins, ce qui montre le caractère fluctuant des paysages dans un passé géologique récent (Planchais *et al.*, 1977; Planchais, 1982).

Par rapport à ces faits, l'élévation du niveau marin telle qu'elle est prévue est faible dans l'absolu (+ 20 cm). La nature même du paysage ne devrait pas être modifiée; la barrière dunaire ne devrait pas être submergée, seuls des réajustements de structure tant géomorphologique que biotique sont à prévoir. Il n'en demeure pas moins que ces faits du passé devront inciter le gestionnaire à une grande vigilance.

(ii) *Evolution du système dunaire*

• Haut de plage: En réponse à l'agressivité de la mer, la précarité des peuplements sera plus grande. Cette destabilisation peut avoir des conséquences non négligeables en raison du rôle de protection que ces communautés peuvent jouer à l'égard des premiers cordons dunaires.

• Système dunaire: plusieurs types de conséquences sont à envisager:

a) arasement des dunes si le cordon est bas et dépassé lors des tempêtes.

b) destabilisation du système dunaire résiduel après démantèlement partiel par les tempêtes. Les sables repris par la mer sont par la suite rejetés à la côte. Le vent les déplace, ensevelissant la végétation préexistante, ce qui destabilise la dune.

c) segmentation des cordons et ensevelissement des dépressions interdunaires. Par voie de conséquence le niveau phréatique sera plus profond. Ces stations qui constituent un refuge pour une faune et une flore sensibles à la sécheresse estivale, biologiquement importantes par leur rareté ou leur endémisme auront une aire plus restreinte.

(iii) *Evolution des terrains salés périlagunaires*

Dans l'hypothèse d'une plus grande agressivité de la mer multipliant l'ouverture des graus et d'une élévation de son niveau favorisant le transit mer-étang, on peut, le long des bordures (niveau proximal), assister à une régression des roselières là où elles existent au profit des chénopodiacées halophiles.

La zone de salant du niveau intermédiaire aura tendance à se renforcer et à se déplacer vers l'intérieur, se rapprochant des zones agricoles, ce qui nécessitera un effort accru dans la procédure drainage/ irrigation.

En Camargue et dans une grande partie du Languedoc, le déplacement vers l'amont du phénomène de salant concernera les terrains de parcours de chevaux et taureaux.

Tant que le maintien des marais d'eau douce pourra être assuré grâce à un régime hydraulique approprié, les mammifères herbivores disposeront de pâturages pour la période estivale.

Les parcours hivernaux sont déjà pauvres. Il faut compter environ 10 hectares pour une jument avec son poulain ou deux vaches et leurs veaux. Le mas de la Bélugue près de Salin-de-Giraud peut nous donner une idée des effets prévisibles. D'après le fermier, l'extension des salines au cours des vingt dernières années a fait baisser de 800 à 400 vaches environ la capacité d'accueil de ce domaine qui couvre 900 hectares.

En ce qui concerne les marais saumâtres temporaires qui environnent les lagunes, le bilan d'eau est essentiellement régi par l'antagonisme des pluies et de l'évaporation. Il est peu probable que le milieu physique subisse des modifications susceptibles d'entraîner un bouleversement radical des biocénoses.

5.4.2 *Le rôle des paramètres climatiques*

Le long du golfe du Lion, les températures minimales du mois le plus froid (m) sont: en Camargue de +1,7°C, à Montpellier de +0,9°C et au Sud, dans les Albères de +4°C.

Cette distribution des isothermes en hiver (Fig. 11.32) laisse apparaître une zone relativement fraîche, centrée sur Montpellier. On peut la mettre en parallèle avec des hiatus dans la distribution de certaines espèces, par exemple *Mesembryanthemum edule*, *Thymelea hirsuta* . . . Si rien ne change, on continuera d'observer leurs oscillations de part et d'autre de leurs limites géographiques au gré des variations interannuelles sinon l'extension ou la régression continue de leur population servira à valider biologiquement les tendances des minimums thermiques enregistrés.

5.5 *L'avifaune littorale*

Pour l'avifaune aussi bien hivernante, nicheuse ou migratrice, les étangs et marais du golfe du Lion constituent un attrait. La Camargue est parmi tous ces sites le plus célèbre. Elle est de plus tout à fait représentative des communautés aviennes de l'ensemble de ces zones littorales. Nous

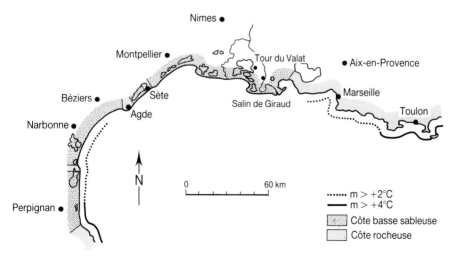

Fig. 11.32 Le Golfe du Lion. Types de côtes – températures minimums du mois le plus froid.

la prendrons comme type d'autant qu'elle est actuellement et depuis longtemps la mieux étudiée.

Un inventaire récent (Blondel et Isenmann, 1981) donne 337 espèces au total dont 111 nicheurs réguliers et 15 nicheurs accidentels. Dans l'ensemble la richesse spécifique varie peu puisque l'on ne compte depuis 1840 que 6 disparitions et 22 acquisitions. Cette tendance à l'enrichissement qualitatif tiendrait depuis un siècle à une meilleure maîtrise de l'eau, la création de nouveaux biotopes (salins, cultures) et une eutrophisation de la Camargue et de ses abords.

La diversité en espèces est telle que nous ne pouvons faire état ici que de quelques communautés aviennes ou d'espèces présentant un intérêt particulier, choisies parmi celles qui seraient les plus concernées par les changements induits par le climat: Grand cormoran (*Phalacrocorax carbo sinensis*), Flamant rose (*Phoenicopterus ruber roseus*), Ardéidés, Laro-Limicoles, Anatidés.

5.5.1 Les espèces

Le Grand cormoran: C'est un hivernant strictement piscivore. Il est présent d'octobre à avril avec des effectifs maximaux de 4000 à 5000 généralement en janvier et février.

Le Flamant rose: C'est un nicheur, la Camargue est le seul site de reproduction connu en France. Il est limité à un îlot (artificiel) dans les salines de Salin-de-Giraud. Cependant la population exploite la totalité des lagunes du golfe du Lion.

Ardéidés nicheurs: Deux espèces nichent isolemment: Grand butor et Blongios nain. Les espèces coloniales nicheuses représentent en Camargue plusieurs milliers de couples.

Une telle diversité d'espèces nicheuses est assez peu fréquente en Europe et place la Camargue parmi les zones de reproduction les plus remarquables. Deux espèces (Héron cendré et Héron pourpré) nichent dans les roselières et parfois dans les tamaris: les autres (Aigrette garzette, Héron bihoreau, Héron garde-boeufs, Héron crabier) établissent leurs colonies dans les arbres. Seuls le Héron cendré et l'Aigrette garzette se nourrissent dans l'eau salée et saumâtre. Les lieux d'alimentation proches des colonies sont préférés. Une analyse plus fine a également montré qu'il existe un choix préférentiel quant au type de marais d'eau douce. En effet les effectifs et la diversité des colonies d'Ardéidés nicheurs augmente avec l'étendue des marais permanents qui les entoure. Les mêmes habitats préférentiels (marais permanents d'eau douce) ont par ailleurs une influence positive sur le succès d'alimentation, ainsi que sur le succès de reproduction de l'Aigrette garzette. Pendant la saison de reproduction, toutes ces espèces sont piscivores à l'exception du Garde-boeufs et du Héron crabier qui préfèrent les amphibiens. Quoiqu'il en soit, la nourriture principale est constituée par des vertébrés qui sont plus abondants dans les marais permanents que dans les marais temporaires.

Laro-Limicoles: Les nicheurs représentent environ 20 000 couples. Pour certains: Sterne hansel, Goéland railleur, la Camargue est le seul point français de reproduction. Pour tous les autres c'est un important centre de nidification.

Dans l'ensemble, le peuplement s'est développé en relation avec l'élévation du niveau des eaux (gestion hydraulique) et l'extension des salins, ce qui crée de meilleures conditions alimentaires et de sécurité.

Une vingtaine de limicoles sont des migrateurs au long cours transitant par la Camargue lors de leur passage prénuptial ou postnuptial; par exemple: Chevalier sylvain, Chevalier cocorli. Toutes sont tributaires de milieux aquatiques de faible profondeur et de vasières riches en benthos.

Anatidés: Ils comprennent 19 espèces hivernantes ou nicheuses (Tamisier, 1987). Leur effectif toutes espèces confondues est très important mais en régression: 130 000 individus en 1970, 100 000 en 1987. Un certain nombre d'entre eux étant classés comme gibier subissent une forte pression de chasse, mais bénéficient d'aménagements destinés à les attirer.

5.5.2 *Les milieux concernés*

La plupart des espèces utilisent alternativement plusieurs types de biotopes pour leur reproduction, leur alimentation (gagnage) et leur séjour (remise). Une restriction due à l'un ou à l'autre peut être un facteur limitant les effectifs.

Les règles qui président au choix de chacun semblent souvent très précises, propres aux caractéristiques biologiques, ou morphologiques de chaque espèce, ce qui leur permet un bon partage des ressources.

Les bosquets

Ce sont des lieux de nidification pour les Ardéidés, sauf Héron pourpré et Héron cendré. Leurs dimensions doivent être suffisamment vastes pour atténuer les effets du vent. La nature du couvert révèle certaines préférences: Orme (*Ulmus sp.*) et Frêne (*Fraxinus sp.*) plutôt que Peuplier blanc (*Populus alba*) et Saule (*Salix sp.*); en Petite Camargue: Pin.

Le bois des Rièges à Genévrier rouge (*Juniperus phoenicea*) n'est pas colonisé.

Lagunes et marécages salés

Elles ont des fonctions multiples:
- Lieu de reproduction: Les îlots ou les digues entourées d'eau suffisamment abritées des submersions servent de lieux de reproduction pour de nombreuses espèces de Laridés, de Limicoles et d'Anatidés.
- Lieu de nourriture: Elles sont utilisées aussi bien par les sédentaires ou les migrateurs comme le Chevalier cocorli (salines).
- Lieu de séjour: Elles jouent un rôle très important en particulier chez les Canards hivernants qui y trouvent des conditions de sécurité satisfaisantes.

Pour ces deux dernières fonctions: salinité, hauteur d'eau, régime des assecs, forme et orientation des rives jouent un rôle important dans le partage de l'espace et limitent la surface réellement utilisée.

Les étangs permanents, temporaires et les sansouires inondées seulement l'hiver ont un rôle complémentaire selon les niveaux d'eau. Les oiseaux peuvent ainsi alternativement les exploiter au cours des saisons ou selon les particularités climatiques de l'année.

Les zones palustres
Ce sont des milieux en eau permanente ou temporaire, doux ou légèrement salés, souvent encombrés d'une végétation aquatique. Le Héron pourpré et le Héron cendré s'y reproduisent installant leurs colonies dans des roselières. C'est de même un lieu de nidification pour le Butor étoilé, le Colvert, l'Echasse et bien d'autres oiseaux n'appartenant pas aux groupes retenus.

Ce sont des lieux de nourrissage importants pour tous les Ardéidés et des migrateurs comme le Chevalier sylvain. Ils y trouvent une nourriture abondante particulièrement en fin de printemps dans les zones en voie d'assèchement où la faune se concentre. Les canards l'utilisent qu'ils soient herbivores ou carnivores. Certains Laridés comme la Sterne hansel fréquentent également ces milieux ainsi que les pelouses et cultures proches.

5.5.3 *Pronostic pour l'évolution de l'avifaune*
Etant en bout de chaîne, les conséquences des changements climatiques pour les oiseaux sont difficiles à prédire.

Hivernants
L'appauvrissement prévisible de la faune de poissons de la partie centrale de la Camargue affectera vraisemblablement l'hivernage des espèces piscivores, notamment du Grand cormoran, du Grèbe huppé et du Héron cendré.

Nicheurs
Parmi les Ardéides fréquentant les lagunes saumâtres, le Héron cendré étant essentiellement piscivore est le seul qui pourrait subir les conséquences de l'appauvrissement de la faune de poissons. Le maintien du peuplement de toutes les espèces d'Ardéidés dépendra du maintien des marais permanents d'eau douce ou peu saumâtre, et de leur faune de poissons et d'amphibien.

Le Flamant, avant toute autre espèce, bénéficiera d'un apport supplémentaire d'eau salée.

N'étant pas piscivores, les Laro-Limicoles ne subiraient pas les effets de changements dans la faune de poissons. En ce qui concerne la Sterne hansel, sa survie dépendra du maintien des marais d'eau douce.

Migrateurs au long cours
Ils ne sont pas piscivores, ils ne devraient donc pas être affectés par le déclin des populations de poissons. Une augmentation de la superficie des milieux salins devrait être bénéfique aux espèces telles que le Chevalier cocorli tandis qu'une diminution de la superficie des marais d'eau douce devrait entraîner une chute des effectifs d'espèces telles que le Chevalier sylvain.

6 Procedes de Lutte Contre les Effets de Recul de la Cote dus aux Elements Naturels

6.1 Les différentes situations rencontrées

Elles seront variées selon la nature des milieux terrestres.

a) Dans les espaces dunaires non limités sur l'arrière (par une route, une station touristique ou des espaces d'agriculture intensive) et en bon état (dune de barrage homogène en granulométrie et en altitude, et continue): l'ensemble du système reculera conformément aux prévisions exposées au chapitre 4 sur tout le linéaire côtier concerné par ces dunes. Il est probable qu'à un pas de temps plus court (10 ans) le recul de la dune sera plus lent, étant donné l'inertie des masses sableuses en jeu.

b) Dans les espaces dunaires dégradés, fortement anthropisés, en général les dunes présentent des altitudes variables, des ruptures dans le cordon dunaire et des pentes de plages anormales (très plates et/ou inversées). Dans ce cas, il y aura accélération des processus de dégradation par évacuation des sables soit vers la mer (repris alors par la dérive littorale), soit vers l'intérieur des terres par franchissement du cordon. On assistera probablement au laminage des plages concernées et à la formation progressive de cordons dunaires continentaux segmentés, orientés parallèlement aux vents dominants de terre, Mistral ou Tramontane. Il est à prévoir alors la transformation du paysage littoral en une succession de zones planes, humides ou non, et de dunes continentales, l'ensemble ayant une orientation générale Nord-Sud et non Est-Ouest comme aujourd'hui. L'acquisition d'un tel paysage nécessitera probablement plus d'un demi-siècle.

c) Dans les espaces dunaires séparant mer et étangs, une augmentation du niveau de la mer et surtout de la violence des tempêtes risque de provoquer une rupture des lidos et l'ouverture de graus.

d) Dans les espaces fortement urbanisés, il y a en général:
– diminution de l'altitude des fonds de plage par arasement et suppression des dunes pour permettre la construction de boulevards de front de mer, d'immeubles, de campings,
– réduction de la largeur de la plage, du fait de la trop grande proximité des enrochements de protection.

Ces deux types de dégradation accentuent les risques de franchissement par les vagues de tempête et le dépôt des sables hors de la zone de plage, surtout lorsqu'elles sont étroites. Ces sables, charriés depuis la zone marine vers la zone terrestre sont perdus pour la zone marine, en l'état actuel des techniques classiques de gestion du littoral.

Dans ces secteurs fortement anthropisés, il est probable qu'en cas de montée importante du niveau de la mer, il y aura aggravation des dégats causés aux équipements, par augmentation de la fréquence des franchissements en tempête.

6.2 Les réponses apportées actuellement en matière de protection côtière

6.2.1 Dans les espaces dunaires

Les techniques légères permettent:
- d'engager un processus de restauration lorsque l'érosion est en partie liée à des anomalies locales de la topographie terrestre, celles-ci ayant

souvent des répercussions sur la topographie marine. Ces anomalies sont souvent liées à une utilisation anthropique incontrôlée des milieux (prélèvements de sables, piétinement, etc.). Dans ce cas les équipements réalisés permettent de ramener l'érosion à des valeurs supportables pour une collectivité voire à enrayer une tendance au recul (Fig. 11.33).

- d'engager un processus de consolidation des cordons dunaires lorsque l'érosion est en grande partie due à des facteurs marins (houles, courants): cette consolidation vise à améliorer les résistances de la frange littorale de façon à éviter une accélération des processus de dégradation. C'est le cas du lido de Sète où les équipements réalisés n'ont pas permis une accrétion du trait de côte mais ont provoqué une amélioration de la "tenue" de la plage vis-à-vis des tempêtes (rectification des profils transversaux et longitudinaux des plages; augmentation des altitudes).

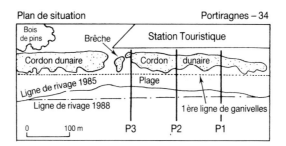

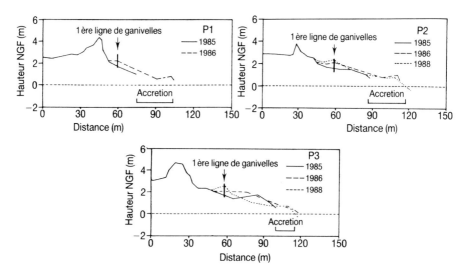

Fig. 11.33 Exemple d'intervention sur le littoral à l'aide de ganivelles. Portiragnes au Sud d'Agde (Hérault).

6.2.2 Dans les espaces fortement anthropiques

Lorsque l'érosion est importante, qu'elle soit d'origine anthropique ou marine, les mesures de protection utilisées depuis une vingtaine d'années visent à provoquer une accrétion du trait de côte à l'aide d'ouvrages lourds (épis et/ou brise-lames) et parfois de rechargements artificiels. Les résultats obtenus font apparaître:
- une stabilisation du trait de côte dans les secteurs protégés à l'aide d'épis en batterie (Fig. 11.34);
- une protection correcte dans les zones protégées par les brise-lames avec accroissement de la surface de plage.

La diversité des phénomènes qui provoquent la régression des plages et leur interaction sur un secteur donné, conduisent à une diversité des moyens de protection.

Des suivis scientifiques de ces ouvrages sont engagés de façon à mettre en évidence les facteurs agissants et à permettre la réalisation des "corrections" nécessaires à apporter.

6.3 Le comportement des gestionnaires

En France, la gestion du Domaine public maritime est confiée aux services du Ministère de l'Equipement qui par le personnel qu'il emploie et les études qu'il mène, participe activement à la protection du littoral.

Cependant, l'Etat n'a pas obligation d'assurer la protection des propriétés riveraines du rivage de la mer. Il ressort au contraire des dispositions des articles 33 et 34 de la loi du 16 septembre 1807 que cette protection en incombe aux propriétaires. L'Etat n'intervenant que par l'allocation de subventions au cas où il le jugerait opportun, notamment au titre de la protection des lieux habités.

Les financements de telles opérations sont donc pris en charge par les riverains, les collectivités territoriales (communes et départements essentiellement, parfois la région).

6.3.1 Les attitudes traditionnelles

L'évolution sociologique du Languedoc au travers du développement touristique sur la frange littorale a conduit à un abandon des pratiques traditionnelles de gestion: remodelage des cordons dunaires et fascinages étaient réalisés régulièrement par les agriculteurs lorsque ceux-ci exploitaient les vignes installées sur l'arrière. L'abandon cultural et l'urbanisation se sont traduits par la disparition quasi-généralisée de ces pratiques et une accélération des processus de dégradation.

Dans les espaces dunaires "de nature", les évolutions du milieu ne sont pas contrôlées tant qu'elles ne participent pas à une dégradation des secteurs qui les bordent (agricoles ou urbains) et tant qu'une structure publique ne prend pas en charge les coûts de restauration.

Depuis une dizaine d'années, les espaces de nature sont progressivement acquis par des organismes publics qui en assument alors la gestion.

Dans les espaces dunaires fortement anthropisés il convient de discerner trois types de secteurs:
- les zones "mitées" où s'alignent le long du littoral des campings, des villas, des cabanons implantés en lieu et place des dunes. Dans ce cas, les techniques de protection ont été prises en charge individuellement

412 *Climatic Change in the Mediterranean*

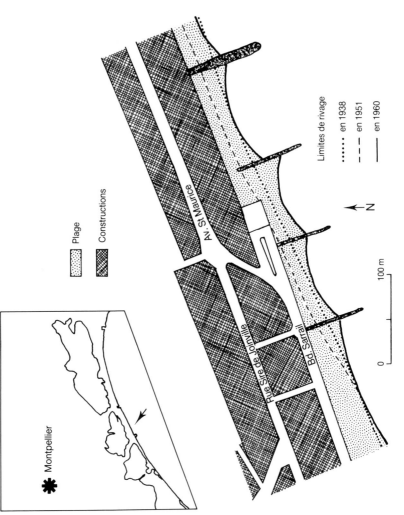

Fig. 11.34 Lutte contre l'érosion des plages au moyen d'épis de protection

(Palavas les Flots, Hérault). L'implantation a eu lieu en 1951. (d'après le Service maritime et de navigation Languedoc Roussillon).

par chaque riverain, tant sur le plan du choix de la solution technique que sur le plan financier. Des enrochements de protection étaient ainsi réalisés au coup par coup, en limite des propriétés. L'absence de coordination, entre les ouvrages et la chronologie suivant laquelle ceux-ci étaient réalisés (chaque propriétaire effectuant son enrochement lorsqu'il estimait sa situation suffisamment grave) a conduit à d'importantes perturbations de la dynamique littorale;
- Les secteurs urbanisés implantés sur des stations touristiques anciennes (Palavas, Carnon, Valras, Le Grau du Roi, etc.): les bâtiments construits après destruction des secteurs dunaires ont fait l'objet d'attaques violentes de la mer. La mise en oeuvre d'ouvrages lourds de protection s'est faite dans un cadre collectif, après constitution d'une association syndicale des riverains de chaque secteur. Les financements sont pris en charge en moyenne à 20 % par l'Etat, 40 % par les collectivités territoriales et locales, 40 % par les riverains.

Toutefois si ces ouvrages ont en général permis une stabilisation de la plage au droit des équipements, il reste encore de nombreux problèmes à résoudre: ensablements éoliens des stations et submersions marines en cas de fortes tempêtes. Des compléments techniques sont encore à mettre en place.
- les secteurs nouvellement urbanisés dans le cadre de la Mission interministérielle pour l'aménagement touristique du littoral Languedoc-Roussillon (La Grande Motte, Port Camargue, Port Barcarès, Cap d'Agde, etc.): la protection côtière et la gestion de la frange littorale ont été prises en charge par l'Etat lors de l'implantation des stations dans les années 1970. Il s'agit généralement d'ouvrages lourds mais on a pu noter un souci de gestion des espaces dunaires dans certaines de ces unités touristiques. La protection et la gestion sont souvent prises en charge par les collectivités, sans faire appel aux riverains.

6.3.2 Les attitudes actuelles
La décentralisation engagée en 1983 a conduit à un transfert des compétences de l'Etat vers les collectivités locales et territoriales impliquant plus fortement celles-ci dans la protection du littoral. Couplée avec le souci de préserver au littoral languedocien une "image de marque" de qualité vis-à-vis du développement touristique, cette responsabilité accrue des collectivités conduit les décideurs à mettre en oeuvre une politique de gestion des littoraux.

Ainsi, un schéma directeur de protection du littoral a été entrepris dès 1983 et se poursuit actuellement. Ce schéma se décompose en trois phases:
1) un diagnostic des évolutions en cours et de leurs causes;
2) un schéma directeur d'intervention adapté à chaque site;
3) la mise en place d'un suivi scientifique et technique de l'ensemble du littoral et des ouvrages réalisés.
1) l'étude diagnostic a consisté à:
- effectuer le constat des évolutions;

- déterminer les causes de ces évolutions;
- analyser les facteurs responsables;
- déceler les tendances des évolutions futures . . .

2) le schéma directeur a défini:
- les zones prioritaires à protéger en fonction des risques décelés ;
- les techniques possibles à mettre en oeuvre;
- les moyens de défense adaptés à chaque site en particulier;
- un programme d'intervention cohérent à l'échelle de l'ensemble du littoral;
- une définition des aménagements par étapes.

3) le suivi: La protection du littoral réalisée par tranches nécessite une analyse du comportement du littoral et des ouvrages réalisés. Les données nécessaires à cette analyse sont:
- des levés réguliers du trait de côte et des profils terrestres de la plage;
- des levés bathymétriques de la plage sous-marine.

Il vise à identifier les dégradations en cours, les réponses techniques à apporter, l'urgence des décisions à prendre et les chronologies à observer.

Dans la mesure du possible on s'efforcera de diversifier les techniques, et pour reprendre un mot du président du Conseil général de l'Hérault, il s'agit d'éviter de "transformer une côte sableuse en côte rocheuse" par la prolifération incontrôlée d'enrochements et d'ouvrages lourds qui hypothèquent définitivement les paysages et grèvent considérablement les budgets publics.

Cette démarche qui passe par une volonté de coordination des différents équipements, est à mettre à l'actif de la décentralisation et du rôle joué dorénavant par la collectivité territoriale (département, région) et est également due à la politique de gestion menée depuis 12 ans par le Conservatoire national de l'espace littoral et des rivages lacustres sur les terrains qu'il acquiert, aux actions pédagogiques mises en place par le Ministère de l'Environnement et l'Université, aux expérimentations techniques menées en Languedoc par le Centre d'études et d'expérimentation pour la protection, la restauration et la gestion de l'espace littoral (CEPREL) en matière de techniques légères de protection côtière et d'analyse des milieux et de leur évolution, aux études et travaux du Service Maritime et de Navigation du Languedoc-Roussillon, ainsi qu'aux efforts déployés par les Associations et Collectivités soucieuses de protection du milieu naturel.

L'ensemble des actions engagées aujourd'hui en Languedoc a conduit les gestionnaires à s'entourer d'un Conseil scientifique et technique destiné à orienter les choix en matière de protection côtière. La prise en compte d'une éventuelle montée du niveau marin se fera donc désormais au travers d'une structure ayant à charge d'évaluer scientifiquement les risques et de coordonner les actions entreprises.

6.3.3 Les coûts de la protection côtière

Au-delà de l'identification rigoureuse des mécanismes en jeu, la protection côtière pose pour les gestionnaires le problème du coût des ouvrages réalisés. Les coûts donnés ci-dessous s'entendent pour 1 km de côte protégée, (évaluation 1987, sans considérer pour simplifier l'analyse, les inévitables interactions entre les deux mécanismes retenus, (attaque frontale des houles et dérive littorale).

Protection contre l'attaque frontale des houles
(i) *Dans les secteurs fortement urbanisés*
- cas où la plage a disparu et où les profils bathymétriques font apparaître un fort creusement des fonds
- ouvrages lourds de type brise-lames
    coût: 6,8 MF[5]
    (toutes taxes comprises, TVA au taux de 18,6%)
- cas où la plage n'a pas complètement disparu et où l'on n'observe pas de creusement excessif des fonds
- ouvrages lourds de type brise-lames coût: 4 MF

Dans les secteurs dunaires: ouvrages légers de restauration des systèmes
- cas graves: renforcement de la résistance du cordon dunaire et/ou reconstitution totale des dunes disparues
    coût: 2,6 MF
- cas moyennement graves: reconstitution progressive du cordon dunaire
    coût: 1,4 MF

Protection contre l'érosion due à la dérive littorale
- ouvrages lourds de type épis en enrochements
    coût: 2,3 MF

Ces chiffres s'entendent hors études préalables et hors opérations (nécessaires) de revégétalisation dans le cas des techniques légères.

Ouvrages expérimentaux
Dans le cadre de l'amélioration de la politique côtière dans le Golfe du Lion, se développent des expérimentations techniques visant à mieux comprendre la dynamique locale et à mettre sur pied des techniques adaptées qui ne grèvent pas le fonctionnement des milieux.
Leur coût s'élève de 1 à 5 MF. km^{-1}.

Mais, face à une montée clairement identifiée du niveau marin il est probable que la protection côtière se focalisera sur les secteurs urbanisés et sur la protection des étangs côtiers, zones économiquement très importantes en Languedoc. Là encore, les recherches actuellement entreprises permettront de limiter les secteurs à traiter aux strictes zones dont la dégradation aura été dûment identifiée et analysée, ce qui permettra de réduire et d'étaler dans le temps les interventions. Le coût élevé de la protection côtière sera alors de peu de poids face aux enjeux économiques en cause.

7. Recommandations
1) Compléter les schémas sur l'évolution récente des climats pour l'ensemble du bassin méditerranéen occidental.

2) Suivre l'évolution du trait de côte en insistant sur les modalités des variations saisonnières et pluriannuelles, afin de vérifier les cycles d'érosion /accrétion.

3) Etablir une classification très fine des indicateurs végétaux ou géomorphologiques de la vulnérabilité à l'érosion marine et (ou) éolienne.

4) Etablir, à l' échelle de l'ordre du 1 /5000, des cartes prévisionnelles de vulnérabilité à court, moyen ou long terme (année, décennie, siècle).

5) Multiplier, en les combinant et les diversifiant, les ouvrages expérimentaux de protection du littoral: défenses depuis la mer (agressions marines), dispositifs sur la plage et la dune (agressions marines et éoliennes).

6) Construire des schémas d'évolution du couvert végétal spontané et (ou) introduit, sous l'action des agressions marines et éoliennes, ainsi que sous l'action des mesures de protection physique du trait de côte, afin de comprendre, puis de favoriser les mécanismes de défense spontanée de la dune.

7) Préciser les schémas d'évolution qualitatifs et quantitatifs des populations halieutiques du milieu lagunaire, en fonction des phénomènes de confinement et déconfinement.

8) Développer les études sur le rôle de la diversité des milieux dans le maintien des communautés animales.

9) Mettre en place un observatoire d'occupation par l'Homme de l'espace littoral, lagunaire et périlagunaire, afin d'identifier les changements dans les pressions socio-économiques et les compétitions dans son usage.

Notes

1 Centre National de la Recherche Scientifique
2 INSEE: Institut national de la statistique et des études économiques.
3 CRPEE: Centre régional de productivité et des études économiques. (UA, CNRS Montpellier)
4 date du dernier recensement général de l'agriculture de la France
5 1MF = 1 000 000 F français

8 Références Bibliographiques

Aloisi, J.-C., 1986. Sur un modèle de sédimentation deltaïque. Contribution à la connaissance des marges passives. *Thèse ès Sciences naturelles,* Perpignan, p. 162 et annexes.

Aloisi, J.-C.et Duboul-Razavet, Ch., 1974. Deux exemples de sédimentation deltaïque actuelle en Méditerranée: les deltas du Rhône et de l'Ebre. *Centre rech. Pau-SNPA,* n° 8–1, 227–240 + 5 figures.

Aloisi, J.-C., Froget, C., Got, H., L'Homer, A. et Monaco, A., 1984. Précontinent et littoral. In Synthèse géologique du S.E. de la France. *Mém. BRGM,* n° 125, **1**, 559–561.

Armi, L., Farmer, D., 1985. The internal hydraulics of the Strait of Gibraltar and associated sills and narrows. *Oceanol. Acta,* **8**, 37–46.

Astier, A., dir. publ., 1970. Camargue. Etude hydrogéologique, pédologique et de salinité. Rapport général. Direct. départ. de l'Agric. (B.du Rhône), p. 352 et annexes.

Baudière, A. et Simonneau, P., 1974. Les groupements à Corynephorus canescens (L.) P.B. et à Ephedra distachya L. du littoral roussillonais. *Vie et Milieu,* **24**(10), 21–42.

Bazille, F., 1974. Nouvelles données sur l'âge des cordons littoraux récents du golfe d'Aigues-Mortes. *Bull. Soc. languedocienne Géogr.,* **8**(3–4), 199–206.

Bazille, F., 1975. Nouvelles données sur l'âge des cordons littoraux du Golfe d'Aigues-Mortes. *Bull. Soc. languedocienne de Géogr.,* **8**(4–5): 199–206.

Bertrand, J.-P. and L'Homer, A., 1975. Le delta du Rhône. Guide d'excursion. IXème Congrés internat. *Sédimentologie,* Nice, p. 65.

Béthoux, J.P., 1979a. Budgets of the Mediterranean Sea. Their dependence on the local climate and on the characteristics of the Atlantic water. *Oceanol. Acta,* **2**, 157–163.

Béthoux, J.P., 1979b. Le régime de la Méditerranée au cours de périodes glaciaires. *Nuovo Cimento,* **2**, 117–126.

Béthoux, J.P., 1984. Paléo-hydrologie de la Méditerranée au cours des derniers 20000 ans. *Oceanol. Acta,* **7**, 43–48.

Béthoux, J.P. et Prieur, L., 1984. Hydrologie et circulation en Méditerranée nord-occidentale. In: Bison, J.J. et Burollet, P.F. (éd.). *Ecologie des microorganismes en Méditerranée occidentale,* AFTP, Paris, 13–22.

Béthoux, J.P., Prieur, L., Nyffeler, F., 1982. The water circulation in the North-Western Mediterranean Sea, its relations with wind and atmospheric pressure. In: Nihoul, J.C.J. (éd.), *Hydrodynamics of semi-enclosed seas,* 129–142.

Bird, E.C.F., 1986. Coastline changes. A global review. John Wiley and Sons, réimpression avec corrections, p. 219.

Blanc, J., 1977. Recherches de sédimentologie appliquée au delta du Rhône, de Fos au Grau du Roi. C.N.E.X.O., (éd.) 75/1193, p. 69.

Blondel, J. et Isenmann, P., 1981. Guide des oiseaux de Camargue. Delachaux & Niestlé, p. 344.

Britton, R.H. and Podlejski, V.D., 1981. Inventory and classification of the wetlands of the Camargue (France). *Aquatic botany,* **10**, 195–228.

Brun, G., 1961. Contribution à l'étude écologique de l'estuaire du Grand Rhône. Thèse 3ème cycle, Marseille, p. 64 + tabl.

Bryden, H.L.and Stommel, H.M., 1984. Limiting processes that determine basic features of the circulation in the Mediterranean Sea. *Oceanol. Acta,* **7**, 289–296.

Busson, G. et Perthuisot, J.-P., 1986. La synthèse des données. In: *Les séries à évaporites en exploration pétrolière, t.1: Méthodes géologiques.* Technip., Paris: 165–217.

Clique, P.M., Feuillet, J. and Coeffe, M.G., 1984. Catalogue sédimentologique des côtes françaises. Côtes de la Méditerranée de la frontière espagnole à la frontière italienne. Partie A, de la frontière espagnole à Sète. *Coll. de la Dir. des*

Et. et Rech. d'EDF, Eyrolles: 1–105.

C.N.R. (Compagnie nationale du Rhône), 1973. Les aménagements d' Avignon et du palier d'Arles sur le tiers aval du Bas-Rhône. Editions de la Navigation du Rhin, Strasbourg, p. 16.

Corre, J.-J., 1976. Etude phyto-écologique des milieux littoraux salés en Languedoc et en Camargue. I. Caractéristiques du milieu. *Vie et Milieu*, **26**(2c), 179–245.

Corre, J.-J., 1987. Les peuplements végétaux et la gestion des côtes basses du golfe du Lion. *Bull. Ecol.*, **18**(2), 201–208.

Corre, J.-J., Bigot, L., Billès, G., Bouab, N., Coulet, E., Gautier, G., Heurteaux, P., Pont, D. et Skeffington, S., 1982. Structure et peuplements des rives d'étangs temporaires en Basse Camargue. *Bull. écol.*, **13**(4), 339–356.

Fernandez, J.M., 1984. Utilisation de quelques éléments métalliques pour la reconstitution des mécanismes sédimentaires en Méditerranée occidentale: apports du traitement statistique. Thèse de 3ème cycle. Univ. Perpignan, p. 230.

Frisoni, G.-F., 1984. Contribution à l'étude du phytoplancton dans le domaine paralique. Thèse Ing. Doct. Univ. Montpellier, p. 171.

Gadel, F. et Pauc, H., 1973. Sédimentation récente à l'embouchure du Grand Rhône: données sédimentologiques et géochimiques. *Bull. Inst. Géol. Bas. Aquit.*, **14**, 127–141.

Got, H. et Pauc, H., 1970. Etude de l'évolution dynamique récente au large de l'embouchure du Grand Rhône par l'utilisation des rejets du Centre de Marcoule. *C.R. Acad. Sci.* Paris, **271**: 1956–1959.

Greslou, M., 1984. Catalogue sédimentologique des côtes françaises. Côtes de la Méditerranée de la frontière espagnole à la frontière italienne. Partie B, de Sète à Marseille. *Coll. de la Dir. des Et. et rech. d'E.D.F.*, Eyrolles, 106–187.

Guelorget, O. et Perthuisot, J.-P., 1983. Le domaine paralique. Expressions géologiques, biologiques et économiques du confinement. *Trav. Lab. Géol. Ec. Norm. Sup.*, Paris, **16**: p. 136.

Guelorget, O., 1985. Entre mer et continent. Contribution à l'étude du domaine paralique. Thèse. Doct. Etat Univ. Montpellier: p. 721.

Guelorget, O., Frisoni, G.-F., Monti, D. et Perthuisot, J.-P., 1984. Contribution à l'étude hydrochimique, sédimentologique de la lagune de Nador (Maroc). Rapport FAO/ MEDRAP, p. 82.

Ibrahim, A., Guelorget, O., Frisoni, G.-F., Rouchy, J.-M., Maurin, A. et Perthuisot, J.-P., 1985. Expressions hydrochimiques, biologiques et sédimentologiques des gradients de confinement dans la lagune de Guemsah (Golfe du Suez, Egypte). *Oceanologica Acta*, **8**(3), 303–320.

Kruit, C., 1951. Sediments of the Rhone Delta. 1– Grain size and microfaune. Verhand. *Konink. Neder. Geol. Mi.*, **15**(2), 357–956.

Lagaaij, R. and Kopstein, F.P.H.W., 1964. Typical features of a fluviomarine offlap sequence. In: Deltaic and shallow marine deposits. 6ème Intern. Sedim. Congr., 1963, Proc. Amsterdam. Elsevier Publ. Cy., 216–226, 4 figures.

L'Homer, A., 1987. Notice explicative de la carte géologique Arles à 1/50000. BRGM (éd.), p. 72.

L'Homer, A., Bazile, F. et Thommeret, J.Y., 1981. Principales étapes de l'édification du Delta du Rhône de 7000 BP à nos jours: variations du niveau marin. *Oceanis*, **7**(4), 389–408.

Le Dain, A.-Y., Barbel, P and Gerbe, A., 1987. Plage et dunes du golfe du Lion. EID et IARE Contrat de plan Etat-Région Languedoc-Roussillon, p. 58.

Millot, C., 1979. Wind induced upwellings in the Gulf of Lions. *Oceanologica Acta*, **2**, 261–274.

Millot, C., 1983. Etat actuel de nos connaissances sur le régime hydrodynamique du golfe du Lion (plateau et pente). Pétrole et techniques, AFTP, **299**, 35.

Millot, C. et Wald, L., 1980. The effect of Mistral wind on the Ligurian current near Provence. *Oceanologica Acta*, 3.

Molinier, R. et Tallon, G., 1965. Etudes botaniques en Camargue. I. La Camargue, pays de dunes. II. Vers la forêt en Camargue. *Terre et Vie, 1ère série,* **19**, 1–192.
Oldham, R.D., 1930. Earth movements in the Delta of Rhone. *Nature,* G.B., **125**, n° 3155, 601–604.
Oomkens, E., 1970. Depositional sequences and sand distribution in the post-glacial Rhone delta complex. In Deltaic Sedimentation. *Soc. of econ. Pal. and Min.; spec. publ.,* n° 15, 198–212.
Parde, M., 1924. Le régime du Rhône. Etude hydrologique. Thèse Univ. Grenoble.
Pauc, H., 1970. Contribution à l'étude dynamique et structurale des suspensions solides au large de l'embouchure du Grand Rhône (Grau du Roustan). Thèse de 3ème cycle, Perpignan, p. 126.
Perthuisot, J.-P., 1971. Présence de la magnésite et de huntite dans la Sebkha el Melah de Zarzis. *C.R. Acad. Sci.,* Paris, **272**, 185–188.
Perthuisot, J.-P., 1975. La Sebkha el Melah de Zarzis. Genèse et évolution d'un bassin salin paralique. *Trav. lab. géol. Ec. Norm. Sup.,* Paris, **9**, p. 252.
Perthuisot, J.-P., 1980. Sites et processus de la formation d'évaporites dans la nature actuelle. *Bull. Centre Rech. explo. Prod. Elf aquitaine,* **4**(1), 207–233.
Perthuisot, J.-P., 1983. Introduction générale à l'étude des marais salants de Salin de Giraud (Sud de la France): le cadre géographique et le milieu. *Géol. médit.* **9**(4), 309–327.
Picon, B., 1978. L'espace et le temps en Camargue. *Actes Sud,* Arles, p. 264.
Picon, B., 1985. Carpes en Camargue, autopsie d'un demi-échec. *Aqua-revue,* Bordeaux, juin-juillet 1985.
Picon, B., 1987. Les conflits d'usage sur le littoral camarguais: protection de la nature et pratiques sociales. *Revue Norois,* **34**(133–135), 73–80.
Picon, B., Tamisier, A. et Dervieux, A., 1987. Le développement récent des activités de tourisme, de loisirs et de chasse dans les exploitations de Camargue. Influence d'une évolution économique sur le milieu naturel. Convention INRA-CNRS, janvier 1987.
Pitovranov, S.E., 1987. A climate scenario based on the Vinnikov/Groisman' approach. European workshop on interrelated bioclimatic and land use changes. Vol. A – Climate scenarios, 24–38.
Planchais, N., 1982. Palynologie lagunaire de l'étang de Mauguio. Paléoenvironnement végétal et évolution anthropique. *Pollen et Spores,* **24**(1), 93–118.
Planchais, N., Quet-Pasquier, L., Cour, P., Thommeret, J. et Thommeret, Y., 1977. Essai de palynologie côtière appliquée au remplissage flandrien de Palavas (Hérault). *C.R. Acad. Sci.,* Paris, sér. D, **284**(3), 159–162.
Pons, A., Toni, Cl. et Triat, H., 1979. Edification de la Camargue et histoire holocène de sa végétation. *Terre et Vie, Rev. Ecol.,* suppl.2, 1979, 13–30 + cartes.
Racine, P., 1980. Mission impossible? L'aménagement touristique du littoral Languedoc-Roussillon. *Midi Libre,* Coll. Témoignages, p. 293.
Ritter, J.,1973. Le Rhône. P.U.F., Paris, Collection "Que sais – je?", n° 1807, p. 126.
Rognon, R., 1981. Les crises climatiques. *La Recherche,* n° 128, 1354–1364.
SOGREAH, 1974. La petite Camargue. Protection du littoral, du phare de l'Espiguette à l'embouchure du Petit Rhône. SOGREAH, R 11 863, R 11 929.
Surrell, E., 1847. Mémoire sur l'amélioration des embouchures du Rhône. Imprimerie cévenole, Nîmes, 1 vol. in 8°.
STSC, 1987. Statgraphics,. STSC Inc, Rockville, p. 600.
Tamisier, A., 1987. – Quartier d'hiver et de transit pour les oiseaux d'eau. *Le Courrier de la Nature,* n° 109, 30–37.
Van Straaten, L.M.J.V., 1957. Dépôts sableux récents du littoral des Pays-Bas et du Rhône. *Géol. Mijnb.,* 19ème Juargang, 196–213.

Vernier, E., 1972. Recherches sur la dynamique sédimentaire du Golfe de Fos. Thèse de 3ème cycle, Marseille, p. 72.

Verlaque, G., 1987. Le Languedoc-Roussillon, Coll. "La question régionale", *PUF*, p. 184.

Wigley, T.M.L., 1985. Impact of extreme events. *Nature*, **316**, 11 July 1985, 106–107.

Wigley, T.M.L., 1987. Climate scenarios. European workshop on interrelated bioclimatic and land use changes. Vol. A, Climate scenarios, 10–23.

ANNEXE I

Cartes des portions du littoral dont la côte est inférieure à 5 m
Elles marquent la limite extrême des zones risquant d'être influencées par une remontée de la mer
d'ici 2025 selon les prévisions les plus pessimistes.

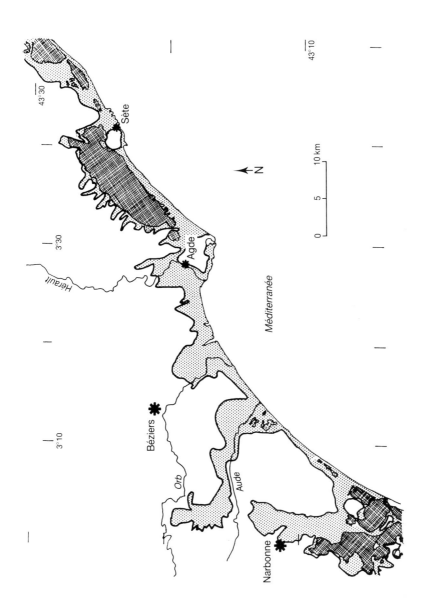

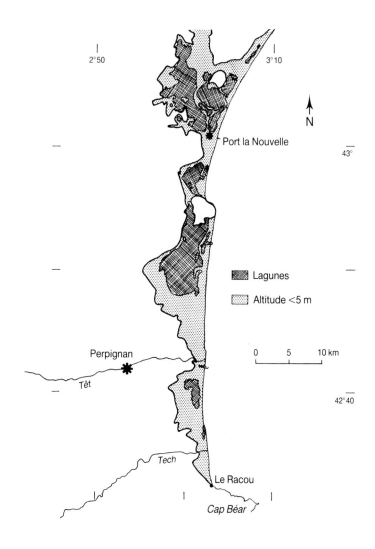

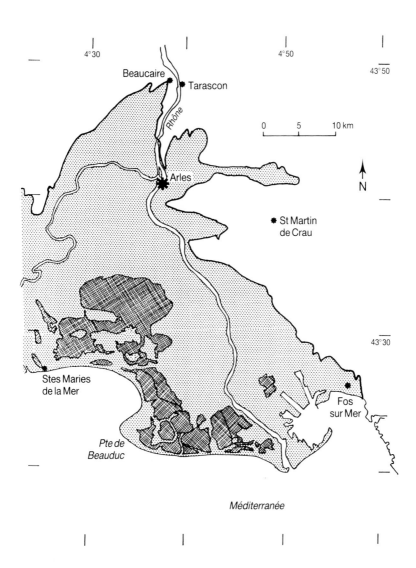

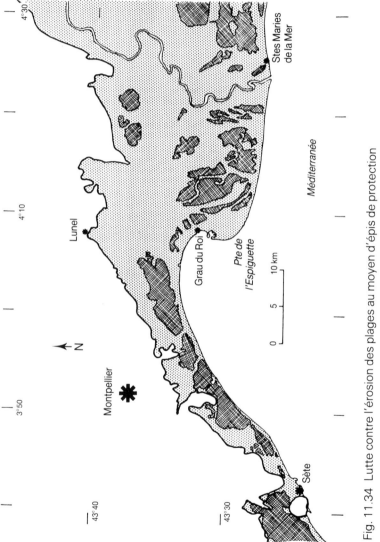

Fig. 11.34 Lutte contre l'érosion des plages au moyen d'épis de protection (Palavas les Flots, Hérault). L'implantation a eu lieu en 1951. (d'après le Service maritime et de navigation Languedoc Roussillon).

Annexe II

Principales activités économiques localisées sur le littoral du Golfe du Lion hors tourisme et agriculture (d'après Verlaque, 1987 et Port autonome de Marseille, 1988)

Tableau 11.9 Productions halieutiques en Languedoc-Roussillon

	1965	1975	1980	1984
poissons débarqués total en tonnes	10269	17000	18375	20372
moules et huîtres total en tonnes	–	8724	8483	7536

Principales industries des villes littorales (1987)

Fos-sur-Mer:
 sidérurgie
 pétrochime
 sous traitance

Sète-Frontignan:
 engrais (superphosphates, nitrate d'ammoniaque, acide phosphorique)
 engrais composés
 pesticides
 raffinerie de soufre
 cimenterie
 pêche industrielle, mareyage
 dépôts de produits pétroliers

Agde:
 produits phyto-sanitaires
 batiment et travaux publics

Port-la-Nouvelle:
 cimenterie
 pêche industrielle
 dépôts de produits pétroliers

Grau-du-Roi:
 pêche industrielle, mareyage

Annexe III

Activités de la zone portuaire de Fos-sur-Mer
(Source: Port autonome de Marseille)

Trafic maritime:
  80 millions de tonnes.an^{-1}
  1822 emplois liés aux activités portuaires

Activités industrielles:	8 927 emplois
Sidérurgie:	2 entreprises, 5858 emplois
Pétrochimie:	10 entreprises, 1569 emplois
Sous traitants:	1500 emplois

Tableau 11.10 Echanges maritimes régionaux Port-la-Nouvelle, Sète

	1960	1965	1970	1975	1980	1984
Total marchandises × 10^3 tonnes	4079.8	4510.6	6110.2	7427.6	9219.5	7583.6
Total voyageurs × 10^3	1.7	1.7	0.6	39.5	88.6	128.6

Annexe IV

Quantification du tourisme par rapport aux autres activités

Faute d'indicateurs statistiques fiables il est impossible de quantifier la part des activités touristiques par rapport aux autres activités (agricoles, industrielles, piscicoles, . . .). Le nombre de visiteurs peut être évalué grossièrement par différence du nombre d'accès au littoral en saison touristique et hors touristique (comptages routiers). La méthode est envisageable en Camargue où les accès sont peu nombreux. Elle serait beaucoup plus difficile ailleurs.

Le comptage des nuitées en hôtel ou en camping ne rend pas compte de l'importance des visites très nombreuses d'une journée, ni de l'importance du camping "sauvage".

Le chiffre d'affaires (indicateur économique) ou l'emprise au sol de l'activité touristique sont difficiles à évaluer. De multiples connexions existent entre habitats permanents (non touristiques) et temporaires (touristiques), entre activités agricoles, d'élevage et touristiques (ferrades*, chasse, gîtes ruraux, promenades à cheval, etc.) au sein des mêmes entreprises et sur les mêmes territoires.

---

*Fête célébrée à l'occasion du marquage du bétail.

Les activités commerciales tendent à s'adapter à la fois à la clientèle permanente de morte saison et touristique. Tous ces phénomènes auxquels s'ajoute le poids de l'économie informelle rendent bien aléatoire toute tentative d'évaluation sérieuse à partir des données disponibles. Un programme de recherche spécifique serait nécessaire pour tenter de mener à bien une telle entreprise.

12

Implications of Climatic Changes for the Po Delta and Venice Lagoon

G. Sestini
(*Applied Earth Science Consultant,
London, UK*)

Abstract

The impact of climatic changes on the north-western Adriatic lowlands in the next decades must be judged in the perspective of the environmental problems that exist there at present. They have resulted not from climatic oscillations nor from the recent low rate of sea-level rise, but from coastal-development activities during the last 50 years that have been carried out with little consideration for the natural processes and their responses. These have included land reclamation, the intense urbanization of the coast, the building of deep harbours and of coastal-defence structures, a decrease of lagoon surface, enhanced land subsidence caused by ground-water extraction, the diversion of fluvial water and solid discharges, and the pollution of lagoons and the sea.

Since the late 1950s, many parts of the shoreline have been retreating, while the lagoons have experienced increasingly higher tide and storm surge levels. Today, the entire coast is in a state of physical instability because of subsidence and sediment starvation, and must be classified as a high-risk zone in respect of the threat of accelerated sea-level rise.

In general, climatic changes could occur gradually, involving a 0.5–1.5°C temperature increase by 2025, and 1.5–3.5°C later in the 21st century. They should not be specifically manifested until about three decades from now. The frequency of abnormal events, however, could increase: hot dry summers, unusually mild and dry winters, erratic precipitation (including heavy rainfall and floods), marine storms, tidal surges, and episodes of water stagnation and eutrophication.

A sea-level rise of 12–20 cm by 2025 generally would not cause flooding of lowlands, but would aggravate beach erosion and the periodically high water levels in the Venice lagoon. Stronger tidal currents and more estuarine conditions in the lagoons would accelerate the degradation of the tidal flats and reed beds, diminishing their value for wildlife and fish breeding.

The main threats of an additional sea-level rise (possibly over 40–50 cm after 2050) would be the serious degradation of the city of Venice and other towns in the lagoons and on the coast, the impairment of the harbour-related activities, and a substantial decline of the beach resorts. The average temperature rise of 3°C would produce a milder winter climate, possibly with greater rainfall, but this also would alter considerably the river regimes, as smaller and shorter-lived snow cover develops over the Alps and Apennines in winter.

As regards agriculture, many crops would benefit from a higher air temperature and CO_2, provided that adequate water supplies are still available. In many parts of the Po Delta, and perhaps elsewhere, it might be more economical to return sub-sea-level lands, now devoted to extensive cereal cultivation, to their original lagoonal state, in favour of fishing, which presently is a more efficient and remunerative activity.

In the Po Delta and elsewhere, the lagoons and marshes could act as a buffer between the open sea and higher land, and also as nature reserves. Industrial and other activities in the areas below 1 m probably could be moved gradually inland without excessive disruption.

Little change is expected to occur in the settlement and economic patterns of the hinterland (elevations between 3 and 5 m) during the next few decades because of low population growth. The main factors controlling regional organization (especially for agriculture and urban development) will continue to be the availability and quality of water resources, but means to control industrial, urban and agriculture pollution must be found soon. New approaches to beach recreation would minimize the impact of sea-level rise on summer tourism.

The role of the present coastal centres of heavy industry (Porto Marghera, Ravenna) and the importance of commercial ports (Venice, Chioggia, Ravenna) will be determined more by general factors of national and international demand than by climatic changes.

Eventually, the basic resources of agriculture and fisheries should change both qualitatively and quantitatively, the former benefiting perhaps from higher temperatures and greater CO_2 in the atmosphere.

A great deal of research in advance will be needed on the impacts of higher temperatures and of sea-level rise, with the systematic collection and analytical treatment of a data base ranging from climatic and oceanographic parameters to hydrology, ecosystems, and social and economic variables.

1 INTRODUCTION

The 300 km-long Adriatic coast of northern Italy, from Monfalcone to Rimini, borders a lowland 15–25 km wide, generally under 2 m elevation, in which remnants of formerly more extensive lagoons, salt and fresh-water marshes alternate with reclaimed lands. Many parts actually lie below sea-level, especially in the area of the Po Delta (Fig. 12.1).

It is a deltaic and lagoonal setting characterized by active geological subsidence and instability of natural environments, subject to river floods, storm surges and coastal shifting. Natural disasters have affected this region since human occupation, as recently as the devastating Po Delta floods of 1955 and 1966, and to the recurringly higher tidal flooding of Venice.

Since ancient times the Po River has represented the main access route from the Adriatic to the interior plains and alpine regions of Italy. The towns of Adria and Spina were already important centres of maritime and land traffic in the early 6th century BC. Commercial activities in Etruscan and Roman times were sustained also by a local coastal economy, based on cattle raising, game hunting and wood exploitation, and important fishing, salt and brick making. The Byzantine capital of Ravenna continued these commercial activities, but devastating floods in the late 6th century AD and the fall of the city in the mid-8th century led to a progressive degradation of the coastal regions.

It was the protection offered by the natural lagoons and their swampy margins that favoured the initial settlement of Venice in the 6th century and later provided the basis for its development into a state and major maritime power. The well-being of Venice depended on a symbiosis with its lagoonal surroundings, in particular on its preservation in the face of siltation from rivers. The result was an increasing necessity to interfere with, or at least to regulate, the natural processes for the benefit of agriculture in the lowlands

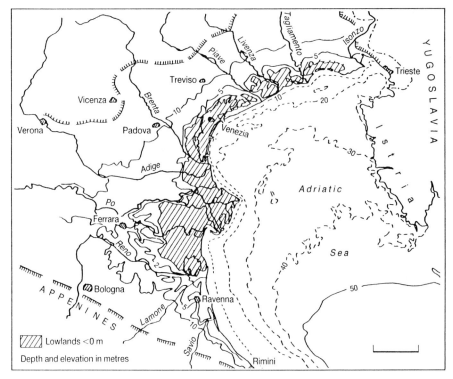

Fig. 12.1 Extent of the lowlands, under 2 m elevation and below sea-level, along the north-east coast of Italy.

and of lagoon access to the sea, which was continuously threatened by the ever-changing passages between the barrier islands.

The legacy of this historical development is the existence today of cultural and artistic centres of inestimable historical significance.

Exploitation of the north-western Adriatic lowlands remained limited, leaving the coastal areas in a fairly natural condition until the first decades of this century. Transformations since the First World War have been dramatic, first by the progressive reclamation of large swampy lagoonal areas (e.g. Comacchio to Ferrara, north-east of Venice), then with the development of the industrial centre of Porto Marghera, and the beginning of the recreational use of beaches. Later, in the early 1950s, it was decided to construct a second large harbour and industrial complex near Ravenna, following the discovery of natural gas fields there.

Coastal use further exploded in the next decade, the years of Italy's economic boom, with the expansion of the industrial centres and the rapid urbanization of the shore, which became one of the largest summer recreational centres in southern Europe. Tourism, both cultural (e.g. Venice) and beach-bound, is now an important resource for the regions of Emilia-Romagna, Veneto and Friuli–Venezia Giulia.

Thus, the north Adriatic coastal plains have reached a stage of advanced land use, with complexly interrelated agricultural, industrial and tertiary activities that are tied to national and supranational markets, and are supported by a well-developed communications network. Resident population within the +5 m contour, however, is not high (1.8 million). There is no growth due to negative birth rates and to migration. Economic development remains limited to a belt marginal to the +2 m lowlands, the exception being the summer beach tourism and the harbour activities.

The postulated climatic changes in time could seriously disrupt the economy and well-being of the coastal zone. A significant rise in sea-level not only would have a direct impact on beaches, lagoons and agricultural lands that lie below sea-level, but also would aggravate environmental problems that have resulted from an unwise exploitation of the coast and lagoons.

Most of the shoreline is in a state of retreat due to reduced river sediment inputs, a consequence of the water management practices upstream and sand mining from river beds. Natural subsidence in the Po Delta and in the vicinity of industrial centres (Mestre, Ravenna) have been magnified by the over-extraction of ground water. Furthermore, the natural flexibility of the coastline has become constrained by the erection of numerous fixed structures, built to protect the agricultural and tourist developments and access to harbours.

The degradation of the Venice lagoon started with the alteration of the tidal flats for the new harbour of Marghera. The later development of base industries, tied to the transformation of imported raw materials, has contributed to the extensive chemical and thermal pollution of the lagoonal waters with consequent serious ecological problems.

But problems are not purely physical. In the wake of the unprecedented concentration of population and capital along the coast, contrasts have arisen between the different economic and political interests involved. Such is the incompatibility of industry and tourism, of land reclamation and the preservation of wetlands for fishing. Lately, the increasing awareness of the degradation of the environment and the necessity to contain the pollution of coastal and inland waters have come in conflict with the maintenance of economic activities. Last, but not least, coastal protection, nature conservation and proper coastal management are hindered by political, financial and administrative wrangles between local authorities, the central government and the different ministries that share responsibility for Italy's coastlines.

2 Geographic Setting

The coastal lowlands considered in this report include the Po Delta, and the lower Romagna and Veneto-Friuli plains and their wetlands, between the shore and the 2 m contour. Sixty per cent of the surface of this zone is wetlands and areas below sea-level.

The coastal belt is composed of beach-dune barriers, lagoons, salt and fresh-water marshes, and reclaimed lands (the remnants of former lagoons and interdistributary bays), separated by the more elevated channel systems of the many rivers that flow from the Alps and the north Apennines. The prominent coastal features are the cuspate deltas of the Isonzo and Tagliamento and the lobate delta of the Po (Fig. 12.1).

2.1 Monfalcone to Tagliamento

On the whole, study of beach morphology and wave parameters (Brambati et al., 1978) indicates a lower wave energy in the Monfalcone–Lignano stretch of the coast, than from Lignano to the Brenta River. The moderate wave energy is due to the presence of offshore bars, and also to an extensive shallow seabed between the emergent and the submarine beach (e.g. the Mula di Muggia bank, a remnant of the former Natissa Delta; and mudflats, from a few hundred metres to 1 km wide, in front of the Isonzo delta).

In the Marano–Grado lagoon, deep tidal and former river channels (5–6 to 10 m deep) connect the outlets to streams and drains at its inner margin. The rivers Aussa and Corno have been canalized for boat access to the Torviscosa and San Giorgio Nogaro harbours. There is also a major east–west navigation canal from Lignano to Grado, part of the Litoranea Veneta canal system. The lagoon is closed by a series of barrier islands and sand banks (Fig. 12.2). The unprotected islands of Martignano and San Andrea are easily washed over during storms and still tend to shift. Otherwise, the islands and the rest of the coast are considerably armoured, with many parallel and normal structures, including the 1.5 km long, 2.5 m high seawall at Grado, which dates back to 1887.

The entire inner margin of the lagoon is lined by a dike not less than 2 m high. It protects a strip of reclaimed land that lies largely below sea-level and is drained by pumps. Canal and road embankments rise generally 2–3 m above sea-level.

2.2 From the Tagliamento to the Piave and Sile Rivers

At the margin of the Tagliamento Delta (Lignano to Bibione) there are still dunes, or sandy high ground, 2–6 m high, but behind lie patches of ground below sea-level. Further west, the coastal beach-dune barrier is only a few hundred (< 500) metres wide; the beaches are narrow, with gentle profiles and offshore slopes (low tide uncovers the coastal sand bars, the high tide reaches the foot of the beach resort buildings). The coastal barrier is interrupted by several river mouths and lagoon outlets (Fig. 12.3) by which are located many pleasure-boat harbours.

The present Caorle lagoon has an area of about 1700 ha with five fish basins, fed by the Nicesolo canal. Sedimentation at the outlet of this canal is causing reduced access of marine tidal waters, with a consequent decrease of salinity in the lagoon. The reclaimed, former Caorle lagoon and wetlands constitute a large area below sea-level, which is crossed transversely by the 1–2 m high fluvial ridges of the Piave and Livenza dikes. Along all canals and road embankments rise 1–2 m above sea-level.

2.3 Venice Lagoon and the Coast to the Adige

The Venice lagoon measures 546 km^3 and is divided into three basins (Lido, Malamocco and Chioggia) by "watersheds": areas of tidal flats (*barene*) which represent high ground historically associated with the former alluvial ridges of the Sile and Brenta rivers (Favero, 1979; Zunica, 1971). Depths vary from 1–3 m in the open basins to 15–20 m in the outlets and canals. The tidal channels are about 4–5 m to 10 m deep, the intertidal areas usually are <1 m (± emergent at low tide). At the inner edge of the lagoon there are salt marshes with fresh-water lake basins (*laguna morta*).

Po Delta, Venice Lagoon and climatic change 433

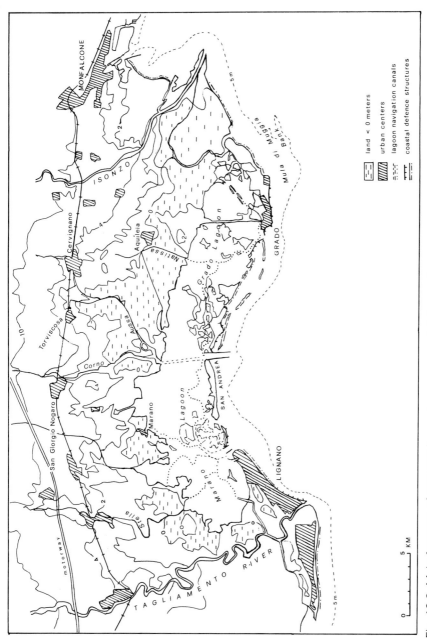

Fig. 12.2 Main features of topography, and occurrence of defence structures between Monfalcone and the Tagliamento River (based on Brambati, 1987; CNR, 1985; and topographic maps).

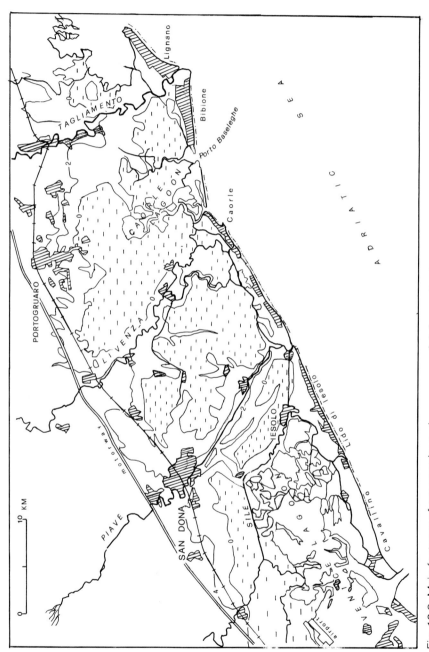

Fig. 12.3 Main features of topography and occurrence of defense structures between the Marano and Venice lagoons (CNR, 1985; Provincia di Venezia; 1983, Regione Veneto, 1987).

The lagoon of Venice and its barrier islands have been much modified (Fig. 12.4). Only a part of the original tidal flats and marshland remains; the rest is reclaimed land and *valli* for aquaculture. The barrier islands have been altered by the erection of the 1.5 to 4 km long jetties and by seawalls (central part of Lido; the *murazzi* of Pellestrina built in 1731–1777, which are 4 m high). The island of Lido has a natural elevation of 2–3 m. At Pellestrina there is no beach and the submerged lower beach face is very steep (1.5–2.3%) with sand bars at 3.8 m depth. The thickness of the barrier sands in the subsurface is 10–20 m.

The Jesolo to Cavallino shore has a narrow (60 m), low-gradient beach made of medium-fine grained sands. It is backed by a discontinuous line of 2–3 m dunes, and the land behind has elevations of 0.5–1 m (roads are 2 m high). There are many groins along the shore.

The inner margin of the north basin of the Venice lagoon is bounded by the levees of the Sile and by road embankments 2–5 m above sea-level. The airport of Tessera, as well as the islands of Venice, Murano and Burano, rise to average 2 m above sea-level; most of the area of Mestre and Porto Marghera has elevations of 2–3 m.

The area of the Brenta-Adige deltas (Chioggia to Porto Caleri) is similar to the Tagliamento Delta, with an external complex of beach ridges and a lagoonal area behind. The south-west margin of the Venice lagoon and the plain between the Adige river and Cavazere are generally below sea-level, but there are various high and continuous transversal embankments, namely the "Strada Romea", the Chioggia–Mira–Mestre road and the Adria–Chioggia railway. Relict beach ridges reach an elevation of 4 m, occasionally 8 m (Favero, 1979).

2.4 The Po Delta

East of Strada Romea, the Po Delta covers an area of 610 km^2 with a 64 km long shoreline from Porto Caleri to Volano. The external delta is a zone enclosed by a series of low, sandy barrier islands, with lagoons, marshes, fish basins and narrow ridges in the north; tidal flats and marshes in the east and south-east, and the Sacca Scardovari bay in the south (Fig. 12.5). Foreshore beach gradients are gentle (0.4–0.8%), steeper only in front of Po di Pila. The present wetlands are almost entirely isolated and hydrologically controlled.

There are no defence structures at the outer edge of the Po Delta, and no tourist developments. The most prominent construction is the ENEL (Electricity Board) thermoelectric plant at Polesine Camerini on the Pila branch. The inner delta consists of reclaimed lands below sea-level (ground depths are quite variable, frequently 2.5–3 m). In the west they are crossed by north–south, north–north-east, north–north-west trending sandy wooded ridges, remnants of ancient shorelines that stand at 0–1 m elevations (e.g. Bosco Mesola). The ridges are discontinuous and have been flattened at many places by sand removal. The margins of the agricultural lands and all river branches are lined by levees and dikes, the highest along the Pila (3.8–4 m), the Gnocca (2.5 to ±3 m) and Goro (2–2.5 m) branches. The elevation of the smaller dikes is 0–1 m to ±1.5 m.

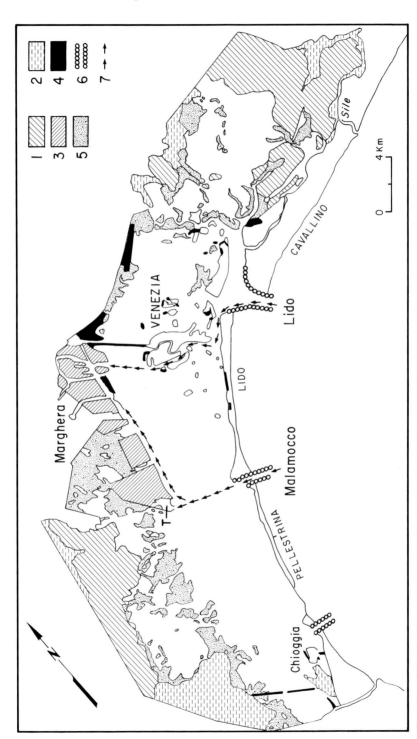

Fig. 12.4 Main land uses around the Venice lagoon. 1. Cities over 100,000 inhabitants, 2. 50–100,000 inhabitants, 3. 25,000–50,000 inhabitants, 4. 10,000–25,000 inhabitants, 5. 5000–10,000 inhabitants, 6. Towns under 5000 inhabitants, 7. Land under sea-level.

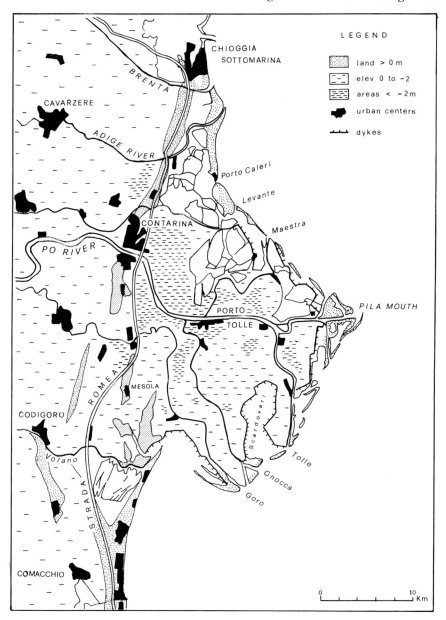

Fig. 12.5 Main features of topography of the Po Delta (based on topographic maps, on 1982 satellite images and data in Marabini, 1985).

2.5 The Lowland Between the Po, Ferrara and the Reno

Much of the region west of Ferrara was swampy and lagoonal until 1870, the beginning of land reclamation. Reclamation between 1870 and 1967 has left few wetland areas near the coast. The Comacchio *valli* system was brackish but not a lagoon (there were no tidal flats); it covered an area three times larger than the present Comacchio lagoon (Fig. 12.6).

Most of the reclaimed area that has replaced the Valle di Mezzano, where water depths were 50 cm, lies now at -1.5 to -3.5 m; most roads and towns (e.g., Comacchio, Lagosanto) are barely above sea-level, locally up to 1 m.

Except near the coast, the remnants of former beach ridges are of negligible elevation, due to subsidence. The Strada Romea is mostly <1 m. Inland, the main elevated structures are normal to the coast, the exception being the road Ariano Polesine to Codigoro which follows a former stream channel.

2.6 Reno to Cervia Lowland

South of the River Reno, the main depressions (<1 m and below sea-level) are inland of Porto Corsini and south-east of Ravenna (Fig. 12.6). From Po di Volano to Cervia the 3–6 km wide coastal belt of Romagna is a series of beach ridges, presently covered by pine woods alternating with fresh-water lakes and marshland areas. In three-dimension, it is a body of prograding beach sands, 20–25 m thick, that passes laterally landward to peats and lagoonal clays (Rizzini, 1974).

The Romagna beaches, 80–100 m wide, are composed of fine to very fine sand, finer grained in the north than in the south (where gravel is present). Sand dunes (2–6 m) were well developed and almost continuous until a few decades ago. Now, the only stretches where they are still preserved lie between Lido degli Scacchi and Porto Garibaldi, and between Comacchio and Ravenna (Cencini *et al.*, 1979).

The coast, as well as the beach-dune ridges, is interrupted by the mouths of numerous rivers of Apennine origin (Reno, Lamone, Fiumi Uniti, Bevano, Savio, Rubicone, Marecchia) and by the outlets of several canals of natural or artificial origin: Porto Garibaldi, Bonifica Destra Reno, Logonovo (Lido Estensi), Bellocchio, Candiano (the canal from Porto Corsini to Ravenna), the Tagliata, Cervia and Rimini canals.

Beach-protection works are numerous, especially offshore breakwaters. The Porto Corsini jetties are >2600 m long to depth of 8 m. Other jetties occur at Cervia, Casal Borsetti, Porto Garibaldi and also at the outlets of Milano Marittima and Lagonovo canals. Urban constructions near the shore occupy 65% of the coast from Volano to Rimini.

3 THE SOCIO-ECONOMIC SETTING

3.1 Historical Introduction

The modern development of the north Adriatic coastal lands of Italy dates back to the first decades of this century, mainly after the First World War when the lowlands attracted the attention of the State, for the economic and partly political necessity to expand agricultural production to increase employment and to resettle population. Land reclamation was carried out

Po Delta, Venice Lagoon and climatic change 439

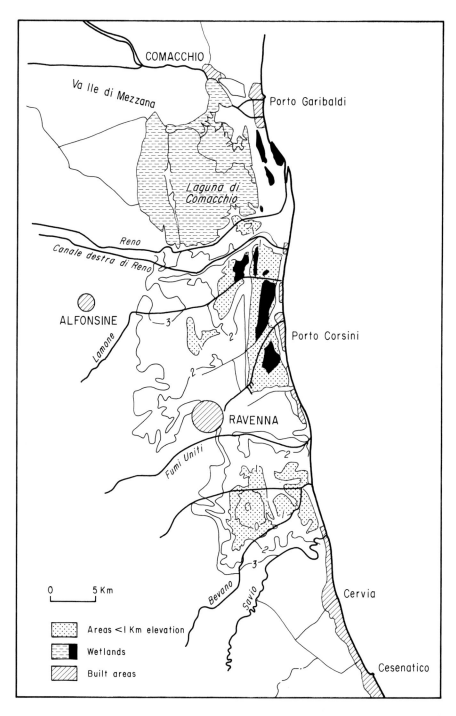

Fig. 12.6 Main topographic features of Romagna region.

in the parts that were still swampy and lagoonal: the Po Delta, near Ravenna, between Comacchio and Ferrara, North and north-east of Venice. As these developments eliminated malaria, population expanded. Cereals were the main crop; later sugar beets and hemp permitted the establishment of sugar and textile mills.

A decision was taken in the early 1920s to create a new harbour and industrial centre near Mestre, to serve as a focus for industrial development at the north-east coast of Italy. The port within the lagoon had the advantage of being sheltered from the sea and only required the excavation of tidal flats.

After the end of the Second World War and the period of reconstruction, a new phase of economic expansion unfolded in the regions of Romagna, Veneto and Friuli–Venezia Giulia (Muscarà, 1979). As in the rest of Italy, the main feature of this development, in the 1950s to early 1960s was a transformation of the economy from agrarian to industrial. In less than 10 years the traditional domination of agriculture was replaced by a productive structure in which industry became the main activity. Economic growth was rapid, spurred by a large number of middle-sized and small enterprises which, taking advantage of labour released from agriculture, were able to compete in international markets, especially the European Economic Community.

In the lower Veneto, this phase of intense transformation especially involved the area from Padova to Venice: Mestre–Marghera saw a radial expansion. The port and industries were enlarged, with new deep access canals to allow the passage of oil tankers, and the Marghera and Venice harbours began to serve as a terminal for trade with central Europe and the north of Italy.

The southern parts of the provinces of Venezia (except for Chioggia, a historical centre with a diversified economy) and the Po Delta, remained, however, relatively depressed. The Po Delta was still a marginal area, outside the main roads and railways, without an urban administrative centre. Population increased until the 1950s, but afterwards declined by 20–25% due to floods, farm mechanization, and the migration of labour towards the industrial centres. The only area of development was from Rosolina to Chioggia, an area favoured by sandy soils for horticulture and by a better commercial infrastructure.

In Romagna the new industrial centre and port of Ravenna was constructed along the Candiano canal, following the discoveries of natural gas fields on- and offshore in the 1950s. Oil refineries and mechanical and chemical industries were established.

A phase of intense urban expansion took place on many stretches of the Adriatic coast, with the rapid growth of new beach resorts, especially between Lignano and Lido, Ravenna and the Po River, and from Cervia to Rimini. This development was a response to the demand of the "economic boom" years for beach recreation by the middle and working classes of Italy and northern Europe. It occurred, unfortunately, with little planning and was accompanied by an extensive alteration of the natural environment (Zunica, 1987; Cencini et al., 1988).

Today, the economy of the coastal belt less 2 m in elevation, is based essentially on tourism and maritime trade, and to a lesser extent on agriculture, which has become more specialized. Fishing is of local importance (Bernardi et al., 1989). Venice is not only a regional and provincial capital; it lives on tourism and is one of the busiest commercial ports in the northern Adriatic. The industrial activities at Marghera and Ravenna, though founded on chemical and metallurgical plants, include a large number of smaller establishments producing consumer goods destined for export.

In addition, lagoonal fishing and aquaculture have expanded and industries have appeared in several of the smaller towns. The communications network includes roads parallel (e.g. the road Rimini–Ravenna–Chioggia–Mestre; the motorway Padova–Mestre–Trieste) and normal to the coast (the motorways Ravenna–Imola, Ferrara–Comacchio, and the roads Rovigo–Po Delta and Venice–Mestre). The international airport of Tessera, at the inner edge of the Venice lagoon, handles 800,000 passengers a year.

More recently, economic growth in the region has slowed down, essentially in consequence of the late 1970s and early 1980s recession and the end of internal migration. Basic industries, especially the government-supported ones, have gone through a state of crisis and of reorganization, while the newer small industries, services, administration and generally the whole sector of tertiary activities have increased considerably.

Inevitably these developments have brought socio-economic problems and negative environmental reactions. While inland a more rational urban and territorial planning is facilitated by a zero population growth and by a reduction of internal migration, in the coastal areas there are conflicts between the expansionist pressures of industries and tourism on the one hand, and environmental protection on the other.

Partly as a consequence of urban and industrial pollution, the quality of the natural environment, paramount to both recreational tourism and to fishing, has deteriorated. The implementation of protective measures for the environment and its natural resources, however, has met with considerable bureaucratic difficulties.

3.2 Population

In 1981 the population of the coastal regions, more or less under the 2 m contour, was about 1 million (Torresani, 1989). High population densities (Fig. 12.7) and towns >10,000 inhabitants (Fig. 12.8) were located along the Portogruaro–Mestre–Padova and Rimini–Ravenna–Argenta–Ferrara axes, i.e. mainly above 2 m elevation. Low densities characterized most coastal areas, e.g. the Po Delta to Comacchio and north-east of Venice regions (the apparent low population density of the Ravenna *comune* was due to the latter's large size; the city has a population of 150,000). In the coastal *comuni*, however, the summer population of the beach resorts usually swells to 10–15 times that of the resident population.

Population trends in the lowland region during the last 40 years are evidenced by comparisons between the 1961, 1971 and 1981 census data (Torresani, 1989) (Fig. 12.9). The main features have been the continuing growth of the economic centres and axes, mainly due to migration, since in this part of Italy (as in France and Spain) the birth rate is low and fertility (1974–1985) has fallen to below the population-replacement threshold.

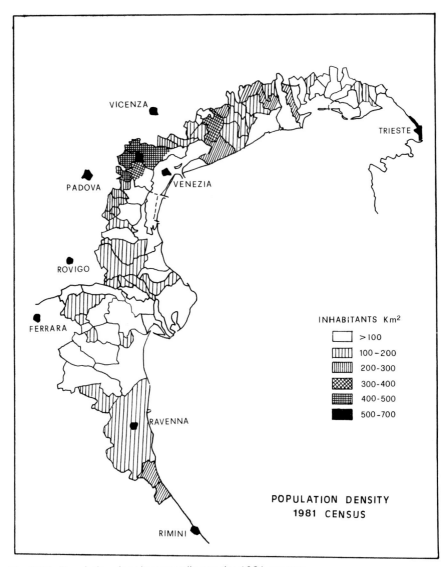

Fig. 12.7 Population density according to the 1981 census.

Negative trends have occurred in rural areas, particularly in the Polesine region east of Rovigo, while population in the Romagna and Veneto–Friuli coastal lands (between San Doná and Monfalcone) has tended to remain stationary or to grow very moderately (with the exception of some sea resorts).

Between 1971 and 1981 growth occurred all around the Venice lagoon, especially in Mestre, but Venice city has and still is experiencing a gradual population exodus (e.g. 79,000 inhabitants in 1989, *vs* 165,000 in 1951), due to lack of employment and to poor infrastructures for some social categories. Blue Plan scenarios (UNEP, 1988) suggest there will probably

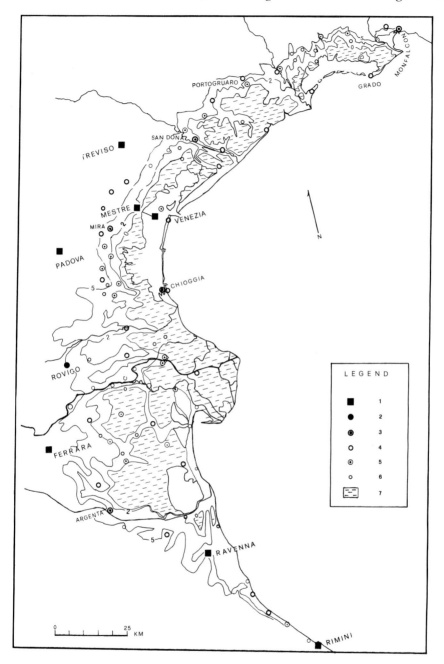

Fig. 12.8 Location of main population centres in relation to elevation. 1. Cities over 100,000 inhabitants, 2. 50–100,000 inhabitants, 3. 25,000–50,000 inhabitants, 4. 10,000–25,000 inhabitants, 5. 5000–10,000 inhabitants, 6. Towns under 5000 inhabitants, 7. Land under sea-level.

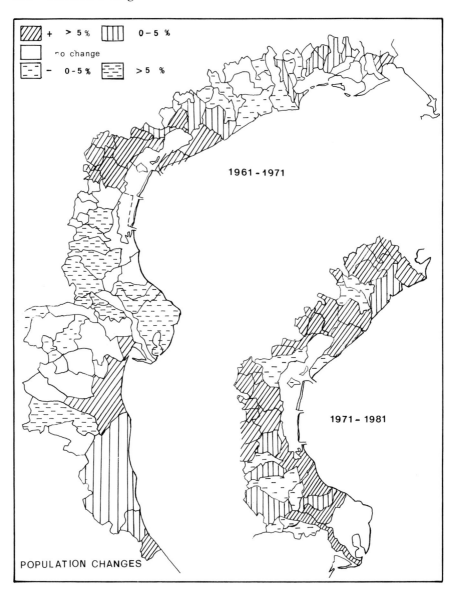

Fig. 12.9 Population changes between the 1961 and 1981 census.

be a 10% overall decrease in population by 2005 due to the continued depopulation of Venice and agricultural areas, and to the low birth rates.

3.3 Agriculture and Fishing

In Veneto agricultural production in the lowland areas is now dominated by corn (> 50% of cropped area), especially in the reclaimed lands, rotating with fodder and soybeans; 30–40% of the cultivated surface is dedicated to wheat and sugar beets, together with corn and soya, in order of importance; locally, there are vineyards (Regione Veneto, 1987). Wheat, with sugar beets and, locally, corn and soyabeans, is the prevailing culture in the Po Delta and in Romagna. Rice is grown in a large area of the Ferrara reclaimed region.

The extent of fodder and "soft" corn cultivation reflects the economic importance of livestock production, while the production of sugar beets has declined in relation to international sugar prices. Overall, agriculture has become very mechanized, market-oriented and dependent on the use of chemicals.

Vegetable gardens (home consumption and export) stretch over the sandy belt from Iesolo to Chioggia, from the Adige to the Po, along the relict beach ridges. Another specialized culture, in Friuli, Romagna and the Polesine, are small stands of poplar, especially along the streams. In Romagna there has been an increase of fruit, grapes and vegetable production in the coastal belt over the sand dunes, partly in response to the development of tourism. In more elevated areas in Veneto as well as in Romagna, between 2–5 m contours, fruit and vineyards are particularly important.

Fishing was traditionally a major economic activity in the northern Adriatic Sea, sustained by a high primary productivity. Fish landings are still the highest in Italy: stimulated by a large market demand, 56.7% of Italy's fish catches between 1971 and 1981 came from the upper Adriatic; 48.4% of that amount was from Emilia-Romagna, 15.8% from Veneto (Orel, 1985). The industry is oriented mainly towards the exploitation of white fish demersal resources. Clam (*Venus*) gathering is important south of the Po Delta in the sandy shallows along the coast. More recently fish production has undergone an overall decline due to environmental, market, organization and financial problems (Bernardi *et al.*, 1989). The future of this activity depends on control of fishing methods (especially trawling), water pollution and catch volumes, which need to be adjusted to biological potential (Jukic, 1985). Organization of fish handling facilities and marketing approaches, and collaboration with Yugoslavia on research, are also deemed necessary.

Lagoonal fishing, on the other hand, has expanded. Open lagoonal fishing involves crustacea (crabs, shrimps), mollusca (*Sepia*, clams), and fishes (eels, mugils, *Dicentrarchus, Sparus, Sgomber, Mullus, Gobius, Platychtys* and *Solea*). In recent decades basin aquaculture has increased in activity and output with respect to open fishing. The basins (*valli*) are lined by levees with special passages for the spring entry of the fry and the autumn capture of the grown fish. In spring and summer the fish remain in the *valli*, either living on natural productivity (extensive aquaculture) or in part artificially fed (intensive aquaculture). Some fish remain in the basins more than one year (eels in particular).

At present professional fishing mainly involves fixed installations in very shallow waters, especially in the upper and central Venice lagoon, and in the Caorle and Comacchio lagoons. There are now, in the Venice lagoon, 25 basins covering 8800 ha. Fish-productivity accounts for 80–150 kg/ha year. The seeding of many basins is done with imported or artificial fry.

Quantitatively and economically important shell fish culture is carried out in the open parts of the Venice lagoon, especially between the

Malamocco and Chioggia outlets. Water pollution is one of the main threats to lagoonal fishing.

3.4 Industries and Harbours

At Marghera and Ravenna the basic industries are metallurgical products, chemicals, petrochemicals, fertilizers and oil refining. At Marghera 200 factories employ 40–50,000 workers. Monfalcone, near the Isonzo delta, is a shipyard town. In the 1960s, 70% of employment was in mechanical industries related to the shipyard. Ship construction has declined significantly in the last two decades; at Monfalcone it survives only for military purposes.

In general, there has been a decline in the large state-supported industrial complexes (Trieste-Monfalcone, Mestre, Ravenna), but a large number of small-scale industries have appeared (e.g. mechanical industries, furniture, clothing) with considerable local specializations. Production has become increasingly oriented towards exports. Activities in the field of construction also have expanded greatly. The location of the main industrial centres in the coastal belt of the north-west Adriatic Sea is shown in Fig. 12.10.

The expansion of North Adriatic harbours before the second World War was related to exchanges between Italy, central Europe, Asia and East Africa through the Suez Canal. Since the early 1950s, major changes in the function of ports have occurred, especially with the growth of oil tanker transport and the establishment of refining and petrochemical plants, and with gradual technical adaptations of harbour installations, forced by greater tanker size and the introduction of roll-on/roll-off and container ships. The ports of the northern Adriatic are lagoonal (Venezia, Marghera, Grado), artificial canal-harbours (Monfalcone, Ravenna, Rimini) and fluvial ports with a canalized access (Torviscosa, Porto Nogaro). Import traffic is predominant (83%), especially for petroleum and related products (60%); exceptions are Chioggia and Torviscosa (a container terminal). Trieste, Venezia and Ravenna are the most important harbours in terms of tonnage handled; if oil traffic is excluded, however, Venezia was (1971–1981) first, handling 50% of northern Adriatic cargo (13.9 million tons/year), followed by Ravenna (27%), Trieste (14.7%), Monfalcone (4.3%) and Chioggia (4%).

External factors have influenced petroleum imports (e.g. economic recession, changes in energy policies, Middle East events); exports are negatively affected by the competition with regard to costs and facilities from Tyrrhenian, Yugoslav and North Sea ports.

Part of the petroleum imports are redistributed by local, coastal shipping. Goods arrive from, or are shipped towards, the interior mainly (95%) by terrestrial transport (inland navigation is limited to cereals in the Venice area); 40% of that by rail, 60% by road.

3.5 Tourism

The demand for beach recreation in the most populated and industrialized part of Italy accomplished an almost total saturation of north-western Adriatic shoreline with its infrastructures, the exception being the Po delta. While figures on the percentage contribution of tourism to the

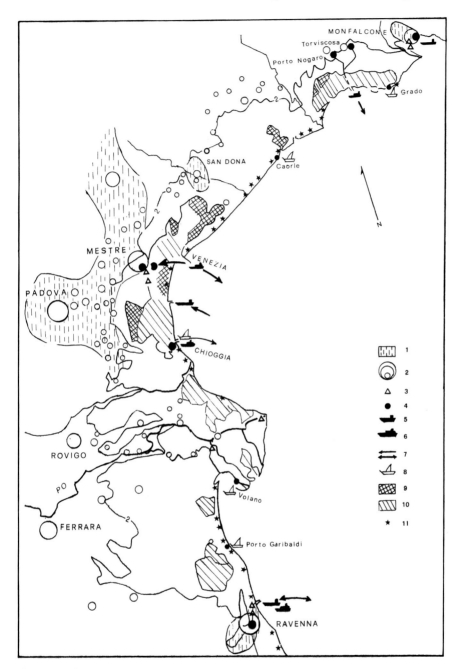

Fig. 12.10 Sketch summary of economic activities in the coastal regions of north-east Italy. 1. Main industrial areas, 2. Industrial centres (size of circles reflects number and importance of plants). 3. Thermoelectric power stations, 4. Ports, 5. Mainly oil imports, 6. Mixed cargo, 7. Maritime trade (arrows stress direction), 8. Fishing centres, 9. Aquaculture, 10. Lagoonal fishing, 11. Beach resorts.

regional GNP are not readily available, some indication of its importance as a commercial activity is provided by the numbers of visitors, as arrivals and presences during three summer months, in proportion to the resident population of the beach resorts. In 1987, for instance, Iesolo registered 0.75 million arrivals, 5.7 million presences, *vs* its 22,000 inhabitants; Caorle had 0.35 million arrivals, 3.6 million presences, and a population of 11,500. The whole coast of Veneto–Friuli Venezia Giulia (Chioggia to Grado) in 1987 had 8.2 million visitors, with 28 million presences, with total population not over 120,000. In 1980 the Romagna coast received 7 million tourists with 36.7 million presences, 70% of which was concentrated in July and August. The main centres of attraction (over 5 million presences) are Rimini, Iesolo and Lignano. The proportions of Italians to non-Italian tourists vary from year to year and along the coast, both in numbers (30–60% foreign arrivals) and length of stay.

On the coast north-east of the Po Delta, visitors tend to stay more in summer houses than in hotels. The reverse is true of the Romagna coast (Muscarà, 1982).

Cultural tourism is one of the main resources of Venice; in 1987 the city had 1.1 million visitors, 70% of which were non-Italians; on the whole they tended to stay briefly (2.8 million presences, Italians accounting for only 28%). A fair number of Venice visitors also stay in nearby Mestre.

Though tourism has become one of the major resources of the north-western Adriatic coastal regions, it is not without pitfalls. As a commercial activity, it is risky, as it is conditioned by fluctuations in consumer demand caused by external economic, social and political factors beyond the control of local tourist organizations. Too little information is available on the causes of such fluctuations, and even reliable statistics on market attitudes and trends are lacking.

As a seasonal activity with heavy demands on use of territory and resources, tourism is threatened by environmental problems (e.g. beach erosion, water pollution) and is itself one of the causes of various problems. The tourism-environment relation is one of conflict. The tourist business has short-term objectives and is generally not interested in the long-term preservation of the environment. In Venice, mass tourism is threatening the gradual destruction of its own objective: the physical stability of the city and its monuments, the continuity of its culture and way of living.

Practical problems created by the seasonal nature of beach tourism involve employment, communications, water and food supplies. There is still a need for studies (e.g. an appraisal of the negative impacts of mass tourism) and for planning the integration of these tourist activities with others that may be compatible with it.

4 The Physical Regime

4.1 Climate

The climate of the northern Adriatic Italian coast is temperate (average annual temperature is less than 23°C). The Alps, pre-Alps and the Adriatic Sea have a considerable influence on climate, the former on rainfall, the latter in moderating temperature (Cantu, 1977).

Relatively mild winters characterize the entire coastal belt, with average temperatures around 2°C. The average summer temperature is 23°C, tending to be higher (>24°C) in the Po Delta and the Romagna hinterland (Fig. 12.11).

Yearly precipitation at the coast is about 600 mm, less in the Po Delta (< 500 mm) and more (700–800 mm) over the Venice lagoon and in Romagna (Fig. 12.12). The rainfall maxima, north-east of the Po River, are in spring (April–June) and in the autumn (October–November), while in Romagna the peaks are in May and in autumn–winter. In the Veneto and Friuli regions the amount of precipitation gradually increases towards the north and north-east to 1000 mm/year (e.g. 1035 mm/year at Portogruaro), and to 2000 mm or more, over the pre-Alps. The precipitation is concentrated in the autumn to early winter months. Snow and rain over the Romagna–Emilia Apennines amount to 1000/1500 mm a year, with maxima in autumn and spring.

Fog is a common winter feature of the coastal lowlands from Ravenna to the Po Delta, with an average of 70–90 days a year in the belt from Venice to the Marano lagoon, and a maximum of over 160 days a year in the lower course of the Piave River.

In winter and spring low-pressure systems characteristically form in the Gulf of Genova, over the Ligurian Sea, and move north-eastwards guided by the alpine highlands. Factors that play a role in the development of these depressions are the thermal contrast between land and sea, which affects the pattern and development of surface pressure; the interaction between the Polar Front and the sub tropical jet streams; the enhancing effect of the Alps on the cyclogentic activity along their southern slopes, under a northerly flow; the effect of terrain features on cyclone formation; and the blocking of cold fronts along the northern rim of the Alps.

In winter, a high-pressure field over central-eastern Europe frequently causes the descent of cold air masses into Mediterranean (or local Adriatic) depressions. These north-easterly winds, very cold and dry (Bora), spill over onto the Adriatic area and northern Italy, funnelled by the alpine passes and valleys. The winds can attain velocities over 100 km/h and last for days.

The winds from south and south-east, which blow humid air up the Adriatic Sea, can reach high velocities, and given the long fetch, can raise substantial waves. Storms occur predominantly in winter and spring, from the north-east, south-west and south-east.

In spring the north-easterly winds become weaker and gradually shift to the south-east. The summer winds are weak, from the east, south-east and south-west. In autumn the wind generally blows from the north-west. There are some variations of the direction of predominant winds in the different parts of the region (e.g. Fig. 12.13).

For the last 100 years Ravenna wind records show no indication of change (Cencini et al., 1979), but at Lido di Venezia predominant wind direction in the last decade has shifted more to the north-east.

Long series of climatic data exist for several localities of north-central Italy (Cantu, 1977). Analyses of temperature and precipitation variations (e.g. Murri et al., 1980; Camuffo, 1984; Papetti and Vittorini, 1988) indicate

450 *Climatic Change in the Mediterranean*

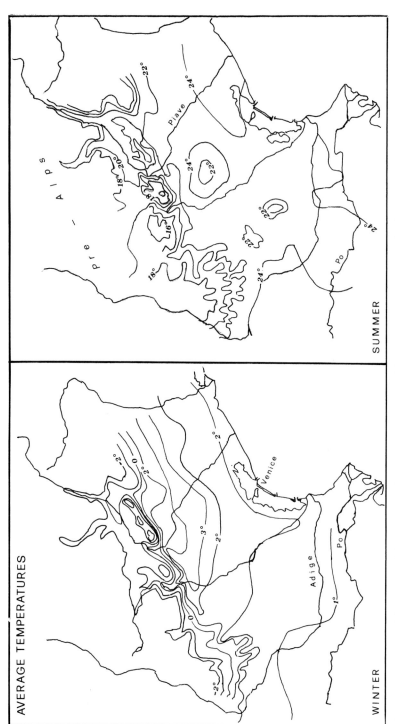

Fig. 12.11 January (a) and July (b) isotherms (1925–1959 averages) (modified from Almagia (ed.) Veneto, UTET, 1962).

Po Delta, Venice Lagoon and climatic change 451

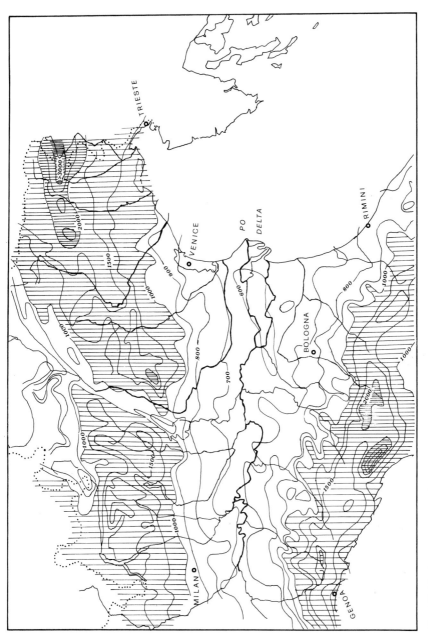

Fig. 12.12 Average annual precipitations 1921–1960 (modified from Ministero dei Lavori Pubblici, Servizio Idrografico, 1975).

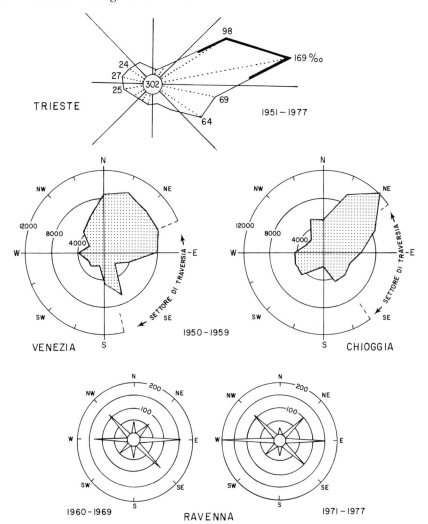

Fig. 12.13 Patterns of wind direction at main localities of north-west Adriatic coast (from Brambati, 1987; Malanotte-Rizzoli, 1981; Cencini et al., 1979).

a general warming trend (especially during the last 15 years) and considerable oscillations in rainfall, without a clear trend, but with important differences between localities (Pinna, 1991).

4.2 Marine Parameters

The northern Adriatic Sea constitutes a rather individual unit of the Mediterranean because of its shallow depths, marked temperature and sea-level variations, and the low salinities due to the fresh-water discharge of many rivers (one-third of all fresh water presently supplied to the Mediterranean). The central and southern basins of the Adriatic are deeper, partially separated by sills, and in general display many of the oceanographic characteristics typical of deep seas.

4.2.1 Salinity and currents

Currents in the Adriatic are mainly gradient currents due to density contrasts, as well as tides and winds (Mosetti, 1985). Water from the Mediterranean through the Otranto channel, with a constant salinity of 38%, becomes progressively diluted to 32% (in summer) near the Po Delta.

Surface waters flow to the south-east, near the west side of the basin (Fig. 12.14). This flow is balanced by an inflow of (intermediate) waters up the eastern side of the Adriatic. Thus, at the surface a general anti-clockwise circulation is developed. The flow of these gradient currents, however, is seasonal, faster in summer (velocities in the order of 25 cm/s, increased to 50 cm/s by tidal movements), slower to almost nil in winter.

In winter, the water density is high, which prevents the spread of the cold, diluted river waters. In early spring the greater stability of the surface layer favours the progressive spread of warmer waters over a large part of the northern basin. In summer this is separated from the underlying water by a pycnocline (Fig. 12.15).

The distribution of water density is strongly affected by surface winds. The water column is overturned and deep water upwells in winter during outbreaks of the cold and dry Bora wind, which causes increased surface evaporation (Zore-Amanda and Gačič, 1987).

Nearshore circulation in the northern Adriatic is locally complex, as it is conditioned by longshore currents, river discharge plumes, tidal currents, and the impacts of fixed structures, especially long jetties at river and lagoonal outlets (Fig. 12.16). The fresh-water plume of the Isonzo flows essentially south-east, secondarily to north-east and south-west, those of the Tagliamento and Piave rivers to the south-west. The current off the Adige moves to the north-east (with a velocity of as much as 80 cm/s), but turns south after a few kilometres.

The high spring and autumn discharge of the Po River heads mainly to the east, with velocities of 18 to 100 cm/s; after losing momentum, the main current turns south. The Po discharge has a strong influence on the oceanographic conditions in front of the Romagna coast by isolating a wide belt of nearshore waters (Fig. 12.15). Summer stagnation is common, broken only by strong north-east winds which cause mixing, or by south-west winds (Libeccio) which may induce upwelling (Accerboni et al., 1982).

4.2.2 Waves

The wave regime on the Italian side of the northern Adriatic is related to wind direction. The most important storms are from the south and south-east (Sirocco), causing high wave-energy concentration on the Venice to Tagliamento coast (ref. the flooding of the November 1966 storm surges; Malanotte-Rizzoli, 1981; Malanolte-Rizzoli and Bergamasco, 1983). One year of measurements 15 km off Venice at 16 m depth indicated half the waves were caused by Sirocco winds (Cavaleri et al., 1981). In Romagna the frequency of south-east waves is double that of north-east waves, but the latter can create wave heights of 3–6 m (Cencini et al., 1979).

Because of gentle gradients, surf conditions in the north Adriatic develop far offshore. During major storms, waves affect the bottom

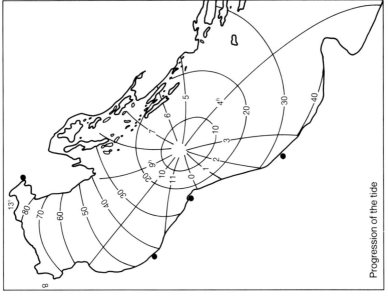
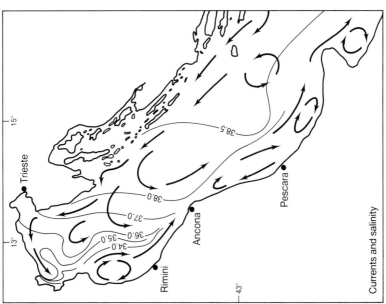

Fig. 12.14 Summary of surface currents, salinity and tides in the northern Adriatic Sea (based on Nelson, 1970; Mosetti, 1985).

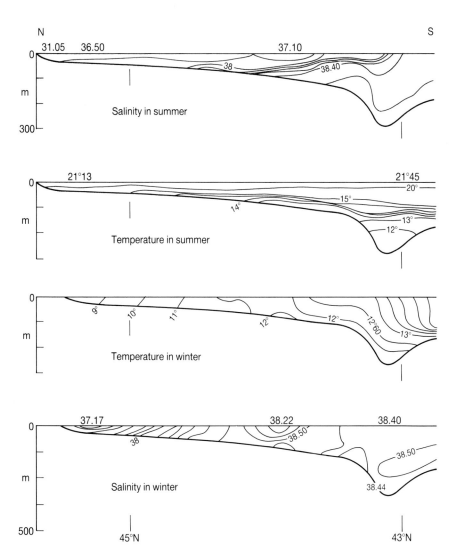

Fig. 12.15 Vertical salinity and temperature distribution in the northern Adriatic Sea (redrawn from: P. Guibout, *Atlas Hydrologique de la Mediterranee*, IFREMER–SHOM, 1984).

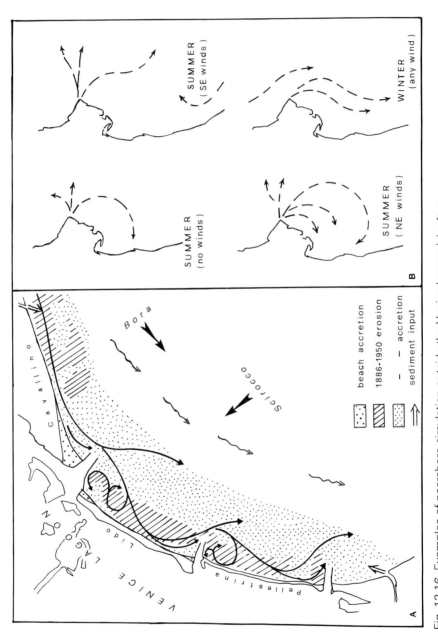

Fig. 12.16 Examples of nearshore circulation outside the Venice lagoon (a), after Gatto, 1984) and off the Po Delta (b), after Marchetti et al., 1982).

3.5–4.5 km off the shore in the north, 4–7 km in north Romagna to a depth of 10 m.

In addition to several wave refraction types due to the interaction of different wave direction and local coastal morphology, the narrow elongated shape of the Adriatic can induce wave reflections from the basin sides, creating confused sea conditions, not always directly related to existing meteorological conditions (Mosetti, 1985).

There seems to have been some cyclicity in the frequency of storm waves from either the Bora or Sirocco directions. For instance, at Lido (Venice) the great storm surges of the 1960s came from south-east, but this direction was less important in the 1970s (Gatto, 1984). A similar situation has been noted at Marina di Ravenna (Cencini et al., 1979). It also has been stated that there has been an increase in storminess, with greater incidence of storm surges in the last two decades (Brambati, 1984).

4.2.3 Sea-level variations

Tides and seiches are the predominant features of sea-level fluctuations in the Adriatic basin (Mosetti, 1985). Because of its enclosed shape, the semi-diurnal tidal amplitudes in the Adriatic increase in a complex manner, from south-east to north-west, with a positive, counterclockwise movement, the amphidromic point located off Ancona (Fig. 12.14). The spring tide elevations are 85 cm at Trieste, 88 cm at Grado, 72 cm at Lido, 60 cm at the Po Delta.

Seiches are always present in the Adriatic Sea, but of variable impact. Those superimposed on rising sea-level (due to wind action) have greater amplitude and are associated with storm surges. In the Adriatic Sea seiches behave in close agreement with the tides and can contribute to severe coastal damage.

In the northern lagoons the elevation and spread of tidal waters is strongly influenced by the depth of the access channels, by the complex network of canals that digitate from them, by depth distribution and especially by the extent of tidal flats. In the Marano–Grado lagoon the flood tidal wedge extends upriver a few kilometres; in the Po Delta it extends several kilometres, especially in summer.

A considerable rise of sea-level due to the combination of tide and other effects occurs generally between October and April, with a prevalence (60%) in November and December, at times of atmospheric depression in the northern Adriatic. Seiches, in association with Bora or Sirocco winds, elevate the high-tide levels to over 1.80 m.

Recently, water surges of 130 cm (even 200 cm) in the Marano lagoon and of 195 cm in the Venice lagoon have been registered. Venice has been increasingly affected by this tidal flooding (*acque alte*) (Fig. 12.17). The incidence of high water had reached 50 times/year in 1985, compared to under 10 times/year in 1925. Since 1972 the average tidal level has increased at least 40 cm (Pirazzoli, 1983). Part of this value (14 cm) has been attributed to an increase in the tidal oscillation, but 27 cm have been caused by a relative local rise of sea-level (see page 465 and Fig. 12.21). In the last 30 years the tidal range inside the Venice lagoon has increased by 10 cm relative to the outside. This is related to a reduction of the surface area of the lagoon and to the deepening of its outlets. In

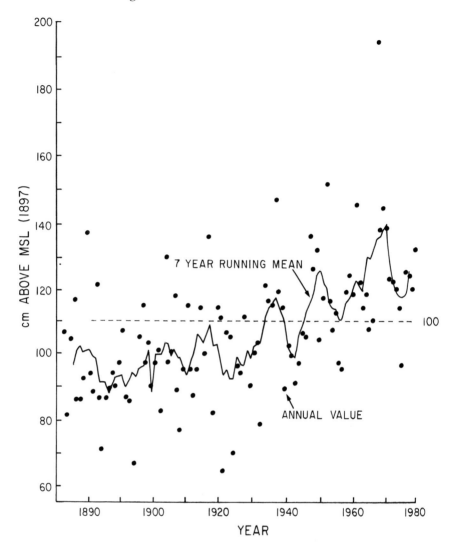

Fig. 12.17 Highest annual tidal in Venice (1872–1981). Data are related to the 1887 msl (after Pirazzoli, 1983).

addition to the general trend of tidal increase there are 20–30 cm high decennial oscillations attributed to the 16.6-year cycle of lunar declination (Pirazzoli, 1983).

4.2.4 Water quality
In the Adriatic, the extensive mixing in the autumn and winter permits good water quality, at least comparable to that in the Mediterranean Sea. Under particular local conditions and in areas close to the shoreline, however, eutrophic conditions arise. The nutrient-rich polluted waters from the Po and the other rivers favour the development of algal blooms,

especially diatoms and dinoflagellates (Zanoni, 1977). A wide area close to the coast of Veneto, Romagna and the Marche is considered as eutrophicated up to 60 km from the shore due to the high concentration of chlorophyll and nutrients.

Red tides have been well known in the Adriatic Sea since the last century but appear to have become more common in the Emilia-Romagna coastal waters since 1975. The general trend is for blooms of diatoms to develop at the end of winter in both coastal and open waters, while dinoflagellate blooms reach maximum intensities between August and October. By the end of summer the red algal blooms can jeopardize the tourist activity at the Romagna resorts, giving rise to anoxic conditions with stinking waters and mass fish kills (Marchetti, 1985).

The effective management of water quality in the upper Adriatic is still an unresolved problem. At present, the contamination of inland waters is increasing and consequently algal blooms have become more frequent and widespread.

In general, conditions of marine water pollution north-east of Venice are considered to be fair (Gasparini *et al.*, 1985). Nevertheless, the stability of the diversified and hitherto stable benthic community has been damaged and disrupted in the Bay of Trieste. South of Venice scattered areas of decaying organisms were reported in 1974, 1977 and 1983; an area of possibly 250 km² is becoming azoic in the North Adriatic (Ghirardelli *et al.*, 1974).

5 GEOLOGY

5.1 The Formation of the Adriatic Lowlands
At the beginning of the Pleistocene the sea formed a wide marine gulf extending far into the Po valley. Active tectonic subsidence contributed to the accumulation of 1500–2000 m of fluvial marine sands, clays and silts. Subsidence was influenced by the continuing deformation of the folds of the outer Padan Apennines, with the axis of major sediment accumulation shifting to the north-east.

The oscillations of sea-level related to glacial periods resulted in a series of marine transgressions and regressions. At the peak of the Würm glacial (20,000 years ago) the Adriatic Sea stood about 100 m below its present level, with a shoreline situated between Ancona and Zara. The northern Adriatic was a large alluvial plain crossed by a palaeo-Po river system into which most of the other rivers flowed. Climate-warming and the ensuing sea-level rise gradually shifted the shoreline northward, leaving a record of its progression in at least 17 submarine platforms, slope breaks and relict beach lines (Colantoni *et al.*, 1979).

During the maximum of the Holocene transgression (6–7000 BP) the sea reached a line 5–20 km inland of the present coast (Fig. 12.18). With a decrease of the rate of sea-level rise and its stabilization (from an average of 15 mm/yr to 4 mm/yr), a phase of active regression began. A prism of Holocene terrestrial, lagoonal and marginal marine sediments accumulated parallel to the shore.

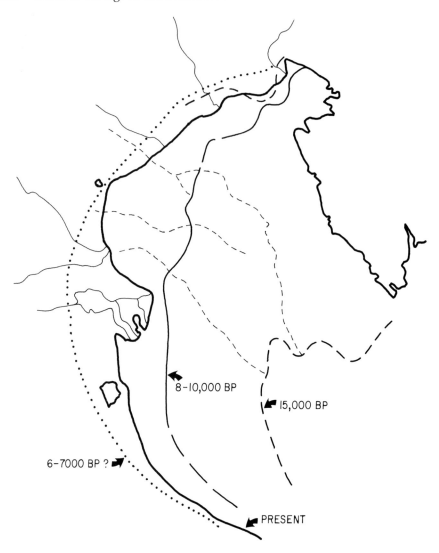

Fig. 12.18 Suggested location of late Pleistocene to Holocene shorelines (source: Favero, 1979; Ciabatti, 1979; Brambati, 1985; Brambati, 1987; Castiglioni and Favero, 1987).

Subsequent coastal changes can be traced in detail from morphological and historical evidence (Veggiani, 1974; Ciabatti, 1979; Fabbri, 1985). In Etruscan and Roman times (3–2000 years BP) several branches of the Po, Adige and Brenta rivers flowed towards the Adriatic. In the lower Po valley, from Chioggia to Ravenna, relict beach-dune ridges indicate the position of former shorelines and of no less than 10 cuspate deltas of Etruscan-Roman age (Fig. 12.19). These deltas were characterized by wide sandy beach zones and had a low rate of advance (450 m/century).

The early phase of delta building continued until the 12th century AD. The development of the modern Po Delta began after a major flood break

Po Delta, Venice Lagoon and climatic change

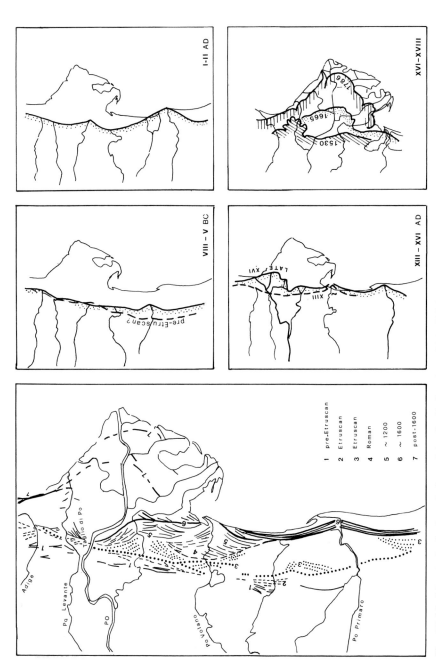

Fig. 12.19 Morphological evolution of the Po Delta since 2000 BC (right – based on Ciabatti, 1979; left – based on Rossetti and Rossetti, 1977).

and channel switch in about 1155, west of Ferrara (the so-called Rotta di Ficarolo). Until that time the Po discharged south of the present delta through the Primaro (now the River Reno) and the Volano branches. The Ficarolo break diverted the main flow to east. In the 14–17th centuries there was a gradual build-up of discharge and bedload related to greater upland rainfall (the "Little Ice Age"), deforestation and increasing flood control (Zunica, 1978).

Another artificial deviation of the main Po channel, the Porto Viro cutoff in 1604, directed the river to south-east, while a second one in 1770 from the northerly Po di Maestra branch gave more importance to the southern Gnocca, Goro and Tolle branches. Finally, further regulation of the channel in 1870 emphasized the role of the present Pila branch as the main outlet of the Po. The post-17th-century deltas were lobate in shape and characterized by rates of advance of 7 km/century.

Thus, the Po Delta, as seen today, is a recent geomorphic development, the product of constricted river advance during the last four centuries. It could be considered to be almost artificial, due to the influence of growing anthropic activities in the basin (deforestation, expansion of agriculture, stabilizing of river beds) and to an increasing control of river discharge.

North-east of Venice major coastal changes were the retreat and eastward deviation of the Natisone–Isonzo Delta (Brambati, 1987) and large-scale artificial deviations of river courses (especially of the Sile and Piave).

Regarding the origin of the lagoons, there is still no conclusive evidence that they were formed because of intermittent subsidence, associated with sea-level rise, or formed by the development of spits and barrier islands related to the advance of the various deltas of Veneto in combination with the westward littoral drift (Zunica, 1971, 1976; Favero, 1979,1985).

5.2 SUBSIDENCE

Subsidence at the edge of deltaic coastal plains is a common feature due to the tectonic sinking of the depositional basin and to the compaction of clays and peats. Pleistocene sediment thickness in the Po Delta, for example, exceeds 2000 m. In Venice geological subsidence has been estimated to be 1.3 mm/yr, but during the Pleistocene it reached 2.6 to 4.5 mm/year at times (Fontes and Bertolami, 1973). Greater rates of subsidence in the Po Delta (5 mm/yr between 1900 and 1950) and in the Comacchio–Goro region (3 mm/yr) was compensated by sedimentation. In the Ravenna region natural subsidence until the early 1950s was 2.7 mm/yr (Bondesan et al., 1986).

Additional sinking, up to 1.3 m, has been recorded in most reclaimed lands in the years immediately following their drainage. This was attributed to a variety of causes, such as compaction after de-watering, the lowering of the water table, chemical and volume reduction in clays, etc. (Montori, 1983; Bondesan, 1988; Schrefer et al., 1977).

Large areas near Venice in the Po Delta and in the Ravenna region have experienced accelerated subsidence since the early 1950s due to excessive extraction of ground water for industrial and urban use. In the Po delta, the extraction of methane-bearing water caused piezometric declines of 40 m

Po Delta, Venice Lagoon and climatic change 463

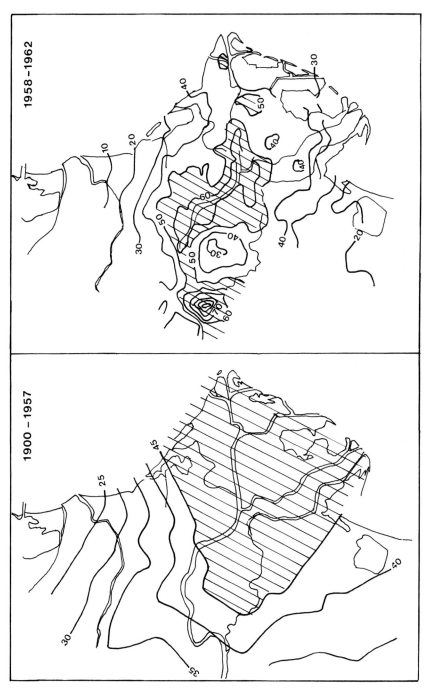

Fig. 12.20 Subsidence in the Po Delta 1900–1962 (Caputo et al., 1970).

in aquifers 100–600 m deep in the late 1950s. Locally subsidence reached 30 cm/year (Fig. 12.20). Total subsidence, between 1958 and 1967 was 3.5 m (Caputo et al., 1970). Considerable sinking occurred also along the margins of the delta, to the north (Chioggia–Adige, 85 cm) and the south (Codigoro, 70 cm).

In Ravenna, land subsidence amounting to 1.2 m between 1949 and 1977 affected an increasingly large area (700 km^2), inclusive not only of the city and its industrial estate but also of the coast and reclaimed marshlands (Carbognin et al., 1981; 1984). In Venice, water overpumping from aquifers 70–350 m deep lowered piezometric levels by 12 m between 1952 and 1969, producing a subsidence of 8 cm at Mestre, 14 cm at Porto Marghera and 10 cm in Venice (Carbognin et al., 1981; Gatto and Carbognin, 1981).

As the relation between aquifers/ground-water surface behaviour and subsidence became known, measures were introduced to reduce or stop the excessive pumping. In Venice, the recovery of the flow field occurred quickly, and subsidence was arrested almost instantaneously. There was even a surface rebound of 2 cm. Presently land subsidence has returned to its natural rates. In the Po Delta and Ravenna subsidence in 1982 was 2 cm/yr; an attenuation was noted in 1986 and it now has returned to the pre–1950 values.

Accelerated artificial subsidence is no longer a problem. It has, however, led to an irreversible lowering of land levels, and produced considerable negative impacts. In the Po Delta the consequences were the reversal of hydraulic gradients in rivers and irrigation canals, with the need of adjustments to beds and embankments, the increased marine flooding of bays, and coastal erosion. In Venice, artificially created sinking has added 12–14 cm to the effects of natural subsidence and to sea-level rise; the relative rise of sea-level being in total 22 cm (Fig. 12.21). As the city is only 0.8 m above mean sea level high tides above 1.3 m can flood 60% of it, those above 1.8 m, 100% (Gatto and Carbognin, 1981).

5.3 Soils

The soils of the Veneto, Friuli and Romagna coastal lowland are essentially alluvial, or alluvial-hydromorphic, with saline tendencies near the lagoons. They may be sandy or argillaceous with (A)C or A–C profiles at least 1 m thick. In the reclaimed areas soils were initially acid (peats) and organic rich, but with low permeabilities.

In the region north-east of Venice (Fig. 12.22) the following pedologic units have been mapped:
1) the sandy soils of the beach zone, with high porosity and a normal organic content;
2) the soils of the former lagoons and swamps, which have a medium-low permeability and can be either sandy (of limited extent, corresponding to former lagoonal channels and outlets), clayey-silty (with high to normal organic matter content), or humic-peaty and rich in organic matter;
3) the alluvial soils, sandy-silty with medium permeability (organic matter: normal to good) that correspond to the distributary channels of the

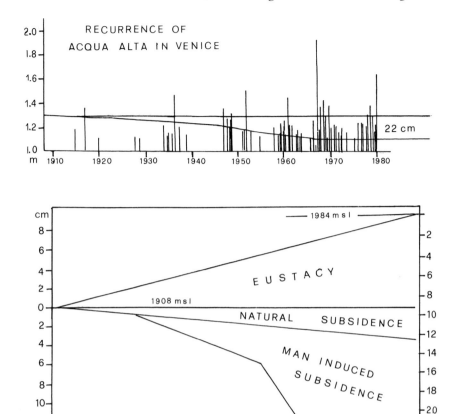

Fig. 12.21 The effects of subsidence in Venice. Top: The increase in time of the occurrence of *acqua alta* in relation to relative rise of sea-level. Bottom: land sinking as related to subsidence and eustatism (from Gatto and Carbognin, 1981).

Piave, Livenza and Tagliamento, as well as the sandy-gravelly soils of older channels of the Tagliamento;
4) above the 3–4 m contour, the soils of the plain are mainly argillaceous, with a low permeability, and are related to modern and past interchannel floods of the main rivers.

In the reclaimed areas of the Polesine and Ferrara–Comacchio region many parts that had peat and clay acid soils were improved by drainage with the Po calcareous waters and by rice paddy cultivation in the first years after reclamation. The soils of the Po Delta and of the Romagna lower plain are essentially similar to those north-east of Venice.

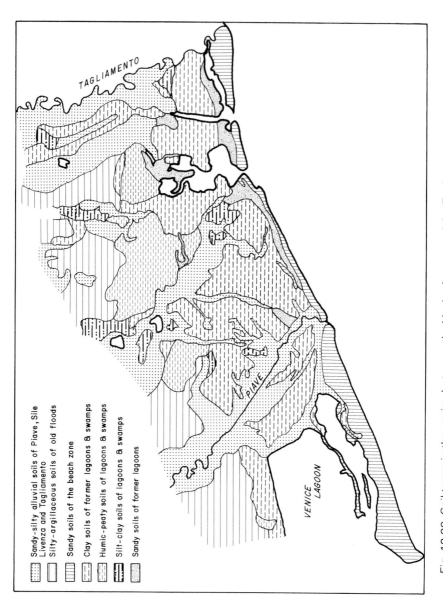

Fig. 12.22 Soil types in the region between the Venice lagoon and the Tagliamento River (based on Provincia di Venezia, 1983).

5.4 Coastal Processes and Stability

There has been a considerable number of detailed studies and reviews about the north-west Adriatic coast, starting with Zunica (1971, 1974). Later publications have dealt with the stretches north-east of Venice (Brambati et al., 1978; Brambati, 1984, 1987), near Venice (Gatto, 1984; Zunica, 1991), the Po Delta (Bondesan and Simeoni, 1983) and the Romagna shoreline (Bondesan et al., 1978; Cencini et al., 1979; IDROSER, 1982).

At present all the Veneto and Romagna beaches are unstable, with retreat along most of the coast since the early 1950s, even at the mouths of the Tagliamento, Adige-Brenta, Po and Reno (Fig. 12.23). At the Adige mouth a previous continuous growth of 6.5 m/yr stopped in 1950. Retreat has been significant especially since the stormy years of 1960–1970.

In Romagna, beach advance was common until about 1935, especially before 1915, with the exception of the stretch Reno mouth to Porto Corsini (Bondesan et al., 1978). Since then there has been a general retreat, especially of the projecting deltas, a straightening of the shore and a steepening of the foreshore. The retreat of the Fiumi Uniti beach has been 3 m/yr. Further north, at Lido Adriano, recession from 1957 to 1977 amounted to 126 m/yr with destruction of pine woods, while at Punta Marina retreat was 70 m/yr. At both places, coastal erosion has coincided with land subsidence, 45 cm in the first case, 35 cm in the latter (Carbognin et al., 1982, 1984).

At the edge of the Po Delta most of the outer beaches and sand bars have been unstable for the last 30 years, either advancing or retreating. In the early 1980s (CNR 1985) the northern coast of the delta was mostly in a state of retreat; in the south advance was registered only downdrift of the (receding) mouths. Moreover, most of the marginal parts, marshes, tidal flats and sand bars, in 1960, had returned to a lagoonal state. Further north, the Pellestrina and Lido shores are affected by erosion, because they are cut off from the littoral sand movement by the Lido and Malamocco outlets jetties (see Fig. 12.16).

In the lagoon of Venice the central basin has been affected by intense bottom erosion (Cavazzoni, 1983). The islands have become unsettled and the remaining tidal flats are eroded. Occurrences of very high, but also low, waters have become more frequent.

The causes of shore instability and retreat are partly natural, partly the consequence of anthropic activities on the coast and in the hinterland. Natural erosion is the consequence of the removal of beach sand by strong littoral drift generated by the oblique south-east or north-east wave approach, of high ($>$ 1.5 m) winter waves and of storm surges associated with high tides, and of eolian beach deflation in winter. This removal of sand has not been balanced by sufficient inputs from rivers, or from offshore sources. A further cause of shore erosion has been the negative effects of fixed transversal structures, such as piers, jetties and groins, that have unbalanced the longshore movement of sand (Zunica, 1991).

One of the principal causes of the observed shoreline retreat is the reduction of river sediment supply, principally due to less slope erosion (re-forestation, abandonment of farming and decreased ploughing in hilly

468 *Climatic Change in the Mediterranean*

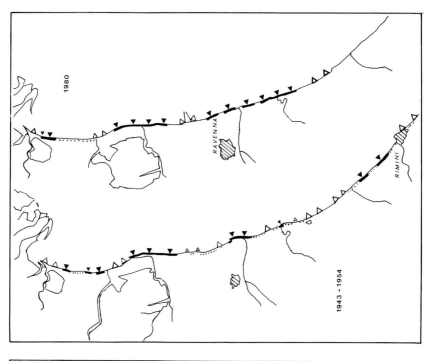

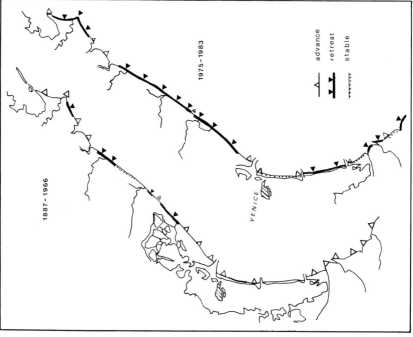

Fig. 12.23 Summary of the state of advance/retreat in the Veneto and Romagna coasts (sources: CNR, 1985; Cencini et al., 1979; Zunica, 1971).

areas), to the retention of sediment in hydroelectric reservoirs, and recently also to industrial dredging of sand from river beds, especially in the plains. In the period 1965–1973 the Po solid load decreased from 16.9 million to 10.5 million t/yr. The bed load, a quarter of these figures, was barely sufficient to satisfy the calculated sand budget of 3.1 million t/yr necessary to keep the delta margin in equilibrium (Bondesan and Dal Cin, 1975; Dal Cin, 1983). Similarly, the solid discharge of the Isonzo, Tagliamento and Romagna rivers has almost halved in the last decades compared to the years before 1950.

The dredging of sand from the Po River bed, officially given as 100 million tons for the period 1958–1981, is estimated to have been up to six times greater. In Romagna the official 8 million tons for the period 1957–1971 is also short of reality (Cencini et al., 1979). The sand still delivered to the sea by the Po comes, in part, from the erosion of the river bed itself, which has been lowered already by 5 ms. Sand discharge will decrease even more when engineering measures are undertaken to prevent the bed from reaching below sea-level. A sand deficit lasting at best 30 years, perhaps 100 years, is forecasted, even if all dredging operations were to stop (Dal Cin, 1983).

To alleviate the direct impact of storm surges, and to maintain beaches, as well as to regulate the access to lagoons and harbours, seawalls, offshore breakwaters, groins and jetties have been built extensively since the 1950s. Generally they represented local solutions of immediate necessity, with no reference to each other within a regional plan. The effect of these structures has been beneficial or neutral in some places, but quite negative in others. They do trap sediment "upstream", but generally sediment starvation and erosion take place in the "lee" (downdrift side). This is a common feature over the long stretches with groins between Iesolo and Lignano. In the Cavallino shoreline, beach retreat is accompanied by wave attack on the dunes, with a general weakening of the shore in the face of storm surges. Downdrift erosion occurs also west of natural offshore build-ups, sandbars that have grown as the indirect result of groins built on the east side of river mouths (e.g. Piave, Tagliamento) to prevent their shoaling (Brambati et al., 1978).

The construction of the jetties at the three Venice lagoon outlets has changed the character of sediment movement, with a strong accretion at the northern and southern ends. The jetties, however, deviate the littoral currents offshore, beyond the breaker zone, and there is no sediment bypass. Similar phenomena occur near all the piers that armour the mouths of rivers and canals in Romagna. The offshore breakwaters on the Romagna coast are only initially successful. They break the waves before reaching the beach, but water tends to stagnate behind and sand accumulates. Gradually, the space behind the breakwaters is filled, the beach is eliminated, and eventually the erosion process is resumed.

It must be concluded therefore, that human "management" of the coast and activities in the hinterland has generally contributed to the current loss of beaches, and at places to their degradation. The fixed structures erected during the last three decades and the deep canals excavated in the lagoons have considerably altered the hydrodynamic regimes of the low north Adriatic coast and its lagoons. The lagoon of Venice, in particular, is suffering from the increasing influence of the sea, e.g. higher tides and faster currents, with greater bottom erosion. The Po Delta is undoubtedly in a precarious state: it is now too prominent, sediment-starved, unable to balance subsidence, and resisting wave action only by the protection of dikes along its entire perimeter.

6 Water Resources

Water in the lower plains of north-eastern Italy is provided by rainfall, surface flow (rivers, canals, lagoons) and underground aquifers. In the past, rainfall over the Apennines and the pre-Alps was sufficient to maintain a flow of surface water through rivers, and to recharge aquifers in Pleistocene strata. Considerable modifications have occurred in the natural flows, in consequence of flood control operations, the retention of water in reservoirs for electric power generation, and in the last decades, to a vastly increased water extraction for irrigation, industry and domestic use (the respective rates of consumption at present are about 56%, 28% and 16%; IDROSER, 1977). The management of rivers, however, has been necessary for centuries because of frequent flooding as well as to meet the needs of agriculture and inland navigation. As river regimes have been altered, dependence on ground water has grown considerably. Pollution problems have arisen during the summer, when much of the surface flow is derived from waste waters returned to the rivers.

Water resources are generally adequate in the Veneto and Friuli regions, because of high precipitation over the eastern Alps and pre-Alps. In Romagna there is a scarcity of water, as rainfall is ill-distributed throughout the year, there are no natural reservoirs in the Apennines, and the streams have torrential regimes with very little flow in summer. What rainwater is not evaporated runs rapidly through the river courses or percolates into the subsurface aquifers. In the Emilia-Romagna region, it has been calculated that out of a total of 9 billion m^3 of precipitation, 49% is surface flow, 25% is lost through evapotranspiration, 25% percolates into the ground (10% reaching the deep aquifers) and 4% is retained by surface ponding, plants and soils (IDROSER, 1977).

6.1 River Regimes
In Veneto and Friuli, surface waters are derived, in addition to local rainfall and runoff, from rivers with different types of regimes:
1) Alpine rivers, like the Isonzo, Tagliamento and Adige, have discharge peaks conditioned by the late spring snow melting and rainfall, and by the autumn rainfall (Fig. 12.24).
2) The rivers from the pre-Alps, like the Piave, have regimes entirely rainfall-dependent, with principal discharge peak in October–November, a secondary one in April–May, and lows in winter and summer. The waters of the Piave are extensively held back by many hydroelectrics reservoirs (in the 1960s, 42% of the drainage system of Veneto was affected by reservoirs).
3) The rivers that originate in the 'risorgive' spring belt, at the edge of the plain (Livenza, Sile, Bacchiglione, Dese, Marzanego, Tartaro, etc.) flow more regularly through the year. Systematic discharge data for

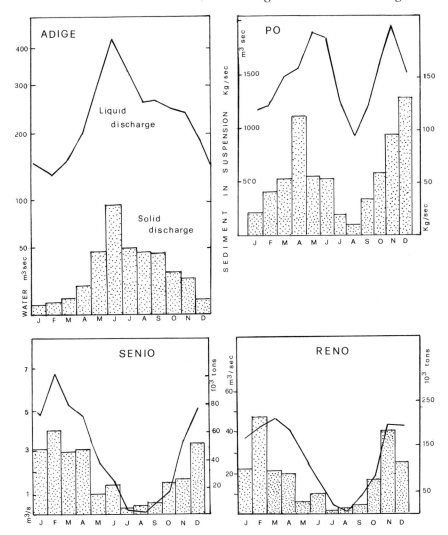

Fig. 12.24 Examples of river discharge regimes (Adige, from Zunica, 1971; Po, from data in Cati, 1981; Reno and Senio, from CNR, 1985).

these, and other rivers, are lacking; furthermore all measurements stations on the major rivers are located no less than 50 km from the sea.

In the coastal lowlands, water flow is maintained by mechanical uplift, and near the coast, the stream regimes are influenced by the tidal oscillation; the inland extension of salt wedges (several kilometres) prevents the use of river water for irrigation and domestic consumption. Parallel with the coast, between the Tagliamento and Po di Levante, there is a network of navigation canals that are also used as agricultural drains.

The middle and lower plain of Veneto-Friuli was considerably affected by the flooding caused by the exceptional rainfall event of 4–5 November

1966 (Fig. 12.25). Flooding appears to have been only partly controlled by topography; an important factor was also the obstruction of water flow by elevated embankments and, near the coast, by an exceptionally high tide (above 1.5 m for 11 hours) (Croce et al., 1971).

The Po's complex regime reflects the partial compensation of the contributions of many effluents, from areas of different types of precipitation. The river basin (70,000 km^2) includes both Alpine (winter lows and May–July maxima) and Apennine (March–April maxima, summer minima) regimes.

As a result, the discharge regime of the Po displays a smaller difference between maxima and minima than other rivers (Cati, 1981). In the lowest reaches the river once had discharge peaks in May–June and November and lows in January–February and August. However, summer discharge has been considerably reduced by retention of water in the mountain reservoirs; the amount "missing" was calculated to be as much as 130 to 210 m^3/s (Bevilacqua and Mattana, 1976). The consequences of this reduction of flow have created navigation problems in the lower course of the river; pollution from agricultural and industrial waste waters; upstream migration of the salt wedge, with impacts on the intakes of fresh water for irrigation and for the regulation of aquaculture basins in the Po Delta.

Because the Po, like the other rivers of the Adriatic coastal plain, has an elevated bed, flooding has been a common occurrence; a break at Pontelagoscuro in 1951 affected two to three of the Polesine region. Despite the stream control works, floods may be increasing in frequency (e.g. 22 flood breaks between 1951 and 1966, as compared to 10–22 floods per century from the 16th to 19th centuries). This could derive from a combination of upstream mismanagements together with a greater incidence of erratic climatic events.

The Reno and Senio and other Romagna rivers have torrential regimes, and are almost dry in summer (Fig. 12.24). The extent of water extraction (156 million m^3/yr for industrial and urban use alone in 1975) is such that in the Romagna plain the streams are now straightened and canalized, and instead of drainage flow collectors have become watersheds mostly used as waste disposal ducts, with deleterious consequences for water quality (Table 12.1). Because of the torrential regimes and the elevated river beds, flooding has been a frequent event, not only in past centuries, but also in recent years. Several river tracts are under active erosion, with levees in danger of undercutting, especially through the parts of the plain affected by land subsidence (Cencini et al., 1979).

6.2 Ground water

Figure 12.26 shows schematically the distribution of aquifers under the Venetian plain. In the foothill region there is a coarse conglomeratic blanket derived from the amalgamation of fans constructed by the Tagliamento, Piave, Brenta, Astico and Adige rivers. This Undifferentiated Freatic Aquifer (up to 400 m thick, average permeability of 10^{-3} m/s) is intensely exploited. The water table is situated at considerable depth near the pre-Alps foothills, but gradually it comes to the surface in the south and outcrops at the Risorgive Line of springs.

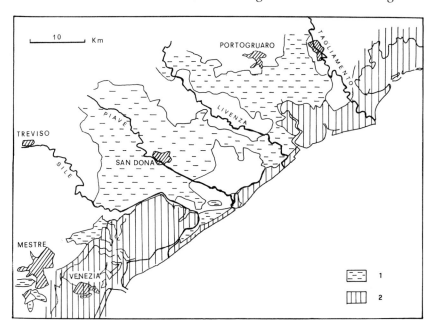

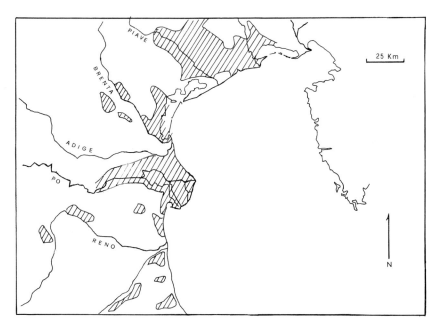

Fig. 12.25 Vulnerability to flooding of the lower coastal plains of Veneto and Romagna. A. Extent of river (1) and tidal surge (2) flooding in November 1966 (based on CNR-Touring Club Italiano *Carta dell'alluvione del Novembre 1966 nel Veneto e nel Trentino-Alto Adige*, 1:200,000, 1972). B. Areas affected by river flooding during this century (from Pinna and Ruocco, 1980).

Table 12.1 Discharge and water quality of Romagna rivers

	Water discharge m³/sec			Solid load kg/m³	Water quality		
	Yearly	Min.	Max.		Moderate	Poor	V. poor
Po di Goro	150	–	–	?	x		
Po di Volano	95	–	–	?		x	
Reno	59	6.66	119	1.4			x
Senio	3.2	–	–	2.3			x
Lamone	8.8	0.90	19.3	1.8			x
Fiumi Uniti	20.5	1.95	39	3			x
Savio	10.8	1.04	21	5.1	x		
Rubicone	2.1	0.13	4.5	?			x
Marecchia	6.1	–	–	?	x		

Source: Servizio Idrografico, *Annali Idrologici*, 1978; Marchetti, 1985.

In the lower-middle plain there is a surface freatic aquifer 50–80 m thick with permeabilities of 10^{-3} to 10^{-4} m³/s. At depth the Lower Confined Aquifer contains alternating fluviatile sands and silt-clays, 100–250 m thick with permeabilities of 10^{-3} to 10^{-5} m³/s and confined water tables.

The basal limits of this aquifer system are Pliocene clays, as well as the top of brackish or salt-water saturation. In deep wells at the coast this discontinuity lies at depth 150–270 m; in the Po Delta at 150–200 m.

The recharge of the Undifferentiated Aquifer is from the main rivers, from rainfall percolation, and from the infiltration of irrigation waters. The flow of the aquifer is closely related to that of the streams, with two peaks, in spring and autumn, and two intervening lows. The natural discharge of the Undifferentiated Aquifer is through the Risorgive springs (average 42 m³/sec) and to the aquifers of the middle and lower plain (Mari et al., 1985).

The Undifferentiated Freatic Aquifer and the Lower Confined Aquifer are the main, and much used, ground-water resources of Veneto-Friuli. However, ground water is suitable for drinking only from the former.

In the Venice-Mestre region six deep aquifers were the most exploited until the late 1970s (Carbognin et al., 1981). The area requirement of ± 7000 1/s (two to three for industrial use) is now almost entirely supplied by aqueducts from the Sile and Livenza rivers.

In Romagna, particularly in the Ravenna area, the aquifer system is known to the depth of 500 m with well-identified silty sandy units between –90 and –430 m (Carbognin et al., 1984). Aquifers in the upper zone are little exploited, as they contain less water and are contaminated by the polluted unconfined surface aquifer. Below 430 m the salt content is very high. Recharge is from the foothills of the Apennines and from the Po River basin, but the relative contribution is unknown.

6.3 Pollution of Surface and Lagoonal Waters

The Venice lagoon is subject to a massive inflow of urban industrial and agricultural wastes. This load of pollutants far surpasses the self-cleaning capacity of the lagoon by tidal flushing, which varies considerably. The

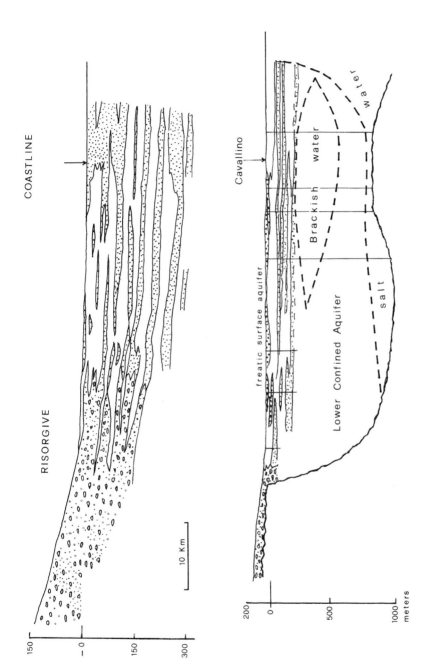

Fig. 12.26 Schematic profile across the subsurface aquifer system of the Veneto plain (based on Mari et al., 1985).

flushing waters derive from the north Adriatic, which is itself affected by various inputs of pollutants.

As regards urban pollution, the amount of oxygen required to oxidize and regenerate the organic matter produced by 10 million people has been calculated 53.6 t/day. In areas with oxygen depletion, ammonia and sulphides produce environments favourable to the development of pathogenous bacteria. Thus large areas of the lagoons have been found to be unsuitable for shellfish farming. Large quantities of nitrogen and phosphorus are also discharged (3970 t/yr and 1096 t/yr, respectively in 1975); 50% of the phosphorus comes from detergents. In certain parts, these inputs periodically cause eutrophic conditions.

Industrial pollution is due mainly to the Porto Marghera industries, where at least 60 plants produce highly polluting waste waters. There is a high concentration of aliphatic hydrocarbons, aromatic polynucleids, cyanide and heavy metals (cobalt, copper, zinc, nickel, chrome, iron). Due to their insolubility the heavy metals accumulate in the fine-grained sediments and in animal tissues; considerable negative effects on the bottom fauna have been noted (Perin, 1975). Thermal pollution is also serious, especially in summer. Discharge of hot waters from thermo electric plants increases water stratification and causes anoxic conditions.

Agricultural waste waters carry large quantities of nitrates, phosphates, herbicides and pesticides. In the 1970s it was calculated that 5–20% of substances used in agriculture were washed by rainwater through drains and canals into the lagoon, with phosphorus and nitrogen amounting to 51.7 t/yr and 2642 t/yr, respectively.

Information available in the 1970s (Perin, 1975) indicated that pollution decreased from the vicinity of the industrial area and the inner margins on the lagoon to the central basin east of Venice, the waters near Malamocco outlet, and the sea. More recent data (Zucchetta, 1983; Cossu and De Fraja Fragipare, 1987) suggest that the lagoon of Venice is polluted beyond acceptable limits (with increasing eutrophication events) and measures to reduce and control pollution are urgently needed.

7 ECOSYSTEMS

Despite agricultural, industrial and recreational developments, remnants of natural environments still exist along the coast of north-eastern Italy. Most of the external zones of the Po Delta (beaches and sand bars, lagoons) have not been affected by tourism. Vegetational oases occur along the margins of canals, rivers and fish basins (Caniglia, 1988). Little remains of the mesophyle forests of oak, ash and carpinus that once covered the Po valley and the Veneto plains. Tiny remnants of woodland exist in Friuli–Venezia Giulia, Veneto (e.g. Cavallino, Nordio, Carpinedo), somewhat larger woods survive in the Po Delta (Mesola) and in Romagna (San Vitale).

Nature conservation is still limited in scale, though the major lagoons and the external parts of the Po Delta are scheduled for restriction as nature reserves. South of the Venice lagoon, only a small percentage of the area of the proposed Po Delta National Park is actually safeguarded (2318 ha *vs* 25,000 km^2 of wetlands). Protection is needed also to preserve the typically zoned biotopes that range from beach to inner lagoon margins and/or old beach ridges (Fig. 12.27).

The deep transformations and alterations of the lagoons and their marginal lands have considerably reduced the survival possibilities of many mammals, those that generally live in or near aquatic, in particular fresh water, ecosystems. The otter, for instance, has entirely disappeared. There also has been a notable decrease of those woodland species that until a decade or two ago lived in the relict woods of the plain or in limited niches amidst the agricultural areas, such as hedges, wooded river and drains margins, and in private parks. Hunting has drastically reduced the number of remaining carnivores, while the escalating use of agricultural pesticides has had a negative impact on bats. Surviving small mammals are insectivores (*Neomys anomalus* and *N. fodiens*), the porcupine (*Erinaceous europeus*), rodents (*Crocidura, Sorex, Soncus, Muscardinus avellanarius*) and the nutria (*Myocastor coypus*), a species introduced from South America (Rallo, 1988).

The number of nesting birds has shrunk due to the disappearance of special environments, such as the oak woods of the plain, and the fresh-water swamps and wet meadows at the margin of the lagoons; within the lagoons, many areas of tidal flats, salt marshes, and reed beds, where many wintering birds feed (Neugebauer *et al.*, 1988). The Po Delta and the other coastal wetlands are also host to thousands of migrating bird species, which converge there through the central European and southern, Carpatho–Danubian routes. The latter is quantitatively the most important for ducks and waterfowl, plovers, ardeids, etc. Many species of columbids, turdids, fringillids and passeriforms also winter there. The most important wintering species are swans (*Cygnus cygnus* and *C. olor*), geese (*Anser fabalis, A. albifrons*), ducks (*Anas, Aythia, Bucehala, Clanqula, Mergus*, etc.) and ardeids (*Ardea cinerea, Egretta alba*). More rare, wintering or nesting, are sterns, avocets, seagulls and *Himanthopus himanthopus*.

8 AN EVALUATION OF THE IMPACT OF CLIMATIC CHANGES

8.1 Introduction

A proper evaluation of the consequences of climatic changes due to greenhouse warming for water resources, agriculture, fisheries, beach tourism, etc., requires knowledge of how climatic elements, like temperature, air circulation and rainfall, would be altered on both local and regional scales. For the Mediterranean regions, there are no existing climatic models that can predict how the local weather will be affected by changed larger-scale circulation patterns (e.g. North Atlantic Ocean, central and eastern Europe). Precipitation forecasts, for instance, need to be refined, as they are still rather contradictory (Wigley, Chapter 1).

Changed annual mean and seasonal temperatures, wind patterns and amount of precipitation would affect:

1) Surface and ground-water flow and river regimes; the incidence of floods and the quantity of sediment transported and delivered to the sea.

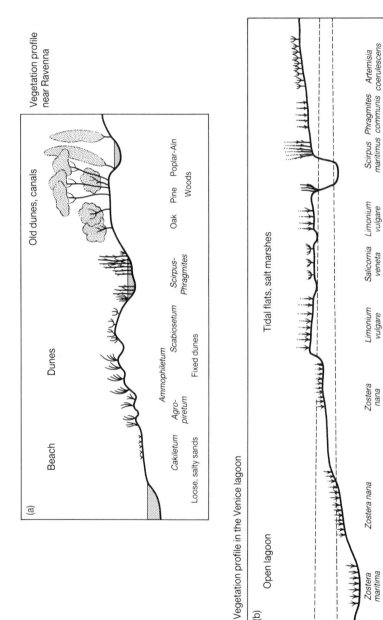

Fig. 12.27 The natural ecological zonation of the coastal belt of the Venice lagoon ((a), from Ratti, 1988) and near Ravenna (b), based on Corbetta, 1968).

2) The movement of marine water masses, especially in terms of erosional impacts on the coasts and of water elevation in the lagoons.
3) The biological resources of the sea and lagoons.
4) The patterns of human occupation and use of the coastal lowlands, particularly with respect to agriculture, fishing, industry, tourism and the quality of the environment.

The effects of climatic changes are likely to be felt gradually and probably could be adjusted to. Nevertheless, even small increases in average temperature, and especially the likely increased frequency of extreme climatic events (Wigley, 1989), could have significant impacts on ecosystems (Warrick et al., 1986). Because of their ecological fragility, related to the land–sea transition, the wetlands would be most vulnerable. Eventually changes will become more drastic. Since temperature oscillations in the last 2000 years have been mostly within 0.5–1°C, increases of 2.5–3.5° undoubtedly would have profound effects on agriculture and marine resources.

The physical impacts of sea-level rise on a lowland coast can be modelled on the basis of morphology, hydrodynamics, sediment budget, land subsidence and the effects of fixed artificial structure. Equally, the impacts of altered rainfall distribution on surface and ground water could be modelled quantitatively. The effects of increased air temperature and changed soil-water parameters on biosystems can be estimated, thus providing some idea of impacts on agriculture and fisheries. What is more difficult to estimate, however, is the impact of the physical and biological changes on the future socio-economic framework of the threatened lowlands.

The economic role of Mediterranean coastal zones is now largely conditioned by external, rather than local, social and market forces. For instance, the future relevance of local industries, agriculture and ports will be largely controlled by worldwide commodity prices, trade trends, competition (e.g. mineral and energy raw materials, cereals and industrial crops) and the future demand for consumer goods in a competitive international community. The role of individual ports may vary in response to altered technology and trade conditions. Markets for consumer goods and services could change in relation to stagnating or declining urban growth. The demand for beach recreation would certainly continue, but social habits could be modified, in particular if the water-edge urban-type resorts of today cannot be physically maintained.

The analysis of the physical impacts of sea-level rise is complicated by the intense human occupation of the coastal zones and their enormously accrued economic value, and by the continued human interference with natural processes. On the north-western Adriatic coast, present land uses are enmeshed with the economy of north Italy and beyond. No return can be envisaged to the pre-development times, and measures to protect the economy are therefore a political necessity. Nevertheless, coastal zone management must be based on cost-effectiveness, an assessment of the value of threatened land uses, not only in terms of their present functions (in the context of local and regional needs) but also those in the decades ahead.

8.2 Changes of Physical Parameters

8.2.1 Climate and precipitation

The published General Circulation Models (GCM) envisage that the doubling of "greenhouse gases" in the atmosphere as early as 2015 or as late as 2050 will eventually produce over northern Italy and the Eastern Alps a maximum warming of 3–3.5°C (see Wigley, Chapter 1). This would cause average winter temperatures in the plain to rise to the level of those of Naples (now 5– 6°C), while average summer temperatures might become more similar to those of Florence (26.2°C) and Palermo (27.2°C). Thus, winters would become milder and summers much warmer than at present.

This level of warming, however, will not be attained in the next three to four decades, considering the lag effect required by the global system to reach an equilibrium. More immediately, the gradual warming of up to 1°C by 2030 would be characterized by an appreciable increase of inter-annual variability (several long summers, sequences of mild or cold winters, more irregular cyclonic events) and increased frequency of extreme events: more droughts, more floods, more heat waves, cold snaps, marine storms, etc. (Wigley, 1985, 1992).

It is not yet established how the temperature increases would affect average rainfall over the Alps and the Apennines. Current GCMs indicate a fair to minor increase of winter and spring precipitation over the northern Adriatic region and the eastern Alps, a lesser one, or possibly no change, over the central and northern Apennines. Evaluation by the "analogues" method (Lough et al., 1985) has suggested that in warmer times yearly precipitation over north-eastern Italy would decrease, particularly in spring and summer. Rainfall depends on the interaction between large-scale air flow and orography. Since changes in the former are virtually certain (Wigley, 1989), an upward shift of main upper westerly flow could reduce the length of the rainy season in the western and central part of the Mediterranean.

Presumably precipitation patterns would remain the same as they depend on the relations between the cyclogenetic depressions originated in the western Mediterranean and the colder air masses that in winter descend from north-western and eastern Europe. Evapotranspiration will probably increase, perhaps by 200 mm/yr (Le Houérou, Chapter 6).

There would still be a contrast between air masses in winter, but a latitudinal decrease of temperature gradient could lead to a weakening of the winter wind regime. On the other hand, if the Mediterranean were to become cooler in winter (according to another scenario) lower atmospheric pressure over the Mediterranean, *vs* higher pressure over central and north Europe (see Lough et al., 1983), would lead to an increased number of depressions and therefore to a stormier climate. Consequently, an increase in winter temperatures will cause a reduction of area and time-permanence of the snow cover, and large areas will be frost-free in winter. For a 1.5°C warming the snowline would rise by 200–300 m (Kuhn, 1987).

The assumption of a qualitatively similar winter air circulation implies the continuation of prevalent strong easterly winds and, secondarily, of south and south-east winds. It is still to be seen whether their relative frequencies would change, especially the westerly winds which have a lesser impact on coastal dynamics.

8.2.2 Impacts on marine parameters

The general opinion is that a rise of temperature (even 1–2°C) and wind frequencies and patterns would eventually alter significantly the surface and deep-water circulation of the Mediterranean basins (Gačič et al., Chapter 7). A shift and/or a weakening of the wind regime (especially the cold winds that blow southwards through mountain gaps, e.g. bora, mistral, etc.) would particularly affect areas (like the central Adriatic) where deep waters are formed and where upwellings occur. Deep-water renewal could become slower. Any appreciable change in the distribution of the bora winds would result, for instance, in the change of volume of deep-water formation, thus affecting the exchanges Adriatic and east Mediterranean (these are influenced by climatic variations over a wide area of western Europe-Atlantic; Gačič et al., Chapter 7).

Large-scale climatic changes will affect not only salinity and sea-level, but also other parameters, such as vertical mixing, horizontal density gradients, evaporation rates, etc. Vertical exchanges are essential to regulate the ventilation of the deep parts of the Adriatic and to facilitate the fertilization of surface waters.

Increased sea temperatures and changes in winter weather could result in reduced deep-water formation, with lower oxygen concentrations in bottom waters, hindering the upward recycling of nutrients. If winter fresh-water discharge increases in the northern Adriatic, the effect would be higher surface temperatures, reduced density and salinity, and therefore a reduction of winter mixing and a poorer oxygenation of bottom waters.

Primary productivity would be favoured by a temperature increase and by vertical stability, but hindered by reduced upward nutrient flux because of less mixing and upwelling, and by greater horizontal transport. The end result would be reduced pelagic primary production. In the coastal areas, on the other hand, the opposite effects could become dominant. A growth in nutrient inputs from anthropogenic sources together with increased stability of water masses could favour an increase of primary production (though this situation could give rise to more frequent episodes of eutrophication and anoxia).

Circulation near the coast, as well as wave directions could be influenced by possibly greater frequency of south-west and south-east winds. In the region south of the Po Delta, summer south-west winds could produce more frequent upwelling. Local circulation close to the coast could change on a seasonal basis due to altered river discharges. The effects of a rise of the sea-level would be felt more significantly near the coasts, where they would be magnified by other factors of sea-level increase: low pressures related to possibly increased cyclonic perturbations from the western Mediterranean and higher tides in the lagoons due to the impact of anthropic factors on lagoon hydrology. However, the effect on the tides by fresh-water plumes near the coast could either be positive or negative, depending on future river discharge.

8.3 Impacts on Coastal Stability

On the basis of recent estimates (Warrick and Oerlemans, 1990) average sea-level could rise globally by 8–29 cm (with a best guess mean of 18 cm) by 2030; and by 21–71 cm (±44 cm mean) by 2070. The relative elevation of sea-level on the Adriatic coast (taking only natural subsidence into account, as man-induced subsidence has by now ceased) could be therefore 13–34 cm in Venice, 28–49 cm at the Po Delta, 19–39 cm at Ravenna (best guess averages respectively of 23, 38 and 29 cm). By 2030 the sea-level rise could be 31.4–81.4 cm at Venice, 61–111 cm at the Po Delta and 41–92 cm at Ravenna by 2070 (averages respectively of 54, 88, 65 cm).

What is of immediate concern is the manner of sea-level rise, whether slow and smooth or in steps, and the way in which the coast will respond, both along shorelines still in a natural state and those strongly modified by man. As the level of the sea rises, beaches and barrier islands are expected to migrate gradually inland (Brunn and Schwartz, 1985), a 30 cm rise may cause a retreat of 30 m, though the actual amount depends on beach gradient. Examples of this recession are indicated by the response of the Caorle lagoon barrier island to a storm in 1977 (Catani *et al.*, 1978); and by the 126 m retreat occurred at Lido Adriano (Ravenna) from 1957 to 1977 because of a 45 cm subsidence (Carbognin *et al.*, 1984). The rate of retreat depends on the condition (continuous *vs* discontinuous) of the dune belt and the thickness of the coastal sand barrier.

Along with increased water level, there would be an increase in the frequency of exceptional high-water occurrences from tidal and storm surges that will exacerbate the present negative aspects of coastal erosion, sediment deficits and lagoon hydrodynamics. Venice and nearby islands will be flooded more frequently, with progressively higher water levels, unless the proposed tide regulation works are built and functioning by the end of the century (Pirazzoli, 1991).

Erosion and flooding of intertidal areas will increase in the lagoons that have an open access to the sea, especially those parts affected by subsidence. The inundation will be augmented by the natural or artificial deepening of lagoons outlets. The lagoon will not be able to migrate because of their inner diked margins.

The edge of the Po Delta will continue to be eroded, at least as far as the foot of the dikes that surround it and line the main river branches. Wave energy in front of sea-walls will gradually increase with the greater water depths. There will be increased erosion of most beaches now used by the summer tourist industry, considering that the river-supplied sand will continue to be deficient for at least another 30 years even if all sand dredging in the lower river courses is stopped soon.

The existing parallel coastal defences (offshore breakwaters, dikes) will have to be raised periodically, perhaps replanned more rationally, to survive periodic high tides and storm surges. The beach nourishment scheme must be intensified. By the years 2010/2020 all present protection works will need adjustment. The cost of beach protection and maintenance will certainly escalate, contributing to the further decline of some beach resorts.

Probably a relative rise of even up to 50–60 cm will not cause the flooding of the lands below sea-level, because most of the Adriatic coast is lined by sea walls, dikes, road and canal embankments, if not at the shore in the immediate hinterland. Many of these structures undoubtedly will have undergone periodic improvements. The sea, however, might overcome all those parts not hydrodynamically adjusted.

Sea-level increases of more than 50 cm (to 100 cm), coupled with more frequent storm surges, could have the following effects:
1) The rapid destruction of the sea-front parts of all exposed coastal towns (Grado, Lignano, Caorle, Lido, Rimini Cervia). Some coastal resorts may no longer be viable.
2) All constructions (piers, groins, low sea walls now no more than 1 m above sea-level) could be seriously damaged if not destroyed.
3) Even low-lying areas of the immediate hinterland could be threatened (e.g. the important orchard belt from Iesolo to Cavallino and Chioggia; the industrial area of Ravenna).
4) All reclaimed areas presently below sea-level would experience increased saline water infiltration, and there would be an increased need of pumping (especially in the Po Delta) with more powerful and costly facilities. Water levels will be higher in several rivers and canals, with consequent salt-wedge penetration.

8.4 Impacts on Hydrology and Water Resources

As mentioned earlier (8.2) current hydrological scenarios are not sufficiently advanced to predict how and where the availability of water could change in respect to the present. Unlike the southern Mediterranean, northern Italy might not experience water limitations, although the Po delta and Romagna already are water deficient.

With more rain than snow in winter, the discharge regimes of rivers from the Alps and pre-Alps would be altered significantly. More persistent higher summer temperatures will cause the disappearance of the eastern alpine glaciers and a reduction of the western glaciers (Kuhn, 1987). At first, the regimes of the rivers that flow from the inner Alps (Isonzo, Adige, etc.) would become more similar to those of the pre-Alpine rivers (e.g. Brenta, Piave). If there is a continuation of the present patterns of catastrophic rainfall every few years, there will be a greater risk of flood in the plains, given the latter's state of vulnerability. The Apennine rivers will continue to carry little or no water in summer, perhaps less water overall.

In later decades, as temperatures become warmer and sea-level rises, increased river flow from the Alps in autumn and winter and more torrential discharge by the Apenninic rivers, together with larger sediment loads due to greater soil erosion and slope instability, would introduce serious problems for the management of: (1) hydroelectric reservoirs and irrigation; (2) flood control; (3) river banks, bridges, canals; and (4) riverine navigation.

In the plains, negative consequences could result from reduced stream flow and greater use of water for irrigation during the longer, drier summers. Irrigation water intakes may no longer be adjusted to the lowered river levels. River and canal pollution would become even more serious problems than they are today.

Since the changes in precipitation and runoff in the Alps and pre-Alps probably would be more qualitative (i.e. timing) than quantitative, the availability of water to recharge aquifers should not be altered in Veneto and Friuli because of the porous nature of the Upper Unconfined Aquifer. Negative changes may be expected, however, in Romagna, with greater problems for the shallow aquifers, especially if excessively used, than at present.

Reduced river flow in summer not only would increase stream pollution, but also salt penetration from the sea, especially with higher sea-level and lowered river beds (due to sand extraction). Saline waters will percolate to the subzero lands, via the surficial ground-water flow. In the coastal areas where soils and ground water are already salty, the effects will be increasingly negative for agriculture. Salinity would reduce the soil capacity to retain toxic materials (pesticides, heavy metals) by making them more soluble. There might be a rise and further inland shift of the subsurface boundary between fresh and brackish water, especially if associated with over-pumping.

Proper water resource management will continue to be a priority, at least to avoid or minimize reduced supplies of adequate quality water for agriculture and industry. The cost of providing good quality drinking water will increase because of salinization. Pollution could continue to be a major problem with surface waters, and can be reduced only if substantial changes are made in the next decade to control it at the source.

Studies of the impact of climatic change on the complex system of surface water origin *vs* utilization by different (and conflicting) users will have to be made as soon as possible. In particular, all industrial and urban developments need to be geared to existing water resources. In the Venice lagoon (and, in the future, possibly elsewhere) long-term hydrological implications or engineering protective measures should be carefully reviewed.

8.5 Impacts on Marine and Lagoonal Ecosystems

The impacts of climatic changes on ecosystems could be considered in their natural evolution, or more realistically, in the perspective of the influence of human activities. Only small parts of the wetland system of the north Adriatic coast are protected at present and likely to be left in a natural condition. The biologic and economic importance of lagoons and marshes is based on the high primary productivity of environments that are conditioned by a delicate balance between brackish and fresh water. The continued existence of a rich and varied migrating and nesting bird fauna, for example, depends on the permanence of such transitional environments: salt marshes and tidal flats, fresh-water ponds and fish basins with specialized vegetation.

The distribution of wetlands in the next 50–70 years will depend on their ability to grow vertically, relative to rates of sea-level rise, since the wetlands cannot migrate inland due to the diked margins of the lagoons.

Aquatic ecology is likely to be profoundly affected by a temperature rise. Shallow sheltered onshore marine and lagoonal areas are likely to become warmer and hypersaline. In the lagoons, water temperatures (maxima and minima) already are higher than those of open sea water (e.g. in the Marano lagoon maxima average 28°C, and reach even 30°C in stagnant areas, while average minima can drop to freezing in shallow parts). Even a 1°C rise might have a marked adverse effect on fish life. While tolerance to temperature variations can be considerable in many aquatic animals (and many also migrate), it is less for spawning and is much reduced in relation to changes in oxygen concentration (reduced with higher temperatures when organisms would consume more), salinity (e.g. greater fresh-water input from rivers in winter), and with the effects of pollutants. Of great importance for vegetation and fauna would be the effects of interannual variability in temperature (hot, dry summers, or very cold winter spells). Migrating species would be adversely affected by alteration of coastal physiography and/or inland hydrology.

In the shallow marine environments of the continental shelf, the proportion of some species may be altered if they are sensitive to small temperature variations. The main impact could be that of a continued discharge of pollutants from the Po and other rivers, enhanced by more frequent eutrophication as well as broad changes in primary productivity consequent upon reduced nutrient distribution related to major oceanographic changes. At a later stage, if the lagoons were to be transformed into bays, the location of natural marshlands and the nature of protected areas would change drastically, with serious effects on migratory birds.

In the open lagoons, migrating species that are omnivorous and adaptable (e.g. mugils, eels), would be favoured, as well as some lagoonal fishes (*Gobius*). More marine fishes would move in (e.g. *Platychtis, Mullus barbatus, Sarpa, Atherina*, etc.). Perhaps the fish fauna in general would be favoured by a rise of temperature, provided primary productivity and bottom fauna have not been impoverished by more marine conditions and/or pollution.

The impacts on the bird fauna rest mainly on physical alterations to their specific nesting and feeding habitats; an increase of salinity, for instance, would be detrimental to fresh-water aquatic plants (*Phragmites, Potamogeton*, etc.) that provide shelter and food to waterfowl. On the other hand, the open lagoons will be used increasingly for wintering by marine birds.

Changes in fish fauna would affect fish-eating birds. Greater extension of brackish or saline areas (*vs* fresh-water areas) would favour (or impede) certain species. The number and types of migrating (staging and wintering) birds also might be influenced by changed conditions in their traditional wintering locations, and in their nesting locations in more boreal regions.

With regard to vegetation, higher temperatures and longer and drier summers could favour the evergreen-oak natural association, but hinder many imported species that characterize the non-natural vegetation. At the level of natural ecosystems, impacts are difficult to assess because of their complexity. The existing balance between species would be altered, the CO_2-richer and warmer atmosphere enhancing the growth rate of some species, but being detrimental to others. CO_2 might alleviate some environmental stresses, such as nutrient deficiencies and drought.

Biotic interactions, e.g. between competing plant species, are quite sensitive to minute changes in metabolic activity. Enhanced effects of environmental changes can be expected in places where they cause the rupture of the synchronization of events which under normal conditions, lead to the activation or suspension of growth (in harmony with the cyclic changes in weather) or to flowering at a time when pollinators are present.

8.6 Impacts on the Economy of the Coastal Zone

The socio-economic system of the north-west Adriatic coastal region is extremely complex, based on innumerable interdependent factors and elements, many of which have little to do with the physical environment. In principle, an evaluation of the impacts of future climatic changes should be based at least on the trends of population, economic productivity in different fields, and social behaviour. Changes over the last three decades, however, may not be indicative of possible future tendencies, witness those of population density, industry *vs* tertiary activities, of cultivated crops, the role of ports, the importance of fishing, and the type and fluctuations of tourism. In fact, socio-economic extrapolations may be hazardous beyond 20–30 years due to the effects of possible modification in economic conditions (e.g. demand for consumer goods, housing, recreation, transport, etc.).

Land uses will gradually adapt to changing physical conditions, spurred by economic necessity. Any investment or commercial activity that would be no longer economically viable will cease to function, or will be transferred elsewhere.

The level of activity of summer tourism, industry, ports and agriculture is already determined by decisional market forces situated outside the coastal lowlands. For instance, the north-western Adriatic ports are one of the main assets of the economy of the north-eastern Italian coastal regions; even assuming that their activities would be streamlined to face competition and changed technical needs, they will certainly be affected by the eventual reduction of petroleum exports and by changes in the world trade of basic raw materials (e.g. timber, textile fibres, metallic ores) due to technological innovations, long before a significant rise of sea-level.

Little change could be expected in the economic and demographic patterns of the >2 m hinterland. Since population in the coastal regions of north-east Italy is likely to decrease, particularly in the lagoonal towns and the low-lying agricultural areas, the basic needs of the region should remain the same, with employment derived from diversified industrial, commercial and tertiary activities; and continued local food production for local use (mainly fruit, vegetables, meat and dairy farming, fishing). The impact of temperature increases may not be direct (except for greater electricity needs due to increased summer air-conditioning), but indirect, affecting fishing and specialized agriculture and the infrastructures of the beach resorts.

Higher sea-level would cause a gradual deterioration of living conditions in the lagoonal towns (e.g. Grado, Chioggia, Venezia and nearby centres). In fact, they have already experienced the effects of flooding due to winter high tides and storm surges, in association with land subsidence. Although the main inland towns lie above 2 m elevation, the uncontrolled rise of lagoonal and canal water levels could also affect them eventually.

There is no doubt that the impact of sea-level rise on beach resorts, as they have developed in the last three decades, would be generally negative, not only because of the shrinking beaches, but also of increased damage due to storm surges and higher costs of protection. The likely deterioration of water, increased cost of summer produce supplies and a possible return of mosquitoes to the marshes could add to the problem. Nevertheless, a degree of adaptation to environmental hazards has already manifested itself, such as the use of swimming pools because of marine pollution and algal blooms. The high demand for beach recreation should remain (UNEP, 1988), and temperature increases would not be detrimental, if only because of the continuing attraction of Mediterranean scenery.

The main problem facing decision makers will be one of policy; which strategy should be adopted by state and local authorities for optimal exploitation of beaches and tourist attractions, in the face of swelling numbers of visitors. Tourism will continue to be an important economic resource, but the approaches to littoral land use should be reviewed. The main issues of tourism and its trends must be investigated, especially the efficiency of high-rise residential facilities in close proximity to the shore. The maintenance of beach recreation facilities will require a more rational approach than that of today. The present state of degradation suggests that the vision of total protection held by local residents, private investors and politicians is not practical nor cost-effective in the long run. A better approach is the introduction of setback lines and zoning, with less urbanized settlements and the adoption of more 'open-space' tourism, wherever feasible, on the model of the Camargue (see Corre *et al.*, Chapter 11).

Regarding the impacts of temperature change on agriculture, cereal production (especially wheat and corn) and associated animal breeding may benefit from higher temperatures and CO_2 concentration. Agrobiological research no doubt will provide adaptations of crops and animal husbandry. Problems, however, would be created by the expected abnormal inter-year variations of temperature and water supplies, by water salinization, by increases in weeds and pests (possibly of diseases), and by changes in soil behaviour.

An increase of evapotranspiration and soil-moisture deficits would occur particularly between late summer and early winter. Weather fluctuations (temperature, availability of water, rains during the ripening stages, etc.) could affect several crops (wheat, soya, beans, sugar beets, tomatoes, vines and fruit trees). Warmer winters and severe water deficits would threaten the existence of those trees (e.g. olive, nut trees) that require a dormant period at relatively low temperatures.

Irrigation could become even more necessary than now, but also more difficult and expensive, because of increased need for soil drainage and greater use of surface and subsurface water resources. Increased use of fertilizers (crops may require more nitrogen in a CO_2-richer atmosphere) and pesticides could increase water pollution.

It is not clear what effect increased salinity and temperature (2–4°C) will have on beach sands and alluvial and clay-peat soils of the coastal lowlands. Some soil parameters could be quite sensitive to temperature and rainfall changes, e.g. the soil composition and salt balance, chemical processes and the supply and breakdown of organic matter (see Imeson and Emmer, Chapter 4).

In the face of certain changes in the physico-chemical conditions of lake and lagoonal waters, continued aquaculture research will be needed in regard to the controlling environmental and biological factors, including the solution of the present difficulties experienced by artificial breeding and feeding in intensive aquaculture. Aquaculture should be considered as a high-priority economic activity, one compatible with the buffer zone function of wetlands in subsident areas, against the impacts of sea-level rise.

9 CONCLUSIONS AND RECOMMENDATIONS

Although climatic changes and sea-level rise would take place gradually, they will first (in the next two to three decades) become manifest through a greater interannual variability, e.g. more frequent, very hot and dry summers, milder winters and a greater incidence of exceptional events (droughts or very heavy rainfalls, winter storms, high tides). Consequently, the Adriatic coast of north-eastern Italy would be exposed to much greater natural stresses than at present, with more frequent marine washovers and a further loss of beaches, despite any engineering measures that have been or will be undertaken to stop them.

Average annual temperature increments of up to 0.5–1.5°C by 2030 would have more of an impact on lagoonal and marine ecology (and therefore on fishing) than on agriculture. There are still several unknowns on the effects of a CO_2-richer atmosphere on crop production. Benefits, in association with agrotechnology adaptations, may be offset by the consequences of abnormal temperature fluctuations and of the scarcity and poorer quality of water.

The social and economic fabric of the coastal plains (>2 m elevation) will bear the effects of climatic change on the lower lands indirectly, as the consequence of the deterioration of primary activities, natural environments, coastal towns and infrastructures of tourism and water availability.

In relation to sea-level rise, the entire coast of the north-west Adriatic should be classified as high risk, though not for the same reasons at all locations. For instance, along most of the shore from Chioggia to the Isonzo, beaches are often narrow with very gentle gradients, mostly retreating. The risk for the hinterland is variable, there are many transversal dikes and embankments, but also several river-lagoon inlets and parts are below sea-level and subsident.

All the islands and margins of the Venice lagoon could eventually be affected by flooding. The Po Delta is a high-risk area for a variety of reasons (hydrology, subsidence, low sediment outputs), but most of it is surrounded by dikes. The Romagna coast already is mostly retreating and in parts subsident; there are too many built-up areas near the shore, many impediments to sediment movement, and numerous waterways giving access to the internal lowlands.

The prospect of a substantial rise of sea-level signifies a choice between a static approach of defending the present *status quo* with local hasty solutions and a flexible view (within a regional plan) of following natural changes with a planned re-development of beach uses and the restoration and management of natural wetlands, both for the sake of their biological resources and as buffer zones against the flooding of inland agricultural areas. In Italy, one of the world's industrial powers, it is probably technically and financially feasible to protect all, or most, of the Adriatic lowlands (including Venice, Marghera and Ravenna) against higher sea-level flooding. It is illogical, however, to envisage the total, future protection of all the contemporary beach resorts by means of engineering measures. Apart from climatic changes, deterioration of environmentally and economically important parameters will derive also from the continuing reluctance of society at large, and of decision-makers in particular, to understand natural processes (e.g. coastal dynamics, lagoon ecology, etc.) and the need to adjust land uses to them. Increased awareness is necessary to appreciate the complex interdisciplinary nature of coastal management.

With regard to the consequences of sea-level rise it will be necessary to investigate the bases for, and introduce a classification of, risk exposure; to investigate which beach resorts no longer will be viable, and how and when to apply concepts of set-back lines; and to control coastal developments, especially the further spread of urban-type beach resorts and pleasure marinas, as well as locating factories and power plants near the shore.

Since by law the State has the responsibility for all activities that relate to the coast, from its physical protection to its commercial use, it should take a leading role in stimulating and implementing all ecologically sensitive planning, above the present local and conflicting interests, and the bureaucratic complexities, for the sake of the well-being of future generations.

The logic of present territorial planning ought to be that of preventing that in a few decades situations will arise in which the only option would be retreat and abandonment.

A series of research activities seem to be advisable at this stage in order to create information systems that would help to cope with possible changes:

1) The coastal studies that resulted in the publication by the National Research Council, of the Atlas of Beaches, should be undertaken again, keeping in mind future evolution.
2) Dynamic model studies of sea-level change for the high- and medium-risk stretches, including a critical investigation of tourist towns that could become undefendable against rising waters.
3) The monitoring of erosion rates, with an effort to relate them to causes, including subsidence.
4) Studies of which coastal protection structures could be most effective, e.g. sloping dikes and beach sand nourishment, as opposed to vertical sea walls and groins.
5) The monitoring of the variations and trends of sea-level (more tidal gauge stations, etc.).
6) Coordinated studies of trends of temperature rainfall and air circulation.

7) The projection of subsidence trends.
8) The mechanics of sea-level rise in relation to the lagoon interiors, including the effects of closing the lagoons with sluices.
9) Analysis of which low-lying reclaimed areas could be returned to a lagoonal state.
10) Study of the ways to restore beach and river sediment budgets, considering future rainfall-induced erosion in the uplands and the needs of inland agriculture and hydropower.
11) Research into manners of waste disposal and of pollution control, in relation to changed hydrological parameters.
12) The systematic study of the ecological parameters of the main components of the lagoonal and shallow marine environments (including data on nutrients and phytoplankton concentration), to permit an analysis of the impacts of salinity and temperature changes.
13) Research into the basic elements that are needed to assess the impact of sea-level rise on the future socio-economic fabric of the regions involved.
14) Research on the legal impacts of climatic change and on the attribution of responsibilities for the counter-measures to be taken.

Abundant basic data already exist in Italy (collected by State and regional institutions) on topography, meteorology, waves and sea-level, river discharge, subsurface waters, pollution and ecosystems. They need, however, to be elaborated and coordinated to form a uniform data-base that will allow the analysis of future trends and the preparation of alternative scenarios. Research dealing with environmental hazards and impacts carried out by different institutes also needs to be coordinated.

10 REFERENCES

Accerboni, E., Manca, B., Michelato, A., Moro, F. and Mosetti, R. 1982. Caratteristiche dinamiche estive dell'Alto Adriatico e loro influenza sui fenomeni di inquinamento. CNR, *Convegno Risorse Biologiche e Inquinamento Marino del Progetto Finalizzato Oceanografia e Fondi Marini*, 891–912.

Bernardi, R., Lands, F., Marinucci, M. and Zanetto, G., 1989. L'Alto Adriatico: articolazione di uno spazio costiero. In: Bernardi, R. (ed.), *Mari e coste italiani*, Patron Editore, Bologna, 123–152.

Bevilacqua, E. and Mattana, O., 1976. The Po River basin: water utilization for hydroelectric power and irrigation. In: *Italian Contributions to the 23rd Int. Geogr. Congress*. CNR, Roma, 181–189.

Boldrin, A., Richards, J., 1988. Mammiferi. In: Comune di Venezia – WWF, *Laguna, Conservazione di un Ecosistema*, Arsenale Editrice, 37–40.

Bondesan, M. and Simeoni, U., 1983. Dinamica e analisi morfologica statistica dei litorali del delta del Po e alle foci dell'Adige e del Brenta. *Mem. Scienze Geol.*, **36**, 1–48.

Bondesan, M., 1988. The Po delta area and its geomorphological problems. In: *Excursion Guide Book, Joint Meeting on Geomorphological Hazards, I.G.U.*, Padova, 131–144.

Bondesan, M., Calderoni, G. and Dal Cin, R., 1978. Il litorale delle province di Ferrara e di Ravenna (Alto Adriatico): evoluzione morfologica e distribuzione dei sedimenti. *Boll. Soc. Geol. It.*, **97**, 247–287.

Bondesan, M. and Dal Cin, R., 1975. Rapporto fra erosione lungo i littorali emiliano romagnoli e del delta del Po, e attivita estrattiva negli alvei fluviali. *Cave e Assetto del Territorio*. Italia Nostra and Regione Emilia Romagna, 126–137.

Bondesan, M., Minarelli, A. and Russo, D., 1986. Studio dei movimenti verticali del suolo nella provincia di Ferrara. *Studi Idrogeologici sulla Pianura Padana*, **2**, CLUP, Milan.
Brambati, A., 1984. Erosione e difesa delle spiagge adriatiche. *Boll. Oceanologia Teorica e Applicata*, **2**, 91–184.
Brambati, A., 1985. Modificazioni costiere nell'arco lagunare dell'Adriatico settentrionale. Antichità Altoadriatiche, XXVII, *Studi Jesolani*, pp. 13–47.
Brambati, A., 1987. *Studio sedimentologico e marittimo-costiero dei litorali del Friuli-Venezia Giulia*. Regione Autonoma Friuli Venezia Giulia. Direz. Regionale Lavori Pubblici, Trieste, p. 67.
Brambati, A., Marocco, R.G., Catani, V., Carobene, L. and Lenardon, G., 1978. Stato delle conoscenze dei litorali dell'Alto Adriatico e criteri di intervento per la loro difesa. *Mem. Soc. Geol. Italiana*, **19**, 389–398.
Bruun, P. and Schwartz, M.L., 1985. Analytical prediction of beach profile change in response to a sea-level rise. *Zeitschr. Geomorphologie, Suppl. Bd.*, **57**, 33–50.
Camuffo, D. 1986. Analysis of the series of precipitation at Padova, Italy. *Climatic Change*, **6**, 57–77.
Caniglia, G., 1988. Il paesaggio vegetale. In: Cumune di Venezia – WWF, *Laguna, Conservazione di un Ecosistema*, Arsenale Editrice, 41–46.
Cantu, V., 1977. The climate of Italy. In: Wallen, C.C. (ed.), *Climates of Central and Southern Europe*. Elsevier, 127–173.
Caputo, M., Pieri, L. and Unguendoli, M., 1970. *Geometric investigations of the subsidence in the Po Delta*. CNR, Ist Studio Dinamica Grandi Masse, Venezia, TR2, p. 37.
Carbognin, L., Gatto, P., Marabini, F., Mozzi, G. and Zambon, L., 1982. Le trend evolutif du littoral Emilien-Romagnol (Italie). *Oceanologica Acta*, Proceed. Int. Symp. on Coastal Lagoons. SCOR/IAB O/Unesco, 73–77.
Carbognin, L., Gatto, P. and Mozzi, G., 1981. La riduzione altimetrica del territorio veneziano. *Ist. Veneto Sci. Lettere, Arti, Rapp. Studi*, **VIII**, 55–83.
Carbognin, L., Gatto, P. and Mozzi, G., 1984. Case history 9.15. Ravenna, Italy. In: Poland, J.F. (ed.), *Guidebook to studies of land subsidence due to groundwater withdrawal*, Unesco, 291–305.
Castiglioni, G.B., Favero, U., 1987. Linee di costa antiche ai margini orientali della laguna di Venezia e ai lati della foce attuale del Piave. In: *Rapporti e Studi, Ist. Veneto di Sci. Lett. Arti*, 17–30.
Catani, G., Marocco, R., Brambati, A., Carobene, L. and Lenardon, G., 1978. Indagini sulle cause dell'erosione nel tratto orientale di Valle Vecchia (Caorle, Adriatico Sett.). *Mem. Soc. Geol. Ital.*, **19**, 399–405.
Cati, L., 1981. Idrografia e idrologia del Po. *Uff. Idr. del Po Min. LL.PP.*, Ist. Pol. e Zecca dello Stato, p. 310.
Cavaleri, L., Curiotto, S., Dallaporta, G. and Mezzoldi, A., 1981. Directional wave recording in the northern Adriatic Sea. *Il Nuovo Cimento*, **4C**, 519–532.
Cavazzoni, S., 1983. Recent erosive processes in the Venetian lagoon. In: Bird, R.C.F. and Fabbri, P. (eds), *Coastal Problems of the Mediterranean Sea*, IGU Commiss. Coastal Environments, Bologna, 19–31.
Cencini, C., Cuccoli, L., Fabbri, P., Montanari, F., Sembeloni, F., Torresani, S. and Varani, L., 1979. Le spiagge di Romagna: uno spazio da proteggere. *Consiglio Nazionale delle Ricerche*, **I**, Bologna.
Cencini, C., Marchi, M., Torresani, S. and Varani, L., 1988. The impact of tourism on Italian deltaic coastlands: four case studies. *Ocean and Shore Management*, **11**, 353–374.
Ciabatti, M., 1979. Ricerche sull'evoluzione del Delta Padano. *Giornale di Geologia*, Sr 2, **34**, p. 26.
CNR (Consiglio Nazionale delle Richerche), 1985. *Atlante delle Spiagge Italiane*. Scale 1:100,000, sheets 40, 40A, 51, 52, 53, 65, 77, 89, 100, 101. *Atti del Convegno Conclusivo*, Roma, p. 403.

Colantoni, P., Gallignani, P. and Lenaz, R., 1979. Late Pleistocene and Holocene evolution of the North Adriatic continental shelf. *Marine Geology*, 33, M41–M50.

Corbetta, F., 1968. La vegetazione delle valli del litorale ferrarese e ravennate. *Notiziario della Soc. Ital. Fitosociologia*, Roma, 67–88.

Cossu, R., De Fraja Fragipane, E., 1987. Stato delle conoscenze sull'inquinamento della laguna di Venezia. Aspetti riassuntivi e conclusivi. *Atti Giornata di Studio: A Vent'anni dall'Evento di Marea del 1966*. Ist. Veneto Sci. Lett. Arti., Venezia.

Dal Cin, R., 1983. I litorali del delta del Po e alle foci dell'Adige e del Brenta, caratteri tessiturali e dispersione dei sedimenti, cause dell'arretramento e previsioni sull'evoluzione futura. *Boll. Soc. Geol. Ital.*, 102, 9–56.

Fabbri, P.C., 1985. Coastline variations in the Po Delta since 2500 BP. *Zeitschr. Geomorph., Suppl. Bd.*, 57, 155–167.

Fabbri, P.C., 1985b. La gestione delle aree costiere. In: *Lo Spazio Spiaggia: usi ed erosioni*. Ed. Autonomi, Firenze, 120–135.

Favero, V., 1979. Aspetti dell'evoluzione recente dell'Alto Adriatico. *Atti Conv. Scient. Naz. P.F. Oceanografia e Fondi Marini*, 2, 1219–1231.

Favero, V., 1985. Evoluzione della laguna di Venezia ed effetti indotti da interventi antropici sulla rete fluviale circum-lagunare. *Convegno Nazionale delle Acque*, Venezia, 402–209.

Fontes, J.C.H. and Bertolami, G., 1973. Subsidence of the Venice area during the past 40,000 years. *Nature*, 244, 339–341.

Gačić, M., Hecht, A., Hopkins, T., Lascaratos, A., 1992. Physical oceanography aspects and changes in circulation and stratification. Chapter 7, this volume.

Gasparini, V. and others, 1985. L'inquinamento marino nell'alto Adriatico: problemi igienico-sanitari. In: *I Problemi del Mare Adriatico*, Atti Conv. Internaz. Alpe-Adria, Univ. Trieste, pp. 367–389.

Gatto, P., 1984. Il cordone litorale della lagune di Venezia e le cause del suo degrado. *Ist. Veneto Sci. Lett. Arti. Rapp. Studi*, IX, 163–193.

Gatto, P. and Carbognin, L., 1981. The Lagoon of Venice: natural environmental trend and man-induced modification. *Hydrological Sciences Bulletin*, 26, 379–391.

Ghirardelli, E., Giaccone, G. and Orel, G., 1974. Evolution des peuplements benthiques du Golfe de Trieste. *Rev. Intern. Océanogr. Méd.*, 35–36, 111–113.

Jukić, S., 1985. Fisheries management in the Adriatic Sea. In: *I Problemi del Mare Adriatico*, Atti Convegno Internaz. della Comunitá Alpe-Adria, Univ. Trieste, pp. 355–361.

IDROSER, 1977. Progetto di piano per la savalguardia e l'utilizzo ottimale delle riserve idriche in Emilia Romagnia. IDROSER, Bologna.

IDROSER, 1982. Piano progettuale per la difesa della costa adriatica emiliano-romagnola. IDROSER, Bologna.

Kuhn, M., 1987. Impact analysis of climatic change in the Central European Mountain Ranges. In: *European Workshop on Interrelated Bioclimatic and land use changes*, Nordwijkerhout, Holland, 17–21 October, 1987.

Le Houérou, H.N. 1992. Vegetation and land use in the Mediterranean Basin by the year 2050; a prospective study. Chapter 6, this volume.

Lough, J.M., Wigley, T.M.L., Palutikov, J.P., 1983. Climate and climate impact scenario for Europe in a warmer world. *J. Climate Appl. Meteorol.*, 22, 1673–1684.

Marabini, F., 1985. Alcune considerazioni sull'evoluzione del delta del Po. *Nova Thalassia*, 7 (Supp. 2), 443–451.

Marchetti, R., (ed.), 1985. *Indagni sul problema dell'eutrofizzazione dell' acque costiere dell'Emilia Romagna*. Dipt. Ambiente-Territori-Trasporti, Regione Emilia Romagna. *Assess. Ambientale e Difesa Suolo, Studi e Documentazione*, 35, p. 310.

Mari, G.M., Sliverti Piuri, E., Ghezzi, G., Marchetti, M.P., Del Giudice, C., Filo, L., Pizzi, G. and Todini, E., 1985. Sistema acquifero regionale e modello matematico di gestione delle risorse idridiche. In: Magistrato alle Acque, *Venezia Atti Convegno Laguna, Fiumi, Lidi: Cinque Secoli di Gestione delle Acque nelle Venezie*, 712–732.

Malanotte-Rizzoli, P., 1981. Le mareggiate di scirocco (vento di Sudest) in Adriatico settentionale ed i loro effetti distruttivi sull'area costiera delle Laguna di Venezia. Ist. Veneto di Sci. Lett. e Arti, Commiss. di Studio dei Provvedimenti per la conservazione e difesa delle lagune e della cittá di Venezia. *Rapporti e Studi*. **VIII**, 19–30.

Malanotte-Rizzoli, P. and Bergamasio, A. 1982. Hydrodynamics of the Adriatic Sea. In: Nihoul, J.C.J. (ed.), *Hydrodynamics of Semi-enclosed Seas*, Elsevier, Amsterdam, 251–286.

Montori, S., 1983. Effetti della subsidenza sui territori di bonifica. *Atti Convegno Subsidenza del Territorio e Problemi Emergenti*, Bologna.

Mosetti, F., 1985. Caratteristiche fondamentali dell'idrologia dell'Adriatico. In: *I Problemi del Mare Adriatico, Atti Convegno Internaz. della Comunità Alpe-Adria*, Univ. Trieste, 155–170.

Murri, A. *et al.*, 1980. Andamento delle piogge nella stazione di Camerino. In: Zanella, G. (ed.), *Atti Primo Convegno di Meteorologia Appenninica*, 1979, 107–127.

Muscarà, C., 1979. *Veneto*. Enciclopedia Europea Garzanti, Milano.

Muscarà, C., 1982. *Gli spazi del turismo (Per una Geografia del turismo in Italia)*. Patron Editore, Bologna, p. 296.

Nelson, B.W., 1970. Hydrography, sediment dispersal and recent historical development of the Po River delta, Italy. In: *Deltaic Sedimentation. Modern and Ancient*. Soc. Econ. Paleont. Miner., Spec. Publ., **15**, 152–184.

Neugebauer, M., Scarton, F. and Semenzato, M., 1988. L'aviofana lagunare. In: Cumune di Venezia – WWF, *Laguna, Conservazione di un Ecosistema*, Arsenale Editrice, 52–56.

Orel, G., 1985. La pesca nella zona costiera. In: *I Problemi del Mare Adriatico*, Atti Convegno Internaz. della Comunità Alpe-Adria, Univ. Trieste, 331–340.

Papetti, F. and Vittorini, S., 1988. Le variazioni del clima a Firenze al 1822 al 1986. *Boll. Soc. Geogr. Ital.*, 73–92.

Perin, 1975. *L'inquinamento chimico della laguna di Venezia, sintesi di sette anni di ricerche*. Consorzio Depurazione Acque dell'Ente Zona Industriale di Porto Marghera. Venezia, 47–89.

Pinna, M., Ruocco, D. (eds), 1980. *Italy, a geographical survey*. Pacini Editore, Pisa.

Pinna, M. (ed.), 1991. Le variazioni elimatiche recenti (1800–1990) e le prospettive per il XXI secolo. *Mem. Soc. Geogr. Ital.*, XLVI.

Pirazzoli, P.A., 1983. Flooding ('acque alte') in Venice (Italy), a worsening phenomenon. In: *Coastal Problems in the Mediterranean Sea*, Bird and Fabbri, E.C.F. (eds), Univ. di Bologna, 23–31.

Pirazzoli, P.A., (1991). Possible defences against a sea-level rise in the Venice area. *J. Coastal Res.*, **7**, 231–248.

Provincia di Venezia, 1983. Studio geopedologico e agronomico della Provincia di Venezia, Map at scale 1:50,000, Venezia, Amministrazione Provinciale.

Rallo, G., 1988. I mammiferi. In: Cumune di Venezia – WWF, *Laguna, Conservazione di un Ecosistema*, Arsenale Editrice, 57–59

Regione Veneto, 1987. Carta della difesa del suolo e degli insediamenti.

Regione Emilia-Romagna, 1981. Uso reale del suolo. Map scale 1:200,000. Serv. Coord. Programmazione, Pianificazione, Ufficio Analisi e Ricerche Territoriali, Bologna.

Regione Veneto, 1987. Carta dell'uso del suolo. Segreteria Regionale Territorio, Venezia. Scale 1:250,000.

Rizzini, A., 1974. Holocene sedimentary cycle and heavy mineral distribution, Romagna-Marche coastal plain, Italy. *Sedimentary Geology*, **11**, 17–37.

Rossetti, G. and Rossetti, M., 1977. Idrografia e idrologia della regione del delta del Po in relazione con la degradazione delle aree tributarie. Ist Geologia, Univ. Parma, Pubbl. p. 53.

Schrefer, B.A., Lewis, R.W. and Norris, V.A., 1977. A case study of the surface subsidence of the Polesine area. *Int. J. Numer. Analyt. Methods in Geomechanics,* **1**(4), 377–386.
Torresani, S., 1989. Il processo di popolamento della costa adriatica (1871–1981). In: Bernardi, R. (ed.), *Mari e coste italiani*. Patron Editore, Bologna, 91–122.
UNDP, 1988. *The Blue Plan, Futures of the Mediterranean Basin.* Executive summary and suggestions for action. Sophia Antipolis, France.
Veggiani, A., 1974. Le variazioni idrografiche del basso corso del Fiume Po negli ultimi 3,000 anni. Padusa, Rovigo, **1–2**, 1–22.
Warrick, R.A. and Oerlemans, J. (eds), 1990. Sea Level Rise. In: *Climate Change*. The IPCC Scientific Assessment, Houghton, J.T., Jenkins, J.G. and Ephraums, J.J., (eds), Press Syndicate of the Univ. of Cambridge, 261–285.
Warrick, R.A., Gifford, R.M. and Parry, M.L., 1986. CO_2, climate and agriculture. In: Bolin, B., Döös, B.R., Jäger, J. and Warrick, R.A. (eds), *The Greenhouse Effect, Climate Change and Ecosystems*. J. Wiley & Sons, Chichester, 363–392.
Wigley, T.M.L., 1989. The effect of changing climate on the frequency of absolute extreme events. *Clim. Mon.,* **17**(2), 44–55.
Wigley, T.M.L., 1992. Future climate of the Mediterranean Basin, with particular emphasis on changes in precipitation. Chapter 1, this volume.
Zanoni, L., 1977. L'inquinamento del mare adriatico attraverso i fiumi. In: *Atti Seminario Intern. Studi, Fenomeni di Eutrofizzazione lungo le coste dell'Emilia-Romagna,* Bologna, pp. 89–105.
Zore-Armanda, M. and Găcič, M., 1987. Effects of bora on the circulation in the North Adriatic. *Annales Geophysicae,* **5B**, 1, 93–102.
Zucchetta, G., 1983. L'inquinamento della laguna veneta. *Ateneo Veneto,* CLXX (n.s XXI), **21**, 5–14.
Zunica, M., 1971. *Le spiagge del Veneto,* CNR, Centro du Studi di Geografia Fisica, Univ. Padova, p. 144.
Zunica, M., 1974. *Documenti cartografici per l'interpretazione del regime litoraneo tra l'Isonzo e il Foglia.* CNR, Padova, p. 47.
Zunica, M., 1976. Coastal changes in Italy during the past century. In: *Italian Contributions to the 23rd Int. Geogr. Congress,* 275–281.
Zunica, M., 1978. Il Delta del Po: elementi per un approccio ecologico. *Boll. Mus. Civico Storia Naturale, Venice,* **24**, 19–30.
Zunica, M. 1987. *Lo spazio castiero italiano.* V. Levi Editore, p. 285.
Zunica, M. 1991. Beach behaviour and defences along the Lido di Jesolo, Gulf of Venice, Italy, *J. Coastal Res.,* **6**, 709–719.

13

Implications of Future Climatic Changes on the Inner Thermaikos Gulf

D. Georgas (*Coastal Environment UNEP Consultant*)
and C. Perissoratis (*Institute of Geology and Mineral Exploration, Athens, Greece*)

ABSTRACT

The city of Thessaloniki, the adjacent lowland plain and the associated gulf constitute the second most important socio-economic centre in Greece. In this coastal area more than 1 million people live, while the fertile plain provides all kinds of crops and the sea sector is extensively used for maritime traffic, fish production and as a recreational area.

All these parameters will be greatly affected by the consequences of the expected future climatic changes. With an initial temperature rise of 1–2°C and a 20–30 cm sea-level rise, the impact will be a gradual one as the present intertidal zone will be flooded. This impact in general will be compensated by the evolution of various natural and other systems as long as they can exhibit high tolerance to the changing environment. However, with a greater temperature (2–3°C) and sea-level (0.5–1.0 m) rise, the impact may be catastrophic and may seriously alter the coastal morphology of the estuary and the unprotected low land. The limited seasonal rainfall will cause problems with the water balance, particularly irrigation, and to hydroelectric power-station reservoirs along the Aliakmon River. Deficiency in soil moisture will decrease fertility, thus requiring better drainage and increased irrigation. Reduced rainfall on the "thirsty" agricultural plain as well as high irrigational needs will lead to the sea-water intrusion into the aquifers, thereby deteriorating the ground-water quality. The bioclimatic shift also may affect the estuarine ecosystems.

Suggested long-term actions include careful monitoring of the local changes (weather, water quality, agricultural yields, insect population, etc.) as well as mapping high-risk areas vulnerable to flooding, protecting the economically important coastal sites, re-adjusting building standards, etc. If planning is inadequate the area will be unprepared for the consequences of the expected climatic changes.

Construction of a barrier in the outer bay is proposed. This barrier will produce a "buffer zone" between the open sea and the vulnerable highly economically important coastal area, but will not affect river discharge or marine navigation.

1 INTRODUCTION

Three interrelated and important parameters make the inner Thermaikos Gulf in northern Greece the second socio-economic centre of the country, after Athens. These are, the city of Thessaloniki to the east; 10,000 ha of fertile low-lying plain in the west formed by sediment from four rivers, and the Gulf itself, rich in fish and intensively used by marine traffic. Today more than 1 million people live in the greater area. The adjacent plain contributes the major part of the crops produced over the entire prefecture, and the port is used for a considerable part of Greece's imports and exports.

Thessaloniki was a major city built during the time of Alexander the Great, but it was also one of the few that not only survived but also developed to become the capital of northern Greece. Built in 316 BC by Kassandros for his wife Thessaloniki (the sister of Alexander), it soon became an important urban and social centre. At that time, the sea extended more than 30 km to the west to the city of Pella, then the capital of Macedonia. Up to the time of the Roman occupation (136 BC), Thessaloniki was a prosperous city because of the gold and silver mines and the agricultural production from the surrounding areas. Meanwhile the sediments deposited to the west by the rivers had made the port of Pella unoperational, which was finally abandoned.

During the Roman occupation, the relative social and economic stability of *Pax Romana* helped the city's progress. In addition to trading, agriculture and mining, an important contribution to the local economy came from the construction by the Romans of the famous "Egnatia Road" which connected the Adriatic and Black seas and passed through the city. During the same period, marshes had covered a great part of the area to the west; the lagoon at the centre had an outlet connection with the open sea in the east. During the succeeding Greek-Byzantine period (4th to 15th centuries), especially between the 9th and 11th centuries, the city became the economic centre of the whole Balkan peninsula and the second most important city of the empire. The population exceeded 100,000 many times as marine trade and industry greatly advanced, along with agriculture. A big fortress built around Thessaloniki extended all along the seashore, protecting the city from pirates and offering a shelter to the ships sailing in the northern Aegean. On the adjacent plain the deposited sediments transformed many parts of the swamps into land which, however, due to the primitive drainage methods, was not yet available for agriculture.

Gradually, especially at the Ottoman occupation period (15th to 19th centuries) agriculture increasingly developed, particularly cotton and tobacco. The area also was used for hunting and fishing. Mining of silver, lead and iron contributed to the prosperity of the Thessaloniki. By the end of the 19th century the whole plain was tillable land with occasional marshes and a small lake at its centre, where the lagoon had existed previously. The coastline was similar to that of today, and the continuous sediment deposition was threatening to cut-off Thessaloniki port, as had happened before at Pella.

The first decades of the 20th century were characterized by a great increase in population, by major agricultural transformations and by large engineering projects. Between 1925 and 1939 the Axios River course was altered, the lake and the swamps were drained, and the whole plain area was given over to agriculture. Former peasants became landowners, and the productivity rose sharply as the new farmers became particularly interested in higher yields from their own farms. At the same time, diversion of the Axios River to a more southward course saved the port from cut-off. As a result by 1951 the total population of the Thessaloniki prefecture reached 450,000, with the city itself having more than 300,000 habitants.

Since the 1950s the area has experienced an unparalleled "boom". Indicative of the city's growth is that between 1967 and 1971 about

50,000 new apartment houses were built. In 1981 the population of the Thessaloniki prefecture was over 850,000, while city inhabitants were about 700,000. Today there are more than 1 million people living in and around Thessaloniki, with 10,000 small and large industries and a total cargo traffic of 10.4 million tonnes/year.

This rapid and often unplanned development and growth, however, had obvious consequences, such as severe pollution of the Gulf due to the domestic and industrial sewage and the agricultural fertilizers, which downgraded considerably the entire natural environment. The city is already feeling the results, as swimming is prohibited in extensive coastal areas of the inner Gulf and living standards in many parts are generally low.

Today the question arises as to the impact of a future temperature and sea-level rise in the inner Thermaikos Gulf and its surroundings. How will the area look then; what will be the impact of non-climatic factors; and how will direct climatic impact be superimposed? Will the bay become a sluggish lagoon with stratified water layers and sapropel formation due to excessive pollution? How will precipitation changes affect seasonal cycles? Will there be an excessive need of water for agricultural and other uses? Could the plain revert to a swamp? What will be the effects on infrastructure, industry, housing, tourism, etc.? And, finally, how can the consequences of human violation on natural settings be minimized?

This chapter, using data from various researchers carried out in the area, attempts to answer the above questions. Moreover, it is an effort to make people, especially decision makers, sensitive to the problems and to urge them to plan as early as possible what measures to take, because when the full verification of the climatic changes occurs, it may be too late for effective action.

2 THE COASTAL ENVIRONMENT

2.1 Geomorphology

The area under study is located in Macedonia, northern Greece, between longitudes 22°30° and 23°00° N and latitudes 40°20° and 40°45° W. Our study also occasionally extends to adjacent areas when additional data are necessary to understand several interrelated mechanisms (geology, geomorphology, hydrology, etc.). The inner Thermaikos Gulf, a shallow embayment with an areal extent of 470 km^2, is bounded by a 160 km long coastline, stretching from the Kitros salt marshes to the Epanomi Cape (Fig. 13.1). Three geographically distinct basins are considered, the semi-enclosed inner Thessaloniki Bay (maximum depth 26 m), the outer Thessaloniki Bay (maximum depth 30 m) and the inner Thermaikos Bay (maximum depth 40 m). The Gulf is characterized by uniform bathymetry, the western part with slightly gentler gradients due to river sedimentation (Fig. 13.1).

The topographic relief of the adjacent land is generally smooth, surrounded everywhere, except to the north-west, by hills and mountains. The land sector to the north-west is occupied by the broad alluvial Thessaloniki plain, which meets the sea in a series of deltas. The plain

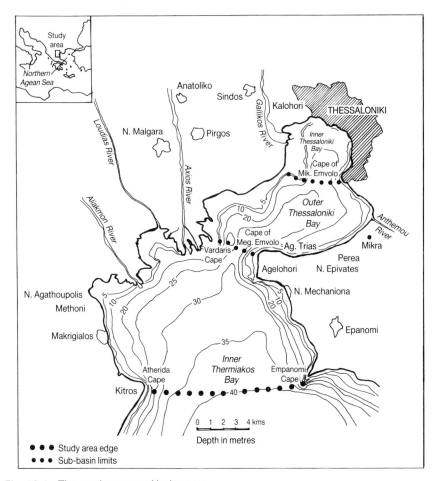

Fig. 13.1. The study area and index map.

itself is also bounded by Mount Paiko (1650 m) to the north-west and Mount Vermio (2050 m) to the west. The Pieria Hills, the northern foothills of the Olympus Massif, stretch eastward to the sea, forming the south-eastern boundary of Thessaloniki Plain. Mount Chortiatis (1200 m) lies almost behind the city of Thessaloniki to the south-east (Fig. 13.2).

Four rivers discharge in the three bays, the Gallikos, Axios, Loudias and Aliakmon rivers. The Axios River is the largest river and is mainly responsible for the geomorphology of the area. It drains a basin of 23,000 km^2 and has a route of 380 km, 300 of which lie in Yugoslavia; only the last 80 km are in Greece (Fig. 13.3). The 350 km long Aliakmon River, the longest river flowing entirely in Greece, drains a basin of 11,000 km^2. Loudias River runs between the two major rivers above and is a 38 km long draining channel with a 1250 km^2 basin. The Gallikos River is presently only a seasonal stream.

The study area belongs geologically to the Axios/Vardar geotectonic zone, which represents the last stage in the infilling of a subsiding basin.

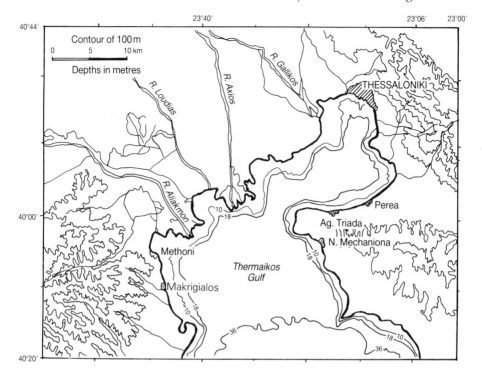

Fig. 13.2. Topography of the area (after Chronis, 1986).

This alpine orogenic basin has experienced intense tectonic stresses and has been infilled by debris and volcanic tuffs from the surrounding mountains. The principal factors in sediment transport, particularly during the Quaternary, have been Axios and Aliakmon rivers (Zimerman and Ross, 1976).

The sediments of the basin reflect the time variability of two major contrasting environments, that is sediment of terrestrial origin brought down by rivers and marine sediments consisting mainly of shell fragments. The interbedded marine and terrestrial facies are particularly evident in the Quaternary deposits of the Thessaloniki Plain, reflecting sea-level changes (Nedeco and Grontmy, 1970).

2.2 Evolution

As mentioned above, the Axios and Aliakmon rivers have been the major controlling factors in the evolution of the progressively expanding Thessaloniki Plain, especially since Holocene time when the sea was close to the alpine marginal boundary (Chronis, 1986). Historic evidence from the 5th century BC indicates that the city of Pella, the capital of Macedonia during the reign of Alexander the Great, was adjacent to the sea. Ruins from this city are now located about 30 km inland (Fig. 13.4). During the 2nd and 1st century BC only an artificial channel existed for the communication to the sea. Struck (1908, Fig. 13.4) believed that from approximately the 5th century AD the progressive filling of the estuary of

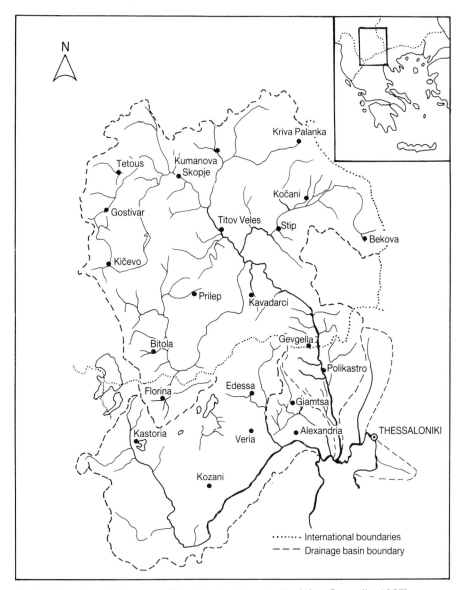

Fig. 13.3. Axios, Aliakmon and Loudias drainage basins (after Ganoulis, 1987).

the western Gulf resulted in the cut-off and final isolation of the shrinking Gianitsa Lake. The lake was connected to the open sea by the Loudias River.

Increasing sedimentation gradually shifted Axios delta to the west where the river was discharging either to Loudias River or to the adjacent Gianitsa Lake. In the 16th century, floods turned the lake into a huge swamp, which periodically received the waters of Aliakmon River. Although the 19th century damming of Aliakmon diverted its course back to the old

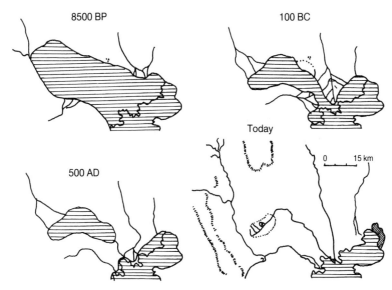

Fig. 13.4. The evolution of Thessaloniki Plain (after Struck, 1908).

channel, a dam break caused catastrophic flooding and malaria became a significant local health problem. Construction projects in 1925–1935 diverted the Axios River and channelized its downstream course to avoid the cut-off of Thessaloniki Bay into a lake. Additionally, the draining of Lake Gianitsa, the reclamation of the adjacent swamp, and other works, prevented the plain from inundation, and malaria subsequently disappeared. Later, reclamation and irrigation projects converted the area into a highly productive agricultural plain of major economical importance (Evmorphopoulos, 1961).

2.3 River Runoff, Sedimentation Rates

The sediment contribution from the major rivers and streams discharging into Thermaikos Bay is influenced by climate, especially precipitation. The local precipitation pattern (see Chapter 3) has the character of the eastern Mediterranean region: high during the winter and low during summer (Fig. 13.5).

Upstream water management of Axios/Vardar River in Yugoslavia has been undertaken through several small projects, whereas in Greece only one irrigation diversion channel exists. However, the present damming is insufficient to prevent flooding. Thus, Axios has a very high mud load, with an annual deposition rate of 8.4×10^6 m³/yr in the delta and a lateral overflow of 0.9×10^6 m³/yr on its way downstream (Therianos, 1974).

In contrast, Aliakmon River has an almost entirely controlled flow through three hydroelectrical dams along its course. Sediment transport, therefore, is relatively low, with an annual rate of deposition in the delta of 5.6×10^6 m³/yr and a lateral overflow of 0.6×10^6 m³/yr (Therianos, 1974). Loudias is a drainage channel with a very low average flow, 2–3 m³/s. Sometimes, during low summer discharge, the sea intrudes in the estuary especially during high tide.

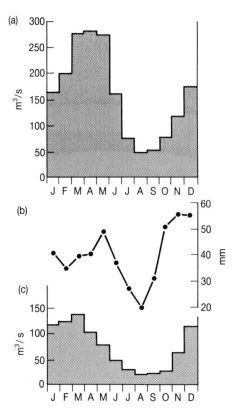

Fig. 13.5. (a) (c), Mean annual variation of the runoff of the Axios and Aliakmon rivers (Therianos, 1974) and (b) mean annual variation of precipitation (1930 to 1972) (Angouridakis and Machairas, 1973).

Gallikos River has a stream-like annual deposition of 0.16×10^3 m^3/yr in the delta and 0.04×10^3 m^3/yr laterally.

2.4 Circulation Patterns, Sediment Deposition, Marine Dynamics

The main factors controlling the local circulation in the Gulf are winds, tides and density currents. Circulation is differentiated in the inner and outer Thessaloniki Bay and the inner Thermaikos Gulf. The prevailing winds are the "Vardaris" north–north-west winter wind and the summer south–south-east etesian winds (see Chapter 3).

Tides are small and semi-diurnal, ranging from 5 to 30 cm during spring. Short and long periods of sea-level oscillation, corresponding to south-west sea breezes and barometric pressures, are also low (Wilding et al., 1980). Significant tidal and density currents occur only during the intensive summer stratification of sea-water masses. The schematic circulation for north–north-west and south–south-east winds is given in Fig. 13.6.

Sedimentation in the Gulf is controlled by the quantity and the grain size of the discharged sediment in combination with the oceanographic

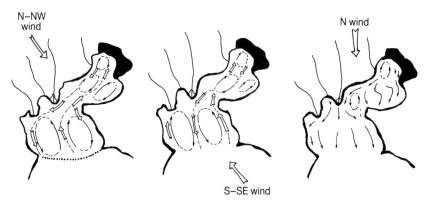

Fig. 13.6. Shallow and deep sea water mass circulation models for N–NW (a), S–SE (b) and surface circulation for N winds (c) (Ganoulis, 1987).

conditions. Thus, the low-amplitude wave energy in the inner Thermaikos leads to a deposition of the coarser sediments in front of the estuaries. "Bird-foot" deltas characterize the Axios and Aliakmon estuaries, resulting in the rapid growth of Thessaloniki Plain during the last 1000 years and the corresponding reduction in the width of the Thessaloniki Bay entrance. Thus, the present distance between the Cape of Mount Emvolo (or Mount

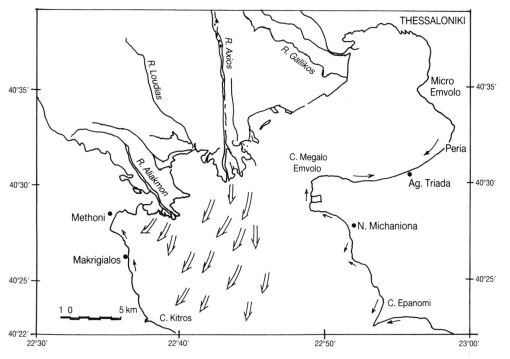

Fig. 13.7. River and coastal sediment movement. Large arrows depict river sediment dispersal while small black arrows indicate coastal sediment movement (Chronis, 1986).

Karabournou) and Axios delta, which is the limit of outer Thessaloniki basin, is only 4.5 km (Fig. 13.7). The limited input of coarse sediment in the east causes the erosion and retreat of the eastern coasts. Therefore, the only coarse debris subject to reworking by wave action is derived from the erosion of the older steep Quaternary sediments at the coast of Emvolo and Mechaniona to the east, and Makrigialos and Methoni in the west. The erosion of the western and eastern coasts is mainly due to the strong wave action generated by the south-east and north-west winds respectively (Chronis, 1986) (Figs 13.7 and 13.8).

The suspended fine material of the river discharges "slides" over the stratified saline wedge. It is deposited off the western coast, where suspended particles flocculate to form larger grains. Changes in this depositional regime usually reflect climatic disturbances (Balopoulos, 1986).

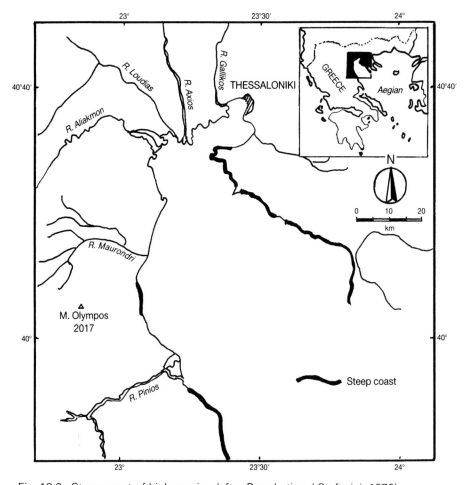

Fig. 13.8. Steep coast of high erosion (after Brambati and Stefanini, 1970).

3 Water Cycle

3.1 Meteorological Setting

The greater Thessaloniki area is characterized by a meso-Mediterranean semi-arid climate with cold winter (min. 0–3°C) (Mavromatis, 1980). However, the local climate of the region does not appear to be typical of its latitude and Mediterranean setting. The elongated north–south Axios valley plays an important role in the weather of Thessaloniki Plain as it drives the eastern European and Balkan peninsula anticyclones to the eastern Mediterranean. The prevailing north–north-west winds frequently blow throughout the year and are enhanced during the winter. During the summer the characteristic southerly Etesian winds of the eastern Mediterranean region are common but weak in Thessaloniki Bay. They follow a diurnal cycle, blowing strongly during the day and abating at night. Smaller-scale winds, like the daily sea breezes, are generated by the land–sea temperature differences. Mean wind speeds have a primary maximum in the winter (December–February) and a secondary maximum during the summer (June–August) (Fig. 13.9).

In spite of the above, the weather of Thessaloniki is characterized by considerable and rapid variability. Sudden summer thunderstorms occur, possibly caused by the annual frontal cool displacement of the warm and dry southerly winds. The annual precipitation average is 480 mm.

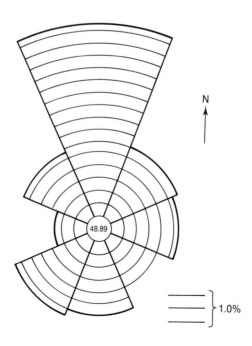

Fig. 13.9. Mean annual frequency of wind direction in Thessaloniki (1930 to 1971) (after Livadas and Sahsamanoglou, 1973).

506 *Climatic Change in the Mediterranean*

Although precipitation is higher during winter, mean monthly values do not appear significantly higher than those in the summer (Fig. 13.10). Air temperatures range from 0–38°C over the year. Therefore sea water is subjected to a strong cycle of summer heating and winter cooling (Angouridakis and Machairas, 1973).

3.2 Sewage and Industrial Output

Industrial wastes, domestic sewage and river discharged pollutants from the adjacent agricultural plain and urban areas are the polluting constituents of inner Thermaikos Bay. The bay receives inputs mainly from the Sindos industrial zone drainage channel, minor disposal from the industrial sites in the east, and from the domestic sewage (Fig. 13.12).

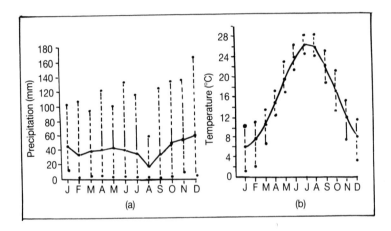

Fig. 13.10a. Meteorological conditions over Thermaikos Bay: (1) mean annual variation of precipitation (1930 to 1972) after Angouridakis and Machairas, 1973), and (2) mean annual variation of air temperature (1892 to 1973, from Flocas and Arseni-Papadimitriou, 1974).

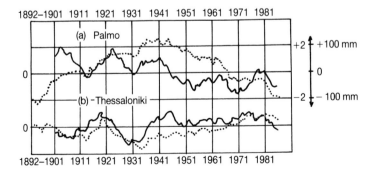

Fig. 13.10b. Palma (a) and Thessaloniki (b) annual precipitation values smoothed with a 10-year running mean (–) and time-series (Palma and Thessaloniki) of annual cumulative deviation from the mean (. . . .) (Maheras, 1987).

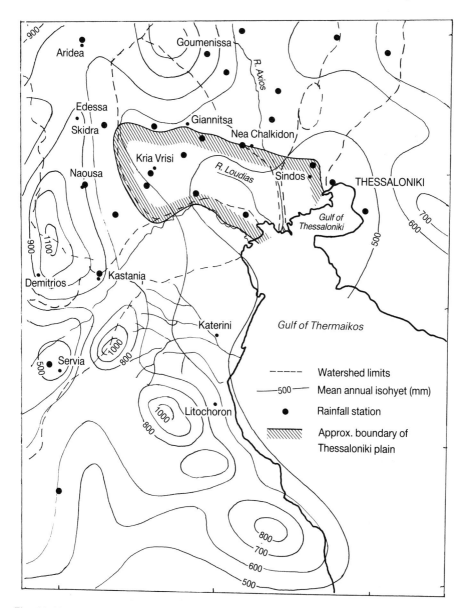

Fig. 13.10c Precipitation catchments of the area (Nedeco and Grontmy, 1970).

A total of over 10,000 industries are distributed along the greater Thessaloniki industrial zone; of them, 300 are considered large. Among the 35 largest are a refinery and a steel mill, a canning factory and papermills. They are primarily responsible for the wastes discharged into the inner bay. However, the industrial plants near the Axios and Gallikos rivers have waste-processing units that discharge either directly into the Bay or to the highly polluted Sindos channel. Additionally, untreated domestic wastes of 1 million inhabitants of Thessaloniki greater area are directly discharged into the Bay. This sewage input is estimated to amount approximately 130,000 m³/day or 36,000 kg/day BOD_5 (Ganoulis, 1987). A new sewage disposal project, which is under construction, will provide a main waste collector and treatment plant that will discharge treated sewage to the Axios estuary.

The pollutants drained from the plain between Gallikos and the Axios delta are derived from the agricultural sector and from several industries. During growing periods and irrigation, the excessive fertilizer and the other additives (pesticides, etc.) are carried through drainage channels and pumping stations to the sea. Consequently, during spring and summer months, the waters of the east and west Chalastra pumping stations are rich in PO_4, NH_4, NO_2, and NO_3 (Fig. 13.12).

Domestic wastes of the coastal and adjacent villages do not affect the outer Thessaloniki Bay, but the Axios, Loudias and Aliakmon rivers and two pumping stations of north Malgara and Klidi do discharge there (Fig. 13.11).

Loudias, which is considered one of the more polluted rivers in northern Greece, is the main drainage outlet for the greater part of the irrigated Thessaloniki Plain area, especially during summer irrigation. It receives additionally a 50,000 m³/day of industrial waste, rich in organic material derived from milk and sugar plants. During the growing season, the values of several pollutants (BOD, COD, PO_4^{3-}, NO_2^-) exceed several times the EEC's higher limits (Ministry of Environment, 1986). Thus, fish have disappeared due to the anaerobic conditions.

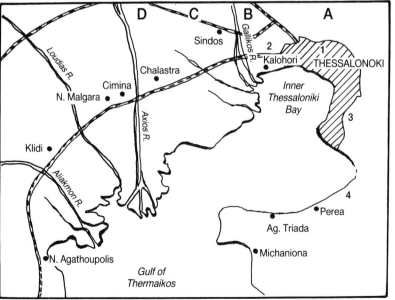

Fig. 13.11. Industrial activities.

Inner Thermaikos Gulf and climatic change 509

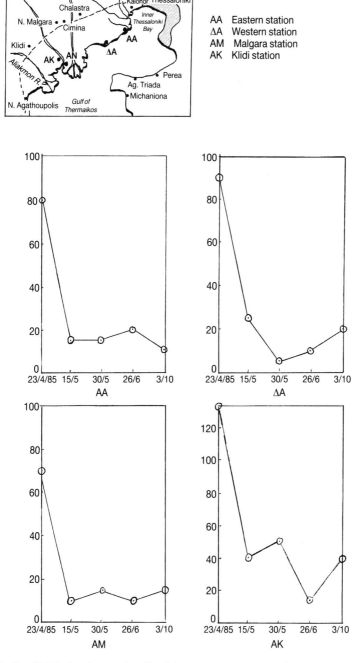

Fig. 13.12a. BOD$_5$ load rates (mg/l) of the pumping station discharges (Ganoulis, 1987).

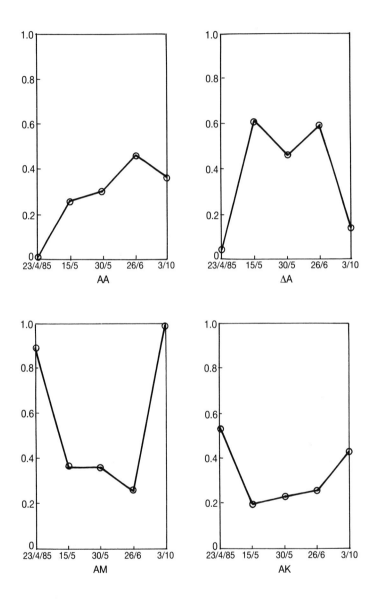

Fig. 13.12b. NO_2^- rates (mg/l) of the pumping station discharges (Ganoulis, 1987).

Inner Thermaikos Gulf and climatic change 511

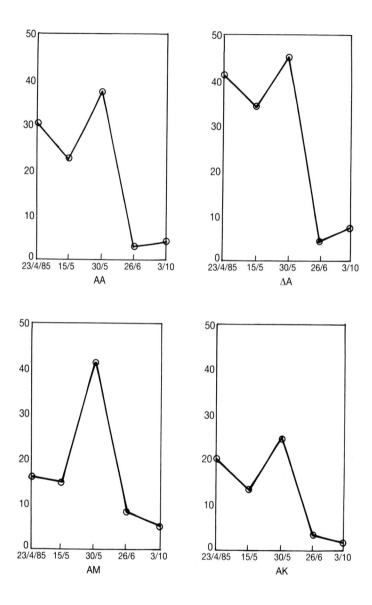

Fig. 13.12c. NO_3^- rates (mg/l) of the pumping station discharges (Ganoulis, 1987).

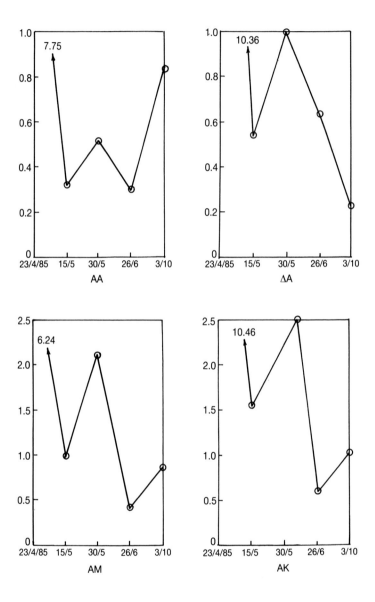

Fig. 13.12d. NO_4^- rates (mg/l) of the pumping station discharges (Ganoulis, 1987).

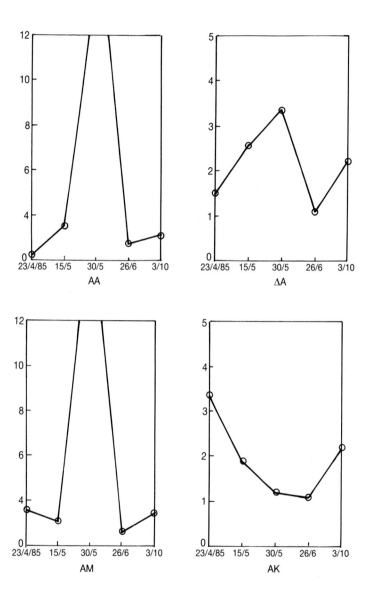

Fig. 13.12e. PO_4^{3-} rates (mg/l) of the pumping station discharges (Ganoulis, 1987).

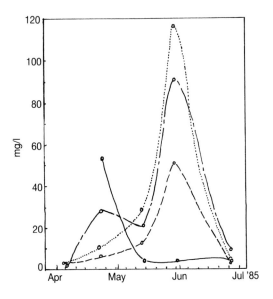
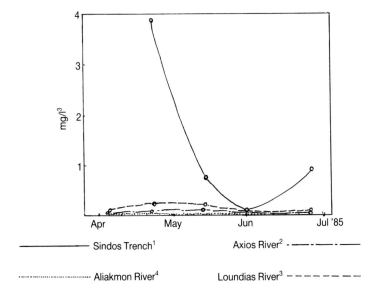

Fig. 13.12f. NO_3^- (upper) and NO_2^- (lower) rates (mg/l) of the Sindos trench[1], Axios[2], Loundias[3] and Aliakmon[4] rivers (Ganoulis, 1987).

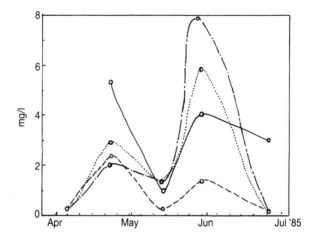

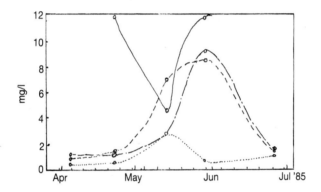

Fig. 13.12g. NO_4^- (upper) and PO_3^{3-} (lower) rates (mg/l) of the Sindos trench[1], Axios[2], Loundias[3] and Aliakmon[4] rivers (Ganoulis, 1987).

Concerning the other rivers, there are no data available for the Yugoslavian part of Axios/Vardar River, although it runs through large industrial towns (Skopje, Titov Veles). Aliakmon River is the cleanest river that discharges in the Gulf (Figs. 13.12f and g), but both it and the Axios waters are rich in $CaMnCO_3$ and $CaCO_3$ due to the karstic limestone and ophiolithic origin of their drained basins, respectively.

3.3 Fresh-water Resources and Ground-water Management

Considering the hydrology of the south-east part of Thessaloniki plain, surface inflow consists entirely of the irrigational diversion of Aliakmon and Axios rivers. Other inflow, such as from rainfall and ground water pumped from artesian wells, adds to the water balance of the area. The outflow drain is managed through a channel network where the drained water is pumped via several stations and the Loudias channel (Nedeco and Grontmy, 1970).

The hydrogeologically important upper sedimentary strata of the whole plain consist of interbedded marine and lagoonal facies represented by clays, sands and gravel. Thus, the top-most 80 m of sediment are of terrestrial origin, gradually changing seaward into marine clays and a few thin sandy horizons (Fig. 13.13).

The ground-water tables can be differentiated into a shallow phreatic water level (at about 10 m depth and a thickness of 6 to 8 m), an intermediate and a deep aquifer. The shallow phreatic zone is supported mainly from the lateral infiltration of the main rivers and other streams, rainfall and irrigation. It is largely controlled, however, by the extensive artificial drainage network. The water table rises and falls within a range of 0.30–0.50 m in response to infiltration from rainfall. Furthermore, it is affected by summer irrigation, mainly by the lateral subsurface drainage and evapotranspiration.

The artesian intermediate aquifer lies at 40 to 60 m (locally 30 to 80 m) depth and is fed from the same phreatic catchments. Also it receives water from the lateral water movement and from the intake marginal areas in the highlands surrounding the plain. The subsurface main flow direction follows the old riverbed course.

The deep artesian aquifer, lying at a 100–200 m depth, is fed by the same source as the intake sources noted above (Knithakis and Tzimourtas, 1987). The marine origin of the sediments produces high salinization-alkalization of the phreatic ground water, as the infiltrating surface water flushes sodium chlorate to the phreatic and artesian aquifers. As a result, the phreatic water is subject to high salinization and alkalization, while the underlying intermediate artesian aquifer is not suitable for drinking. Mean annual fluctuation of the water table is 0.5–4 m. The deep artesian ground water exhibits satisfactory quality, and is used mainly by the Water Authority of the city of Thessaloniki and its suburbs. Water-table fluctuation on the order of 7 m is common due to the intensive use of this aquifer.

4 Ecosystems

4.1 Soils

The pedological features of the study area are characterized by interbedded marine, lagoonal and river deposits. The first two are classified as entisols, whereas the third is considered as inceptisols. The high mica and the relatively low quartz content is typical of the soils occurring in this area. The poor clay crystallization and sand disintegration are also typical of young soils. According to silt content, several soil families are distinguished, ranging from coarse silty soils along the coast to more clayey types landward. Permeability depends largely on the top-soil structure and tends to increase with depth. The silty soils exhibit high drainability

Inner Thermaikos Gulf and climatic change

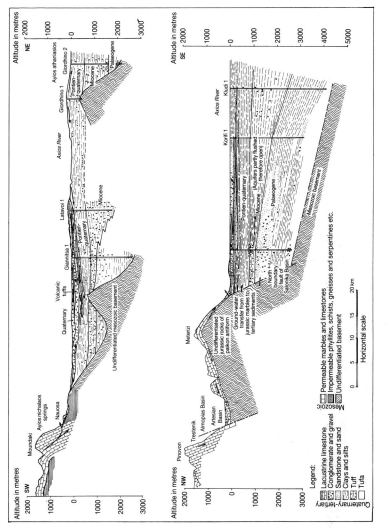

Fig. 13.13. Hydrogeological cross-section across Thessaloniki plain (Nedeco and Grontmy, 1970).

and are usually used for rice growth. The construction of the drainage and irrigation systems along the Axios and Aliakmon rivers, which carry waters rich in Ca^{++}, led to a decrease of the soil salinity and alkalinity. The application of gypsum-based techniques were also beneficial to the desalinization of the reclaimed plain soils. Although de-alkalinization is a slower process than desalinization, the occurrence of non-saline/alkaline soils indicates a great improvement of the reclaimed soils. The persisting salinity in a few local areas is due either to the insufficient irrigation and drainage conditions or to upward seepage and concentration of saline ground water in a locally high water table (mainly along the Loudias River and south of the Aliakmon River). In general, salinity increases with depth, indicating a leaching process. (Nedeco and Grontmy, 1970).

4.2 Biota

4.2.1 Flora

Terrestrial vegetation of the coastal environment is characterized by the following vegetation types: sandy beaches, rice fields, saline soils, marginal saline soils, lagoons and river forest. The river-drained basins exhibit broad bioclimatic zonations. The following is a brief examination of these types:

Vegetation of the sandy beaches. Along the west–south-west coast, between northern Agathoupolis and the Aliakmon estuary as well as at several sites on the Axios estuary, the loose sand and sediment zone extends 3–6 m from the sea, and is colonized by nitric- and nitrophile-type vegetation. On the intertidal zone, especially where organic detritus has accumulated, the *Salsola kali-Xanthium strumarium* plant communities grow, typical and characteristic species of all Greek coasts. On the coasts with less slope, such as south of Aliakmon and between the Axios River bank and Kalohori, the area is occupied by the *Antipicethum tatarici* community. Although a few data are available, it does not appear to be any typical "sand dune" plant community, (Appendix I, Ministry of Environment Report, 1986). In the densely populated eastern coasts and beaches, chlorophycaea and phenophycaea species characterize the aggravated marine environment and all biotopes of the sector. Consequently, there is negative evolution of the benthic flora which is contaminated from the industrial and domestic wastes (Nicolaides, 1985).

Vegetation of rice paddyfields. Rice fields are artificial temporal wetlands, characterized by shallow stagnant water, not unlike an artificial lake environment. The algal-rich floor, mainly due to the high eutrophic level of the water, plays an ecologically important role in rice growth. Several nitrogen-blocking blue-green algae species in cultivated alkaline soils supply nutrients and oxygen to the rice. On the other hand, several types of fish living in the rice paddies (i.e. *Tilapia nitolica*) feed on blue-green algae.

Saline soils vegetation. Where cultivation was established after reclamation, marine saline soils predominate. Typical Mediterranean communities are found locally, the dominant of which are: the coastal *Salicomiethum european*, spreading in the extended submerged areas that are flooded most of the time (lagoons, saltmarshes); *Salsola soda-Sueda splendens* in limited humid areas in the estuaries; *Halocnemum strobicei* occurs in recently reclaimed areas between the rivers; *Juncetum maritimi*, are in areas very close to the sea; *Hallmionetum portulacoides* occurs near channels and creeks on saline soils.

Vegetation of the marginal saline soils. Estuarine environments, not under direct influence of sea water, are considered marginal. Although most of these marginal areas are now cultivated, locally the dominant species is *Juncus acuctus* which grows in huge communities, especially between the Loudias and Aliakmon rivers. High-growing shrubs (*Tamarix hampeana*) occur in elongated strips along the banks downstream of the Axios and Aliakmon rivers. *Tamarix* shrubs grow better in humid conditions, while in saline places its growth is limited. These species are considered an ideal biotope along both rivers, and are protected by the Ramsar contract.

Vegetation of the lagoons. Reed families find favourable growing conditions in lagoons and swamps, forming two communities, *Bolboschoenetum maritimi* on saline fresh water and *Phragniterum* in small channels with direct access to the sea.

River forest vegetation. Relict river forest vegetation is present along the downstream banks of the Axios and Aliakmon rivers, where dominant species are *Salix pedicellata, Salix alba, Populus alba* and *Alnus glutinosa*.

4.2.2 Fauna

Although land reclamation and other recent projects have limited the natural biotopes, the deltas of Axios and Aliakmon rivers are still considered as wetlands of major importance. The Ramsar contract will protect the forests along the downstream banks of the Axios and Aliakmon, the small sandy river islands and the salt marshes and lagoons. Summer rice paddyfields are all ideal places for birds which nest or temporarily rest.

Most of these significant bird colonies are unique in Greece and among the largest in Europe (except USSR). Birds are spread from Aliki lagoon (20 km south of the study area) up to Gallikos River, and from the coast to the marginal cultivated land. The wetland is characterized by the widespread salt marshes, lagoons and swamps, with abundant mudflats. The low tidal range and the low-amplitude seasonal prevailing north wind (Vardaris) permit a great deal of the coastal zone to provide a hospitable environment for thousands migration birds (Ministry of Environment, 1986).

A significant number of eagles, herons and cormorants nest along the banks of downstream Axios, between the main motorway and the railroad, as well as in the reed beds along the channels of the Loudias. A big community of avocets rests on the Aliki (near Kitros), and pelicans are in the mouth of the Aliakmon River. Marsh harriers often feed from the rice fields, whereas knots, plovers, avocets and curlews are found on the mudflats during winter. Gulls seem to have benefited from the increasing pollution in the deltas, while the already limited numbers of duck and geese have been seriously reduced by hunting.

4.3 Agriculture

The productive estuarine area, as well as the rest of the plain, was formed after reclamation and subsequent soil improvements. When the area was given to cultivation, the complex ecosystem benefited from irrigation and was soon developed into a more simple system with high fertility and productivity. Thus, by enhancing nutrients and by adding fertilizers, better yields were achieved. The action of soil bacteria, however, was critically reduced, thus introducing the use of insecticides, which are essential to reinforce the reduced resistance of the plants. Although, the link of the ecosystem was already broken as a result of the agricultural development, all crop cultivations are characterized by the extremely high use of fertilizers which then increase the eutrophic level of the main collector. Also insecticides and pesticides negatively affect the marine ecosystem.

Today the highly productive Thessaloniki plain has both major agricultural and economical importance. In an area of about 10,000 ha, the sector between the rivers is cultivated mainly with maize, rice, cotton, cereals and sugar beets. The rest of the lowland is used for pasturage and animal breeding (Fig. 13.14).

Corn is considered one of the most valuable crops, easily grown in a wide variety of soils, ranging from sandy to clayey silts. Yields are around 800–1200 kg/1000 m^2. Saline coastal areas between rivers, having a high water table, are conducive to intense rice growth. This crop culture, with high needs in water (3 m^3 water per m^2 area), yields 600–8300 kg/1000 m^2. Cotton is mainly cultivated between the Loudias and Aliakmon rivers, yielding 300–400 kg/1000 m^2. Sugar beets are cultivated to the north-west plains near a sugar factory, yielding 5000–7000 kg/1000 m^2. Cereals are mainly between the Aliakmon and north Agathoupolis (Fig. 13.14). The 1986 agricultural production of the coastal plain area and its percentage contribution to the total production of Thessaloniki district (in brackets) for the various crops was: cereals: 48,400 tons (21%); corn: 61,000 tons (75%); rice: 76,000 tons (100%); cotton: 31,600 tons (80%), sugar beet: 27,400 tons (81%) (Prefecture of Thessaloniki, 1986).

4.4 Aquaculture

The Gulf of Thessaloniki was known widely for its excellent quality of its fish, but pollution has gradually degraded the marine environment. Fish farming has taken place in the area since the last decade, while between the Loudias and Aliakmon a big fish hatchery was employed. Today the inner Thessaloniki Bay is seriously polluted by industrial and domestic sewage and, as a consequence, fishing activities have been banned. However, due to high eutrophicity, estuaries are still attracting fish, especially at the Loudias and Axios estuaries.

At the outer Thessaloniki Gulf, in the estuaries of the Axios River and at the opposing eastern coast, there is a number of primitively equipped oyster and mollusc farms (Fig. 13.15) with a total production of 1000–2000 tons/year.

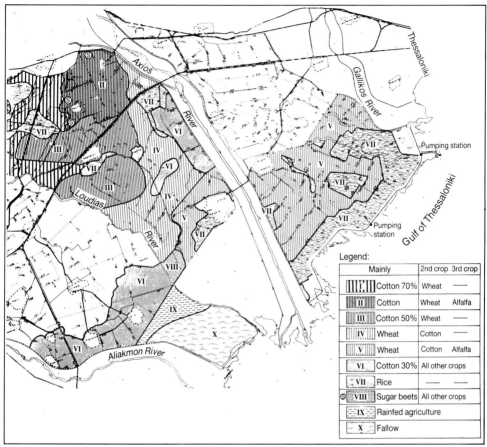

Fig. 13.14. Crop distribution on the coastal Thessaloniki plain (Nedeco and Grontmy, 1970).

5 COASTAL ACTIVITIES

5.1 Industry
Thessaloniki is the second economically most important city in Greece. The fast growing industrial activity of the area (see Chapter 3.2) is distributed in the four major industrial zones in the north–north-west suburbs and also in four minor urban industrial sites at the edge of the city (Fig. 13.11). Some of the largest industries are a brewery and a bottling plant (Gallikos); paper mills and canning factories (Monastirion and Axios); a chemical and a steel industry and a refinery (Diavata). Some of them are quite close to the coastline and are developed down to the shore, especially east of Kalohori, where there is a refinery (expanded down to the jetties), a cement industry, slaughter houses, etc.

5.2 Infrastructure
Due to its geographical setting, the city of Thessaloniki exhibits a well-developed transportation network. Railroad and motorways run across

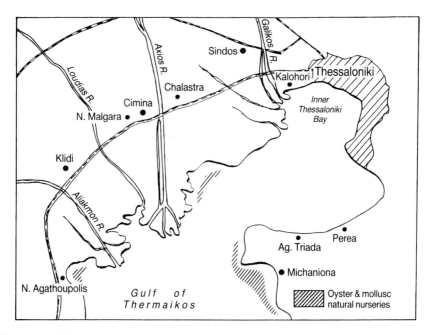

Fig. 13.15. Sites of natural nurseries of oysters *(Ostria edulis)* and molluscs *(Mytilus galloprovincialis)*.

the upper estuaries, linking Thessaloniki with Athens (520 km), with the Yugoslavian borders (70 km), and with central and eastern Macedonia and Thraki. It is also the nearest European approach to the eastern Mediterranean. The port of Thessaloniki has a total cargo maritime traffic of 10.4 million tones/year (Thessaloniki Port Authority, 1987). The direct airlink to some main European capitals and domestic airlink is done through the Mikra Airport which is also used for the summer tourist charter flights. Most of the lowlands of the reclaimed estuarine area are protected by a 1–1.50 m high dyke along the sea, whereas a cement sea wall of similar height protects Thessaloniki itself.

5.3 Recreational and Tourist Facilities

Increased pollution of the area from the city's unprocessed domestic discharges has become the dominant problem for the recreationally used eastern coast of the inner and outer Thessaloniki Bay. Walking in the city along the sea was once pleasant and enjoyable. Now the repulsive odour in the sea breezes drives people away. The situation is better at Agia Triada Beach in central Thessaloniki Bay, and further at the outer bay, where the beaches of Nea Michaniona and Epanomi still attract tourists during the summer. These are the nearest beaches to the city, just 30 km away, highly populated at summer season. Camping facilities are organized in two sites, at Agia Triada and at Epanomi, run by National Tourist Organization (NTO). Additionally, Methoni, Nea Agathoupolis and Makrigiakos are camping beaches at the west coast, south of the Aliakmon estuary (Fig. 13.16).

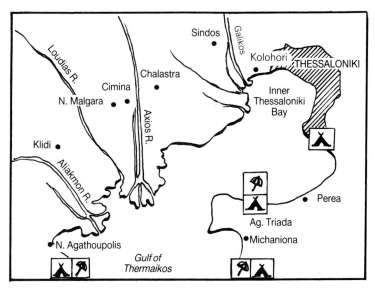

Fig. 13.16. Camping sites and tourist beaches.

5.4 Population Distribution

Thessaloniki is the second largest city in Greece and the capital of Macedonia. Growing fast, from 320,000 inhabitants in 1951 to 702,000 in 1981, it now has reached 850,000 inhabitants, 1 million with the surrounding suburbs. It is a densely populated coastal city which has expanded southward to the coastal suburbs and villages. Tourist villages at the eastern and western coasts exhibit a summer seasonal population of over 45,000 people (Fig. 13.20).

6 Impacts

6.1 Coastal Environment

Sea-level changes and river-transported debris have been the major factors in shaping the geomorphology of the studied area. Since any natural environment is subject to continuous change, one way to outline the expected future climatic changes is to monitor lesser amplitudes of sea-level change.

According to the proposed climatic changes scenario, the expected 1–2°C temperature rise will be followed by a gradual mean sea-level rise of 20–30 cm in the next 50 years, as a consequence of thermal expansion of the sea-water column and partial melting of the polar icecaps. However, local sea-level changes could increase significantly because of land subsidence and soil compaction. In such a case the excessive sea-water masses might slowly flood the low-lying unprotected coastal and estuarine areas, as well as the lowlands and the reclaimed areas of Thessaloniki plain (Fig. 13.17). This will not be pronounced in the next 50 years, but as storm surges gradually become more intense, catastrophic events will occur.

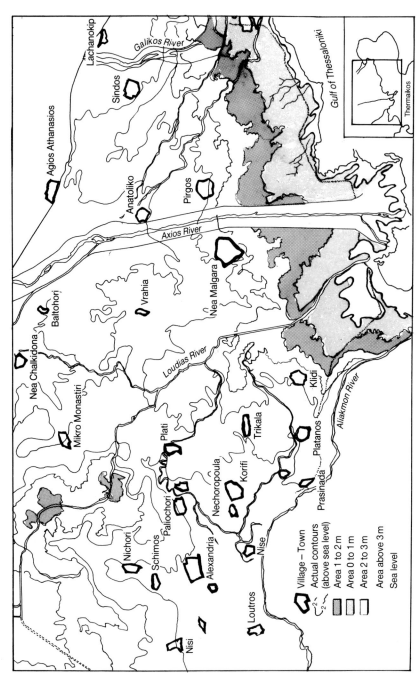

Fig. 13.17. The Axios plain landscape showing the low-lying areas of high-flood risk (Nedeco and Grontmy, 1970).

Inner Thermaikos Gulf and climatic change   525

The intrusion of salt water into the low-flow river mouths of the Loudias and Gallikos rivers, as well as the Aliakmon and Axios, is expected to take place, especially during high spring tides reinforced by south-west winds. This could possibly lead to a decreased sediment deposition at the upper delta, causing a reshaping of the present-day lower deltas. Lagoons with direct access to the sea also will be subject to flooding, while marshes will be converted to lagoons. When mean sea-level rises by 0.8–1.0 m (40 to 75 years from now, depending on local subsidence), this effect would become noticeable and catastrophic with major economical impact as the coastline shifts landward (Fig. 13.18). Most of the coastal areas of

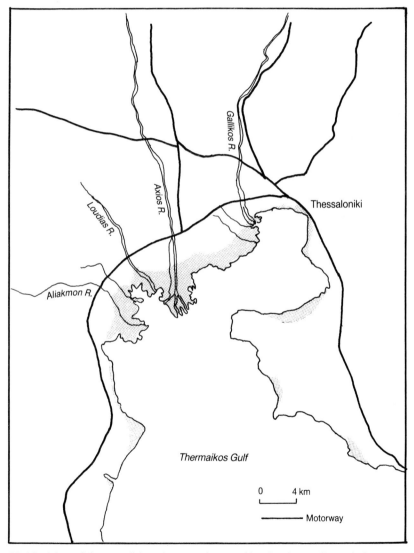

Fig. 13.18. Map of the possible submerged coastal lowlands at a 3 m relative sea-level rise. The total inundated area is approximately 152 km².

the reclaimed agricultural Thessaloniki plain are today defended by a 1.5–2.0 m high dike, and at the city of Thessaloniki by a 1–1.5 m high concrete sea wall. In this case, the height of the constructions might have to be increased. Also engineering measures will have to be carefully evaluated for their negative and positive effects, as they are not likely to represent a realistic long-term solution.

Low-lying sensitive areas, such as the developing industrial zone of Sindos (from the port to Kalohori) (Fig. 13.1), the Mikra airport, the area next to Aliakmon River and the east coast, are totally exposed to even a minor sea-level rise. Therefore, it is urgent to identify and monitor the present coastal lowlands that are vulnerable to flooding, and to map the high-risk areas.

However, it must be noted that no significant changes are expected in the circulation and the tidal regime of the studied area. Thus, due to the added watermass, short period mean sea-level oscillations caused by south-westerly sea breezes may slightly exceed the present tide value of 10–15 cm.

A temperature rise will possibly result in a warmer sea, reinforcing the sea-water stratification. This will help the offshore transport of fine sediment, as the suspended matter and the river water will be transported further away. On the other hand, the oxygen depletion and the reduction of the eutrophic zone are expected to cause severe consequences to marine biota (see section 4). This might establish bottom anaerobic conditions, especially in the presence of excessive organic carbon environment in the sluggish and polluted waters of inner Thessaloniki Bay. Thus, under a reduced sediment input, the deposition of sapropelic black mud could begin.

High evaporation, due to the warmer climate, will increase the future needs of fresh water and the hydrological balance will be affected. The Axios and Aliakmon rivers' runoff is expected to decrease due to the greater water used for irrigation and hydroelectric energy. Therefore bilateral international arrangements between neighbouring countries should specify the quantity of the river water to be used by each country in case of international river paths like the Axios. These arrangements should include future diminishing conditions due to the climatic impacts.

Faster sediment filling of water reservoirs and the irrigation channels of the Aliakmon River could result from higher sediment loads, due to more torrential rainfall onto drier desertified slopes. The lack of coarse sediment input will increase the erosion of the already eroded eastern and south-western coasts. As a result, the tourist beaches of Emvolo, Mechaniona, Macrigialos and Methoni eventually may be destroyed.

The Loudias River is expected to become polluted as it drains part of Thessaloniki plain during winter. The Gallikos River on the other hand, will become a highly polluted, more temporal stream, as precipitation might be reduced into fewer but heavier storms.

6.2 Water Cycle

Based on the revised scenario of Wigley (1989), a future northward climatic shift is expected. According to this, the autumn and spring seasons will be slightly shortened in length, whereas winter will be gently

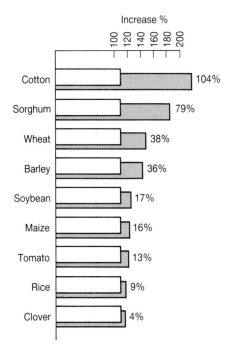

Fig. 13.19. Predicted increase in yields of nine major crops caused by a doubling of carbon dioxide concentration (UNEP/GEMS, 1987).

mild and rather moist. Although the future local cyclonic regime of the southern end of the Balkan peninsula is unpredictable, Vardaris, the important prevailing north winds, are not expected to be affected. Summer Etesian winds, on the other hand, might be increased due to the temperature rise.

Local annual precipitation will be rather reduced, characterized by increased variability and patchiness of the rainfall; warm season rainfall will almost disappear, transferred to winter. Aridity will increase not only as a result of decreased annual precipitation, but also due to the change in frequency and magnitude of the rainfall events. Because of temperature rise, evapotranspiration will be greater. On the other hand, a chain of impacts associated with the probable deterioration of soil properties and structure will take place which might influence the balance between infiltration and runoff especially in the reclaimed areas. If overland flow increases at the expense of infiltration, excessive erosion and flooding could occur throughout the drainage basins and the lower plain, also causing hydrological disturbances. Therefore, a general decline in percolation and eventually in the ground-water supply must be expected.

Saline water will intrude further landward into the plain's subsurface, thus deteriorating the phreatic water and soil quality. On the other hand, there will be no direct impact on the phreatic intermediate and deep artesian water. There also is no evidence that coastal artesian aquifers have direct access to the sea, as they exhibit almost constant annual salt variations throughout the year, due to the dominance of marine sedimentary strata. However, quality might be influenced by the fact that the overlying phreatic water will be deteriorated due to lateral percolation.

Under a rising sea-level the irrigation and drinkable fresh-water needs will rise, especially in the summer. Thus the intensive use of artesian water might gradually result in sea-water intrusion. In this case first the intermediate and then the deep coastal aquifer will be subjected to salinization, deteriorating the quality of drinkable water sources (Kalohori, Axios) used by the city population.

Lower discharge of surface flow will reduce the rivers' ability to carry and dilute waste materials. Sea and water pollution is therefore likely to increase unless effluents are properly treated. The already highly polluted Sindos drainage channel and later the Loudias River, as well as the other minor drainage channels, must be incorporated into a central cleaning sewage system of the whole Thessaloniki area. The design of the central cleaning outfall will be concerned, due to the diminished river runoff. The reuse of water biologically cleaned, might be useful for agricultural irrigation.

Finally, bacterial growth will be stimulated by high temperature, and water-borne health risk will be aggravated. Additionally, warm anoxic stagnant waters and sediments, mainly during long warm periods, will present ideal growing conditions for a number of pathogens, increasing the risk of infections, especially malaria, in "swampy" areas. Diseases, which are now either absent or exceptional, might become recurrent and significant, possibly in epidemic form, unless hygiene standards are re-adjusted.

6.3 Ecosystem

Soil, considered as an open system, will respond immediately to the dynamic processes that might be induced by a warmer and rather dryer environment. These processes will affect the subsurface hydrology and soil chemistry as well as the nutrient balance of the soil.

Small changes in terms of rainfall, frequency and magnitude, in combination with increased evapotranspiration, might lead to intensification of certain pedological processes. Under a moisture-deficient regime, porosity and permeability will be reduced, thus degrading soil structure and texture. Particularly, the Aliakmon karstic surroundings may exhibit a tendency towards crust formation that gradually could lead to surface pore sealing and consequently to lower infiltration rates, especially in cultivated areas with higher runoff and erodibility. Irrigation would then become difficult, requiring better soil drainage.

In addition, the problem of seasonal salt accumulation in the top soil of the reclaimed plain could become serious again. This is because salts would not be sufficiently flushed from the soil but would increase their residual contents, affecting soil fertility and productivity at the crop fields. Also, the sea-level rise and the subsequent rise of the saline phreatic water table could produce a secondary partial salinization of the plain. Therefore the recently vanished local ponding in several places might reoccur.

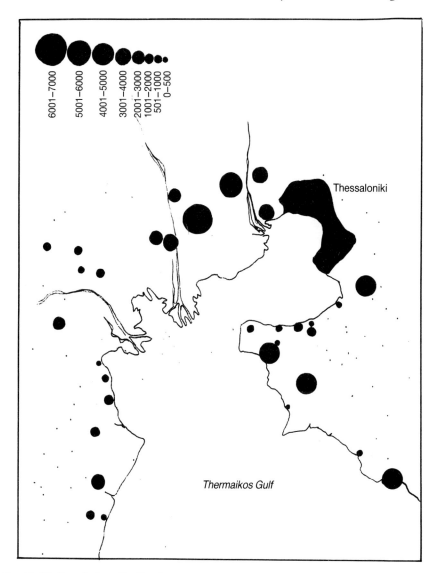

Fig. 13.20 Population distribution.

In the estuarine environment, organic matter plays a vital role in soil fertility because it retains nutrients and moisture favourable to soil microfauna. Forest fires, in contrast, lead to a loss of the upper soil horizon rich in organic matter. High temperature could cause soil degradation, reinforced in cultivated areas by increased mineralization of the organic matter, especially where organic input is high. In places where the high temperature combines with drought to cause a greater susceptibility to slaking, there also will be a negative effect on the soil microfauna, particularly when the soil eventually becomes moistened by rain.

The induced Mediterranean climate, as far as the impact on plants and the agricultural production is concerned, will be characterized by a general water shortage, more evident during the prolonged dry seasons. Occurrence of extreme weather phenomena and accelerated plant-growing cycles also will be stimulated by CO_2 emission. A combination of these factors, in conjunction with soil degradation, could modify plant growing periods. Net biomass, especially weeds, could benefit at the expense of grass in grazing fields. Weather variabilities also could be critical for other field crops on the Thessaloniki plain. Although cotton grows quickly in an enriched CO_2 environment (Fig. 13.19), its industrially important fibre quality is vulnerable to warm summer winds and heat-stress conditions during a late growing season. Precipitation decrease and patchiness will negatively affect hard wheat production, which is highly dependent on spring rainfall and also on the prevailing weather conditions during harvest. In addition, the supplementary irrigation for the excessive water needs of rice and other crops would lead to salinization, negatively affecting maize production. On the other hand, for the other adaptable crops like sugar beets, earlier planting methods could be used to avoid extreme summer events. Thus, the remaining crop cultures will gradually be replaced by new varieties that exhibit better adaptation to the changed environment, or by new subtropical crops, as bioclimatic zonation is shifted to higher latitudes and/or altitudes.

Early warm and moist springs can favour a high level of fungus and bacterially induced diseases, in spite of prolonged dry seasons in a CO_2-rich environment. Increased temperatures also may selectively benefit insects. Temperature rise will allow extensive migration of air- and water-borne species from southern environments. Under the gradually changing climate, plant-protection practices and technologies for agricultural pests have to be adapted to the new conditions in order to meet future requirements.

Due to a rising sea-level, marshes will change to lagoons. As a result the most sensitive lagoonal organisms of the lower food web will be easily adapted, as they exhibit a high salt tolerance to increased saline environments. Thus, temperature in the closed marsh systems will possibly control the growing cycles, especially where higher temperatures could lead to subtropic conditions and finally to thermal pollution.

Benthic biomass also will exhibit a seasonal differentiation. Certain species will particularly benefit from the climatic changes, such as epiphytic algal flora and sea grasses. Increased organic input will lead to higher eutrophic levels and consequently to increased primary production in estuaries. But the well-stratified marine environment in the inner Gulf will subsequently result in oxygen depletion that may lead to bottom anaerobic conditions especially where sewage inputs are high.

Birds and a few other animals have shown a high degree of adaptability to changing terrestrial environments during reclamation projects in the estuaries. Thus, although the projected climatic changes would force small populations of rare protected species to disappear, most species probably will continue to live there, as long as their biotopes are well protected.

6.4 Economic Impacts

Human beings and their constructions exhibit a high adaptation and tolerance to temperature variations. Thus, the effects of sea-level rise seem to be the major threat of the expected climatic change. However, these impacts, superimposed onto other future problems, are not expected to be noticeable during the first 10–20 cm rise, when the sea will gradually cover the intertidal zone. In the enclosed Gulf regime this could be manifested during storm surges, reinforced by strong south winds where waves will be prolongated, washing the city's exposed sea-wall pavement.

However, the effects would become significant under a 30–50 cm or higher sea-level rise, with important economic repercussions. Sea water will cover broad coastal areas, especially at sites of high vulnerability, such as between the Aliakmon River estuary and New Agathoupolis, which is not considered an economically important area (mudflats and swamp). It will also cover the area between Kalohori and the port where the sea could easily reach the nearby coastal industries. In the steeper eastern shoreline, under prolongated wave action, beaches as well as other recreational facilities could be seriously eroded and destroyed. At the area of Mikra airport, sea surges could damage the low-lying runways built on a former lagoon, with serious economic impact. Most of the 1–1.5 m high protection constructions, like dikes and sea walls, which can be easily overflowed during storm surges unless other compensating action is taken, would also be affected. Motorways and railroads are expected to be unaffected because they lie above the sensitive low-lying plain areas.

7 Conclusions – Suggested Actions

Impact should be first sensed about 2030–2040 when a 1–2°C rise in average temperature is expected, possibly accompanied by a 20–30 cm rise of mean sea-level. However, later impacts of a 2–3°C and of a 0.5–1.0 m gradual sea-level rise might be catastrophic if no action is taken.

The coastal environment of the semi-enclosed embayment of the inner Thermaikos Gulf will accept the excessive water that will first gradually flood the intertidal zone and then part of the unprotected lowland (east of Kalohori, north of Agathoupolis). As a consequence, properties and low-lying industries might be damaged. Storm surges also will gradually cause significant damage along the whole coastline, as waves will easily overwash the present sea barriers along the reclaimed agricultural and the cement sea wall along the city of Thessaloniki. The coast at the eastern tourist area might gradually disappear because of the high erosion, producing significant economical effects. The low-lying Mikra airport will be seriously threatened if no action is taken. The seasonal reduction of both the Axios and Aliakmon River discharge, due to the limited temporal rainfall, will cause problems to irrigation and to hydroelectrical power stations along the Aliakmon River.

Temperature rise will enhance stratification of the Gulf waters, especially during the summer, which may affect the sediment depositional regime. Fine suspended-sediment load will be deposited further away from its sources causing the shallowing of the navigation channels. Stratification also could increase eutrophication.

A probable northward shift of climatic zone will result in longer summers. Additionally, increased variability and patchiness in rainfall might extend summer aridity. Thus, irrigation needs of Thessaloniki agricultural plain, in combination with the lower ground-water infiltration, will cause a critical fall of the water table and lead to the increased salinization of the ground water. Sea water will progressively intrude into the water table, thus deteriorating the quality of the phreatic and intermediate ground water. The risk of contaminated drinking water will also increase, as conditions will be ideal for several water-borne pathogens and bacteria.

Reduction of rainfall during the hot summer might cause deficiency in the soil moisture, thus degrading soil structure and agricultural fertility. Moreover, reduced wash-off could possibly cause seasonal salt-accumulation in the top-soil of the reclaimed lowlands. As a result better drainage of the plain will be required.

Water shortage during the growing season could diminish agricultural production, especially for crops that depend on spring rainfall (i.e. wheat). Also, salinization of the irrigation water would have negative consequences on grain yield. Consequently, new varieties of crops have to be introduced. On the other hand, fast-growing plant cycles, due to increased CO_2 and temperature, might increase fertilizer consumption, as the needs in NO_3 will be higher. Generally, marine and land weeds are expected to benefit from these conditions. As bioclimatic zonation gradually shifts northwards, several species may disappear and new ones appear. Finally, there will be an increased risk of agricultural pests, insect populations, bacteria and diseases, especially in the swamps.

The action to be taken should include a careful monitoring of local long-term changes of the following parameters: weather pattern and seasonability; underground water quality and water-table elevation; soil quality and salinization of reclaimed areas; soil fertility and agricultural yields; insect population and movements, pests, bacterias, diseases; forest fires; floods.

The policies needed to prevent possible catastrophic impact include: mapping of high-risk coastal sites vulnerable to flooding and re-adjustment of the present flood defensive constructions (dikes, sea-bound sea walls); careful protection design of coastal sites and the infrastructure of high economic importance (airport, port, industrial zone, city); re-adjustment of the building standards for low-lying areas.

From the above-mentioned data and estimations it is obvious that the predicted climatic changes could cause, large-scale and radical consequences in the greater Thessaloniki area. These enormous changes require equally remarkable and daring future steps. Such steps might be the damming and isolation of the Thermaikos Bay and its use as a "buffer zone" in order to diminish or remove the impact of a sea-level rise on the low-lying coastal area. Construction of a barrier across the 4.5 km long stretch between the north-east of the Axios delta and Mount Envolo Cape (maximum depth of 27 m) depth may be within the capabilities of tomorrow's technology. The damming of Thessaloniki Bay and its transformation into a controlled lagoon, with essential navigational outlets, will not negatively affect the greater Thermaikos Gulf marine environment, as the sea-water circulation and sewage output regime would remain almost unaffected.

8 REFERENCES

Angouridakis, V.E. and Machairas, P.C., 1973. Precipitation (I) *Sci. Annals, Fac. Phys and Mathem.*, Univ. Thessaloniki **13**, 347–380.

Balopoulos, Th. E., 1986. Physical and dynamical processes in a coastal embayment of the northwestern Aegean Sea. *Thalassographica*, **9**, 59–78.

Brambati, A. and Stefanini, S., 1970. Sedimentological research of the coastal sediment of the Gulf of Thessaloniki. *Boll. Soc. Geol. It*, **899**, 383–400 (in Italian).

Chronis, Th. G., 1986. The modern dynamic and recent Holocene sentimentation at the Inner Thermaikos Plateau. PhD thesis, Dept of Geol., Univ. Athens (in Greek).

Evmorphopoulos, L., 1961. The Changes of Thessaloniki Gulf. *Techn. Annal*, No. 205–208, 51–76.

Flocas, A. and Arseni-Papadimitriou, A., 1974. On the annual variation of air temperature in Thessaloniki. *Sci. Annals. Fac. Phys. and Mathem.*, Univ. Thessaloniki, **14**, 129–150.

Friligos, N., 1977. Seasonal Variation of Nutrients Salts (N, P, Si) dissolved oxygen and chlorophyll-a in Thermaikos Gulf (1975–76). *Thallasia*, Yugoslavia, **13**(3/4), 327–342.

Ganoulis, J., 1987. Oceanographic elements and environmental investigation of the impacts of the draining project of Thessaloniki Gulf. Dept of Hydraulics, Univ. Thessaloniki, Rep. Minist. of Environment (in Greek).

Knithakis, E. and Tzimourtas, S., 1987. Chemical study of the underground water of the Axios-Loudias River basin. I.G.M.E., Thessaloniki, Unpubl. (in Greek).

Livadas, G.C. and Sahsamanoglou, C.S., 1975. Wind in Thessaloniki, Greece. *Sci. Annals. Fac. Phys and Mathem.*, Univ. Thessaloniki, **13**, 411–414.

Maheras, P.C, 1987. Temporal fluctuations of annual precipitation in Palma and Thessaloniki. *J. of Meteo.*, **12**(123), 305–308.

Mavromatis, G., 1980. The bioclimate of Greece, relations to the natural vegetation and bioclimatic maps. *Inst. of Forest. Research*, Athens (in Greek).

Ministry of Environment, 1986. Boundary determination programme of the Wetlands of Ramsar contract. Deltas of Aliakmon-Loudias-Axios (in Greek).

Nicolaidis, C.G., 1985. Qualitative and Quantitative Study on the benthic macrophycae at the polluted areas of Thessaloniki Gulf. PhD thesis, Univ. Thessaloniki, **23**, sect. 46.

Nedeco and Grontmy, 1970. Regional development project on the Salonica plain. Min. of Agriculture (OECD).

Prefecture of Thessaloniki, 1986. Agricultural developing program, year 1986. Direct. of Agriculture, Thessaloniki (in Greek).

Struck, 1908. Mazedonisch Fahrten II. *Die Makedonischen Miederlander*, Serjevo (in German).

Therianos, D.A., 1974. The geographical distribution of river water supply in Greece. *Bull. Geol. Soc.*, **XI**, 28–58 (in Greek).

Thessaloniki Port Authority, 1987. Statistical data, year 1986.

UNEP/GEMS, 1987. The greenhouse gases. Nairobi, Env. Lib. No 1.

Wigley, T.M.L., 1989. Future Climate of Mediterranean Basin with particular emphasis on changes in precipitation. Chapter 1, this volume.

Wilding, A., Collins, M. and Ferentinos, G., 1980. Analyses of sea water level fluctuations in Thermaikos Gulf and Salonica Bay NW Aegean Sea. *Eustar. Coast. Mar. Sci.*, **10**, 325–334.

Zimerman, J. and Ross, V.J., 1976. Structural evolution of the Vardar root zone in northern Greece. *Geol. Soc. Amer. Bul.*, **87**, 1547–1550.

14

Implications of Climatic Changes for the Nile Delta

G. Sestini
(*Applied Earth Science Consultant, London, UK*)

ABSTRACT

The impact of an atmospheric warming and of sea-levels 30–70 cm higher than at present on the coastal lowlands of the Nile Delta during the next 50 to 100 years will depend not only on the level of population and economic activity, but also on the degree of coastal development during the next two to three decades. Intensification of land use in the coastal region is inevitable, due to the continued growth of population and the consequent need to augment food production through the further extension of land reclamation and of lagoonal fishing, and of new industrial and commercial activities.

The lower Nile Delta has large areas under 1 m elevation, with parts below sea-level, including sizable coastal lagoons. Natural protection is afforded by high sand dunes on some stretches, but other areas are vulnerable to flooding by winter storm surges. Higher sea-level will accelerate the present retreat of the shores, which increased substantially after the building of the Aswan High Dam. Protection of essential coastal uses and of agricultural lands will be necessary.

A sea-level rise of 10–20 cm by itself would be of little consequence, but it would certainly aggravate local problems such as wave attack on harbour installations, particularly on the Nile-mouth promontories and shoreline stretches that are subject to flooding by storm surges. Augmentations over 30–50 cm would have more serious effects, imposing extensive measures of protection. A plan of coastal management would have to be considered at an early stage, and reasonable steps taken during the next 10–20 years; otherwise major disruption is to be expected on the Alexandria to Abuqir coast, at Burg el Burullus, Damietta and Port Said. It is also possible that the Burullus, Manzala and Bardawil lagoonal barriers might be broached by the sea. Though coastal retreat will continue, the building of fixed defence structures would have to be carefully evaluated for possible negative counter-effects.

A flooding of the coastal lowlands is not expected. In theory, a sea-level rise (or relative rise, to include the effects of land subsidence) of 100 cm could flood land within 30 km of the coast or more, affecting 12–15% of Egypt's arable land and 8–10 million people. However, the cultivated lands are already bordered by dikes, and more protection works are likely to be constructed along with new road, reclamation or water preservation schemes. Yet higher lagoon and estuary water levels would mean, if not checked by sluices, easier penetration by salt wedges. Thus, the impact of sea-level rise will be mainly on the financial resources to carry out new protection schemes, to make adjustments to harbour infrastructures, new land use, or reconstruction.

The main consequences of higher average temperatures would be increased rates of evaporation, and therefore greater salinity of waters and soils, with negative impacts on ground water and on reclaimed land agriculture (e.g. continued or greater need for salt flushing by fresh water). In general, the greater temperature and CO_2 concentration could favour agriculture (probably with some shift in crops) and a changed system of crop rotation. Agriculture, however, could be negatively affected by increased evapotranspiration and soil salinity and may require more irrigation and fertilizers (especially nitrogen). A possible increase of weeds and pests could heighten insecticide use, thus magnifying the pollution problems of surface waters.

The lagoonal ecosystems, and hence fish resources, would probably adjust to gradually changed conditions of salinity and water temperatures. They would be affected, however, sooner than climatic changes, by artificial schemes of lagoonal management. Careful consideration would have to be given, therefore, of their environmental impact two to three decades from now.

The socio-economic structures of the lower Nile Delta probably will be affected more by population increase and urbanization, and by economic trends external to Egypt, than by climatic changes; nevertheless water supply and food production could be altered. Beach tourism, especially at some of the present resorts (Alexandria, Baltim, Ras El Barr, Port Said), will certainly suffer from a rise of sea-level; but other resorts with wider and more stable beaches, if properly developed, might take over as the main recreational centres of the Nile Delta. Present regional planning that involves population planning and the use of natural resources and land, and covers time-spans of two to three decades, will have to take into account the longer-term climatic changes.

1 Introduction

The surface of the Nile Delta (20,000 km^2) represents only 2.3% of the area of Egypt, but as much as 46% of the total cultivated area (55,040 km^2), and it accommodates approximately 45% of Egypt's 50 million inhabitants (1986). An extensive part of the delta lies below 2 m in elevation, comprising a zone, 30–55 km wide and 4840 km^2 in area, containing the cities of Kafr el Dawar Rosetta, Damietta and Port Said (Fig. 14.1). The importance of this zone is further underlined by its high level of agricultural and fish production. Besides being the site of expanding aquaculture activities, the coastal lagoons represent ¼ of total Mediterranean coastal wetlands; as such they provide both staging and wintering areas for seasonal bird migrations.

Due to limited rainfall, the Delta region is totally dependent on the Nile River for its water needs for urban, industrial and agricultural uses. Variations in rainfall in the Nile catchment areas of East Africa have had, historically and recently, a profound effect on water availability in Egypt. Nevertheless, there are still unused and under-utilized sources of water, such as ground and waste-water and the winter surplus of surface water, when irrigation requirements are reduced.

For many decades, especially since the final closure of the Aswan High Dam in 1965, the coast of the Nile Delta has been in a state of disequilibrium, with pronounced beach erosion at the Rosetta, Damietta and Burullus promontories. So far this erosion has only involved a few resorts and local infrastructures, but its consequences would be accentuated by a significant rise of sea-level. To the effects of sea-level rise one must add those of geological land subsidence, which locally can amount to as much as 5 mm/yr. Reclamation of former lagoons and marshes has caused additional and faster sinking in some areas; others could be affected if ground-water extraction were to become important and excessive, which might be the case with increased agricultural and urban development.

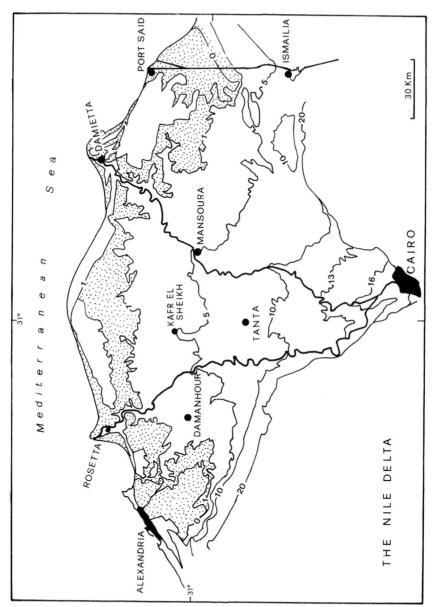

Fig. 14.1 Main topographic features of the Nile Delta (adapted from Sestini, 1989).

The socio-economic outlook of the Nile Delta, therefore, is dominated by the need to feed, house and employ Egypt's fast-growing population. Thus, far the extension of arable land has been accomplished by reclamation of the desert end of lagoon margins; further increase is limited by encroachment on fishing and by soil salinity. Increased development also could accentuate coastal pollution as well as add to the existing shoreline instability. It is essential, therefore, that long-term development take into account the effects of climate alterations and the need for the conservation of natural resources.

The aim of this chapter is to evaluate the likely impacts of climatic changes on the Nile Delta coast and the bordering lowlands, given their morphology and their present and projected level of land use. It also evaluates near-term needs, i.e. in the next 10–20 years, in order to minimize or to adapt to the changes that may be caused by a possible atmospheric warming of 1.5–3°C and a sea-level rise of about 30–70 cm by the year 2070.

2 Geographic Setting

The Nile Delta is very flat, only 18 m above sea-level at Cairo, 150 km south of the coast. One-third lies below 3 m elevation, and several parts of the latter area are below sea-level (Fig. 14.1). The coast has an arcuate shape with the protrusion of the two cuspate subdeltas built by the Rosetta and Damietta branches of the River Nile, and has a total length (between Alexandria and the Bay of Tineh) of 315 km.

The Nile Delta physiographic provinces of concern in this study are:
1) The lower delta plain, with its wetland belt, characterized by active geological subsidence. The lagoons cover large areas, although their size has been considerably reduced in recent decades because of land reclamation.
2) The delta front, with its 1–10 km wide sand belt made of beaches, dunes and backshore sandy plains, and shaped by the coastwise movement of sediment by littoral currents and winds.

The lowland belt of the Nile Delta under 2 m elevation is 30 to 55 km wide. Detailed morphological descriptions have been published by Frihy et al. (1988), Sestini (1976, 1989), and, as regards the coastline, by UNDP/Unesco (1978) and Tetratech (1986). The coastal zone can be subdivided into six sections on the basis of the higher-ground watersheds that are more or less normal to the coast.

1) *Alexandria to Maadia.* The Alexandria coast is bordered by a fossil dune ridge 5–10 m high (up to 20 m south of Dakheila); there are 2–3 m cliffs and numerous small pocket beaches between rocky capes (Figs 14.2 and 14.3). Besides the harbours of Alexandria–Dakheila and Abuqir, the coast is intensely built-up, with a major 20 km road by the sea (the Alexandria "corniche") protected by a bulkhead wall. Most of Alexandria lies above 2 m; low areas are present, however, between the west and east harbours, as well as at the outlets of the Mahmoudia and Mex canals (El Sayed, 1991).

South and east of Alexandria lie Lake Mariut and a large agricultural area, both 1–3 m below sea-level, that extend over the sites of former lakes

Nile Delta and climatic change 539

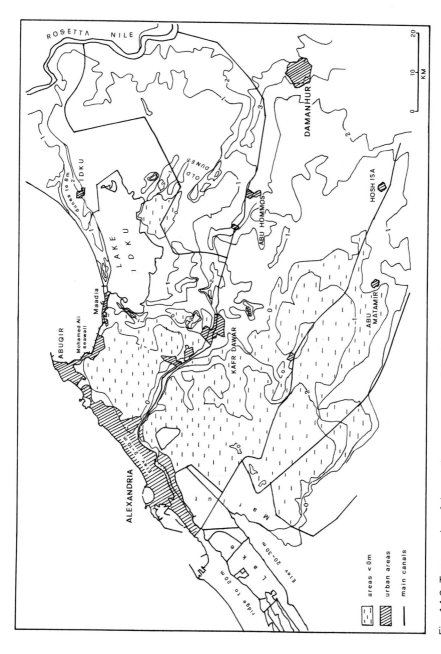

Fig. 14.2 Topography of the north-western part of the Nile Delta. Main features from interpreted 1989 satellite images, elevation contours from survey of Egypt 1945 maps – both at scale 1:100,000.

540 *Climatic Change in the Mediterranean*

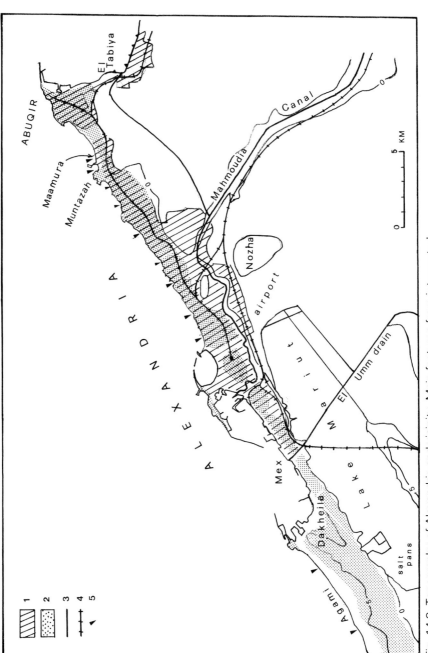

Fig. 14.3 Topography of Alexandria and vicinity. Main features from interpreted 1989 satellite images, elevation contours from survey of Egypt 1945 maps – both at scale 1:100,000. 1. Urban areas 2. Areas above 2 m 3. Main canals 4. Railways

Mareotis and Abuqir (Fig. 14.2). Drainage is provided by pump stations at Mex (Alexandria) and El Tabya (Abuqir), and marine flooding is prevented by the 10 km long, 4 m high Mohamed Ali sea wall. The Abuqir and Mariut lowlands are separated by the ridge of the Mahmoudia Canal, on which are the Alexandria-Tanta road and railway and several towns, including Kafr El Dawar (population 250,000). Lake Mariut is divided into several basins by the El Umm Drain and by road and railway embankments. The western end of Mariut is entirely cut-off; it is a saline lagoon used by salt works.

2) *Maadia to the Rosetta Nile.* The Abuqir lowland and Idku lagoon are separated by a ridge, 1 m (in parts 2–3 m) above sea-level, that follows the course of the former Canopic branch of the Nile (Fig. 14.2). The outlet of Lake Idku is 250 m wide and is presently protected from silting by two jetties, 250 m long. The plain south of the lake contains very low ground, but is interrupted in the western part by a series of ancient (<5000 BP) dune ridges, 4–7 m high, that are stabilized by plants. The main town is the fishing centre of Idku (population 77,000).

The eastern shore of Abuqir Bay is flanked by a sandy plain up to 6 km wide, with scattered low dunes (Fig. 14.4). Parts of the plain are below

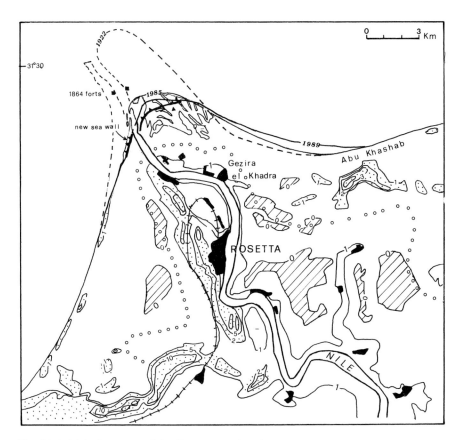

Fig. 14.4 Main features of the Rosetta region. (1989 satellite images).

sea-level and become flooded during rain storms. There is a discontinuous narrow strip of 1–2 m dunes by the beach.

Much higher and more extensive are the Idku and Rosetta dunes, in several places 10 to 20 m high. Although the dunes are active, they are cultivated (orchards and date palms), especially on their south and east sides. At the western margin of the Rosetta dunes the El Rashidya Canal (with the Abuqir–Rosetta railway embankment) has permitted a certain westward expansion of agriculture.

In the Rosetta promontory the Nile has formed a ridge of land >1 m (the river levees are generally 2–3 m high) that protrudes between the depressions of eastern Abuqir Bay and of the western part of the Burullus lagoon (Fig. 14.4). The tip of the Rosetta promontory is the site of active erosion (the two forts built about 1860 and the 1945 lighthouse have been overcome by the sea in the last few years). The shape of the Nile mouth is ever-changing, with cross-channel bars and spits attached alternatively to the eastern and western shores (Frihy, 1987). A sea wall 6 m high has been erected in an attempt to slow down retreat and sea encroachment on the agricultural area of Gezirah El Khadra.

3) *The central region (Rosetta to Gamasa)*. The central region comprises the barrier between Lake Burullus and the sea, the south shore and islands of Lake Burullus, and the coastal belt from El Burg to Gamasa, including the lowlands of former Biheiret El Bessar (Fig. 14.5). The barrier tapers from about 4 km width in the west to about 1 km in the east, near the lake outlet at El Burg. It is formed by a 1–2 m wedge of sand over lagoonal clays. Between Abu Khashab and Hanafi the coast is flat and the barrier between the sea and Lake Burullus is mostly a backshore plain (0.5–1 m elevation), with low dunes (1–2 m) on the south side isolated or in continuous belts. In winter the plain tends to be flooded by storm surges.

The beach sands are fine-grained at the Rosetta promontory and increase in size eastward. They are much coarser-grained (0.250–0.500 mm) at Burullus (as far as Gamasa) than either at Rosetta or at Ras el Barr (Unesco/ASRT/UNDP, 1976; Frihy *et al.*, 1988) Between Burullus and Gamasa the shore is characterized by large asymmetric beach cusps with an average length of 400–600 m, a width of 30 m, occasionally reaching 200 m. Because of their eastward movement, the position of the shoreline at Baltim Beach Resort varies yearly.

The outlet of the Burullus lagoon has always been unstable, but before the 1940s its tendency to migrate eastwards, due to littoral drift, was balanced by strong in- and out-flows between sea and lagoon. The need to keep the channel open for fishing needs prompted the erection of a series of jetties, the first of which (250 m long on the west side) stimulated erosion on the east side. Consequently, a sea wall (500 m long and about 3 m high) was built to protect the town of El Burg (Fig. 14.6).

East of El Burg, the coastal zone is a distinctive morphological unit, averaging 10 km wide, characterized by sand dunes and sandy depressions. In the Baltim peninsula, between Lake Burullus outlet and the Kitchener Drain, there are two belts of dunes, a main one along the coast (up to 18 m high), and a smaller and lower one (2–5 m) near the lagoon. Between the two belts are several basins, some of which are permanent lakes and others (below sea-level) flooded during winter and used as salt pans. Date palms

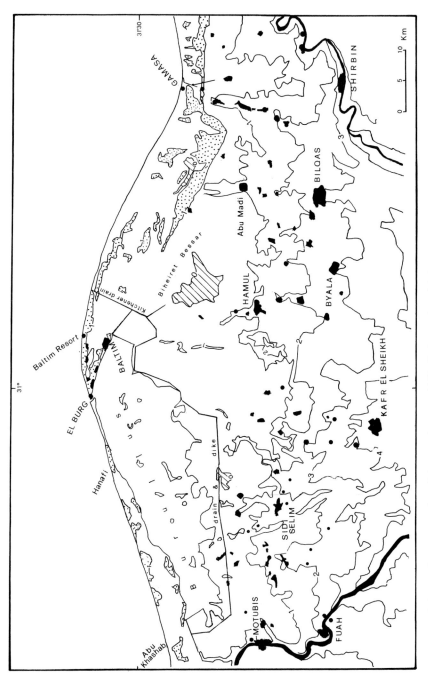

Fig. 14.5 Schematic topography of the north-central Nile Delta region. Main features from interpreted 1989 satellite images, elevation contours from survey of Egypt 1945 maps – both at scale 1:100,000.

544 *Climatic Change in the Mediterranean*

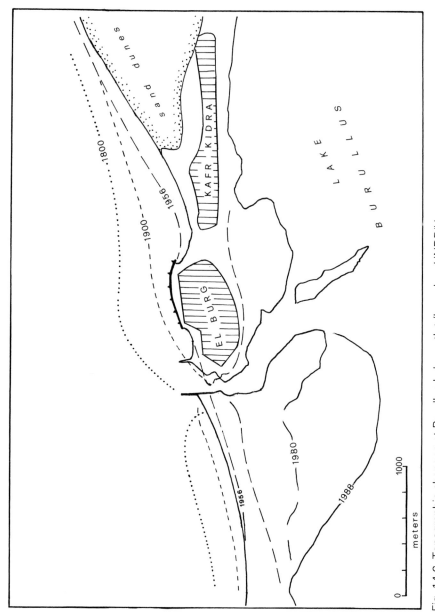

Fig. 14.6 Topographic changes at Burullus Lake outlet (based on UNDP/Unesco, 1978; Tetratech, 1986; and satellite images).

and vegetables are grown on the sandy soils; the main resource of El Burg and Baltim (total population 75,000), however, is fishing.

Between Kitchener Drain and Gamasa the dunes rise over a backshore plain, more inland. The dune belt is more continuous and higher in the south, although there are some large gaps and many parts are below sea-level (Fig. 14.6). This area is still largely uninhabited.

Lake Burullus is very shallow (1.6 to 0.5 m), especially in the eastern part, which has a tendency to be muddy and more brackish, due to agricultural drain discharge. As many as 50 islets dot the lake; some form alignments which may represent elevated relict river channels. The plain below 1 m elevation is very extensive. The only significant relief is provided by thin sinuous ridges, generally 2 m (locally 5–6 m) high that are the remnants of past Nile channels (the Sebennytic and Thermoutiac branches; Sestini, 1976; Tousson, 1934). The ridges have been considerably altered by land-reclamation operations. The cultivated plain, which at several places lies below sea-level, is separated from the belt of marshes near the lakeshore by a continuous dike, generally 3–4 m high. Most villages above 5000 inhabitants are situated above the 1 m contour (Fig. 14.6).

4) *Gamasa to the Damietta Nile branch*. This coastal stretch is a zone 4–5 km wide with a flat backshore plain and dunes; the dunes (maximum elevation 2–3 m) form two belts, a discontinuous northern one with north–south trending seif dunes and a southern continuous one oriented east–west. Between the dunes at several places, especially north-west of Damietta where the new harbour is located (Fig. 14.7), there are very low areas. At the Damietta mouth of the Nile various protective structures (a pier, a sea wall and several groins) have been built partly to stabilize the estuary, partly to protect the Ras El Barr beach resort.

East of the mouth, the shore is presently oriented almost east–west. Eight km to the east the coast turns south-east, with a large protrusion, resulting from the growth, since the 1950s, of a series of new curved spits (Fig. 14.7). This part of the Nile subdelta is made of sets of beach ridges with lagoons in between. The ridges, composed of very fine sand a few metres thick, rest on marine silty clays (Anwar *et al.*, 1984). It is a region without agriculture and almost uninhabited. By contrast, the Damietta Nile embankment is intensively cultivated and has a high population density (in 1986 the Damietta region contained about 300,000 inhabitants). The alluvial ridge is 2–3 km wide, 1 m high (the 2 m contour is located 24 km south of Damietta) except for the levees of the Nile (± 2 m) and the town of Damietta, where the ground is about 1–2 m above sea-level.

5) *Damietta to the Suez Canal* (Fig. 14.8). The lagoonal barrier east of El Diba is narrow (a few hundred metres) and low, without dunes. It is crossed by the two, now canalized, outlets of Lake Manzala, which are armoured by short piers. Port Said, originally built over an area of only 4 km^2, has recently expanded to the south-west and south; the excavation of the new Suez Canal Bypass, has doubled the available land for further urban and industrial use in the east. The natural elevation of the city is no more than 1 m.

Lake Manzala is large (1350 km^2) but no more than 1 m deep. Numerous chains of islands subdivide it into several basins; the islands are rapidly

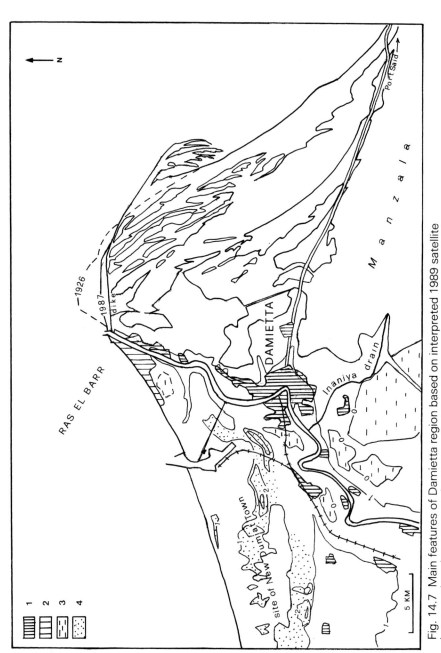

Fig. 14.7 Main features of Damietta region based on interpreted 1989 satellite images, elevation contours from survey of Egypt 1945 maps – both at scale 1:100,000. 1. Urban areas and villages. 2. Summer resort. 3. Areas below sea-level. 4. Dunes.

Nile Delta and climatic change 547

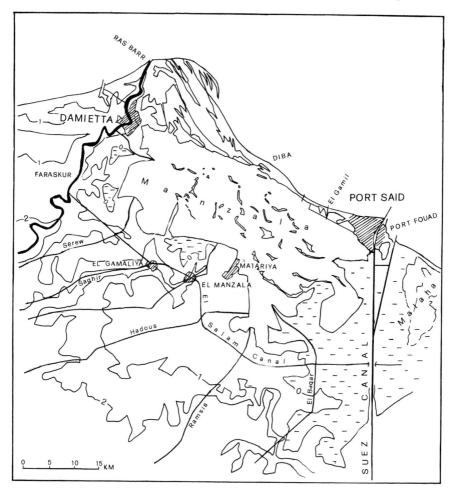

Fig. 14.8 Main topographic features of the Lake Manzala region.

widening due to reed growth and siltation. The extensive lowland under 2 m elevation adjacent to the western shore of the lagoon is divided into lower (<1 m and <0 m) and higher (1–2, even 3 m) sections. The outstanding alluvial ridge is that of the Saghir Drain, which follows the trace of the ancient Mendesian Nile branch. It is an important communication and population axis (e.g. El Gamalyia, 45,000; El Manzela, 55,000; El Mataryia, 76,000 inhabitants). The lands north and south of this ridge are well drained and cultivated; the lake margin is diked along much of the western part.

The south and south-eastern margins of Manzala still have in parts extensive area of marshes, brackish lakes and salt pans. The land below sea-level, extends as far south as a line from Qantara to Tineh, and to a distance of 15 km north of Ismailia.

6) *East of the Suez Canal to North Sinai.* Between Port Said and Tineh there is a cut-off portion of Lake Manzala (El Malaha lagoon) and a salty sand

plain. The Malaha lagoon is mostly a few tens of centimetres deep, and can be totally dry at times; in the plain of Tineh to the east, flooding is common in winter from rains and sea washovers.

East of Tineh, there are large salt flats and gently undulating dunes, 5–10 m high. The Bardawil lagoon (85 km by 22 km, 0.5–3 m deep) is closed by a narrow sand barrier (300–1000 m) that frequently is overtopped by waves. There are three outlets through this barrier, two are artificial and in constant need of dredging; the natural outlet is at the eastern end of the lagoon (Por and Ben Tuvia, 1981). Salt marshes line the south and eastern shores. Apart from the lagoon, there is no low-lying land in northern Sinai, because of the development of high coastal dunes (up to 50 m).

3 THE SOCIO-ECONOMIC SETTING

3.1 Economic Setting and Population

The Nile Delta contributes 30–40% of agricultural production and 60% of fish catch (marine and lagoonal). Half of Egypt's industrial production comes from the delta, mainly from Alexandria, which is also the country's main commercial port, and the Suez Canal at Port Said, one of the main sources of foreign exchange (Anonymous, 1985). It has been estimated (Walker, 1988) that approximately 15% of Egypt's current gross domestic product (GDP) originates in the lowlands under 3 m. Also, large areas targeted for priority development and reclamation lie under the 1 m contour. Major future projects (Fig. 14.9) include a coastal road, industrial centres near Rosetta, Baltim, Damietta, more fish farming, new towns and the possible transformation of parts of the Burullus and Manzala lakes into fresh-water reservoirs (Harris, 1979; IFAGRARIA, 1984; Tetratech, 1986; ARICON, 1989).

Agriculture is still the largest sector of the national economy, but due to the expansion of industrial activities in the last three decades, its share in the GDP has decreased from 40% in the 1950s to 29% in 1974, and to 17% in 1984–1985 (Beaumont et al., 1988). Today, agricultural production, especially cereals, has become insufficient to satisfy domestic needs. In the early 1970s Egypt was still a net exporter of agricultural products. Now, it is forced to import over 50% of its food requirements and 70% of its wheat.

The availability of water resources remains a major worry. In 1987 Lake Nasser was at its lowest level since its ultimate filling in 1978, because of the recent drought in the Blue Nile catchment area. Other problems of agricultural expansion are the deterioration of water and soil quality due to salinization (Assen, 1983).

The highest population densities are found in the central delta, along the Nile branches (e.g. Damietta), and the Mahmoudia Canal (Kafr Dawar, Mahmoudia districts) (Fig. 14.10). Population is much lower and more widely distributed in the areas with elevations less than 1 m (e.g. Bilgas, Sidi Selim). Apart from Kafr el Dawar, Damietta, Port Said and of Alexandria, all main population, commercial and industrial centres are situated above the 3 m contour.

The Egyptian population is growing at the rate of 2.3–3% per annum due to high fertility as well as recent improvements in health and a

Nile Delta and climatic change 549

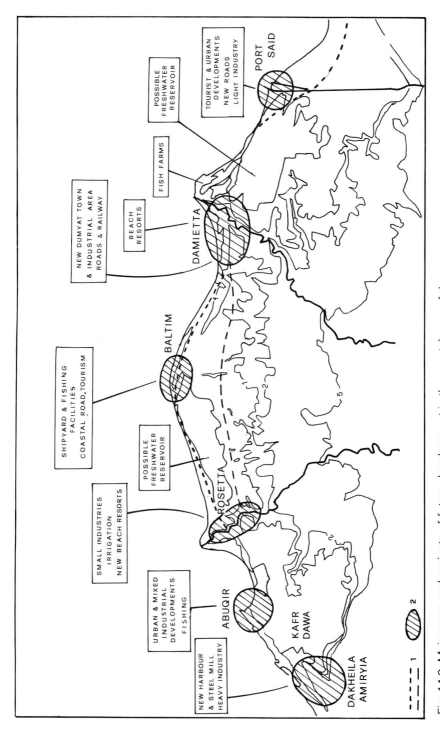

Fig. 14.9 Main areas and projects of future development in the coastal zone of the Nile Delta (from data in ARICON, 1979). 1. Main transversal roads. 2. Development poles.

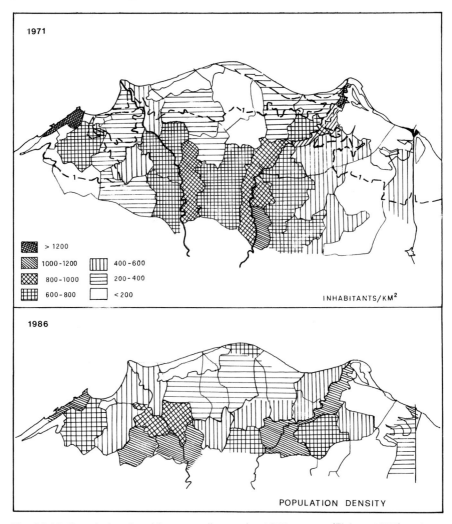

Fig. 14.10 Population densities according to the 1971 census (Fisher, 1978) and to 1986 data (ARICON, 1989).

corresponding decrease in mortality. Projected population by the year 2000 ranges from 62 to 67 million and by 2025, from 85 to 95 million (UN, 1979; UNEP, 1988); in the coastal area < 3 m, it could be over 12 million (about 7 million by 2020, east of Rosetta Nile) (ARICON, 1989).

During the last 15 years urbanization has been intense, especially in Cairo and Alexandria (PADCO, 1981). Alexandria's population has swollen from 0.5 million in the 1950s to 3.5 million at present, with the addition of 0.5–1 million summer visitors. Urban and industrial expansion in Alexandria is physically restricted, by the sea on one side, by Lake Mariut and agricultural land on the other. This narrow strip is now a continuous sequence of high-rise buildings. As a consequence of its rapid population and industrial growth, Alexandria is faced with a waste-water problem that has created health hazards on the beaches and has been detrimental to fishing.

The urbanization of Damietta and Port Said has been on a lower scale; nevertheless, all towns and villages in the belt <3 m (Figs 14.2, 14.5, 14.7, 14.8) are shown in satellite images of 1989 to have expanded by one-third to one-half *vs* 1982. Port Said (population 450,000) has a great potential as a commercial and industrial centre at the entrance of the Suez Canal, but expansion has been limited by the lack of space, squeezed between the sea and two lagoons.

Given that tourism is one of the main resources of Egypt (LE338.3 million in 1985), the beaches of the Nile Delta are little used and mainly on a national scale, serving Cairo and the cities of the lower delta. The erosion of Alexandria's small beaches already has stimulated a rapid development of the western coast (Agami to Marsa Matruh) and of North Sinai (Arish). Government plans for future development are directed to these areas, as well as to South Sinai, the Red Sea coast and Upper Egypt, rather than the Nile Delta.

3.2 Agriculture and Land Reclamation

In the Nile Delta the most important crops by cultivated area are clover, maize, cotton, wheat, rice, fruits and vegetables (Fisher, 1978). Fodder, maize, rice, sunflower, sugar beets and soya beans are the predominant crops of the lower (<2 m) coastal zone, especially in the reclaimed areas. Considerable market gardening is developed near the larger towns. Clover is widely grown as animal fodder and to increase the nitrogen content of the soil. Broad beans are also a common crop for animal and human consumption, as well as for soil regeneration. Rice production has increased with perennial irrigation and lowland reclamation, where it is a favoured crop because of high yields even in moderately to highly saline soils. Cotton production, traditionally the main cash export crop for which Egypt is famous, has decreased from 400–500,000 tons (1973–1983) to 350,000 tons at present (1988); in the coastal delta it is cultivated only in the Damietta region.

One of the reasons for the inadequate performance of agriculture and the rise of food imports is the government's control on cropping patterns and farm prices (Beaumont *et al.*, 1988; Walker, 1988). The decade 1984–1985 was responsible for a 20% decrease of the cotton cultivated area, 10% of the wheat and rice areas. On the other hand, at the same time, the area dedicated to fruit and vegetables, which are not subject to quotas and fixed prices and are increasingly exported, rose by 23%.

With the shift to perennial irrigation, which has greatly reduced the deposition of alluvium in the delta, soil fertility has declined, imposing the use of artificial fertilizers, particularly superphosphates and nitrates. More fertilizer is used in Egypt than in the rest of North Africa combined, a considerable amount of it being imported.

The building of the Aswan High Dam, which stores the annual Nile flood waters, not only has permitted the year-round irrigation of all cultivated lands and a more intensive use of the soil, but also the extension of farming by the reclamation of wetland areas near the coastal lagoons and the desert fringes of the delta. About 1 million feddans (1 feddan =

552 *Climatic Change in the Mediterranean*

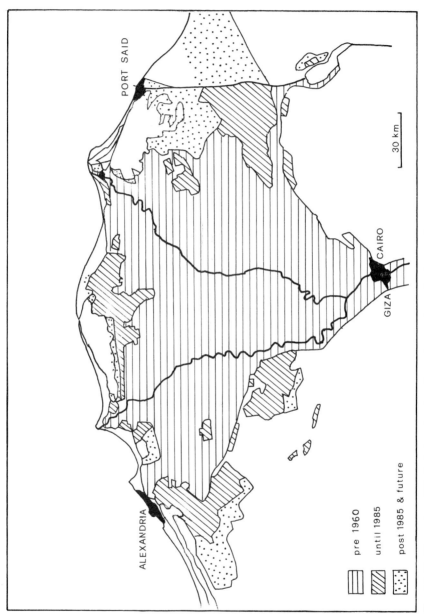

Fig. 14.11 Recent and likely future land-reclamation projects in the north Nile Delta (based on Beaumont et al., 1988; and on 1989 satellite images).

0.43 ha) have been reclaimed in the past 20 years, but according to the Ministry of Irrigation, 2.8 to 3 more million feddans (40% of the total land now under cultivation) could be reclaimed by the year 2000. This would require at least 8.8 km^3 of irrigation water annually. Over one-half of that area is north of Cairo, mainly in the eastern delta and north of Sinai (Fig. 14.11). The most significant (recent and future) delta projects are:

1) A scheme in the Mariut region which includes the additional use of Alexandria sewage water for irrigation.

2) Large lagoonal-swampy areas (120,000 feddans) at the edges of the Idku and Burullus lagoons were drained at great cost; but in the south-east part of Burrullus the soils still contained 10% salt 10 years after reclamation (Boumans and Machali, 1983). A leaching programme is under way to remove excessive soil salinity by means of subsurface drainage, using surplus winter water (the so-called 1986–1995 Drainage Project Five).

3) In the Suez Canal region the plain of Tineh, east of the Suez Canal, is to be watered by the El Salam Canal, which is derived from the Nile south of Damietta and, by underpassing the Suez Canal, should deliver 13 km^3 of water for a 1680 km^2 reclamation that includes a 576 km^2 agro-industrial project to produce vegetables, fruit, livestock and dairy products, and the creation of numerous new villages (ARICON, 1989).

Present irrigation efficiency is estimated at 50%. Egypt could make do with much less water than it presently consumes if present irrigation methods and traditional crop husbandry, as practiced in the old lands of the delta, were changed (Assen, 1983). The Ministry of Irrigation has suggested that the present average mean water use of 8000 m^2 per feddan per year should be reduced to 7420 m^2 in clay soils, and to 5250 m^2 in sandy soils. Irrigation practices need to be changed in response to future reduced water supply and/or increased losses due to evaporation; the cultivation of water-intensive crops, such as rice, could be jeopardized.

Finally, there is the problem of an actual loss of arable land. Although 1 million feddans of new land were reclaimed from the Nile valley and the delta between 1952 and 1980, considerable amounts of existing land were lost: by 1980, as much as 20,000 feddans were wasted because of urban development and brick making (now forbidden by an Act of 1985). In addition, problems of waterlogging and salinity, exacerbated by the perennial irrigation methods, have increased with more intensive cropping. Less concern with drainage has resulted in a general rise of the water table.

3.3 Fishing

Marine fishing (Mediterranean and Red Sea) accounts for 25% of Egyptian fish catches (125,000–160,000 tons a year), versus 60% from the coastal lagoons of the Nile Delta and 15% from the Nile River. The Mediterranean fishing has been strongly influenced by environmental conditions, such as the cessation of Nile nutrient discharge after 1964 (Georgy, 1966; George, 1972; Bishara, 1985) and in the last two decades by the increase of water and sediment pollution (Safeh and Karashili, 1985). Marine catches in the early 1970s had decreased to 20% of the pre-1964 amount (17,000 tons in 1969 *vs* 90,500 tons in 1962); most affected were the sardines (*Sardinella*, especially *S. aurita*), previously 30–50% of total fish landings, and prawns due to the sharp decrease in primary productivity (Bishara, 1975). Marine fish production in the period 1975–1983 varied from 12,000 to 20,000 tons per year, the most active marine fishing centres being Alexandria (west and eastern harbours), Abuqir, Rosetta, Burg el Burullus, Damietta and Port Said. The main fish species marketed are: *Mullus barbatus, Mugil cephalus, Pagellus, Sardinella, Solea solea, Temnodon saltatus, Box salpa*; squids, crabs and shrimps.

In spite of the use of advanced fishing equipment in recent years, statistical reports indicate a decline in catches (Fig. 14.12). One cause is considered to have been the spread of pollution (Halim *et al.*, 1985; Saleh and El Karashily, 1985) which has affected especially the bottom fauna (see below). Some species, however, may have benefited from increased organic matter in the sea; for example, sardines (5631 tons landed in 1981) and *Mullus barbatus*, a bottom feeder.

Of much greater importance for fishing are all the lagoons of the delta coast and Lake Mariut. Manzala is said to produce 16–35% of all Egypt's fish; 12–15,000 tons are produced yearly from Burullus. Fishing activities sustain a large number of people: 35–40,000 fishermen in Manzala and 47,000 in Burullus.

Fish production in the lagoons increased with the greater year-round drainage water discharge after 1964, which freshened the lagoonal waters and consequently altered the ratio of fresh-water to euryhaline fishes. Only in Lake Mariut has there been a decline of fish production, due mainly to pollution. Quantitatively the main species caught are mullets (*Mugil capito, M. cephalus*) and *Tilapia* (*T. nilotica, T. galilea*), but also crabs, shrimps and eels (especially in Idku and Manzala lagoons). The mullets spawn in the sea and migrate into the lagoons in winter-spring, where they attain maturity. *Tilapia*, in contrast, is an entirely fresh-water fish, and catches are generally higher in summer-autumn. In Bardawil lagoon, where salinities range from 40 to 100%, all commercial fishes spawn in the sea: *Solea vulgaris, Sparus aurata, Liza ramada, Dicentrarchus abrax*. Fisheries are entirely dependent on the maintenance of the two artificial outlets, without which the lagoon would quickly return to the state of a large salty mudflat.

Fish farming is growing in importance (Bishera, 1983) and is sponsored both by local government and the Ministry of Agriculture (Aquaculture Development Programme; General Authority for Fish Resources Development). Important fish farms are near Alexandria (Nozha hydrodrome), by the south shores of Idku (a 2000 feddan-farm producing nearly 5000 tons per year) and in Manzala. In Burullus and Manzala there is also a considerable number of private farms, consisting of seasonally opened and closed basins (hoshas). Limitations to aquaculture are the scarcity of mullet fry, traditionally caught near the lagoon outlets and in the Nile estuaries, due to pollution and disturbance by mechanized navigation; the build-up of aquatic macrophytes (Beheira Governatorate project to clear the Idku lagoon of weeds and vegetation); and conflicts between agriculture and aquaculture interests about reclamation of the lagoon margin. At present, considering the poor returns from the salty soils of the newly reclaimed lands, aquaculture should be given a priority. According to the Ministry of Agriculture, in fact, the economic value of fish farming (LE 1600 per feddan) is superior to that of any crops except cotton and citrus fruits.

Nile Delta and climatic change 555

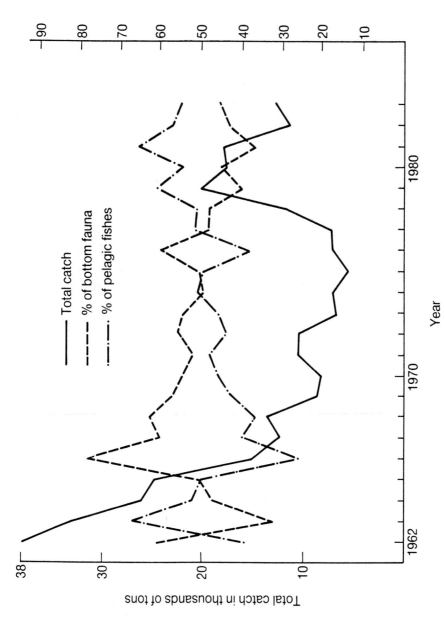

Fig. 14.12 Annual variations of total fish catch, of bottom fauna, and of pelagic fishes in the Mediterranean waters near the Nile Delta (from Saleh and Karashili, 1987).

3.4 Industries and Communications

Nile Delta industry is concentrated in Alexandria (40% of Egyptian industries), where it includes food processing, textiles and clothing, shoes, chemicals fertilizers, cement, small-scale metallurgical and mechanical manufacturing. A steel plant is being developed at Dakheila. Other industrial centers, in the western delta (Fig. 14.13) totally or partly dependent on natural gas (Abuqir Bay gas field), are at Kafr El Dawar (especially chemicals) and at El Tabia, near Abuqir (cotton mill, rayon fibre plant, paper and cardboard, dying materials, food canning, cotton dying and finishing). Small shipyards are located at Abuqir, Rosetta, Damietta and Port Said. Other industries of local importance concern mainly food processing; notable exceptions are a few rice and sugar mills.

Industry is expected to expand in the next decades by 2.8 to 7.7% annually, according to different economic and political scenarios as well as population and internal market growth (UNEP, 1988). The future of industry also will depend on internal development policies and international exchanges (north–south, east–west), and on availability of energy; diversification in the production of consumer goods will entail seeking export markets, with consequent exposure to international competition.

Population and urban-industrial expansion will require better communications in the lower delta than those existing today (Fig. 14.13). To the Alexandria harbour, the main commercial port (16 million tons imported, 1.5 million exported in 1980) have been added a new harbour at Dakheila, the new port of Damietta, and new facilities at Port Said. While communications between the ports and centers of production and Cairo are satisfactory, there is a lack of major east–west road and rail links. Two main transversal highways across the delta, one to run near the coast, are at an advanced stage of planning (ARICON, 1989).

Water transport is less developed than might be expected; in the delta it is limited to the Nubarya canal, the Rosetta branch of the Nile (where high water level is guaranteed by the Edfina barrage) and the Damietta branch, upstream of Zifta. Other canals are quantitatively of lesser importance (including the Mahmoudia canal and waterways towards and through the Manzala lagoon). The Damietta branch will probably acquire a greater role in the near future, due to its connection with the New Dumyat harbour.

4 THE PHYSICAL REGIME

4.1 Climate

Despite the fact that the coast of Egypt is semi-arid, its climate can be considered Mediterranean. The weather is highly seasonal in nature and is strongly related to the high-pressure systems whose limits overstep the boundaries of the Mediterranean area and extend towards the north Atlantic, Eurasia and Africa (Birot and Dretschk, 1956; Wigley and Farmer, 1982; Bucht and El Badry, 1986).

During winter, a semi-permanent low-pressure area known as the Cyprus Low is usually located over the eastern Mediterranean. These months are the windiest, with dominant winds from north-west and west–north-west, less frequently from north and north-east. From

Nile Delta and climatic change 557

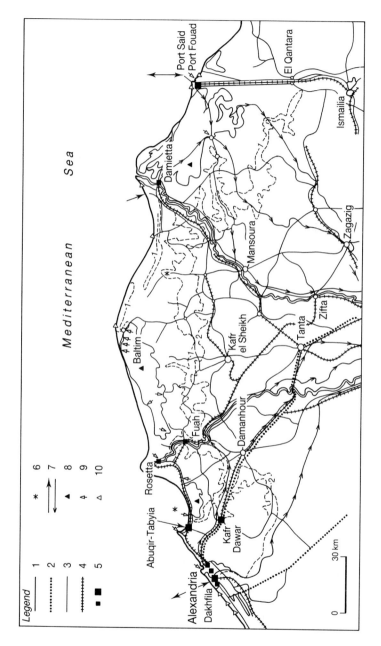

Fig. 14.13 Communications, ports, industrial centres and natural resources in the lower Nile Delta. 1. Main roads 2. Two-lane motorways 3. Navigation canals 4. Railways 5. Industrial centres 6. Natural gas fields 7. Commercial ports 8. Fishing (include aquaculture) 9. Fishing centres 10. Summer resorts.

558 Climatic Change in the Mediterranean

November to March–April, cyclonic storms, associated with moving depressions, appear regularly (Fig. 14.14). Low-pressure centres generally move along the northern side of the eastern Mediterranean. Such storms are responsible for the generation of the highest waves on the Egyptian northern shores. Weaker storms between Crete and the Nile Delta usually last about two days; they are less frequent, shorter-lived, and have shorter fetches. Thunderstorms occasionally affect the Egyptian coastal area, accompanied by sustained winds of 43 to 60 knots for short periods and instantaneous wind gusts in the range of 70 to 90 knots. These thunderstorms occur particularly between October and May, less frequently in the summer and early autumn.

During the spring months (April to May) a gradual weakening of the Cyprus Low coincides with development of a high-pressure ridge over the Mediterranean, and a low-pressure zone over the Arabian peninsula and the north-central Sahara. These weaker pressure features result in a decrease of the average wind speeds over the Mediterranean. When the depressions are counteracted by strong blasts of polar air, the south-west and south hot and dry winds (Khamessin, Ghibli) become violent, raise ground temperatures, lower the relative humidity, and transport sand and dust.

In the summer (June to September) the high-pressure ridge that runs east–west across the central and eastern Mediterranean, and the low-pressure area over the north-central Sahara reach their maximum development. As a result average wind speeds during summer are greater; in July the dominant winds are from the west–north-west. From mid-September, the Arabian-Persian depression disappears progressively, and high-pressure belts from south-east Europe and North Africa extend over the central part of the eastern Mediterranean, producing a very calm sea in October.

Temperatures in northern Egypt vary considerably with season as well as diurnally. Mean maximum temperatures range from 23–27°C, and mean minimum temperatures vary between 14.2 and 19.5°C. Temperature extremes range from 6 to 32°C.

From May to August it is entirely dry over the south Mediterranean coast and the Middle East. Humidity, however, is high with frequent nocturnal condensation (dew). The first rains may appear at the end of September. Annual rainfall, which varies considerably locally, falls mainly between early October and March. At other times it occurs as occasional thunderstorms. Significant precipitation is limited to the coastal belt, especially in the north-west; Alexandria receives 150–200 mm of rain per year, as opposed to 70 mm at El Arish, near Gaza (Table 14.1).

Rainfall decreases rapidly south of the coast. The erratic distribution of rainfall and its insufficiency are worsened by the high insolation and evaporation. Relative humidity is high at the coast, with values highest early in the morning and lowest at noon; it drops south of Cairo. The recorded rates of evaporation are high, from 3.9 to 7.9 mm a day.

4.2 Marine Parameters

The continental shelf off the Nile Delta extends to a depth of 200 m; west of Abuqir and Alexandria it is only 10 km wide, but widens to 50 km north of Rosetta and 70 km north of Port Said (Fig. 14.15). The marine environment

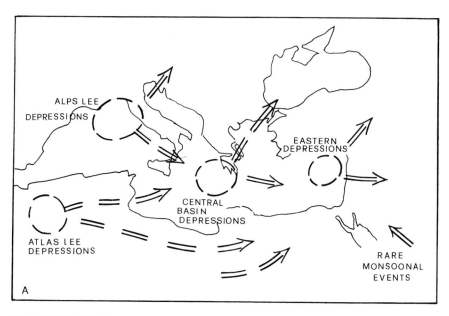
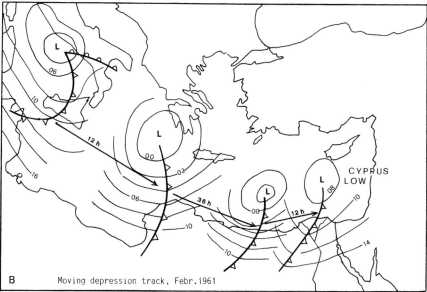

Fig. 14.14 Schematic patterns of depressions migration in the eastern Mediterranean (A, based on Wigley and Farmer, 1982; B, from ASRT/Unesco/UNDP, 1976).

560 *Climatic Change in the Mediterranean*

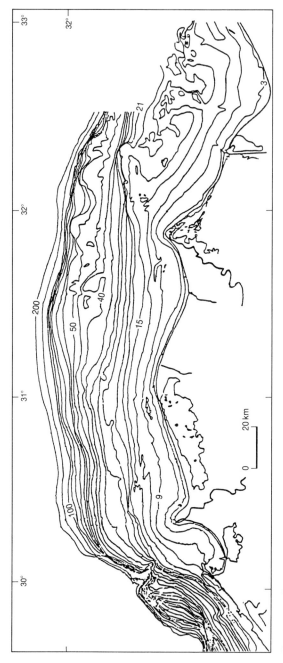

Fig. 14.15 Bottom topography of the Nile Delta continental shelf (adapted from Misdorp and Sestini, in Unesco/ASRT/UNDP, 1976).

Table 14.1 Egypt – General climatic data

	Average annual temp. °C	Average January temp. °C	Average July temp. °C	Average annual precipitation mm	Relative humidity %	Evaporation mm/day
Alexandria	20.2	13.6	26.2	166	70	5.3
Baltim	19.0	15.0	26.5	154	68	–
Port Said	21.2	14.5	27.1	173	70	–
El Arish	20.1	14.2	25.0	97	70	–
Sakha	20.1	14.8	27.0	55	69	3.9
Cairo	21.9	13.8	28.6	42	50	2.8

Source: Balba (1981)

of the Nile Delta shares the general characteristics of circulation, salinity and density of the eastern Mediterranean (Unluata, 1986), but variations occur near the coast due to local shoreline and submarine configurations as well as to the influence of fresh-water discharge.

4.2.1 Coastal waters
Before the cessation of Nile River discharge, the physical parameters of water over the continental shelf were strongly influenced by the Nile flood (Halim *et al.*, 1967; Sharaf El Din, 1973, 1976; Halim, 1987). Since then, of course, salinity distributions have changed considerably. A characteristic feature in summer is the occurrence of a surface layer of high salinity (max. ≈ 39.4%), about 20–30 m thick, overlying a subsurface layer of lower salinity (Fig. 14.16). This water mass may originate somewhere in the eastern part of the Mediterranean Sea, formed under the influence of the special circulation pattern in that region (Sharaf El Din, 1976).

Since the Aswan High Dam closure, transparency also has increased and dissolved oxygen levels have decreased (but still remain high, as characteristic of the eastern Mediterranean); water temperature has not changed substantially. In summer, waters are evenly stratified, from 26–27°C at the surface to 18°C at 75 m; in winter, they are well mixed at 17–18° to 100 m depth (Fig. 14.17). In Abuqir Bay surface water temperatures follow air temperature, the annual average is 22.6°C (28.6°C August–16.1°C February).

4.2.3 Currents and waves
Surface water circulation in the Mediterranean is generally characterized by a counter-clockwise gyre (Morcos and Hassan, 1976). The velocities of the surface geostrophic currents off the Egyptian coast are stated to be 25–50 cm/sec, though velocity and direction are much influenced by wind (Sharaf El Din, 1973). Off the Nile Delta the surface current was responsible for transporting much of the fine silt-clay of the Nile flood throughout the Levantine basin (turbid flood water could be traced as far as Lebanon a few weeks after discharge). Near the delta, a weak clockwise eddy in Abuqir Bay and a stronger one (60 cm/s) (Murray *et al.*, 1981) off Damietta are present. In Abuqir Bay the main clockwise eddy has a return

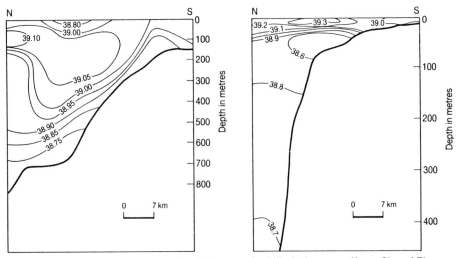

Fig. 14.16 Vertical distribution of salinity near the Nile Delta coast (from Sharaf El Din, 1977).

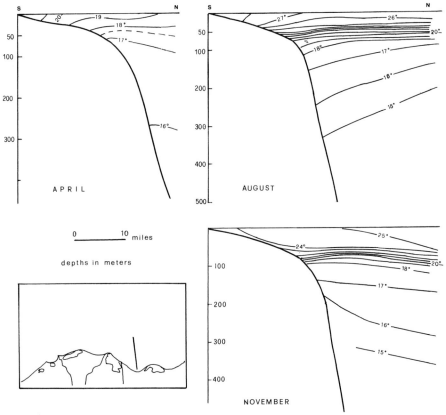

Fig. 14.17 Vertical distribution of temperature near the Nile Delta (from Morcos and Hassan, 1976).

flow through the sea bottom valleys in the middle of the bay; a secondary counter-clockwise eddy exists in the south-western part of the bay (Inman and Jenkins, 1984).

Information about bottom currents and the extent and pattern of sediment movement on the continental shelf is still limited, as are measurements, coverage and duration. Manohar (ASRT/Unesco/UNDP, 1976) concluded that summer swells and winter waves should be capable of resuspending sediment at depths as great as 60 m. In fact, north-east of Damietta strong bottom currents are said to have created a field of actively migrating sand ridges at depths 25–60 m (Coleman et al., 1980; Murray et al., 1981). Elsewhere (e.g. Abuqir Bay, north of Burullus, north of Bardawil), however, long-period measurements have shown, however, that near bottom current velocities exceed 10 cm/sec (max. of 20–30 cm/sec) only 10–15% of the time (ASRT/Unesco/UNDP, 1976; Gerges, 1981; Manohar, 1981).

The constant north-west surface winds over the central and eastern Mediterranean in summer generate swells with periods 9–10 seconds and heights of 0.8–1.5 m, lengths 100–200 m. Wave fetch can be as much as 1000–1500 km from the Ionian and Aegean seas. From November to April storm waves have periods of 7–8 seconds, are 1.5–3 m in height and approach mainly from the north-west, secondarily from the north (Manohar, 1981; Nafas et al., 1991) (Fig. 14.18).

4.2.4 Sea-level variations

The tides along the Nile Delta coast are semi-diurnal with a mean range of 12–30 cm (Fig. 14.19), but extreme ranges occur cyclically (Table 14.2). Maximum annual water levels recorded at Burullos (five years), Ras El Bar (five years), Alexandria (23 years) and Port Said (25 years) (Sharaf El Din and Moursy, 1977; Manohar et al., 1977; Sharaf El Din et al., 1990) suggest 50-year recurrences to be 126 cm and 100 cm at Ras El Bar and Burullos, respectively, with a 20-year recurrence of 95 and 80 cm high tides.

Normally, the main effect of tidal oscillations is on the dynamics of the lagoon outlets. During storms, however, sea-level can rise up to 1.5 m (wave set-up in conjunction with high tides), flooding several stretches of the coast. On average, spring tides and storms should coincide about twice a year.

4.2.5 Water quality

Pollution in Alexandria and in Abukir Bay has become a serious problem (Unesco/UNDP, 1987; Dowidar et al., 1976). About 6 million m³/day of

Table 14.2 Tidal ranges on the Nile Delta coast (source Manohar, 1981)

Location	Extreme range (m)	Mean range (m)
Alexandria	1.17	0.26
Maadia	0.59	0.12
Rosetta	1.05	0.20
Burullos	0.93	0.14
Ras El Bar	1.18	0.17
Port Said	1.32	0.33

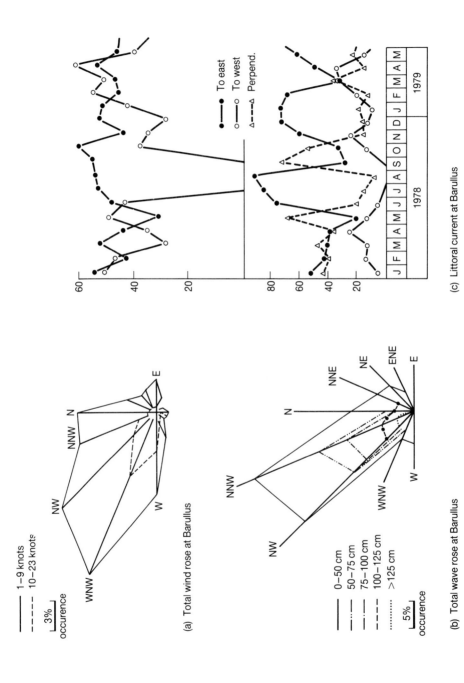

Fig. 14.18 Waves, direction and height; wind and littoral currents directions and speed at Burullus (based on Manohar, 1981).

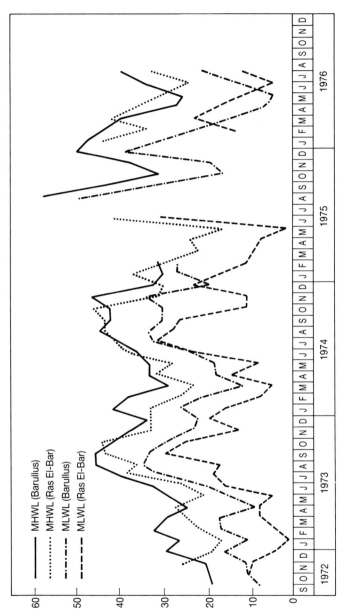

Fig. 14.19 Variations of tidal levels, Burullus and Ras El Barr (courtesy of N. El Fishawi).

agricultural drainage water, carrying agro-chemicals and trace metals, are disposed of west of the city, together with industrial wastes. Untreated domestic waste-water is still released through a number of outfalls along the coastline. Mex Bay, near Dakheila, is subjected to heavy mercury contamination from two sources, the wastes of a chloro-alkali plant and agro-chemicals; concentrations in water and sediments exceed the background levels.

The mercury level in the fish frequently rises above the standards recommended by WHO. It is highest in demersal fish such as the flatfish and the mullids, and in scombrids. It is also in the Mex fish that DDT and its major degradation products reach their highest levels.

One-year monitoring of the bathing beaches of Alexandria showed that of the 12 main beaches, seven greatly exceeded all internationally agreed standards of microbiological contamination, and five remained heavily contaminated at all times. A steady aggravation appears to have taken place over the last few years, undoubtedly due to increased population density.

The effects of pollution in fish resources appear to be both locally positive, by the organic input through domestic waste-water effluents, and negative from the threat of industrial wastes. In Abuquir Bay the fish life-cycle is at present seriously endangered by the smothering effect of the "black liquor" released by the El-Tabya paper mills (2 million m^3/yr, heavily loaded with particulate cellulose and lignin). Landings of mullet and shrimp, both from the bay and from the connected Edku lagoon, have dropped in recent years, and flatfish, such as *Solea solea*, from the bay have decreased in average size as a result of the deterioration of bottom fauna (Halim *et al.*, 1985).

There is a high flux of chlorinated hydrocarbons into the bay from waste through the Edku lagoon outlet and the mixed industrial El-Tabya effluent. Hexachlorocyclohexane (HCH), DDE and DDD were found to be the major compounds in sea water. It is in the Abuqir sediments that the highest levels of chlorinated hydrocarbons are encountered. Sediments along the whole coastal zone appear to be a sink also for PCBs and for herbicide derivatives, trichlorophenols and PCPs. DDT, absent in water samples, is present in all sediments.

The pollution of Abuqir Bay is still more or less limited to the south-western part. The polluted water, however, may extend under certain conditions eastward and northward, and thus affect other parts of the bay. Its spread is easily detectable and traceable because of low oxygen contents in the polluted waters. Although mass mortality of fish and other animals have not been reported in the bay, the plankton content of the polluted area is very low and benthic life in front of the El-Tabya outfalls is completely lacking. It is expected that continued and increasing discharge of industrial wastes will damage marine life in the entire bay.

5 GEOLOGY

5.1 The Formation of the Nile Delta
Unlike other major deltas of the world (e.g. Mississippi, Niger) the delta built by the Nile River is of relatively recent geological age. A paleo-Nile

started to advance across a marine embayment in the Late Pliocene, and developed especially in the Pleistocene through major sea-level changes associated with glacial periods. During low sea-level stands, large quantities of sand and mud were transported and dispersed far into the eastern Mediterranean, forming a large submarine fan.

In the period between 8000 to 5000 BP, when sea-level began to approach its present situation, the marine transgression had reached its maximum landward extent, as far as 10–20 km inland of the present lagoons. The modern progradation started with the development of small delta lobes (Arbouille and Stanley, 1991; Coutellier and Stanley, 1987; Sneh *et al.*, 1986; Sestini, 1989), related to several Nile River distributaries, which are named with reference to the known river channels and mouths of the 5th–3rd centuries BC (accounts by Herodotus and Strabo) (Fig. 14.20). In the east the lobes built by the Tanitic, Mendesian and Pelusiac and later the Damietta branches overlapped one another as they evolved. The lobes, typically deflected eastward, were probably cuspate subdeltas with a notable asymmetric growth of beach ridges.

Two thousand years ago the main flow was through the western (Canopic) and the central (Sebennytic) Nile branches (Fig. 14.20). The Rosetta and Damietta branches were then no more than canals. The disappearance of the older distributaries occurred mainly in the 2nd–5th centuries AD, the eastern branches (Tanitic, Pelusiac), and perhaps the Sebennytic branch, persisting until the 9th century (Tousson, 1934).

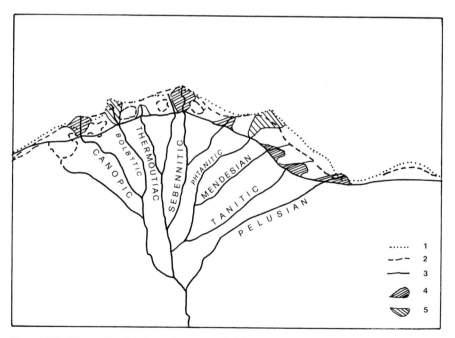

Fig. 14.20 Nile Delta distributaries and subdeltas in pre-historic times to about 2000 BP. 1. Present coastline 2. Coastline 2000 BP 3. Coastline about 7–8000 BP 4. Ancient subdeltas 5. Younger delta lobes.

The position of the coast about 2000 years ago can be estimated with reasonable approximation from historical information. In Abuqir Bay the extent of the Canopic delta lobe is well documented by archeological and geological data. The growth of the Rosetta and Damietta cuspate heads can be calculated by reference to the position of the sea by the location of towns at the time of their founding in the 9th century AD. Off Rosetta, advance from 900 to 1800 AD was 10 km, at Damietta 15 km. South-east of Port Said, similarly dated progradation between 850 and 1800 AD was about 9 km. A series of old shorelines near Tineh (Sneh and Weissbrod, 1973) can be traced through the relict beach ridges of the Manzala lagoon islands. The coastal bulge at Port Said appears to be the remnant of the progradation by the ancient Mendesian and Tanitic branches between 3000 to 1400 BP (Coutellier and Stanley, 1987).

Although in Late Dynastic-Ptolemaic times much of the northern delta was lakes and swamps (as suggested by archeological records), the Burullus and Manzala lagoons were not as extensive as today (Butzer, 1976). Smaller lagoons or lakes, possibly represented as interdistributary bays, were closed-in by the longshore growth of spits from the various cuspate subdeltas. They also may have developed out of flood basins, especially during stages of large Nile floods. The lagoon expansion during the last 2000 years has been irregular, mainly due to subsidence behind a stable beach barrier. There is a record of a rapid spread of swamps and lagoons since the 5–10th centuries, also in coincidence with major earthquakes (Tousson, 1934; Ben Menachem, 1979).

Coastal changes can be followed from maps since the early 18th century, and with more precision from topographic surveys since early 1800 (UNDP/Unesco, 1978). Throughout the last century there was a steady advance of the Rosetta and Damietta promontories (Fig. 14.21), respectively, averaging 30 and 10 m/yr, and accretion in all the embayments towards the east. The shores east of Baltim Beach Resort, as far as Gamasa, and east of the Port Said protrusion were advancing. The only exception was a slow retreat of the Burullus promontory, from west of El Burg to Baltim Beach (800 m in 100 years?), a tendency probably dating back to the disappearance of the Sebennytic mouth, which was located farther north. Stretches of the Manzala barrier were also retreating.

About 1910, an overall 25% decrease of Nile discharge, due to reduced monsoonal rainfall over eastern Africa (Rossignol-Strick, 1983), started the present coastal instability and headland recession. From 1910 to 1965 the Rosetta promontory receded by 2.5 km (with the destruction of the first lighthouse); the Damietta promontory by about 2 km.

5.2 Subsidence

Considerable long-term subsidence of the coastal zone is indicated by the 10–80 m thick layer of post-8000 BP nearshore marine, lagoonal and deltaic sediments (Fig. 14.22), with average rates of deposition as much as 5 mm/yr in the north-east part of the Nile Delta (Coutellier and Stanley, 1987; Stanley, 1990, and of 4 mm/yr in the central part (Arbouille and Stanley, 1991). Subsidence appears to have been more intense in the lagoonal belt than near the present coast. At the margin of Abuqir Bay, 15 m of compacted lagoonal/beach deposits represent sedimentation since 4800

Nile Delta and climatic change 569

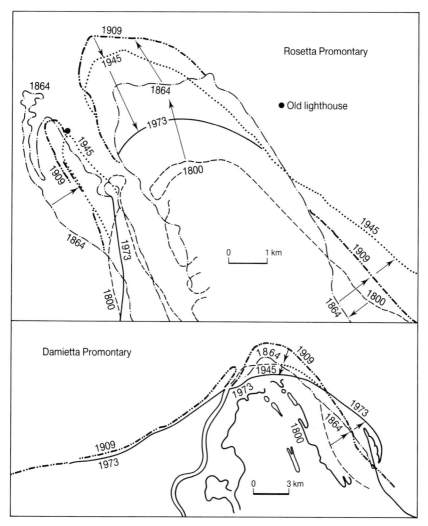

Fig. 14.21 Coastal changes at the Rosetta Nile mouth, 1800–1945 (UNDP/Unesco, 1978).

± 150 BP (UNDP/Unesco, 1978). Other evidence for continued subsidence during the last 2000 years is the following:
1) Roman ruins (Alexandria to Abuqir), today at 5–8 m below sea-level, lie deeper than accounted for by a sea-level rise of 1.5 cm/100 years.
2) The occurrence in the lagoons of drowned river banks, beach ridges and submerged ruins.
3) The many areas presently 1–3 m below sea-level (Figs 14.2, 14.4, 14.5) include reclaimed lands south of the Burullus lagoon, where drainage waters have to be pumped upwards to the lagoon.

In the Suez Canal region, north of Ismailia, a sinking of 50 cm in 80 years was documented by Goby (1952). The eastern part of Lake Manzala

570   *Climatic Change in the Mediterranean*

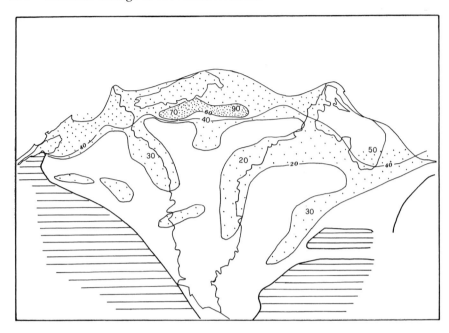

Fig. 14.22 Thickness of Late Pleistocene-Holocene silts (post-8000 BP), as an indirect indication of geological subsidence. Thickness in metres. Horizontal ruling: raised Pleistocene deposits (from Sestini, 1989, modified).

(Port Said and the northern part of the Suez Canal zone) appears to be subsiding at the rate of 5 mm/yr (Emery *et al.*, 1987; Stanley, 1988). This subsidence could be a contributing factor to the frequent flooding of the region east of the Canal. There are no recent nor on-going measurements with regard to subsidence. It is a fact, however, that natural sinking is no longer compensated by sedimentation from the yearly floods.

5.3 Soils

The soils of the Nile Delta coastal belt are genetically related to the major morphologic units of alluvial plain deposition: sand dunes and coastal sandy plains, Nile levees and channels (silt to fine sand), Nile flood basins composed of silt, areas of marshes (Fig. 14.23). Modifications have been introduced by continued agricultural activity (Balba, 1981).

The sandy soils of the coastal littoral belt are rich in $CaCO_3$. The thickness of the "A" zone is generally 40 cm in the areas between the sand dunes. Deposited in the brackish water of lagoon-lakes and swamps, the clay soils of the fluvio-marine marshlands contain some lime but much salt and gypsum. The exchangeable Na and Mg are high, and the soils are therefore saline/sodic. They are all heavy clays, but locally loam subsoils may provide better drainage. The clay soils of former swamps are rich in organic matter.

South and between the lakes, recently reclaimed areas have soils generally made of fine-textured clay and silt, normally poorly drained unless

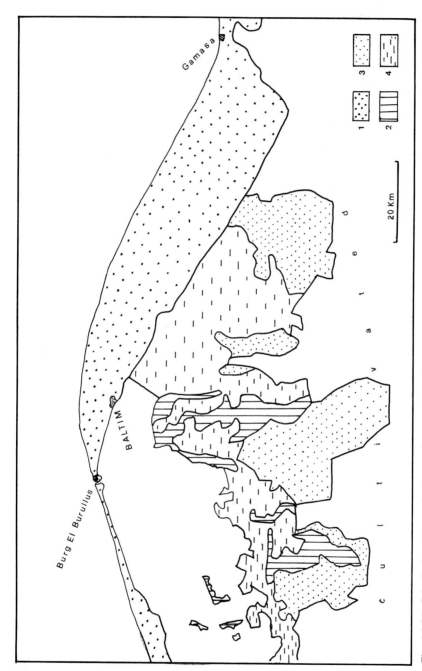

Fig. 14.23 Principal soil types in the north-central Nile Delta. 1. Calcareous sands 2. Silts-fine grained sands of fluvial channels 3. Silts (floodbasins, ± salty) 4. Acid soils of former marshes: clays, loams, peats. Based on 1956 pre-reclamation, morphological units (Sestini, 1976, 1989).

mixed with shells. The water table is shallow, about 50–100 cm from the surface. The soils are saline sodic with a high concentration of salts on the surface before reclamation. Vast areas had to be drained and irrigated with Nile water, frequently applied to improve the physical properties of the impermeable clay soils and to decrease their exchangeable sodium content.

5.4 Coastal Processes

In the Nile Delta wave energy concentration is particularly high on all north-east-trending coastal stretches and the promontories (Inman and Jenkins, 1984; Manohar, 1981). The north-west-oriented stretches are attacked more by north-west and north waves as a consequence of refraction, than by the less frequent north-east and north–north-east waves (Fig. 14.24). The eastward littoral drift drives the beach and nearshore sands of the Nile Delta to and beyond North Sinai. Conservative estimates are that 1,000,000 m^3 of sand are moved yearly by littoral currents, which are quite strong (Fig. 14.18); in addition, large amounts of sand have been and continue to be removed from the beaches by the offshore winds. Estimates of sand losses are in the order of 200,000 m^3 yearly west of the Rosetta mouth and 400,000 m^3 a year in the Burullus-Ras El Barr coast (Manohar, 1981; Smith and Abdelkader, 1988).

Prior to the final 1964 closure of the Aswan High Dam, the Nile discharged about 85–90 million tons of sediment into the sea in late summer, 65% through the Rosetta, 35% through the Damietta mouth. One-third was fine to very fine-grained sand (0.125–0.065 mm) which was deposited in the immediate vicinity of the mouths, and then reworked in winter by waves and littoral drift. The other two-thirds of the sediment was carried offshore and alongshore in suspension. Because of the pronounced density gradients and regional circulation, a surface plume of suspended clay and silt was carried eastward along the delta front. Such turbid plumes are still formed today, as shown by satellite images (UNDP/Unesco, 1978; Klemas and Abdel Kader, 1982). At a few points on the coast (e.g. in the middle of Abuqir Bay between Rosetta and Burullus, east of Port Said) there may be offshore sinks, with sands being moved off the coast to the continental shelf (Inman and Jenkins, 1984).

During winter storms, the deeper fine-grained bottom sediment (silty clay) off the Rosetta and Damietta capes is stirred into suspension and moved seaward, mainly to the east (Summerhayes et al., 1978). At present, the only source of sand is the shore itself and the inner shelf to 30 m depth, where areas of relict (Holocene) sands have been mapped (Summerhayes et al., 1978; Frihy et al., 1990). The actual movement of these sands, however, is not known.

Retreat is continuing on all the most exposed parts of the coast, and instability has been noted elsewhere (Fig. 14.25). There are a number of stretches where shoreline variability has created problems in relation to existing or planned land uses.

Twenty-five years after the cessation of Nile discharge, the following conditions had been noted along the shorefront (UNDP/Unesco, 1978; Tetratech, 1986):

Nile Delta and climatic change 573

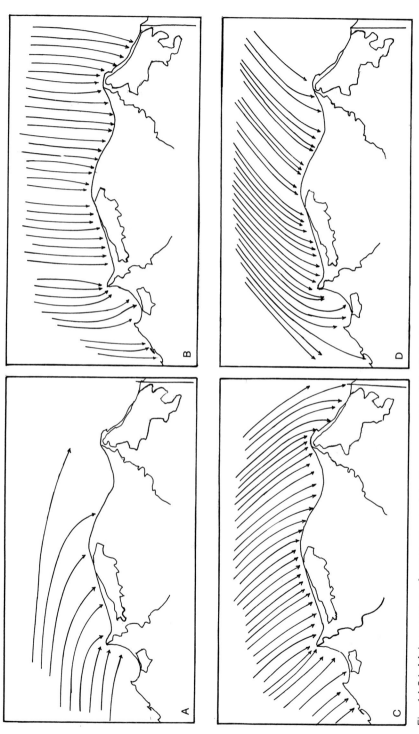

Fig. 14.24 Main patterns of wave approach to the coast of the Nile Delta (based on Tetratech, 1986). A. Westerly waves B. Northerly waves C. North-westerly waves D. North-easterly waves. ACL: T–10 seconds.

574 *Climatic Change in the Mediterranean*

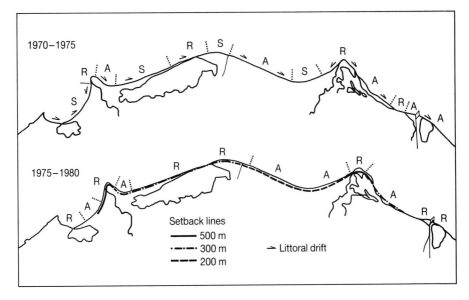

Fig. 14.25 The state of erosion-accretion of the Nile Delta coastline, 1970–1980 (top from UNDP/Unesco, 1978; bottom from Tetratech, 1986).

(1) At Alexandria, as the coast is directly exposed to the north-western and north waves, the eastern and western harbours' outer sea walls are in constant need of attention due to frequent overtopping by waves. The small beaches of the city are retreating because there is no littoral drift supply of sand. Sediment movement is complicated by the rocky capes, islands and shallow submerged rocks.

(2) In the west part of Abuqir Bay, the Mohamed Ali sea wall may be undermined by scouring. The wall, which is subjected to violent action by storm waves > 2 m in height, had to be renovated in 1981–1984 (Tetratech, 1986). The Idku outlet has been stabilized by jetties, but flooding is common in the town of Maadia; a tendency of erosion to the west may be increased by the new fishing harbour. The eastern Abuqir Bay shore is maintained by the erosion of the Rosetta promontory, but seaward sediment transport about 18 km east of Idku outlet has resulted in increased beach steepening.

(3) Retreat of the Rosetta headland between 1970–1987 increased to 80–120m/yr (Frihy, 1987). Erosion of the beach barrier sands has exposed the underlying finer-grained marine sediments to wave action; the submerged delta cone is thus being eroded to 25–30 m depth or more. Intense erosion occurs especially at the river mouth and is already weakening the sea wall. This structure will eventually function as a detached breakwater, following the gradual progress of erosion at its ends; thus, it should slow down the retreat of the promontory.

(4) East of the Rosetta promontory, for 30 km, the shoreline tends to be stable and presumably will continue to be so for another two to three decades. East of Hanafi to 6 km west of the Burullus outlet, however,

the shore is narrow and the swash zone has tended to steepen with some retreat, perhaps because of accretion at the outlet western jetty. Burg El Burullus is protected from erosion by a renovated sea wall, but at the east end of the latter, a serious erosion of the sand dunes (6 m/yr) has occurred for the past 30 years. Land, however, has extended towards the lagoon, both east and west of the lake outlet (Fig. 14.6).

Further east, the Burullus headland is retreating an average 10 m/yr with erosion of the dunes. At Baltim Beach Resort, where the shore had been fluctuating periodically, there is now a slow retreat, averaging 3 m/yr. Short of beach protection (i.e. sand renourishment, offshore breakwaters), the present resort town will be destroyed by 2000 AD. Active longshore and seaward sand transport prevails eastwards to Gamasa; the coast is generally stable or accreting.

(5) The stretch from Gamasa drain to Ras El Barr is one of moderate longshore activity, but 7 km west of Ras El Barr retreat has occurred (250 m from 1945 to 1975), in association with the retreat of the Damietta headland. The New Dumyat Harbour jetties have stimulated accretion on both sides, but have probably modified the local wave and current regimes, with silting of the access canal. Dredging is already necessary and artificial sediment bypassing may be needed by 2000 AD.

(6) Due to the predominant north-west wave approach, the entire Damietta promontory is under erosion, and the Ras El Barr protective structures have been only partially or temporarily effective. Circulation in the Nile estuary is presently entirely tidal (there is no river flow from the Faraskur Dam) and dredging is required. East of Ras El Barr an earth dike extends as far as the spit (Fig. 14.7) (Tetratech, 1986); its envisaged revetment into a stronger structure could eventually increase the erosional problems of the Nile mouth area and cause retreat at its eastern end.

(7) The whole stretch from Damietta to Port Said is very unstable due to a very active longshore sand migration. The spit has been growing by 90–100 m/yr, and the shoreline for 15 km eastwards is under its dynamic influence, with a series of erosion-accretion patterns; the Diba embayment and hump are features that move as much as 100 m/yr; eventually they, and the spit, might encroach on the new Manzala lagoon outlet.

The section 19 km west of Port Said is a unit in itself. It is a weak area, except for the 4 km west of the Port Said western jetty, which has been accreting for over 100 years, though at slower rate in the last decades. It is characterized by long-term retreat (the barrier between lake and sea was reduced from 1000 to 200 m width between 1810 and 1945, and some retreat has occurred since then; UNDP/Unesco, 1978) and by frequent flooding from the sea.

East of the Suez Canal, erosion of the Port Fouad beach has resulted from the lack of sediment supply from the north-west. Long-term erosion of the formerly advancing coast, east of the new Canal Bypass jetties, is inevitable; this and the frequent flooding of the plain, as far as Tineh, could have a negative impact on new land reclamation schemes there.

6 HYDROLOGY AND WATER RESOURCES

The Nile Delta coastal zone is almost totally dependent on the Nile for its surface and ground-water needs. The building of the Aswan High Dam has permitted the storage of annual flood water and the year-round irrigation of all lands presently cultivated; in addition to electricity production, a substantial amount of water has become available for agricultural expansion. However, there is still a pressing need to pay close attention to the efficient use of the stored water, especially to allow new land reclamation.

Measures to conserve water (Master Water Plan, 1984) regard increased usage efficiency and the reduction of "non-beneficial" flows into the Mediterranean. Water used for irrigation and urban-industrial waste waters are returned to the river and drained through the coastal lakes and canals to the sea. Extra water also flows to the sea in winter, when irrigation requirements are reduced.

Egypt now receives, by international agreement, 55.5 km^3 of Nile water through the High Dam Lake Nassar. In the future this quantity might be increased by various water-saving projects in the Sudan and neighbouring countries (Kashef, 1981a; Haynes and Whittington, 1981). However, variations in rainfall in East Africa can have a profound effect on Egypt's water resources and well being; witness the recent low water situation in the Lake Nasser reservoir in consequence of years of drought in Ethiopia (Smith, 1986).

6.1 The Nile Today

The course of the River Nile is divided roughly into three parts: the lower Nile to Aswan (First Cataract); the middle Nile, between Aswan and Khartoum; and the two upper branches, White Nile from Uganda, and Blue Nile from Ethiopia (Lake Tana) (Fig. 14.26). At present the Blue Nile furnishes on average 68% of the flood discharge and 72% of the sediment load; it is fed by the summer monsoon rains that fall over Ethiopia. The discharge minima in winter reflect the equatorial rains in East Africa supplied more sluggishly by the White Nile, the flow of which is strongly retarded by the swamps of the Sudd, in southern Sudan.

In this century, from 1902 (first Aswan Dam) to 1964, aqueous discharge at Aswan averaged 84 billion m^3/yr, mostly (93%) between July and November, with a peak in August-September (Halim, 1991). The range was 45.5 to 140 km^3/yr. Between Aswan and the sea, as much as 60% of water discharge was lost to irrigation, evaporation and seepage. Thus, before 1964 actual water discharge into the sea averaged 50–60 km^3, 65% of which through the Rosetta branch (42.9 km^3 in 1959–1963). During the dry season, the estuaries were kept closed by the Faraskur (Damietta) earth dam and the Edfina (Rosetta) weir dam.

The first Aswan Dam, built in 1902 and further elevated in 1921 and 1933, had a storage capacity of 5 km^3. The High Aswan Dam reservoir is 164 km^3, but evaporation losses are high: 15 km^3/yr. The High Dam provides a controlled mean annual discharge of 58.60 km^3 of water downstream, which has enabled expansion of the cultivated area by 1.3 million feddans, and a conversion of 700,000 feddans from basin to perennial irrigation. About

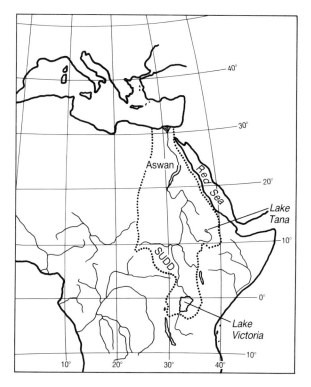

Fig. 14.26 The Nile Basin.

10,000 MW of electricity are generated annually, navigation conditions along the Nile have improved, and flood protection is now guaranteed.

To all intents and purposes the Nile below Aswan has been reduced to the status of an irrigation canal. Barrages at Esna, Asyut, Naga Hammadi (south of Cairo) keep the level of the river high enough for perennial irrigation in the valley. Today the Nile Delta region receives 35 km^3 of surface Nile water; 33% of this amount is lost due to evapotranspiration and infiltration to ground-water aquifers (21%). The remaining surface water is channelled across the delta through a system of feeder canals originating from the Delta Barrage and from the Ismailia Canal (Fig. 14.27).

Discharge of water to the sea takes place mainly through the lagoon outlets to the Nile mouths, the Mex (Alexandria) and El Tabyia (Abuqir) pumping stations, and the Kitchener and Gamasa drains. In total, about 13–16 km^3 are discharged to the sea, six of which flow through the Rosetta and Damietta mouths. The largest outflow is through the Rosetta estuary (±25 m^3/s, increasing to 550 m^3/s in January–February), the Idku (134 m^3/s), El Burg (145 m^3/s) and El Gamil (125 m^3/s) outlets. Maximum flows through the Kitchener and Gamasa drains and the Damietta branch are less than 50 m^3/s.

6.2 Nile Discharge Variations

The most significant aspect of Nile discharge for water resources in Egypt has been its variability. In the last 100 years natural flow has varied from

578 *Climatic Change in the Mediterranean*

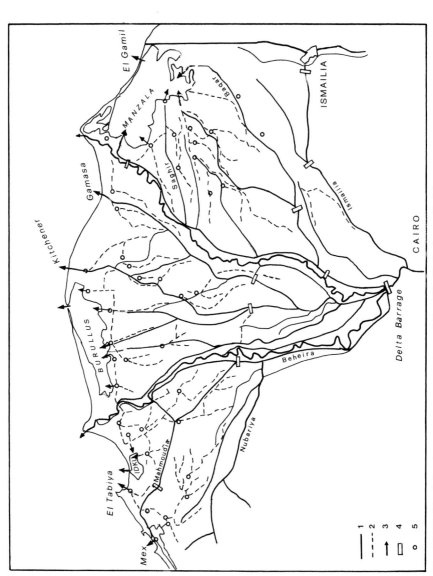

Fig. 14.27 The irrigation-waterways system of the Nile Delta (compiled from Drainage Research Center, *Report on Quality of Drainage Waters*, Ministry of Irrigation, Water Research Center, Cairo, 1986). 1. Main canals 2. Irrigation drains 3. Main discharge points 4. Barrages 5. Main pumping stations

$45.5 km^3$ (1913) to $150 km^3$ (1978) (Fig. 14.28). In general, floods were 25% larger in the 19th century due to heavier rainfall in Ethiopia (UNDP/Unesco, 1978; Rossignol-Strick, 1973). By contrast, seven dry years (1979–1986) in the watersheds of the Nile caused the level of Lake Nasser to fall by 21 m, forcing a series of emergency withdrawals, with deleterious consequences for power generation (Fig. 14.29).

The cyclicity of Nile discharge is well established (Fairbridge, 1984; Hassan, 1981; Hamid, 1984). Examination of discharge records has indicated both short (8–20 years) and long (60–90 years) periods of flood cycles. In the present century there have been five episodes of extremely low Nile flow, corresponding to the great Sahelian droughts affecting the region from western Sahara to Somalia: 1911–1915 (when the total discharge of the Nile was only 45% of the average flow), 1918–1928, 1941–1944, 1969–1972 and 1973–1976 (Fairbridge, 1984).

6.3 Hydrology of the Lagoons

At present, inflow of drainage water dominates the water balance (and hence the water quality) of the coastal lakes (Table 14.3). In spite of

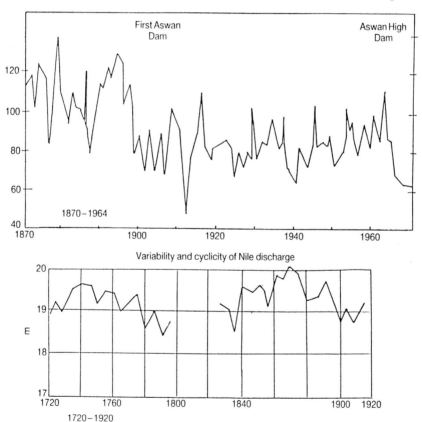

Fig. 14.28 Variability and cyclity of Nile flood discharge (from R.E. Quelennec, in UNDP/Unesco, 1978 and ASRT/Unesco/UNDP, 1976).

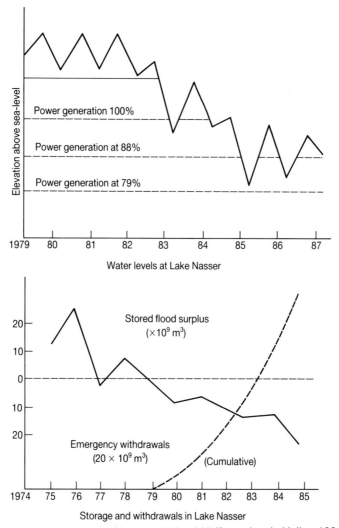

Fig. 14.29 Water levels in Lake Nasser, 1979–1987 (from data in Halim, 1991).

an evaporation rate of 170 (in Burullus) to 200 mm/month (in Manzala) (June–July) and a total calculated loss of 2 km³/yr for the two lakes, the outflow is much greater than the tidal and occasional storm-surge inflow of Mediterranean water; rainfall contribution is negligible.

Water-level changes are due to seasonal differences in drain and canal discharge (Fig. 14.30); tidal level changes are minimal. Three main drains flow into Edku Lagoon, one canal and five pumping stations discharge into Burullus, while six major canals and drains reach Manzala (Figs 14.8, 14.27).

In general, Nile fresh water has low salinity (300 ppm). Drain waters have salinities of 1000–2500 ppm, but some are greater in the periods of reduced flow. Salinity values in the lagoons, therefore, are related primarily to the

Table 14.3

	Present area (km²)	Drain discharge (m³/yr)	Outflow (m³/yr)	Salinity ‰	Temperature winter (°C)	summer (°C)
Mariut	63	?	?	1.09–2.63	–	–
Idku	71	+1.8 billion	?	4.3–14.8	12.1	28.5
Burullus	402	+8.1 billion	6.2 billion	0.5–21	13	29
Manzala	1350	+4.4 billion	?	1.44–10.6	11.3	30.5
Bardawil	700	–	–	41–68 (up to 70–120)	12.7	30.5–34

(*Sources*: Dowidar et al., 1976; Halim and Gerges, 1986; Krumgalz et al., 1980; Maiyza, 1992; Delft Hydraulics, 1991)

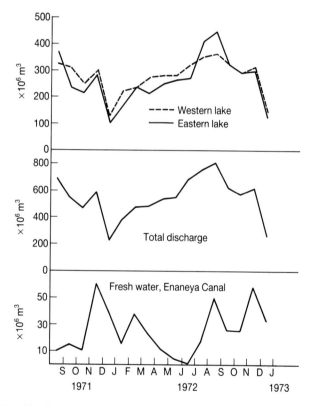

Fig. 14.30 Monthly discharge of drainage and fresh water into Lake Manzala (Halim and Gerges, 1981).

balance between drain water and the influx of marine water; thus salinities increase towards the outlets, where they reach values of 15–20‰. Higher values in the eastern part of Burullus also are due to saline drain water from the Baltim pumping station. In Manzala the northern part of semi-enclosed shallow lagoons is more saline due to evaporation. But salinity is also related to water temperature and evaporation; it is generally higher in summer, with a maximum in July. Because the lakes are shallow and the waters are well stirred by the frequent winds (from south-west in winter, north-west in spring-summer, occasionally from north and north-east), there is no temperature and salinity stratification.

The hydrology of the Bardawil lagoon is totally different. For centuries it was a dry salt and mud pan, occasionally flooded by storms, with local development of salt marshes. The two present outlets (Boghaz I, II) were artificially excavated in 1950. Salinities in the lagoon are normally 40 to 70%, but rise to 120% in sheltered parts. Salinity increases from the surface to the bottom, and from the northern to the southern parts; there is a definite stratification, and water levels vary according to marine inflow (which is high after north, north-west storms).

The degree of pollution of the coastal water bodies of the Nile Delta varies from high (in Lake Mariut) to minor (but increasing) in the main lagoons. Water quality has been badly affected by agricultural waste waters carrying pesticides and fertilizers (Unesco/UNDP, 1987). Consequently, in Manzala and Burullus during the last two decades, there has been an increase in phosphate and a corresponding decrease in dissolved oxygen (El Rayis et al., 1979). A part of Lake Mariut receives waters, untreated sewage and industrial discharges from Alexandria and through El Kalaa Drain. Phosphorus, which increased 15 times between 1963 and 1978, is now 13.1 mg/l/yr, and there are strong eutrophication tendencies. Eutrophy problems also have recently appeared in Burullus, once one of the cleanest lagoons (Beltagy, 1985); organisms also have revealed high concentrations of Hg, Cd, Cu and PCB's.

6.4 Ground water

A recent water resources management programme conducted by the Egyptian Ministry of Irrigation for the purpose of land reclamation (Kashef, 1983) has indicated that the delta has large ground-water resources in the Main Delta Aquifer (Pliocene-Early Pleistocene coarse sands at 200–800 m depths, Said, 1981; Fig. 14.31). The aquifer increases in thickness from about 250 m in the south to about 900 m at the coast, is limited to east by the Suez Canal, and tapers off towards the Western Desert. The recharge of the aquifer is mostly from the Nile (the contribution of rainfall is minimal). The aquifer is recharged through the infiltration of excess irrigation water and seepage from the extensive network of canals and drains.

In the recent past, the potentiometric (piezometric) levels in the Delta aquifer fluctuated in response to the river stages. After the construction of the High Aswan Dam and the closure of the Nile by this dam in 1964, potentiometric levels reached a more or less stable pattern. The present potentiometric levels are slightly higher than those before 1964 due to the low irrigation efficiency and the slight increase of the almost stationary and regulated water levels of the Nile. The hydraulic gradients of ground-water flow in the Delta increase from about 8 cm/km in the South to about 12 cm/km in the North. Near the river and the canals, the gradients increase to more than 50 cm/km.

The annual infiltrated recharge to the aquifer could amount to 6.4 km^3/yr. In 1980 the total withdrawal of ground water for both

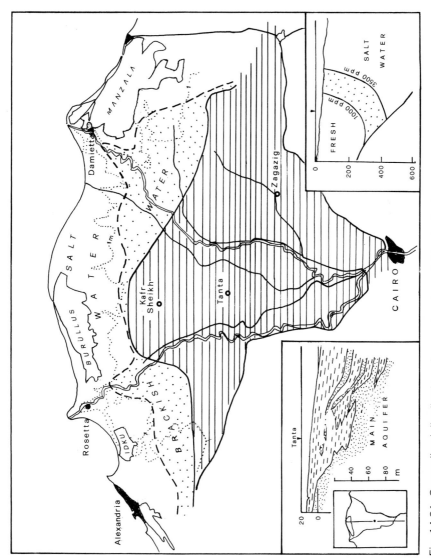

Fig. 14.31 Generalized distribution of ground-water salinities in the Nile Delta (based on Kashef, 1983).

agriculture and urban-industrial purposes was about 2.4 km³/yr, Cairo and Giza use about 19%; most is used in the western (40%) and eastern delta regions (35%). The balance of 4 km³ should represent the potential water surplus for future development. This number is approximate, but if correct, it would indicate an annual gain to the aquifer. If this were to be repeated yearly, it would lead to water logging. On the other hand, increased water withdrawal would reduce water logging, leading to better agricultural production, but with the danger of increased salt-water intrusion.

Ground water in the northern Nile Delta is generally unsuitable for irrigation due to the presence of high amounts of soluble salts derived from sea water intrusion (Shata and El Fayoumi, 1970). Special care is needed to prevent lime accumulation and soil alkalinity where ground water is used; more frequent irrigation and leaching at frequent intervals are needed to prevent soil drying (Ghandour et al., 1985). The quality of ground water is therefore a serious hindrance to its extensive use in the lower delta. The salt-water wedge intrudes deep in the interior of the aquifer, extending landward about 130 km from the coast (Fig. 14.31).

6.5 The Utilization of Water Resources

The water available at present to Egypt, by international agreement with Sudan, is at least 55.5 km³/yr. Usage demands on Nile water (Fig. 14.32) were as follows in 1985 (Ministry of Irrigation, 1985): hydropower generation and navigation, at least 1.6 km³/yr; urban and industrial, 2.43 km³/year; agriculture, 29.4 km³/year. About 6.5 km³/year are returned to the river after irrigation use, mostly in upper Egypt. Evaporation losses are high, perhaps 3.6 km³/year, downward losses due to underground percolation may amount to 6.4 km³/year, and an estimated 2.4 km³ are extracted from the aquifer (Kashef, 1981).

The water needs of industries and cities are expected to grow to 11 km³/year by the year 2000. By the same date, the irrigation of all the potentially reclaimable land (2.4 million feddans) would require an extra 20 km³/year. The 1964 Water Master Plan of the Ministry of Irrigation mentioned that water would be sufficient to irrigate 1 million ha. above

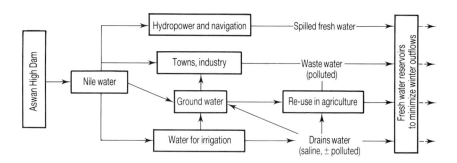

Fig. 14.32 Schematic representation of the utilization of Nile waters in Egypt.

the present 2.4 million ha. Other reviews of water resources (Assen, 1983) have indicated that no more than 0.3 million ha could be irrigated.

It is clear that if future industrial and agricultural developments are to meet the demands of an increased population, measures must be taken to conserve water by increased efficiency of usage. Reclamation efforts will have to be reduced, unless some combination of the following is enacted: (a) increased efficiency of the irrigation system and improvement of farm management practices; (b) utilization of new irrigation technologies, such as drip and sprinkler methods; (c) substantial re-use of drainage water and of city waste waters; (d) increased use of ground water; (e) a limitation of the amount of water that flows to sea in winter, when irrigation needs are less.

Proper ground-water resource management would be necessary (e.g. location of withdrawal stations, continuous monitoring of ground water systems and salt water intrusion, etc.) and developments could be safely started only in the southern part of the delta and along the Nile branches, while extra usage in the northern areas would be delicate, requiring careful investigation, including possible subsidence effects.

In the months of low irrigation demand (especially during the winter closure of the irrigation system) water releases for power generation at Aswan exceed the aggregate demand for irrigation and all other uses. The capture of the superfluous Nile water flow and its storage in a portion of the Burullus and Manzala lagoons would permit the saving of at least 2–3 km^3 of fresh water to irrigate 200,000 to 300,000 feddans in the vicinity of the two reservoirs (Ministry of Irrigation, 1985). One project envisages the bounding of 670 km^3 of the Manzala lagoon by a 121 km long, 2.5–5 m high dike that would keep the lake level 0.5 to 1.5 above msl. Water would be brought in from the Damietta branch, upstream of the new Damietta Barrage, by a 8 km long feeder canal; an additional 0.82 km^3 of drainage water could also be fed into the lake. 0.5 to 1 km^3 of water could be stored; without increased evaporation salinity is expected to range 400–1000 ppm. The lake water would then be pumped into the El Salam canal for irrigation use. In Burullus, an area of 470 km^2 would be encircled by a 114 km long dike, and 1–2 km^3 of water would be diverted into the lake by means of a 15 km feeder canal; 1 km^3 could be added from drains. Water would be used to irrigate an extra 41,000 feddans to be reclaimed along the south shores of the Burullus lagoon. Salinities would vary from 400 to 1100 ppm.

Although these large-scale projects are feasible from an engineering viewpoint, they should be considered only within the general system of water and land management for Egypt. Usage (i.e. fisheries vs. agriculture), management and environmental aspects still need to be addressed. With the additional possible impact of climatic changes, a number of points would need to be considered: optimal size and water levels in relation to sea water infiltration, rates of evaporation, seepage; the amount of water to be stored in relation to the amount and timing of the Nile surplus; the effects of future salinities and distribution of nutrients, and water depths on vegetation and natural fish breeding; the reservoir impacts on health.

7 ECOSYSTEMS

7.1 The Marine Environment

The quality of Mediterranean water remains good (Emara *et al.*, 1973), but biological productivity has been severely affected by the over 95% decrease in phosphates and nitrates that followed the cutoff of the Nile flood (Halimet *et al.*, 1967; Halim and Gerges, 1986). Today, minor phytoplankton blooms still occur in winter.

Biomass near the coast is 2–20 times larger than in deeper water, due to water mixing, drainage and Nile discharge. The shallow Abuqir Bay near Alexandria, an important area for fish spawning (e.g. *Mugil, Sparus*, etc.), is increasingly affected by industrial and agricultural pollution (Bishara, 1983). The adsorption of pollutants by bottom sediments has produced a marked change in the proportion of the pelagic vs. benthic fauna (Saleh and Karashily, 1985). Decreased fish catches are related to these impacts on the bottom fauna.

7.2 The Brackish Lagoons

Although much reduced by reclamation, subject to many threats and nowhere officially protected (the only natural reserve in Egypt is near El Arish in North Sinai), the lagoons and wetlands of the Nile Delta represent vast natural ecosystems of considerable biological importance. Many stretches of shoreline and dunes are still in a natural state, and the shallow lagoons present a variety of morphological and vegetational biotopes that support a rich aquatic and bird fauna. The lagoons are extremely important for migrating and wintering birds, particularly as they are the only extensive brackish water bodies on the north-eastern coast of Africa.

The ecology of the Egyptian coastal lakes is reasonably well studied (Aleem and Samaan, 1969; Samaan, 1977; Wahby and Bishara, 1977; Shaheen and Yousef, 1980; Dowidar and Hamza, 1983; Halim and Gerges, 1986). Productivity in the lagoons is much higher than in the adjacent Mediterranean Sea; fishing yields for instance average 24 t/km^2 in the lagoons vs. 1 t/km^2 in the sea. Idku, Burullus and Manzala lagoons present similar features in being shallow, with a flow pattern controlled by fresh-water discharge, and subject to wind influence. The environment provides optimal conditions for primary producers (phyto- and zooplankton): the water is well mixed with regard to salinity, temperature and nutrients. Productivity has doubled in the last 20 years due to increased amounts of drainage water. A considerable percentage of the water bodies are covered by aquatic plants (Fig. 14.33), especially near the south shores and near the islands, with high productivity near the mouths of the drains.

The phytoplankton is characterized by continuous blooms that increase at the time of high drain discharge. In the Idku lagoon phytoplankton production is estimated to be 0.604 gm C/m^2; the lagoons are considered to be mesotrophic. There are significant phytoplankton variations related to the distribution of nutrients. The zooplankton is characterized by a mixture of estuarine and various coastal Mediterranean species.

Apart from Lake Mariut, which is not connected to the sea, the fauna of the brackish lagoons is characterized by a mixture of marine and euryhaline to fresh-water fish. *Mugil*, eels, *Solea*, *Sparus* and several other fish enter the lagoons for feeding; the fry of these fish aggregate in very large numbers near the lagoon outlets (as well as in the Nile estuaries) at different times of the year, but especially in winter and spring.

The highest bottom biomass was recorded (Samaan, 1977) in areas devoid of hydrophytes; the lowest in the plant areas and in the region of lake–sea connections. The bottom fauna is made of brackish-water forms that can tolerate wide ranges of salinity, mainly the polychaete Nereis, amphipods (*Corophium*, *Gammarus*) and pelecypods (*Ancylus*). The areas covered by *Potamogeton* and *Ceratophyllum* are benthos-poor, especially where the plant cover is dense (6 gm dry wt/m^2).

Although a considerable number of fish are found in the lagoons, the predominant species are tilapias (especially *Tilapia nilotica* and *T. zillii*), followed by mugils. The proportion of fresh water to marine fish depends on salinity, hence on the seasonality of drain discharge versus marine inflow. In Manzala lagoon Ezzat and Hosny (1983) found the following proportions: *Tilapia* (86%), mugils (9.5%), *Morone labrax* (2.1%), eels (1.6%). In the Burullus lagoon, mullets make 15–30% of fish catches.

The tilapias thrive in the plant belts because of the high oxygen saturation values, and because they feed on the epiphyte algae that cover the leaves of *Potamogeton* and *Ceratophyllum* and on the amphipods and snails that feed on the algae and diatoms. The mullets, in contrast, are associated with the pelagic ecosystem of herbivores (copepods, cladocerans, rotifers, shrimps, mysids, snails). In Lake Manzala carnivorous fish constitute only 3% of the total.

The shallow and saline Bardawil lagoon has a high productivity (Por and Ben Tuvia, 1981). The bottom is covered by a seagrass (*Ruppia spiralis* and *Cladopora*); vegetation near the shore is poor due to the high salinity. The phytoplankton is also poor, but there is a rich zooplankton and bottom fauna (copepods, ostracods, tintinnids, nematode worms, molluscs, shrimps and crabs).

All the fish migrate between the lagoon and the sea; of a total of 65 species, one fifth are east Mediterranean fish and many Red Sea species are present. The most common species are *Sparus auratus*, mugils (six species), which are particularly adapted to the high salinity, *Dicentrarchus*, *Argyrosomus*, *Umbrina*, *Epinephelus*, *Atherina* and *Aphanius*.

The coastal lakes present a variety of biotopes suitable to wintering, migrating and nesting birds: e.g. the sandy flats, the reed beds and marshes, fish culture basins and the mudflats. It has been estimated (Meininger and Mullie, 1981) that no less than 800,000 birds (mostly water, wading and shorebirds) winter in the Burullus. Mariut, Idku and Bardawil are important localities for staging birds, though active hunting (especially in Mariut), land reclamation, fishing activities and pollution have reduced the typical habitats of migrating birds. It has been suggested that part of the lagoon be made into a wildlife reserve.

8 An Evaluation of the Impact of Climatic Change

8.1 Introduction

An analysis of the consequences of the possible future rise of air temperature and of sea-level for the Egyptian coastal region of the Nile Delta involves the difficult task of projecting both contemporary physical and socio-economic parameters. The review of the present characteristics of the coastal region indicates that impact analysis should be based on the following premises:

1) The generally retreating state of the coast, caused by the lack of Nile sediment input since the building of the Aswan High Dam. Although erosion is particularly active in the most prominent and sediment-starved parts, it also has been enhanced by artificial fixed structures, which have disrupted the natural movement of sand.
2) The rapid urbanization of the coast, especially east of Alexandria towards Abuqir Bay; west of Port Said and near Damietta. Continuing demographic growth forecasts a population over 12 million in the 0–3 m coastal region by 2020.
3) The probable intensification of land use in the 0–2 m belt. Development projects include new roads, towns and industries (Fig. 14.9), land reclamation and possible use of parts of the Manzale and Burullus lagoons as fresh-water reservoirs. This will result in the strengthening of the economic role of centres other than Alexandria, such as Damietta and Port Said (first order), Rosetta-Motubis and Baltim-El Burg (second order). It also will increase water and energy needs and bring further pressure on the environment (e.g. on the quality of surface waters, on the lagoon margins, on the weak parts of the shoreline).
4) Agriculture in the 0–2 m coastal region by and large makes use of poor soils affected by salinity in reclaimed areas. To the traditional crops have been added industrial cash crops (other than cotton) such as sugar beets, sunflowers, fruits for export. Cereal cultivation has declined; local fruit-vegetable areas will expand with the rise of new towns.
5) Water resources are entirely dependent on the Nile, and both surface and subsurface waters are quite saline. Because of this, ground water is not much used in the coastal zone.
6) The economically very important lagoonal fishing (including aquaculture), as well as lagoonal wildlife, relies on a specific physical system (shallow water, salinity, temperature, water circulation, plant cover) that would be dramatically affected by human-induced changes (pollution, reclamation, damming).

The foreseeable consequences of an average temperature increase of 1.5°C and of a rise of sea-level of up to 25 cm by 2030 (Milliman, Chapter 2; Warrick and Oerlemans, 1990) could be:

1) Greater direct physical impacts of the sea on the harbours and on shoreline uses (including summer resorts), the accelerated retreat of the promontories and a possible break-through of the low-lying coastal sand barriers.
2) The direct effects of a higher sea-level on the hydraulic and biological

functioning of the lakes, either as fresh-water reservoirs, or as brackish lagoons.
3) The indirect effect of higher sea-level, in addition to raised temperature (i.e. with higher evapotranspiration) on the further salinization of coastal soils and aquifers, which would magnify the problems of reclaimed lands agriculture.
4) The effects of overall climatic changes on hydrology, that is on Nile water supply and aquifer recharge, due to possible changed precipitation patterns in East Africa and along the Mediterranean coast.
5) The consequences of higher temperatures, in association with human induced changes, on fishing and wildlife in the lagoons.

The impact of climatic change on the social and economic structure of the lowland region would derive from the effects of 1–4 above, but would also have to be evaluated in the context of:
1) population and economic growth;
2) the extent to which development according to government plans will have saturated the 0–2 m coastal zone by 2000–2010;
3) the magnitude of environmental problems (e.g. pollution);
4) external factors, such as world trade and political trends (Blue Plan scenarios; UNEP, 1988).

The future management of the coastal region will have to be based on 'cost effectiveness', that is, on an assessed value of the threatened land uses, not so much in terms of present functions but more in terms of those in future decades. It will require either extensive physical armouring of the shores, or choosing between: (1) irreplaceable coastal uses (e.g. national and military harbours, lagoonal fishing, etc.) to be defended, (2) uses that could be shifted elsewhere (e.g. industries, roads, airports) or (3) the adoption of different approaches to the exploitation of the shoreline and wetlands, that take environmental impacts into account.

8.2 Impacts on Weather and Marine Parameters

The four general circulation models examined by Wigley (Chapter 1) suggest that a doubling of CO_2 in the atmosphere could increase average temperatures in the Nile Delta region by 0.5–2.5°C, perhaps as early as 2030; the increase could be larger in winter than in summer. Summer temperatures would become more similar to those of the upper Nile valley south of Cairo, with a considerable increase in the rate of evaporation, but probably with a greater relative humidity than at present.

Wind directions and frequencies will depend on circulation over the Mediterranean and Near East and climatic changes over Saharan North Africa and East Africa. At present GCM models, however, are a rather crude, generalized form of prediction, being particularly unable to reproduce present weather conditions (especially rainfall) at a regional scale. Will north-west winds continue to be dominant? Will the centres of low pressure of the central Mediterranean continue to move towards the Middle East or will they be shifted more frequently northward? What alterations, if any, would affect onshore wind frequencies, especially the Khamessin and its dust storms? Changes of wave regimes and energy, and frequency of storms, might arise if there are significant modifications in the general air circulation.

According to present GCM scenarios, precipitation over coastal Egypt could remain the same, or maybe decrease slightly (Wigley, Chapter 1). Much will depend on the pattern of interaction between the central and eastern Mediterranean depressions and the polar continental air from the Iranian plateau in the winter-spring, which brings rain to Egypt and nearby countries. Cyclogenesis either could decrease (due to less contrast between lands and sea) or increase because of greater evaporation and atmospheric moisture. Rainfall probably will continue to be very irregular. Another possibility is that monsoonal rains could reach the south-east Mediterranean less occasionally than in the past (Wigley and Farmer, 1982). The intensity of monsoon rainfall is expected to increase, hence more precipitation over the Nile catchment areas; however, if the tropical easterly jet, which is an essential part of the monsoon system, moves northward, East Africa rainfall could be reduced instead.

Climatic changes could have a significant impact, in the long run, on water circulation in the Mediterranean (surface and deep layer, e.g. the Atlantic Water, and the Levantine Intermediate Water), because both evaporation and wind are responsible for surface circulation (see Gačič et al., Chapter 7). Water temperature and salinity should increase, with a more marked summer stratification, a decrease of dissolved oxygen (especially in summer) and a greater turnover rate of nutrients.

8.3 Impacts on Coastal Stability

Consideration of coastal evolution with the added factor of sea-level rise can be made in regard to either the present state of the shore and lowlands, or as they will be two to three decades from now when, presumably, more fixed structures will have been built by the shore (Coastal Protection Master Plan; Tetratech, 1986). Different levels of impact can be envisaged in relation to different sea-level rise scenarios (Delta Hydraulics, 1991). At the present rate of rise on the Egyptian coast (2.5 mm/yr, El Fishawi and Fanos, 1989), and assuming an average of 1.5 mm/yr rate of subsidence by the coastline (although in the lagoonal areas it could be higher), by 2030 sea-level could be on average 18 cm higher. Estimates of global rise due to greenhouse warming (Warrick and Oerlemans, 1990) are in the range of 8–29 cm (best guess average of 18 cm) by 2030, and of 21–71 cm (±44 cm average) by the late 2000s.

Thus, relative rise at the Nile Delta could be of 12–34 cm (average 24 cm) by 2030, and of 33–83 cm by 2070. Along with the elevation of water level there would certainly be an increase in the frequency of exceptional high water occurrence resulting from storms (Sharaf El Din et al., 1990).

Today, the degree of risk exposure of the different sections of the coast varies (Fig. 14.25). Artificial or natural protection already exists in critical areas, but many weak stretches are unprotected against an increased frequency of washovers. The threat to new urban and other developments will probably necessitate the installation of coastal defences in those areas.

Along the Alexandria to Abuqir coast increased wave overtopping of harbour structures and the "corniche" road, and the gradual shrinking of the city's beaches can be expected with even a mere 10–20 cm rise of sea-level. There would be increasing damage to tourist and other housing and a gradual deterioration of the more exposed, seaside parts of Alexandria.

In Abuqir Bay, erosion of the Rosetta promontory could maintain coastal shape in the western part, but the Mohamed Ali sea wall could become increasingly exposed to undermining, which also may pose a threat to the adjacent shores and harbour fixtures and therefore the low-lying lands behind them.

Retreat of the Rosetta promontory will undoubtedly be retarded by the sea wall, even as a detached breakwater. At the current rate of sea-level rise the shore is estimated to retreat close to Gezirah el Kladra by 2020 (UNDP/Unesco, 1978). Unless the wall is extended, erosion could proceed inland from its extremities (Figs 14.4, 14.25). A second sea wall may become necessary.

The western part of the Burullus barrier is not likely to be breached, but it will be flooded more easily and frequently. The high-risk stretch is between Hanafi and Burg el Burullus. With 24–34 cm relative rise of sea-level by 2030 the continuing erosion and steepening beyond the breaker zone will repeatedly cause the undermining of the sea wall and jetties. The village of El Burg eventually will have to be moved. If extensive precautions are not taken (beach sand nourishment, for instance) the sands of the barrier to the west will be lost, the sea could break through, and a new outlet could be formed. The proposed coastal road might act as a protective seawall (if adequately constructed), but eventually could also contribute to erosion.

Erosion also will increase along the coast between Burullus and a point 15 km east of Kitchener Drain, and the shore will eventually recede. Artificial beach renourishment may be the only sensible way to control the Baltim Sea Resort stretch, at least on a short-term basis.

The fate of the Damietta promontory and adjacent areas will depend upon the progress of offshore erosion, and possibly upon the thickness of the more easily erodible recent sediments, as stiffer lagoonal clays occur at shallow depth. The tip of Ras El Barr and the Nile estuary cannot possibly continue to be efficiently armoured against direct wave attack. Particularly vulnerable would be parts of the Ras El Barr peninsula. The coast west of the New Dumyat jetty may remain stable due to the impounding of sand drift, but as the shore is low, flooding could occur over sections of the New Dumyat town.

The coastal stretch north-west of Port Said is likely to continue to retreat gradually due to the active littoral sand drift, without breakthroughs except in the El Gameel stretch of the lagoon outlets. Particularly threatened by flooding would be the proposed tourist resort of Diba and the Port Said beach and westward town extension. The coastal road may need to be elevated independently of building a retaining dike for the proposed fresh-water reservoir.

In conclusion, the main problems to be envisaged as a result of sea-level rise are damages to the cities of Alexandria, Port Said and Dumyat New Town, especially to port structures, urban seafronts and beaches; the storm surges flooding of the presently, naturally or artificially, unprotected low-lying stretches, and the risk of breakthroughs in the Burullus and Bardawil barriers. The coastal barriers would probably survive, retreating and encroaching on the lagoons like they have done in the past, so long as sand supply is maintained by longshore drift, from the erosion of the Rosetta and Damietta promontories. A grave danger lies in a reduction of sand supply in consequence of the protective structures erected at the Rosetta and Damietta headlands.

The lagoon outlets, open as they are at present, however, would experience greater current scouring and greater tidal flow. After a relative rise of 25 cm of sea-level, the outlets would need to be regulated by sluices to prevent higher water levels in the lagoons; which would seriously change their ecologic parameters.

It is not inconceivable that the most vulnerable parts of the Nile Delta coast will have been protected, though at considerable cost, against the effects of a greater sea-level rise four to five decades ahead. The cost will become much larger later on and could impose agonizing choices, considering the priority that will have to be given to the vital economic structures of Alexandria, Damietta, Port Said and the Suez Canal. The major impact, therefore, will be financial. Technical problems will be no less relevant. In principle, all the Nile Delta coastal uses could be 'defended'; the question is how to do it without the negative side-effects that usually follow the erection of fixed coastal structures. Otherwise the effects of sea-level rise, instead of being checked, could get disastrously out of control.

8.4 Impacts on Surface and Ground Water

The future of the Nile Delta water resources is dominated by two variables: (1) increased surface evaporation, and (2) the behaviour and trends of Nile supply *vs* the need for increased water use. If the 7–9, 19–20 and 60–80 year cycles constitute any basis for prediction, the next century (and the decades 2020–2030 in particular) could be periods of higher floods. In this case, greater Nile water supply could offset evaporation losses.

Greater rates of evaporation would badly affect the lowlands, with regard to increased salt accumulation in soils and water salinities in the shallow and unflushed parts of the lagoons. At present semi-arid hypersaline areas occur (or did occur before reclamation) in several parts of the lower delta. There will be a greater need of fresh water to flush the salts; otherwise drains into the lagoons and/or coastal reservoirs will become more saline. Surface-water pollution, already a serious problem near Alexandria, could become acute because of urban and industrial expansion. Quality, and perhaps availability, of ground water could become a serious problem in the lower delta. All areas of the plain below 5 m elevation have saline ground water, making it poor for irrigation and unsuitable for drinking. Accelerated water extraction from the deep aquifer to answer population and economic growth demands, even if carried out with caution, would require treatment facilities and might pose the threat of land subsidence, especially in parts that already are sinking naturally.

In the coastal zone and wetlands, sea-level rise would have at least moderately serious consequences for the shallow aquifers. The coast-parallel sand ribbon is only a few to several meters thick (except near former Nile outlets). Seepage through this layer, after a rise of 1 m, could be up to 10 million m^3 of sea water a year into Lake Burullus, i.e. about 30% of the direct seepage loss of the entire aquifer to the sea (Kashef, 1983).

Not only must the entire water management scheme be improved, but a more powerful drainage system would be necessary to cope with regional infiltration of sea water towards the wetlands. In this context, the proposal to convert large parts of the Manzala and Burullus lagoons into fresh-water reservoirs at 0.5 to 1.5 m above sea-level would require a serious re-thinking in regard to saline penetration, as well as on other impacts on the hydrologic balance.

8.5 Impacts on Marine and Lagoonal Ecosystems

Marine ecosystems on the inner continental shelf would be influenced mainly by temperature changes, and by whatever changes that may take place in the water chemistry and nutrient abundance (natural or induced by pollution) that condition primary productivity.

The lake and lagoonal ecosystems (aquatic vegetation, euryhaline to fresh-water fish, migrating fish and birds) would be directly affected by increases in water salinity and temperature; the latter due to greater evaporation and more saline drain waters. Higher water temperature would favour primary (hence fish) productivity, but as life in the lagoons depends on complex interrelated environmental conditions, local excessive temperatures, high salinity and (maybe) less dissolved oxygen would be dangerous for many species. Temperature not only affects the survival and distribution of fish, but also their growth, rate of development, activation of reproductive processes and susceptibility to diseases. A salinity increase in the inner parts of the lagoons would have a negative impact on *Potamogeton pectinatus*.

More marine fish would migrate into the lagoons, and the more salinity-tolerant lagoonal species could benefit (e.g. *Tilapia zillii* and *T. aurea*, as opposed to *T. nilotica* and shrimps). In part, however, this depends on the fate of the plant areas, which are the main nursery and feeding grounds.

On the whole, the natural impacts on ecosystems (marine and lagoonal) could be minimal, and mostly gradual and self-adjusting. However, the ecosystems of 2025 may not be the same as today, because of the anthropic impacts of pollution and altered hydrological conditions due to lagoonal management schemes. Fauna and flora and the proportions of the different types of commercial fish would certainly change in the next three to four decades, especially if the lagoons are transformed into fresh-water lakes. The consequences for nesting and wintering birds would be drastic due to the loss of many environments that are vital for them.

Were it not for sea-level rise, it would be desirable that the lagoons are kept in communication with the sea, even by artificially opening new outlets, to facilitate fish migrations, the flushing of excessive plant cover and of agricultural pollutants, and for water renewal. In future, the quantities of drainage water are expected to decline, due to increasing re-use for irrigation, but the salinities of the remaining flows most likely would increase.

8.6 Impacts on the Economy of the Coastal Zone

Approximately 15% of Egypt's current GNP originates in the coastal area under 2 m elevation. Expected development in this part of the Nile Delta over the next two to three decades will include urban expansion, new industries, new roads (especially east–west), more reclaimed land and aquaculture farms, perhaps also enclosed fresh-water reservoirs in parts of the major lagoons. The GNP share will probably rise to 20%, owing to the increasing commercial and industrial role of Damietta and Port Said. Population will have increased by 30%, with local densities over 1000 inhabitants/km^2 in several parts.

With higher temperatures, the more numerous, larger and more crowded urban centers could experience serious difficulties with fresh-water supplies and waste disposal. If ground-water use increases, treatment plants may be required. More energy would be needed for air conditioning. Problems could arise also with public health. Malaria and schistomatosis occur in the regions south of the lagoons; climatic change might spread them further.

Maintaining the efficiency of port structures and of the inland drainage system in the face of a sea-level rise of even 20–30 cm will require extensive adjustments. To these costs will have to be added the expense of protecting lagoonal inlets, new settlements and infrastructures by the shore, if built in locations of high risk, as well as paying for the negative consequences of fixed structures erected to satisfy purely local needs, but not in accordance with an overall plan of coastal management.

The impact for the economy of Egypt of a sea-level rise over 50 cm would be even more considerable. By 2025 technical and financial assessments of these consequences must have been made. Failure to envisage appropriate counter-measures in regional planning will result in serious economic losses through direct destruction, flooding or gradual deterioration.

The consequences of sea-level rise on beach tourism undoubtedly would be negative for several Nile Delta summer resorts, like Alexandria, Ras El Barr and Port Said. However, those outside the delta, on the western and eastern Mediterranean coast, will not be affected, and other delta locations, with wider and more stable beaches, if properly developed, would undoubtedly take over as alternative resorts.

Agriculture, at least some crops (e.g. rice, cotton, corn, and fruit trees like bananas, mangoes and dates) may benefit from higher temperatures, but also would be affected by increased evaporation and soil and water salinities. Agricultural management could become more expensive, and a greater application of fertilizers and pesticides (if imposed by changed soil conditions and by a possible increase of weeds and pests) could increase the pollution of surface waters.

Fishing, if gradually adjusted to changes, should not be much affected, except qualitatively, in the types of species caught or reared in aquaculture basins. Impacts on fishing by anthropically changed hydrological conditions in the lagoons will be felt well before those of atmospheric warming or substantial sea-level rise.

9 CONCLUSIONS

In the context of the present climatic conditions of the Mediterranean coast of Egypt, a gradual increase of average temperature of 0.5–1°C by 2030 would probably be of little consequence. The future of water resources, however, is uncertain, depending as they are on Nile supply, and this in turn on climatic changes in East Africa. Greater rates of evaporation will certainly increase soil salinities, as well as the temperature and salinity of water in the lagoons (even if parts of them would be transformed into fresh-water reservoirs). Besides evaporation there would be salt-water intrusion from the sea and an influx of salts from (probably reduced) agricultural drains outputs.

The impact of a relative rise of sea-level, even 20–30 cm, will be primarily on the financial resources to adjust all fixed coastal structures, especially those of the present harbours, and those built to protect new coastal roads and urban areas. The lagoonal inlets may require additional regulation and probably, with higher sea-levels, also sluice systems; likewise, the internal lagoonal margin dikes and pumping stations would need adjusting.

Overall, however, the state of the environment in the coastal region of the Nile Delta in 20–30 years time will have been influenced more by direct anthropic intervention than by (as yet) small climatic changes. A land-use saturation of the coastal belt is to be expected, owing to population and economic growth. Considerable pressure will be borne on the environment, with the encroachment of urban areas on agricultural land, sedimentation of lagoon margins to the detriment of fish resources, new demands on water supplies, water pollution, and the disruption of the coast by even more fixed structures.

It should be reasonable therefore that the current territorial development planning for the northern delta region fully acknowledges the interdisciplinary nature of coastal management. The mistakes made elsewhere of coastal overdevelopment and concrete 'fixation' ought to be avoided, as the main consequence of climatic change due to global warming would be the greatly increased costs of protection and of adaptation, or else very substantial losses.

10 RECOMMENDATIONS

The most urgent need is to elaborate scenarios of the lower Nile Delta in 2020 that incorporate the interactions between social and economic development on the one hand and the environment on the other. The following future trends should be considered:
1) rates of coastal erosion and land subsidence;
2) actual and predicted changes of climate;
3) impacts of pollution and of water management schemes on ecosystems;
4) cyclic fluctuations of Nile floods;
5) population and industrial growth, with attendant food supply and commercial/financial implications.

These estimates would assist present territorial planning that involves population and natural resources management for the next 20–30 years.

A data-base should be set up to hold both existing and future information so that the physical and socio-economic models can be processed. The following investigations are recommended:
1) *Topography*. The assembly of data from the latest surveys; new repeated surveys, especially of the high-risk areas and of those of suspected subsidence.
2) *Subsidence*. Initiation of studies on subsidence from ground-water data (comparable to those done in Venice and north Adriatic coast of Italy); namely, south of Alexandria, near Rosetta, Damietta, around the Manzala lagoon to Port Said.
3) *Monitoring of coastal dynamics*. Periodical surveys with beach profiles, controlled aerial photography, monitoring the continued rate of headland retreat, modelling risk areas and the impacts of new fixed structures according to various increments of sea-level rise.
4) *Subsurface data*. Extend a programme of borings to the coastal area, especially the lakes and sand barriers, both to study ground water and salinization and to evaluate mechanisms of barrier retreat.
5) *Dune studies*. Evaluation of the effectiveness of dune protection, especially documenting the rates of dune growth.
6) *Detailed soil survey*. With experimentation on the reaction of clays and loams to salt increases.
7) *Waste disposal and pollution*. Studies on the relation between increased agricultural and industrial activities and the pollution of surface waters, especially with less water flushing and/or higher water salinities.
8) *Ecosystems*. Collate and quantify all existing data on lagoonal ecosystems (physical parameters, productivity, bottom and fish fauna, vegetation, bird fauna) that could be fed into alternative models for sea-level rise, increased temperatures and artificial alterations to the lagoonal hydrology.
9) *Climate*. Research into existing climatic data (temperature, evaporation, rainfall, wind directions) to evaluate cyclic trends, number, frequency and intensity of storms, wave approach and intensity.
10) *Water supplies*. Evaluation of published and unpublished data regarding climatic changes in the catchment area of the River Nile and their effects on water supplies in Egypt, with an estimation of their future trends. Evaluation of the state of ground-water information.
11) *Population*. Analysis of present and projected demographic distribution in the coastal areas and projected urbanization needs, especially water.
12) *Economic data*. Analysis of present economic parameters and trends to assess the economic performance of all the parts of the coastal lowlands during the next two to three decades (e.g. agricultural crop capacity for domestic/export use; fresh-water and marine fisheries; light/heavy industries; transportation needs; tourism potential).

11 REFERENCES

Aleem, A.A. and Samaan, A.A., 1969. Productivity of Lake Mariut, Egypt. Pt. I, Physical-chemical aspects. Pt. II, Primary production. *Int. Rev. Res. Hydrobiol.*, **54**, 315–355, 491–527.

Anonymous, 1989. *Quarterly Economic Reviews of Egypt*. Annual Supplement, The Economist Publications Ltd., London.

ARICON, 1989. *Structural Plan of the Comprehensive Development for North Delta Area and the Northern International Road*. Cairo Arab Intern. Consultants, Report for Ministry of Development, New Communities Housing and Utilities, 3 vols. (Arabic, Engl. Summ.).

Arbouille, D., Stanley, D.J., 1991. Late Quaternary evolution of the Burullus lagoon region north-central Nile Delta, Egypt. *Marine Geology*, **99**, 45–66,

ASRT/Unesco/UNDP, 1976. *Proceedings of a seminar on Nile Delta Shore Processes*. Alexandria, p. 630.

Assen, J.A., 1983. Land reclamation in Egypt. *Land & Water Intern.*, **58**, 3–10.

Balba, A.M., 1981. Sources and protection of soil and water of the Mediterranean coast of Egypt. *Advances in Soil and Water Res.*, Alexandria, **6**, p. 73.

Beaumont, P., Blake, G.H. and Wagstaff, J.M., 1988. *The Middle East, a geographical study*. Chapter 19, Egypt: Population growth and agricultural development. David Fulton, London, 511–530.

Beltagy, A.I., 1985. Sequences and consequences of pollution in northern Egyptian lakes. I. Lake Burullus. *Bull. Inst. Ocean. Fisheries ARE*, **11**, 73–97.

Ben Menachem, A., 1979. Earthquake catalogue for the Middle East (92 BC–1980 AD). *Boll. Geofisica Teor. Applicata*, **21**, 245–313.

Birot, P. and Dretsch, 1956. *La Mediterranèe et le Moyen Orient*. Presses Universitaires de France, Paris, pp. 257–291.

Bishara, N., 1983. The ecology of Abuqir Bay and its impact on the fisheries problem. *Bull. Higher Institute of Public Health*, **XIV**, 3.

Bishara, N.F., 1983. A review of aquaculture in Egypt. *Rapp. Comm. Int. Mer Med.*, **28**(6), 91–96.

Bishara, N.F., 1985. The problem of sardine fisheries in Egypt. *Arch. Hydrobiol.*, **103**(2), 257–265.

Boumans, J.H. and Mashali, A.M., 1983. Seepage from Lake Burullus into the reclaimed Mansour and Manisa polder area. *Agricultural Water Management*, **7**, 411–424.

Bucht, B. and El Badry, M., 1986. *Weather in the Mediterranean*. Vol. I. General Meteorology, 2nd edn. Her Majesty's Stationery Office, London, p. 362.

Butzer, K.W., 1976. *Early Hydraulic civilization in Egypt, a study in cultural ecology*. The University of Chicago Press, Chicago, p. 139.

Coleman, H.H., Murray S.P. and Salama, M., 1980. Morphology and dynamic sedimentology of the eastern Nile Delta shelf. *Marine Geology*, **41**, 325–339.

Coutellier, V. and Stanley, D.J., 1987. Late Quaternary stratigraphy and paleogeography of the eastern Nile Delta, Egypt. *Marine Geology*, **27**, 257–275.

Delft Hydraulics, 1991. *Implications of relative sea-level rise on the development of the lower Nile Delta, Egypt*. Delft Hydraulics, Report H 927.

Dowidar, N.M., Morcos, S.A., Saad, M.A. and El Samta, M.E., 1976. Hydrographic observations on pollution in Abuqir Bay, Alexandria, Egypt. *Acta Adriatica*, **18**, 381–396.

Dowidar, N.M., Abdel Moabi, A.R., 1983. Distribution of nutrient salts in Lake Manzala (Egypt). *Rapp. Comm. Int. Mer Medit.*, **28**(6), 185–188.

Dowidar, N.M. and Hamza, W.R., 1983. Primary productivity and biomass of Lake Manzala (Egypt). *Rapp. Comm. Int. Mer Medit.*, **28**(6), 189–192.

El Askary, M.A. and Frihy, O.E., 1986. Depositional phases of Rosetta and Damietta promontories on the Nile Delta coast. *J. Afric. Earth Sciences*, **5**, 627–633.

El Fishawi, N.M. and Fanos, A.M., 1989. Prediction of sea-level rise by 2100, Nile Delta coast. Subcommission Med. Black Sea Shorelines, *Newsletter*, **11**, 43–47.
El Ghandour, M.F.M., Khalil, J.B. and Atta, S.A., 1985. Distribution of carbonates, bicarbonates and pH values in ground water of the Nile Delta region, Egypt. *Ground Water*, **23**(1), 35–41.
El Rayis, O., El Din, S.H., Abu-El-Amayem, M., 1979. Hydrography and distribution of heavy metals and pesticides in Lake Manzala. IAPSO, General Assembly, Canberra, Australia.
El Sayed, M., 1991. Implications of relative sea-level rise on Alexandria. In: Frassetto, R., (ed.), *Impacts of Sea Level Rise on Cities and Regions*. Marsilio Editori, Venezia, 183–189.
Emara, H.I., Halim, Y. and Morcos, S.A., 1973. Oxygen, phosphate and oxidizable organic matter in the Mediterranean waters along the Egyptian coast. *Rapp. Comm. Int. Mer Medit.*, **21**, 345–347.
Emery, K.O., Aubrey, D.G., Goldsmith, V., 1988. Coastal neo-tectonics of the Mediterranean from tide-gauge levels. *Marine Geol.*, **81**, 41–52.
Ezzat, A. and Hosny, C.F.H., 1983. Etudes sur les pecheries du lac Manzala, Republique Arabe Unie. *Rapp. Comm. Int. Mer. Mediterr.*, **28** (6), 119–124.
Fairbridge, R.W., 1984. The Nile floods as a global climatic/soil proxy. In: Mörner, A. and Karfen, W. (eds), *Climatic Changes on a Yearly and Millennial Basis*, Reidel, 181–190.
Fisher, A.B., 1978. *The Middle East: A physical, social and regional geography*. 7th edition, Methuen, London.
Frihy, O.E., 1988. Nile Delta shoreline changes: aerial photographic study of a 28-year period. *J. Coastal Res.*, **4**, 597–606.
Frihy, O.E., El Fishawi, N.M. and El Askary, M.A., 1988. Geomorphological features of the Nile Delta coastal plain: a review. *Acta Adriatica*, **29**(1–2), 51–65.
Frihy, O., Khafagy, A., El Fishawi, N., Fanos, A.M., 1990. Nile Delta coast identification and evaluation of offshore sand sources for beach nourishment. In: Quelennec, R.E., Eroulani, E. and Michon, G. (eds), *Littoral 1990*, Eurocoast, Marseille, 724–728.
George, C.J., 1972. The role of the Aswan High Dam in changing the fisheries of the Southeastern Mediterranean. In: Tagi Farvaz, M. and Milton, P.J. (eds), *The Careless Technology*, The Natural History Press, New York, 159–178.
Gerges, M.A., 1981. Recent observations of currents from moorings in the Egyptian Mediterranean waters off the Sinai coast. *Ocean. Management*, **6**, 159–171.
Georgy, S., 1966. Les pecheries et le milieu dans le secteur de la Mediterranée de la RAU. *Rev. des Travaux*, Inst. des Pêches Maritimes, **30**, 23–92.
Goby, J.E., 1952. Histoire des nivellements de l'Isthme de Suez. *Bull. de la Societe d'Etudes de l'Isthme de Suez*, **5**, 23–43.
Halim, Y. and Morcos, S.A., 1966. Les rôles des particules en suspension dans les eaux du Nil en crue dans la repartition des sels nutritifs au large de ses embouchures. *Rapp. Comm. Int. Mer Medit.*, **28**, 733–736.
Halim, Y., Gerges, K. and Saleh, H.H., 1967. Hydrographic conditions and plankton in the SE Mediterranean during the last normal Nile flood. *Int. Revue Ges. Hydrobiol.*, **52**, 401–425.
Halim, Y., Gerges, S.K., 1981. Coastal Lakes of the Nile Delta. Lake Manzala. *Symp. on Coastal Lagoons*, Unesco, *Tech. Papers in Marine Science*, **33**, 135–172.
Halim, Y., Saleh, H.H. and Salim, A., 1985. Environmental conditions in Abuqir Bay, east of Alexandria, downstream from El Tabia effluent. Impacts on the fish associations in the coastal zone. In: FAO *The Effects of Pollution on Marine Ecosystems*, Fisheries Dept., No. 352, 105–111.
Halim, Y., 1991. The impact of Man's alterations of the hydrological cycle on the oceans margins. In: Mantoura, R.F.C., Martin, J.-M. and Wallast, R. (eds), *Ocean*

Margin Processes in Global Change, Dahlem Konferenzen, John Wiley and Sons, Chichester.
Hamid, S., 1984. Fourier analysis of Nile flood levels. *Geophysical Res. Lett.*, **11**, 843–858.
Harris, 1979. *Master Planning and Infrastructure Development for the Port of Damietta.* R. Harris, Inc., Consulting Engineers, Report to Ministry of Reconstruction and New Communities.
Hassan, F.A., 1981. Historical Nile floods and their implication for climatic change. *Science*, **212**, 1142–1145.
Haynes, K.E. and Whittington, D., 1981. International management of the Nile-Stage Three. *The Geographical Review*, **71**(1), 17–32.
IFAGRARIA, 1984. *Lake Burullus Area Development Project (First Stage).* Prepared for Ministry of Development, Governorate of Kafr-El-Sheikh.
Inman, D.L. and Jenkins, S.A., 1984. The Nile littoral cell and man's impact on the coastal littoral zone in the SE Mediterranean. *Proc. 17th Int. Coastal Eng. Conf.*, ASCE/Sydney, 1600–1617.
Kashef, A.I., 1981a. The Nile, one river and nine countries. *J. of Hydrology*, Elsevier, 53–71.
Kashef, A.I., 1981b. Technical and ecological impacts of the Aswan High Dam. *J. of Hydrology*, **53**(1–2), 72–84.
Kashef, A.I., 1983. Salt water instrusion in the Nile Delta. *Ground Water*, **21**(2), 160–167.
Klemas, V. and Abdel Kader, A.M., 1982. Remote sensing of coastal processes with emphasis on the Nile Delta. *Proc. Int. Symp. on Remote Sensing of Envir.*, Cairo, Egypt, p. 27.
Krumgalz, G.S., Hornung, H. and Oren, O.H., 1980. The study of a natural hyper-saline lagoon in the desert area (the Bardawil lagoon in northern Sinai). *Estuar. Coastal Mar Science*, **10**, 403–415.
Maiyza, I.M., 1992. Water budget of Lake Burullus, Egypt. *Estuarine Coastal and Shelf Science*.
Manohar, M., 1981. Coastal processes at the Nile Delta coast. *Shore and Beaches*, **49**, 8–15.
Meininger, P.L. and Mullis, W.C., 1981. The significance of Egyptian wetlands for wintering waterbirds. *The Holy Land Conservation Fund*, New York, p. 110.
Ministry of Irrigation, 1985. Conservation of Nile outflows to the Mediterranean. A preliminary report. ARE Ministry of Irrigation, Cairo, p. 35.
Morcos, S.A. and Hassan, H.M., 1976. Water masses and circulation in the SE Mediterranean. *Acta Adriatica*, **18**(13), 200–218.
Murray, S.P., Coleman, J.M., Roberts, H.H. and Salama, M., 1981. Accelerated currents and sediment transport off Damietta Nile Promontory. *Nature*, **293**, 51–54.
Nafas, M.G., Fanos, A.M. and Elganainy, M., 1991. Characteristics of Waves off the Mediterranean coast of Egypt. *J. Coastal Res.*, **7**, 665–676.
PADCO, 1981. *Egypt, Urban Growth and Urban Data.* Report of the National Policy Study, Advisory Committee for Reconstruction, Ministry for Reconstruction. A report prepared by PADCO, Inc.
Por, F.D. and Ben Tuvia, A., 1981. The Bardawil lagoon (Sirbonian lagoon) of North Sinai. A summing up. *Rapp. Comm. Int. Mer Medit.*, **27**(4), 101–107.
Rossignol-Strick, M., 1983. African monsoons, an immediate climate response to orbital insolation. *Nature*, **3030**, 46–49.
Said, R., 1981. *The River Nile.* Springer-Verlag, Berlin.
Saleh, H.H. and El Karashily, A.F., 1985. Effects of pollution on fish populations in Egyptian waters. In: FAO *The Effects of Pollution on Marine Ecosystems*, Fisheries Dept., No. 352, 216–228.

Samaan, A.A., 1977. Distribution of bottom fauna in Lake Idku. *Bull. Inst. Ocean. Fisheries*, Cairo, 7(1), 59–90.
Sestini, G., 1976. Geomorphology of the Nile Delta. In: *Proceedings of the Seminar on Nile Delta Sedimentology*, Unesco/ASRT/UNDP, Alexandria, 12–24.
Sestini, G., 1989. Nile Delta depositional environments and geological history. In: Whateley, K.G. and Pikering, K.T. (eds), *Deltas. Sites and Traps for Fossil Fuels*, Geol. Society Spec. Publs., 41, 99–127, Blackwell Scientific Publication, Oxford.
Shaheen, A.M. and Yousef, S.F., 1980. Physiochemical conditions, fauna and flora of Lake Manzala, Egypt. *Water Supply and Management*, 4, 103–113.
Sharaf El Din, S.H., 1973. Geostrophic currents in the south-eastern sector of the Mediterranean. *Symp. on Eastern Medit. Sea, IBP/PM–Unesco*, Malta, 1973.
Sharaf El Din, S.H., 1976. Effect of the Nile flood on the estuarine and coastal circulation pattern along the Mediterranean Egyptian coast. *Limnol. Oceanogr.*, 22, 194–207.
Sharaf El Din, S.H. and Moursy, Z.A., 1977. Tide and storm surges on the Egyptian Mediterranean coast. *Rapp. Comm. Int. Mer Medit.*, 24, 33–37.
Sharaf El Din, Ahmed, K.M., Fanos, A.M. and Ibrahim, A.M., 1990. Extreme sea-level values on the Egyptian Mediterranean coast for the next 50 years. In: *Proceedings Int. Seminar on Climatic Fluctuations and Water Management*, Cairo, Paper II–6, (in press).
Shata, A. and El Fayoumi, I., 1970. Remarks on the hydrogeology of the Nile Delta. In: *Hydrology of Deltas, IASH/Unesco, Bucharest Symp.*, I., Unesco, Paris.
Smith, S., 1986. Effect of Ethiopian drought on water resource management in Egypt. *J. Soil and Water Cons.*, 4(5), 297–300.
Smith, S.E. and Abdelkader, A., 1988. Coastal erosion along the Egyptian delta. *J. Coastal Res.*, 4, 245–255.
Sneh, A. et al., 1986. Holocene evolution of the north-eastern corner of the Nile Delta. *Quaternary Research*, 26, 194–206.
Stanley, D.J., 1988. Subsidence in the northeastern Nile Delta: rapid rates, possible causes and consequences. *Science*, 240, 497–500.
Stanley, D.J., 1990. Recent subsidence and northeast tilting of the Nile Delta, Egypt. *Mar. Geology*, 94, 147–154.
Summerhayes, C., Sestini, G., Misdorp, R. and Marks, N., 1978. Nile Delta: nature and evolution of continental shelf sediment system. *Marine Geol.*, 24, 37–47.
Tetratech, 1986. Shore Protection Masterplan for the Nile Delta. Report to Shore Protection Authority, Ministry of Irrigation, Cairo.
Tousson, O., 1934. Memoire sur les anciennes branches du Nil. *Mem de la Soc. Geogr. d'Egypte*, Cairo, 4, p. 144.
UNDP/Unesco, 1978. *Arab Republic of Egypt: coastal protection studies*. UNDP/EGY/73/063 Final Report, FNR/SC/OSP/ 78/230.
UNDP, 1988. *The Blue Plan, Futures of the Mediterranean Basin*. Executive summary and suggestions for action. Sophia Antipolis, France.
Unesco/ASRT/UNDP, 1976. Proceedings of the Seminar on Nile Delta Sedimentology. Alexandria, Oct., 1975. *Acad. Sci. Res. Technol.*, Cairo, p. 250.
Unesco/UNDP, 1987. Aquatic Environmental Pollution Project, University of Alexandria. UNDP/EGY/73/958, Final Report.
UNITED NATIONS, 1979. Population distribution and urbanization in selected countries in the Middle East. In: *Studies on selected development problems in various countries of the Middle East*, New York, 59–78.
Unluata, U., 1986. A review of the physical oceanography of the Levantine and Aegean basins of the Eastern Mediterranean, in relation to monitoring and control of pollution. Middle East Technical University. *Inst. Mar. Sciences*, Erdemli, p. 55.

Wahby, S.D. and Bishara, N.F., 1977. Physical and chemical factors affecting fish distribution in Lake Manzala, Egypt. *Acta Ichthyologica et Piscatoria*, **7**(1), 16–29.

Walker, T., 1988. Egypt. *Middle East Review*, 55–60.

Warrick, R.A. and Oerlemans, J. (eds), 1990. Sea level rise. In: Houghton, J.T., Jenkins, J.G. and Ephraums, J.J., (eds), *Climate Change, The IPCC Scientific Assessment*. Press Syndicate of the Univ. of Cambridge, 261–285.

Wigley, T.M.L. and Farmer, G., 1982. Climate of the Eastern Mediterranean and Near East. In: Bintliff, J.L. and Van Zeist, W. (eds), *Paleoclimates, Paleoenvironments and Human Communities in the Eastern Mediterranean Region in Later Prehistory*. BAR Intern. Series, **133**, 3–37.

15

Implications of Climatic Changes in the Mediterranean Basin

Garaet El Ichkeul and Lac de Bizerte, Tunisia

G. E. Hollis[1]
assisted by
C.T. Agnew[1], F. Ayache[2], M. Crundwell[1], R.C. Fisher[3],
A. Millington[4], K. Selmi[5], M. Smart[6], A.C. Stevenson[7]
and A. Warren[1]

([1]Department of Geography, University College London
[2]Direction de l'Environnement Agricole, Tunis
[3]Department of Biology, University College London
[4]Department of Geography, University of Reading
[5]Direction Général des Forêts, Tunis
[6]Ramsar Convention Bureau, Gland
[7]Department of Geography, University of Newcastle)

Abstract

This case study of the wetland National Park at Ichkeul, the Lake of de Bizerte and the associated coast of north Tunisia examines the impact of a temperature elevation of 1.5°C and a sea-level rise of 20 cm by the year 2025. The aim is to assess the likely changes to natural systems and the consequent socio-economic impacts. When appropriate regional climatic models have been refined, the forecasted changes in rainfall may prove to be more significant than the changes investigated here. For this study it was not possible to find a spatial analogy with the same rainfall as Bizerte and a mean temperature 1.5°C higher.

The study area is in the process of adjusting to a long period of apparently accelerating change. Post-Pleistocene swings, the entry of sea water through the cutting of a sea canal in 1895, deforestation, channelization and agricultural improvement have all had demonstrable effects on the Ichkeul–Bizerte lakes. A scheme to construct at least three dams on the rivers flowing into Ichkeul between 1983 and 2000 could dramatically change the hydrology and ecology of the National Park. A management scheme involving a sluice and reservoir releases may offset most of the deleterious changes to the National Park which is so important that it is protected under three international conventions.

There may be slightly more soil erosion in the headward valleys but the sediment will be trapped there with a modest reduction in the sediment yields in the larger rivers, reservoirs and the downstream lakes. The reduction of sediment flowing out to sea, because of the climatic changes and the trap effect of reservoirs, may accelerate coastal erosion.

The actual evapotranspiration in the region will increase by around 10% when mean air temperature rises by 1.5°C. This will result in, at least, a 10% decline in riverflow. The salinity of the latter will also increase. Potential evapotranspiration and open water evaporation will rise by a minimum of 12%. The demand for irrigation water will rise by at least 12% when reservoirs are depleted by reduced river flow and increased evaporation. The average storage in the reservoirs will fall by up to 26% and they will be nearly empty for up to 19% of the time. An expected 25% filling of the reservoirs with sediment will seriously increase the water-supply problems, with mean storage falling to around 60% of the levels under present conditions. The sea-level rise, acting on its own, is not likely to have a significant impact on either the Lac de Bizerte or Garaet El Ichkeul.

The combined effects of the dam scheme and the rise in temperature will effectively turn the Ichkeul National Park into a saline sebkha. It is likely to lose all of its food plants for its wintering and breeding waterfowl. Some groups of birds, notably flamingos and waders, may benefit marginally. The nationally important fishery will disappear from Ichkeul. Some fresh-water species and some marsh plants may survive in isolated remnants in the rivers and river mouths. The impact of global warming on bird migration and wintering areas may offset some of these deleterious trends.

Forests are likely to suffer from the increased temperature and aridity. The coastal plantations will suffer from salinization and, probably, remobilized sand dunes. There may be some shifts in the limits of present natural vegetation types but winter rainfall is the main determinant here. Agriculture is likely to change towards even more intensive irrigation in some areas and a greater reliance on grazing generally. The largest impact on infrastructure will be foundation problems with roads and buildings arising from saline ground water and swelling clays. Sewage flooding may follow the sea-level rise because of inadequate slopes of sewers. Sea-flooding itself is not likely to be a major problem.

Overall, existing environmental problems are likely to be exacerbated; agriculture will suffer; inland and lagoon fisheries may have already disappeared through the impact of the dam scheme; sea fisheries may benefit a little; industry will be largely unaffected; water resources will decline in both quantity and quality; transport will be unaffected save for shipping which may benefit slightly; settlements will suffer through their foundations and sewer systems; the quality of urban life may decline through an accelerated influx of farmers leaving the countryside; geomorphological systems will change only marginally; and the hydrology and ecology of the Ichkeul wetland will change dramatically through both the dam scheme and the temperature change because the conservation management measures are not likely to be equal to the developing problems.

1 INTRODUCTION

1.1 Terms of Reference

The UNEP sponsored International Conference on Changes in Sea-Level in Villach (9–15 October 1985) adopted the conclusions of a global rise of mean temperature of 1.5 to 4.5°C and a sea-level rise of 20 to 140 cm by the end of the 21st century. This case study examines the likely impact of a temperature elevation of 1.5°C and a sea-level rise of 20 cm by the year 2025.

Many experts have suggested larger changes by 2025, and some studies have examined rises in sea-level of between 1 and 5 m by 2100. A speculative study of these large sea-level changes would have had to be hedged with uncertainty. This would not have made it as effective as an instrument for policy formulation.

This study's modest assumptions and the derived conclusions are very firmly based and can be a springboard for policy. The case study involves the National Park at Garaet El Ichkeul and the Lac de Bizerte in northern. Tunisia and necessarily considers changes to the catchments of the influent rivers and to the coastal area around Bizerte.

1.2 Objectives
The objectives of the case study are:
1) to examine the effects of the sea-level change on coastal ecosystems;
2) to examine the possible effects of temperature elevations on the terrestrial and aquatic ecosystems, especially those of economic importance;
3) to examine the possible effects of climatic, physiographic and ecological changes on the socio-economic structures and activities; and
4) to determine areas or systems which appear most vulnerable to the likely changes.

1.3 Assumptions
The adoption of the assumption of a 20 cm sea-level rise presents few problems, and it is possible to speculate on its impacts with some freedom. However, the assumed 1.5°C temperature rise presents serious problems because such a change could not occur without possibly highly significant changes in rainfall amount and timing, wind direction, humidity, etc. However, the various global climatic models have not reached a stage of refinement where they agree in the likely direction and extent of change in regional climates (Wigley, 1988). Therefore this study has had to proceed simply on the basis of +20 cm and +1.5°C.

1.4 Methodology
Four methodologies have been used:
1) A group of eight researchers, with years of experience at Ichkeul and elsewhere in Tunisia, formed a think-tank. Each prepared a written estimate of the likely impact of the assumed changes at Ichkeul on the systems that were familiar to them. The results were debated by the whole group. After each of the papers had been considered and the whole situation debated; the workers redrafted their papers in the light of the discussions. In summary, scientific environmental principles were used for informed speculation.
2) Past work at Ichkeul has yielded a set of computer routines and data to calculate potential and actual evapotranspiration, and a large computer model of the hydrology of Ichkeul and the rivers draining into the wetland. The latter simulates the lake level and lake salinity on a daily basis and has daily weather data, daily hydrological data and some lake data for the period 1952–1982. The simulation model allows for the possibility of constructing dams in the river basins and for various types of sluice to be operated in the channel that links Lake Ichkeul to the sea water of the Lac de Bizerte. These computer procedures have been heavily used to make quantitative estimates of the impacts of the changes on the Ichkeul wetland and the existing and proposed water-resource reservoirs.

3) Maps of the whole area with detailed topographic information below 5 m NGT were sought so that formal hypsometric curves, analyses of land use in the 0–20 cm, 20–50 cm and 50 cm–1 m zones, and estimates of the areas inundated by the sea and affected by the sea-level change could be made. Several series of maps with the requisite information exist but they are classified. Within the Ichkeul National Park, some preliminary maps of topography with spot heights at a 10 cm interval were available for the planning of the agricultural improvement of the plain to the south of Ichkeul.

4) A final methodology was to make a spatial analogy with another site similar in character to Ichkeul–Bizerte that already enjoys the elevated mean annual temperature projected for Bizerte. In seeking such a site it was essential that it should have a coastal location and that it should have around 600 mm of rainfall. It is clear from Table 15.1 that once Bizerte has a mean annual temperature of 19.6°C and an annual rainfall of 650 mm, it will represent a new climatic type which does not presently exist in Tunisia. An examination of the volume on *Climates of the World* (Griffiths, 1980) reveals that suitable analogues are not available elsewhere in the Maghreb. For instance, Agadir (18.5°C/224 mm), Sirte (20.0°C/187 mm), Benina (19.5°C/258 mm) and Tripoli (19.5°C/286 mm) all show that, at present, a rise in mean annual temperature is associated in these latitudes with a greatly reduced rainfall. This methodology was not pursued further.

Table 15.1 Mean annual temperature and rainfall for coastal towns in Tunisia (after Service de la Météorologie Nationale, 1967; and Kassab and Sethom, 1980)

Town	Mean temperature °C	Annual rainfall mm
Bizerte	18.1	650
Tabarka	17.9	900
Tunis – Manoubia	18.0	450
Sousse	18.6	325
Sfax	18.8	210
Gabès	19.2	180
Djerba	20.1	200

2 COASTAL TUNISIA

The Gouvernorat of Bizerte in northern Tunisia contains "the industrial north" of the country with its oil refinery and steelworks, a major agricultural zone in the Plain of Mateur, a series of existing and projected dams, major military installations, and one of the Mediterranean's most important wetlands within the Ichkeul National Park (Fig. 15.1). Bizerte, a town of over 100,000 people, has an important tourist industry directly adjacent to the fine sandy beach and a major commercial port with a cement works. Menzel Bourguiba, a town with over 50,000 inhabitants and located on the

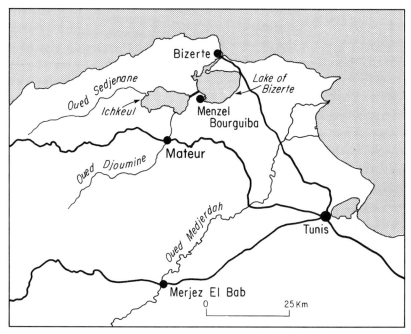

Fig. 15.1 Location map of coastal Tunisia.

inland side of the marine waters of the Lac de Bizerte, has the steelworks, naval dockyard and a host of small factories specializing mainly in textiles. Mateur, with around 40,000 people, serves its agricultural hinterland with engineering, agricultural supplies, marketing facilities and food processing plants.

The region has one of the highest rainfalls in Tunisia thanks to the mountains that form the headwaters of the Oueds Sedjenane and Djoumine (Table 15.2). Under natural conditions around 350×10^6 m^3 of fresh-water runs off into the sea from the area. All of the water resources of Tunisia, save those of the Bizerte region and a more remote area around Tabarka in the extreme north-west, are now fully exploited. Consequently, it has been decided to implement a water-resources scheme which will take around 65% of the runoff eastwards to Tunis and beyond and to use some of the reservoir water for local irrigation. This scheme will have major implications for the wetland National Park at Ichkeul which is of importance for its wintering and breeding grounds of many bird species, its commercial fishery, and its grazing marshes used by thousands of cattle, sheep and goats.

2.1 The Blue Plan Scenarios and Tunisia

Barić and Gašparović (1988) cite data for Tunisia showing that agriculture contributed 15% to Gross Domestic Product (GDP) in 1984 which was a reduction from 21% in 1976. Industry grew in the same period from 30 to 35% of GDP, whilst services remained at 49 to 50%. The area of irrigated agriculture in Tunisia grew by 71% between 1975 and 1984 but it still represented only 4% of the total arable area in the country.

Table 15.2 Climate statistics for Bizerte (after Service de la Météorologie Nationale, 1967)

	J	F	M	A	M	J	J	A	S	O	N	D	Year
Rain days	16	13	12	9	6	4	2	3	8	11	13	16	113
Rainfall (mm)	112	79	56	44	24	12	4	7	34	70	92	119	653
Mean temp. °C	11.3	11.6	13.4	15.4	18.4	22.5	25.2	25.9	24.4	20.4	16.4	12.6	18.1
Mean min. temp. °C	7.7	7.8	9.4	10.9	13.7	17.8	20.2	21.0	19.6	16.0	12.0	9.2	
Mean max. temp. °C	15.5	15.4	17.5	20.0	23.1	27.8	30.2	30.9	29.2	24.9	20.6	16.0	

The Blue Plan for the Mediterranean has envisioned five planning scenarios (Barić and Gašparović, 1988):

Reference-trend scenario	T_1
Worse-trend scenario	T_2
Moderate-trend scenario	T_3
Alternative reference scenario	A_1
Alternative integration scenario	A_2

The projections for the various Blue Plan scenarios for Tunisia are summarized in Table 15.3.

This outline suggests that in 2025 Tunisia will have twice as many people as it has today; a growth rate lower than and maybe only half that of the recent past; a massive increase in the demand for water by a burgeoning tourist trade and industry; and a modest amount of its GDP coming from agriculture which has a significant element under irrigation.

2.2 The Environment in Tunisia

The environmental changes which will be wrought by the rise in mean global temperature and the rise in sea-level will therefore affect an environment already under great pressure from the projected doubling of population. However, as the US National Parks Service and MAB Secretariat "Environmental Profile of Tunisia" shows, the environment in Tunisia is already stressed (National Parks Service, 1980). The report lists the major environmental problems of Tunisia as follows:

1) *Loss of agricultural land*, existing or potential, through erosion caused by damaging cultivation practices and inadequate management.

2) *Degradation of range and forest land* through uncontrolled grazing, extraction of fuelwood, and erosion that follows the removal of the vegetative cover.

3) *Overcrowding of the urban centres*, caused by migration of people fleeing the deteriorating situation in the countryside, and producing problems of pollution of the public water supply and spread of diseases related to poor sanitation.

4) *Industrial pollution*, apparently so far limited in extent, but noted as factory effluents are discharged into the Lake of Tunis, and the refinery at Bizerte and the numerous factories in Tunis lower air quality in the region.

Table 15.3 Blue Plan scenarios for Tunisia (after Barić and Gašparović, 1988)

	T_1	T_2	T_3	A_1	A_2
GDP growth rate (2000/1985) 1985/1960 = 5.2%	2.5	2.2	3.2	4.2	4.7
Population growth (2025/1980)	201%	219%	201%	189%	189%
Water use by tourists (2025/1984) for Mediterranean region	270%	200%	330%	390%	450%

5) *The potential of increasing ground-water salinity* by saline irrigation waters percolating down to the water table.

6) Finally, *lack of adequate protection for native flora and fauna*, thus impoverishing the national and global pool of genetic resources.

2.2.1 Garaet El Ichkeul

The Tunisian National Park at Ichkeul is probably the most important site in North Africa for wintering waterfowl and is comparable to the better known wetlands in the Camargue, France, and Coto Doñana in Spain. It is one of a handful of sites in the world whose international importance is recognized under the Unesco World Heritage, Ramsar, and Biosphere Reserve Conventions. The 90 km^2 lake has a minimum salinity between 3 and 14 g/l in winter when the rivers are in flood. In summer, because of evaporation and the inflow of sea water, the salinity normally rises to over 40 g/l. The 30 km^2 of fresh-water marshes around the lake are maintained by winter inundation from the lake and the continual input of river water (Fig. 15.2). Up to one-third of the lake's area is taken up with massive beds of *Potamogeton pectinatus* which is the major food for the overwintering pochard (up to 120,000 birds), wigeon (112,000 maximum) and coot (188,000 maximum). The marshes are dominated by a bulrush, *Scirpus maritimus*, whose new shoots and bulbous root parts provide the staple diet of the greylag geese. They have numbered up to 20,000 and it seems likely that Ichkeul holds the whole of the north-east and central European populations in some years. The recruitment of young fish to Ichkeul is greatest during March when there are normally high-water levels, low salinity and good flows. The out-migration of fish normally occurs against the inflow of sea water in August and September. The invertebrate fauna of Ichkeul is constrained by the annually alternating regimes of water level and salinity and is impoverished by comparison with true fresh-water and marine habitats.

An interpretation centre for this watery paradise is complete and it is envisaged that it will develop into both an environmental education centre for Tunisians and an excursion halt for the thousands of foreign tourists who seek more than Tunisia's sun, sea and sand.

However Ichkeul is more than a wildlife reserve. It has an important fishery based on eels and mullet. Around 200 tons of fish are taken each year with a value of more than $800,000. Local people and farmers from the surrounding agricultural land graze animals on the marshes which, because of their wetland character, stay green much longer than the surrounding fields. There are often 1500 cattle and over 1000 sheep and goats to be seen on the marshes. The local people cut the reeds and rushes for thatch and take the woody stems of *Tamarix* for fuelwood. Ichkeul was an important hunting area before the declaration of the National Park. Finally, thousands of Tunisians visit the hot sulphurous springs that emerge from the base of the faulted limestone massif of Djebel Ichkeul which rises some 500 m directly out of the lake and marshes. These early summer human migrants to Ichkeul believe that the "Hammams" have both medicinal and sacred values.

The wetland ecosystem is, however, threatened by a scheme to construct dams on each of the major rivers and to divert about 65% of their

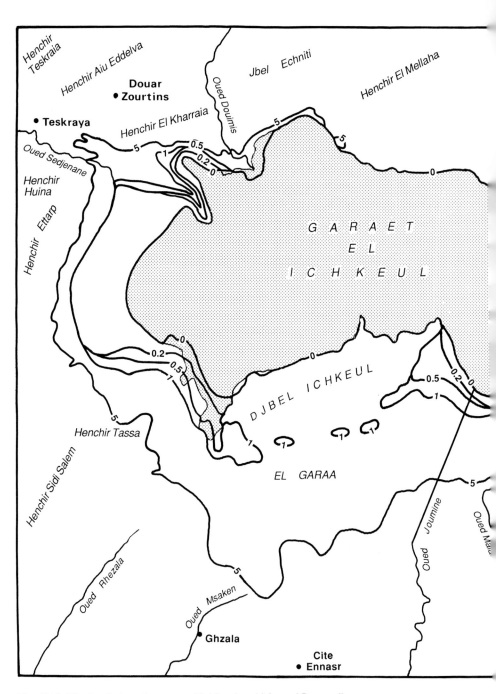

Fig. 15.2 The land elevation around Ichkeul and Menzel Bourguiba.

Garaet el Ichkeul/Lac de Bizerte and climatic change 611

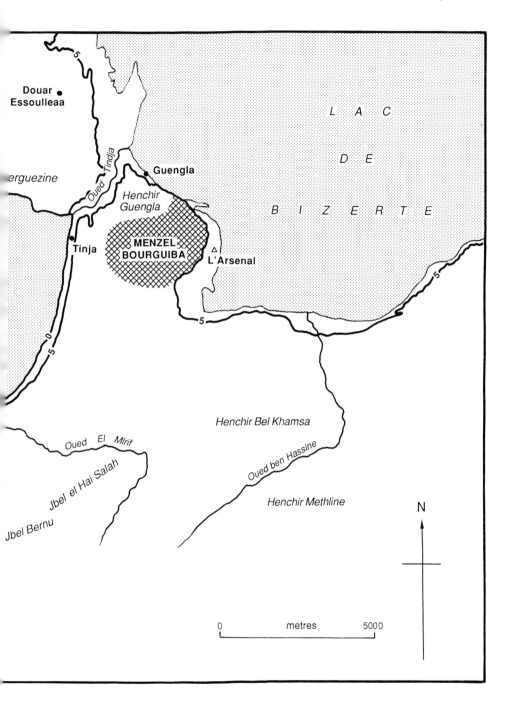

waters for urban, industrial and irrigation requirements. The transfer may eventually extend as far south as Sousse and Sfax. The Ichkeul rivers are being tapped because the water resources of the whole of the country, save for these relatively moist north-western mountains, are completely developed.

A computer model of the hydrology of the catchment, vegetation mapping from LANDSAT satellite imagery and fieldwork, and contemporary and historical waterfowl counts have been linked to provide a quantitative ecological model. This has translated the computer's forecasts of hydrological conditions after the completion of the dam scheme into ecological forecasts. In addition it has been possible to experiment with alternative conservation measures to examine the options for the conservation of Ichkeul. The management strategies that have been evaluated include a sluice on the channel linking the lake and the sea operated to suit either the waterfowl or the fishery; a simple weir to exclude sea water; the pumping of saline lake water to permit the inflow of less salty sea water; releases of water from the reservoirs; and the reduction of the lake area in order to reduce evaporative losses.

The model has shown that a staged response is needed to offset the anthropogenic impacts on the Park. If a full scheme is implemented, the hydrology of the lake will change little, the *Potamogeton* and the *Scirpus* will be conserved and the resultant carrying capacity for wintering waterfowl is not likely to diminish significantly. The first two dams are already complete and a sluice is being constructed to retain water in winter, to flood the marshes and to exclude sea-water inflows in the summer except when the fish are migrating out to sea in August and September. When the current construction of the third dam has been finished, the operating rules of the sluice will have to be changed to exclude all sea-water inflow and the fish will have to use the projected fish pass. This, alone, however, will not entirely safeguard Ichkeul from ecological changes after the third dam.

Although protected by three prestigious international conventions and Tunisian National Park legislation, the area already suffers from some of the environmental problems enumerated by the United States National Parks Service (1980). The vegetation of the Djebel and of the marshes is heavily grazed to the point of serious degradation in many places. The salinity of the ground water will rise significantly as a direct result of the dams and the irrigation scheme but also as a result of the increase in salinity in the lake that can be expected after the water-resource scheme has been completed. Similarly, whilst protection for the flora and fauna exists on paper, Ichkeul itself fully merits its listing as one of the most threatened Ramsar sites in the world (Ramsar Bureau, 1987) and as one of Africa's 24 most threatened protected areas (Thorsell, 1987).

2.2.2 Lac de Bizerte

The Lac de Bizerte has a surface area of 150 km^2 with an average depth of 8m (Unesco, 1986). It is fed with an average of 370×10^6 m^3 of fresh to brackish water each year from Lake Ichkeul via the Oued Tindja, and eight small streams bring approximately 25×10^6 m^3 of fresh water directly into the lake (Fig. 15.2). The lake is connected to the sea by a canal 1500 m

Table 15.4 Temperature and salinity of Lac de Bizerte (after Gimazane, 1981)

	Jan	Feb	Mar	Apr	May	Jun	Jul	Aug	Sep	Oct	Nov	Dec
Mean temperature °C 1977/81	9.1	11.1	12.8	13.8	17.1	20.8	23.6	24.1	22.3	18.8	14.7	11.2
Mean salinity g/l 1979	38.0	36.3	34.7	34.5	35.0	35.8	37.2	38.3	38.6	38.5	37.6	36.4

Table 15.5 Ichkeul before the sea canal was cut in 1895 (the salinity of Lac de Bizerte was assumed to be 10 g/l)

Simulation	Lake level – Ichkeul (cm NGT)					Lake salinity – Ichkeul (g/l^{-1})				
	Mean	Max.	Min.	Med.	UQ	LQ	Mean	Max.	Min.	Med.	UQ	LQ
Present conditions 1952–1982	32.5	330	−6.1	19.1	52.9	2.5	17.2	49.8	2.6	15.0	24.5	8.2
pre-1895 conditions (using 1952–1982 flow, rainfall etc.)	32.5	330	−6.1	19.1	52.9	2.5	6.5	30.4	1.4	5.5	8.6	3.4

Notes: Med. = Median. UQ = Upper quartile. LQ = Lower quartile. NGT = Tunisian datum.

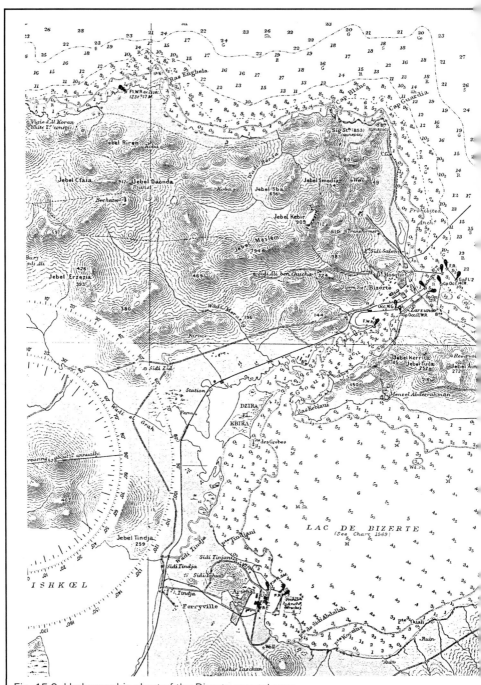

Fig. 15.3 Hydrographic chart of the Bizerte coastal zone.

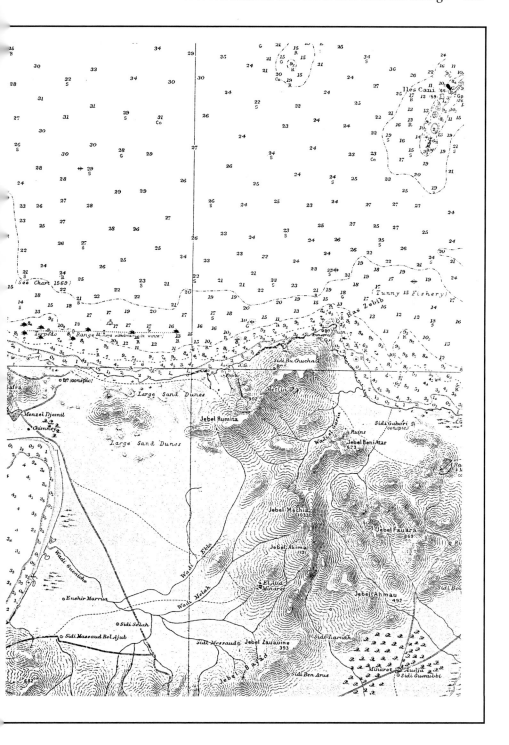

long, 300 m wide and 12 m deep that was cut in 1895 (Fig. 15.3). At present the lake has a small tide of around 30 cm and a salinity in the range 29 to 38 g/l^{-1} (Table 15.4). Before the cutting of the canal it is believed that the lake was brackish in character with a seasonally reversing flow via a narrow channel through what is presently the old port at Bizerte. The lake used to rise in level during large inflows of runoff.

Caulerpa prolifera is the most abundant species among the benthic communities. The lake has a fish production of about 4 kg/ha/yr. 30 species, mainly *Sepia ssp.* and *Pagellus spp.*, are caught using trammel nets (Unesco, 1986). There is also an important shell-fish production near to Menzel Jemil which yields 25 tons/year of *Crassostrea gigas* and 175 tons/year of *Mytilus galloprovincialis*.

A shipping channel 12 m deep was dredged across the lake floor in 1950 to allow warships to use the Naval Arsenal at Menzel Bourguiba. The naval installations and the port remain of considerable significance at Menzel Bourguiba and much of the iron ore for the nearby steelworks arrives by ship. The port at Bizerte which lines part of the canal at its seaward end is of great national significance. Other important installations around the lake include several military establishments, a major cement works and an electricity generating station (Fig. 15.4). The rapidly growing population of Bizerte and the industrial pollution of air, water and land are further local examples of environmental problems listed by the United States National Parks Service (1980). Lake de Bizerte, being deep does not support important bird populations, and agricultural land comes right down to the edges in most areas.

2.2.3 The coastal zone

To the north of Bizerte as far as Cap Blanc the coast consists largely of hard rocks, sandstones and limestones, with cliffs and wave-cut platforms. There is an extensive sand beach just to the north of Bizerte and this is backed by low dunes that have been heavily developed with a complex of tourist hotels. Around the mouth of the Bizerte canal and near to the breakwaters there are a number of informal land-fill sites using rubbish. To the south of the ship canal there is long sweep of sandy beach backed with dunes up to 20m high. The Bizerte oil refinery has been built on a levelled site on the landward side of these dunes. To the south of the refinery around Rmel there is a very large plantation of pines covering the whole dune system. A forestry school and a military establishment have been built within the forest plantation. The dunes are presently stabilized except for the seaward edge. These dunes and the beach at Rmel are extensively used by local people for recreation. The old port of Bizerte is crowded with small fishing boats that derive their catch from the coastal waters.

The coastal area round Bizerte is of minimal importance to birds. The main interest in the coastal area centres on storm-driven birds along the beach, or on trans-Mediterranean migrants arriving in dune vegetation or pinewoods.

2.3 Contexts of Change: Historical, Current, Future

The region has undergone enormous changes in the post-Pleistocene era; all human induced change is superimposed upon these longer-term

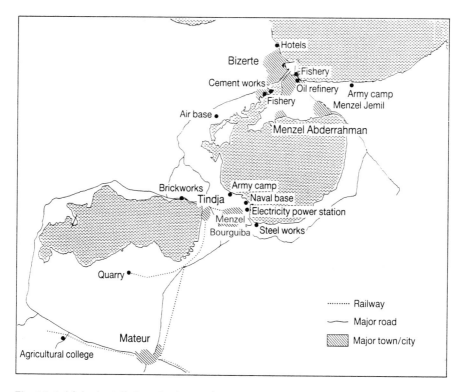

Fig. 15.4 Major installations in the study area.

changes. Massive forest clearance and other vegetation changes in the last 2000 years have had a significant effect on the wetland and coastal systems.

The present floral component of Lake Ichkeul, *Potamogeton pectinatus* and *Ruppia cirrhosa* with a limited amount of the complex alga *Lamprothamnium papulosum* and the marine *Zostera noltii*, is very recent. Pollen analysis of a ^{210}Pb-dated sediment core from the lake demonstrates that *P. pectinatus* and *Ruppia* only assumed their present role about 1900 (Fig. 15.5). Before this date, the lake appears to have been fresher with no evidence of a marine influence in the form of marine ostracods or foraminifera for the previous 500 years (Stevenson *et al.*, 1986). The more recent sediments suggest that the present lake ecology is characterized by increasing marine influences since bands of marine foraminifera, marine ostracods and occasional bands of *Cerastoderma edule* occur. These major changes in the ecology of the lake are probably the result of the construction of the Bizerte canal in 1895 (Unesco, 1986) which altered hydrological conditions within Lac de Bizerte from a brackish/fresh system to a totally sea-water dominated system. This inevitably had an a knock-on effect on salinity conditions within Ichkeul (Bonniard, 1934) involving an increasing marine influence. The extent of these changes was simulated by the hydrological computer model and Table 15.5 suggests that before the ship canal was

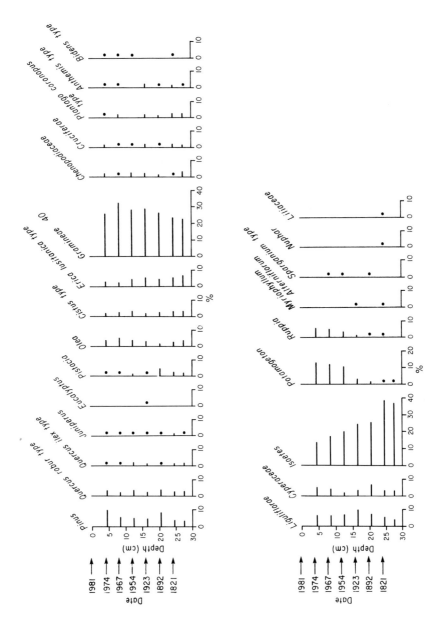

Fig. 15.5 Pollen analysis of a core from Lake Ichkeul with ^{210}Pb dating.

cut, raising the salinity of Lake Bizerte from perhaps 10 g/l to the present 36 g/l, the lower quartile salinity of Ichkeul was only 3.4 g/l and the mean was 6.5 g/l compared to the equivalent values of 8.2 and 17.2 g/l for the period 1952–1982. It is possible that the system is still adjusting to this massive change.

Deforestation, agricultural investment in drainage and reclamation, urbanization and industrialization have brought a huge change to the region during this century. Table 15.6 shows that the rate of sedimentation in Ichkeul has accelerated from 0.28 mm/yr early in the 19th century to around 1.5 mm/yr in the middle of the 20th century and it is now around 6 mm/yr. The explanation for the accelerating rate of sedimentation is not certain but it almost certainly lies in human activity in the catchment areas around Ichkeul. Land-use changes, channel straightening and "improvement", and changes in husbandry practice are all likely to have played a role. For instance, until the 1950s the Oued Djoumine discharged into an inland delta to the south and east of Djebel Ichkeul. Since then it has been increasingly channelized and by 1981 its water and sediments passed directly into Lake Ichkeul.

Table 15.6 Sedimentation rates in the centre of Lake Ichkeul from ^{210}Pb dating of a sediment core

Depth (cm)	Date	Age (years)	Sedimentation rate (mm/yr)
0.0	1980	0.0	6.0
0.5	1979	0.6	5.1
4.5	1972	7.6	7.0
8.5	1966	14.7	1.5
12.5	1952	28.5	1.5
16.5	1923	56.9	1.0
20.5	1890	90.2	0.3
24.5	1819	161.2	

The construction of dams for the diversion of 65% of the fresh-water inflow to the wetland National Park at Ichkeul during the period 1983–2000 has and will have a significant effect upon the hydrology and ecology of Ichkeul, Lac de Bizerte and probably on the coastal zone too (Table 15.7 and Fig. 15.6).

Projected conservation measures for the National Park, such as a sluice in the outflow/inflow channel will permit the manipulation of the hydrology and ecology of Ichkeul (Table 15.7 and Fig. 15.6).

The hypothesized rises in temperature and sea-level are simply additional perturbations to the system. The system also depends to a very large extent upon the amount and seasonality of rainfall, the rate of evaporation (determined as much by the direction of the prevailing wind as ambient air temperature), the economic and political relations between Tunisia, the Arab World and the EEC, the world price of energy (particularly oil), the rate of national population growth, etc. It is likely that major changes, both those forecast and those unpredicted, in the social, economic and

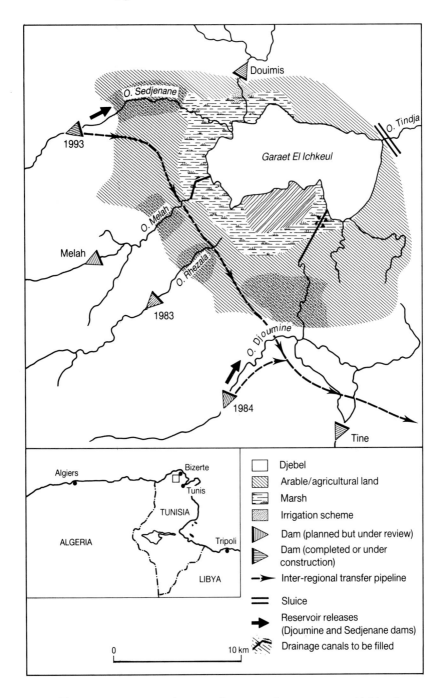

Fig. 15.6 The water resource scheme and conservation measures at Ichkeul.

Table 15.7 The effects of the dam scheme and sluice-operating regimes on the forecasts of parameters of ecological significance at Ichkeul (data for 1952–1982)

Condition simulated	Lake level (cm NGT)						Salinity (g/l)					
	Mean	Max.	Min.	Med.	UQ	LQ	Mean	Max.	Min.	Med.	UQ	LQ
Natural	32.5	330	−6.1	19.1	52.9	2.5	17.2	49.8	2.6	15.0	24.5	8.2
DD, RD, AS and SD	18.5	282	−4.7	12.7	25.1	2.4	38.1	86.9	5.3	36.2	48.9	25.0
DD, RD, AS, SD and Fish Sluice Regime	43.1	333	−27	29.8	68.0	7.4	31.5	86.8	5.6	28.4	41.1	19.5
DD, RD, AS, SD, 1988 sluice and reservoir releases	52.6	337	−11	46.6	82.2	17.6	23.3	65.0	5.4	21.6	24.5	15.3
	(Reservoir releases average 4.4 × 10⁶ m³/yr, max. 23.2, upper quarter 6.5)											

Notes: Abbreviations
DD Djoumine Dam
RD Rhezala Dam
SD Sedjenane Dam
AS Agricultural improvement scheme in the Plain of Mateur
Fish sluice Regime of levels (cm NGT) Sept.–Aug of 0, 20, 35, 60, 100, 80, 20, 20, 20, 20, 0
The 1988 Sluice follows a modified regime, with the controlled top water level (cm NGT) at −Sep. 50, Oct. 50, Nov. 60, Dec. 80, Jan. 100, Feb. 80, Mar. 70, Apr. 65, May 60, Jun. 55, Jul. 50. During the period from May onwards the actual water level in the lake is often lower than this controlled level because of evaporation. Indeed, when the lake falls below −10 cm, the sluice gates are opened to allow sea water to flow into Ichkeul. In addition there are reservoir releases of 0.4 × 10⁶ m³ per day when the lake is below −7.5 cm, but these releases cease during drought years with a frequency of at least one year in five.
Med. = Median. UQ = Upper quartile. LQ = Lower quartile. NGT = Tunisian datum.

political arena will have a far greater impact than the predicted changes in temperature and sea-level.

3 Manifestations of Rises in Temperature and Sea-Level by 2025 and Beyond

It is important to consider the manner in which the expected changes in temperature and sea-level will come about in order to appreciate how they may, or may not, affect policy and practice. Three considerations may be appropriate:
1) The changes are likely to be imperceptible by the general public. The changes are likely to be masked by normal seasonal and yearly fluctuations for the directly affected professionals, e.g. farmers, water engineers and harbour masters.
2) The changes are likely to have their impacts through a greater frequency of severe events rather than the initiation of new extreme conditions. For instance, the rise in mean temperature is likely to result in more frequent droughts for farmers, a prolongation of runs of years with above-average demands made on reservoirs and an increased frequency of very hot summers.
3) There are not likely to be any catastrophic events in the Ichkeul–Bizerte area that can be linked directly to the projected rises in temperature and sea-level. Consequently, there may not be any major triggers for substantial changes in public policy.

3.1 Water Resources, Rivers, Reservoirs and Ground Water

3.1.1 Evaporation and evapotranspiration

The first method employed in this section is to calculate potential evapotranspiration (PEt) and evaporation (PE) using the Penman formula. This is suitable as it incorporates the variables that may be affected by the projected changes, i.e. temperature, vapour pressure, radiation balance and windspeed. A second approach uses the Complementary Relationship which has been forcefully advocated in the 1980s by Morton (1983a, 1983b, 1986).

Monthly observations are available from Bizerte 1970–1977. Fig. 15.7 shows the importance of the radiation budget compared to the aerodynamic function reflecting the maritime influences on the atmosphere in this area. Initially PEt is calculated using an albedo of 0.25 for grassland cover and the analysis below considers changes to each of the climatic variables in turn.

Potential Evapotranspiration: Temperature
Fig. 15.8 shows the impact of increasing air temperature by 1.5°C is most apparent in the summer months when an increase of 0.5 mm/day results. Although this does not appear to be very significant in Fig. 15.8 the difference does accumulate over the year as will be shown. For comparison, a 25% increase in temperature is also plotted with a dramatic rise in PEt evident.

Garaet el Ichkeul/Lac de Bizerte and climatic change 623

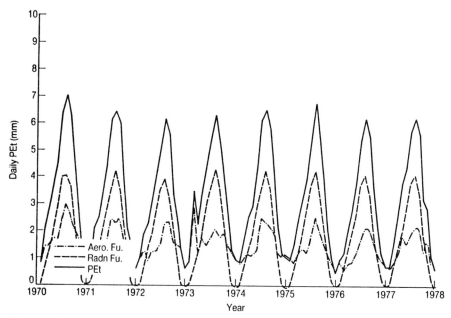

Fig. 15.7 Potential evapotranspiration at Ichkeul for 1970–1977 using the Penman method and showing the relative contributions of the radiation flux and aerodynamic elements in the equation.

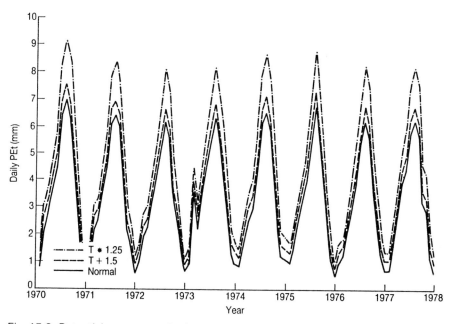

Fig. 15.8 Potential evapotranspiration at Ichkeul using modified temperature in the Penman equation.

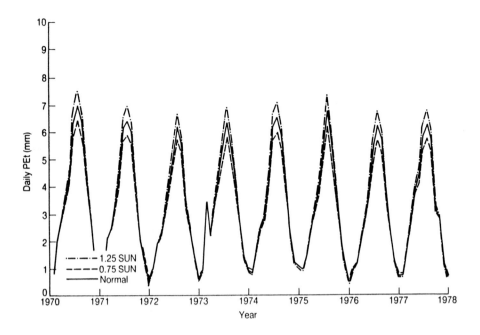

Fig. 15.9 Potential evapotranspiration at Ichkeul using modified sunshine data in the Penman equation.

Potential Evapotranspiration: Sunshine
Given the importance of the radiation budget in determining PEt, changes in the amount of solar radiation reaching the Earth could significantly alter evaporative losses. This variable is influenced by the amount of cloud cover and the moisture content of the atmosphere. A rise in

Table 15.8 The effects of changes in climatological parameters in Penman estimates of potential evaportranspiration and evaporation. Summary of results (mm/day)

Condition	Potential evapotranspiration (Albedo = 0.25)		
	Mean	Max.	Min.
T NORM	3.24	6.99	0.47
T + 1.5	3.68	7.55	0.74
T * 1.25	4.49	9.13	1.01
SUN * 0.75	3.07	6.45	0.57
SUN * 1.25	3.42	7.53	0.37
T + 1.5 and e * 0.75	4.34	8.21	1.21
T + 1.5 and e * 1.25	3.00	6.84	0.22
Potential evaporation (Albedo = 0.05)			
T NORM	4.13	8.62	0.79
T + 1.5	4.59	9.22	1.08
T + 1.5 and e * 0.75	5.25	9.88	1.57

temperature could lead to dissipation of clouds as they are evaporated. Hence Fig. 15.9 shows the effect of a 25% increase in the duration of sunlight. Alternatively, an increase in evaporation associated with a rise in temperature could lead to higher humidities and an increase of cloud cover, so the effects of a 25% decrease in sunlight has also been plotted. On average the effect of these two changes is 0.18 mm/day.

Potential Evapotranspiration: Vapour Pressure
The Penman formula is not particularly sensitive to changes in windspeed but vapour pressure (VP) influences both the energy budget and aerodynamic function. A rise in temperature will cause a decrease in relative humidity if VP remains constant but increasing PEt rates could result in a corresponding increase in VP thus maintaining humidities. Fig. 15.10 shows that there is little change to PEt even if VP increases by 25%. If the area became desiccated, however, and VP was reduced by 25%, then there is an increase in PEt of 1.22 mm/day in the summer.

Actual Evapotranspiration
Table 15.9 shows that PEt will increase by 14% because of the increase in temperature of 1.5°C and by 34% if this is accompanied by increased desiccation. These increases will tend to occur in the drier, summer months and hence may have little impact on actual rates of evapotranspiration. A monthly water balance was therefore computed for vegetation that is

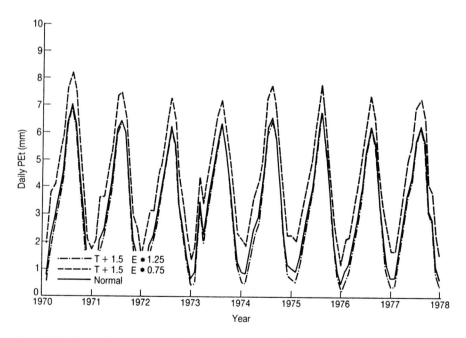

Fig. 15.10 Potential evapotranspiration at Ichkeul using modified vapour pressure data in the Penman equation.

not inundated following the Grindley procedure with a root constant of 75 mm. The calculation uses mean values for PEt for Tunis for the years 1970–77, monthly mean rainfall for Bizerte for 1900–1980 and commences in January by assuming soil-moisture deficits are zero.

Table 15.9 Mean annual evaporative losses (mm)

Condition	Mean	Difference	% increase on T NORM
Potential evapotranspiration			
T NORM	1186	–	–
T + 1.5	1347	161	14
T + 1.5 and e * 0.75	1585	399	34
Potential evaporation			
T NORM	1507	–	–
T + 1.5	1680	173	12
T + 1.5 and e * 0.75	1916	409	27

The results in Table 15.10 show that severe desiccation is experienced from June to September in normal circumstances with an increase in soil-moisture deficits during April and May following an increase in temperature. Table 15.9 indicated that PEt rates increased by 14% whilst AEt rates show a corresponding rise of 8.6% (546.5 to 593.6 mm). Clearly, the rise of 1.5°C has a smaller effect upon AEt rates because there is so little evapotranspiration during the summer months from the dry upland land-use types. It is noticeable that at the end of the first year the soil-moisture deficits have not returned to zero and that only the results from the second year should be considered because they are essentially free of the effects of the assumed starting conditions. The rise in AEt has the effect of reducing the water surplus (effectively runoff from the catchment). The second year's results show a reduction of water surplus (effectively runoff and ground-water recharge) from 56.4 mm to 10.0 mm, a fall of 82.4%.

The calculated percentage runoff (56.4 mm out of 603.3 mm of precipitation – 9.3%) accords well with Kallel's (1986) calculation of an overall runoff coefficient of 27% for the whole Ichkeul basin and only 10% for the Oued Ben Hassine that flows directly into Lake de Bizerte. The former figure includes the higher rainfall and relatively moist mountain areas whilst the 10% for the Ben Hassine is directly comparable with the figures in Table 15.10 which are for the station at Bizerte. The higher and moister parts of the Ichkeul basin are likely to suffer a larger increase in actual evapotranspiration as a result of the temperature rise. They have more moisture available for evapotranspiration and so their runoff coefficients will fall by a larger absolute value than that calculated for Bizerte but by a much smaller percentage amount. The lack of data for the calculation of a Penman estimate of evapotranspiration for these higher and moister parts of the catchment makes a quantification of these ideas impossible at present.

Table 15.10 Grassland monthly water balance (mm) for preset temperatures (TNORM) and a 1.5° rise (T + 1.5)

	TNORM					T + 1.5				
	Rainfall	PEt	SMD	AEt	Water surplus	Rainfall	PEt	SMD	AEt	Water surplus
Jan.	79.5	27.1	0.0	27.1	52.4	79.5	38.8	0.0	38.8	40.7
Feb.	104.8	54.5	0.0	54.5	50.3	104.8	67.0	0.0	67.0	37.8
Mar.	75.8	75.5	0.0	75.5	0.3	75.8	88.8	13.0	88.8	0.0
Apr.	43.4	100.3	56.9	100.3	0.0	43.4	114.9	84.5	114.9	0.0
May	29.8	131.6	113.0	85.9	0.0	29.8	146.7	117.0	62.3	0.0
Jun.	14.5	173.0	124.0	25.5	0.0	14.5	188.3	125.0	22.5	0.0
Jul.	2.5	200.6	125.0	3.5	0.0	2.5	216.7	125.0	2.5	0.0
Aug.	10.3	177.1	125.0	10.3	0.0	10.3	191.7	125.0	10.3	0.0
Sep.	31.1	114.1	125.0	31.1	0.0	31.1	127.4	125.0	31.1	0.0
Oct.	78.8	75.0	121.2	75.0	0.0	78.8	76.1	122.3	76.1	0.0
Nov.	70.7	36.2	87.1	36.2	0.0	70.7	47.4	99.0	47.4	0.0
Dec.	62.1	21.6	46.6	21.6	0.0	62.1	31.9	68.8	31.9	0.0
TOTAL	603.3	1186.6		546.5	103.4	603.3	1347.0		593.6	78.5
Jan.	79.5	27.1	0.0	27.1	5.8	79.5	38.8	27.8	38.8	0.0
Feb.	104.8	54.5	0.0	54.5	50.3	104.8	67.0	0.0	67.0	10.0
Mar.	75.8	75.5	0.0	75.5	0.3	75.8	88.8	13.0	88.8	0.0
Apr.	43.4	100.3	56.9	100.3	0.0	43.4	114.9	84.5	114.9	0.0
May	29.8	131.6	113.0	85.9	0.0	29.8	146.7	117.0	62.3	0.0
Jun.	14.5	173.0	124.0	25.5	0.0	14.5	188.3	125.0	22.5	0.0
Jul.	2.5	200.6	125.0	3.5	0.0	2.5	216.7	125.0	2.5	0.0
Aug.	10.3	177.1	125.0	10.3	0.0	10.3	191.7	125.0	10.3	0.0
Sep.	31.1	114.1	125.0	31.1	0.0	31.1	127.4	125.0	31.1	0.0
Oct.	78.8	75.0	121.2	75.0	0.0	78.8	76.1	122.3	76.1	0.0
Nov.	70.7	36.2	87.1	36.2	0.0	70.7	47.4	99.0	47.4	0.0
Dec.	62.1	21.6	46.6	21.6	0.0	62.1	31.9	68.8	31.9	0.0
TOTAL	603.3	1186.6		546.5	56.4	603.3	1347.0		593.6	10.0

Evaporation
Actual rates of evaporation may differ from potential rates depending upon the size of the evaporating surface and the salinity of the water. Both factors will influence losses from Lake Ichkeul but it is difficult to quantify the effects. Lake evaporation will be affected directly by temperature rise and this can be estimated using the Penman equation with an albedo of 0.05 although it must be noted that the climatological observations from Bizerte are made over a land surface. Table 15.9 shows that the rise of 1.5°C produces a 12% increase in lake evaporation with a 27% increase if the air is drier.

The rise in sea-level will extend the area of the lake thus increasing the volume of water evaporated but it will also result in greater inundation of the marshes and surrounding vegetation. A comparison of PEt and lake PE from Tables 15.8 and 15.9 suggests that there will be an increase in evaporation of 27–25% in these areas with or without a temperature rise. An increase of 42% is found by contrasting PEt in normal circumstances with lake evaporation and a 1.5°C rise. The difference is even greater if actual evapotranspiration is used. Inundation clearly produces a large increase in evaporation of 20–40% compared to non-inundated regions.

The Complementary Relationship Approach
The Grindley approach in calculating regional actual evapotranspiration is limited. It has been described by Morton (1983a, p.3) as, "conventional . . . and based on assumptions that are completely divorced from reality". Indeed after a recent field trial by Mawdsley and Ali (1985) the Grindley approach was described as (p.390), "poorer than one would like, so that the search for a suitable model to satisfy the need (of calculating regional actual evapotranspiration) remains".

There are several conceptual falsehoods implicit in the Grindley approach. The model is one-dimensional and calculates only point evapotranspiration. The role of plants in responding to an "evaporative demand" is complex, controversial and is unlikely to be correctly simulated by the adoption of "root constant" principles (Morton, 1986). Furthermore, the use of the Penman equation as a forcing function in calculating evapotranspiration from large areas is wrong as the potential evapotranspiration calculated is more the result of the actual evapotranspiration occurring in the surrounding area than an upper limit of evapotranspiration that the Grindley model assumes (Morton, 1986). Finally, the Grindley approach considers only one evaporative surface, which in a catchment as diverse as the Ichkeul–Bizerte area, will lead to serious errors in the calculation of regional actual evapotranspiration.

The Complementary Relationship approach, proposed by Bouchet (1963) and developed by Morton, potentially overcomes these shortcomings. This approach is based upon an assumed negative complementary relationship between potential point evapotranspiration and actual regional evapotranspiration induced by changes in the availability of water. For example, if air is passing over a dry continuum, the regional evapotranspiration will be low, causing the overpassing air to be hotter and drier. This air will therefore have a high potential rate of evapotranspiration but a low actual rate. Obviously, the opposite is true for air passing over a wet continuum. Morton (1983a) suggested this relationship could be expressed as:

$$ET + ETP = 2ETW \text{ so that}$$
$$ET = 2ETW - ETP$$

where, ET is the actual regional evapotranspiration, ETP is the potential point evapotranspiration, and ETW is the wet areal evapotranspiration – the evapotranspiration that would occur if the area was saturated and not short of water.

This approach has the advantages of calculating the regional evapotranspiration, so that all the differing evaporative surfaces are accommodated. Consequently, actual regional evapotranspiration using the Complementary Relationship Areal Evapotranspiration model (CRAE) as devised by Morton (1983a) was calculated for the Ichkeul–Bizerte catchment using the data obtained from Bizerte 1970–1977 with "normal" data and with an increase in temperature of 1.5°C (Tables 15.11 and 15.12).

Table 15.11 Mean monthly CRAE model estimates of actual regional evapotranspiration for Ichkeul–Bizerte catchment in mm

Month	Normal	+ 1.5°C	% change	PEt from Penman	AEt from Grindley
Jan.	18.1	18.3	1.3	27.1	27.1
Feb.	29.8	30.8	3.5	54.5	54.5
Mar.	57.4	59.1	3.0	75.5	75.5
Apr.	86.0	88.2	2.6	100.3	100.3
May	97.2	99.1	2.0	131.6	85.9
Jun.	102.6	104.7	2.1	173.0	25.5
Jul.	92.0	94.0	2.2	200.6	3.5
Aug.	57.4	59.1	3.0	177.1	10.3
Sep.	29.5	31.3	6.2	114.1	31.1
Oct.	19.7	20.8	5.5	75.0	75.0
Nov.	16.4	16.9	3.1	36.2	36.2
Dec.	16.0	16.2	1.2	21.6	21.6
Total	622.1	638.5	3.0	1186.6	546.5

Before any discussion on the effect of an increase in 1.5°C can occur, the validity of the CRAE model estimates have to be established. Although Morton (1983b, p. 1350) has described this diagnostic checking as, "a surefire formula for scientific sterility", it is, nevertheless, a useful technique in exposing the weakness of the Grindley approach and emphasizing the CRAE's superiority.

The CRAE model estimates of actual evapotranspiration follow the expected seasonal cycle, with low values in the autumn and winter months, coupled with high values in the spring/early summer, with a decrease in actual evapotranspiration in the height of summer (Table 15.11).

Table 15.12 Annual CRAE model estimates of actual regional evapotranspiration for the Ichkeul–Bizerte catchment in mm

Year	Normal	+ 1.5°C	% change
1970	667.7	685.9	2.7
1971	644.3	662.6	2.8
1972	640.1	657.0	2.6
1973	610.4	629.4	3.1
1974	616.8	632.7	2.6
1975	625.1	641.0	2.5
1976	554.0	568.0	2.5
1977	646.5	663.1	2.6
Total	625.6	642.4	2.7

Comparison with the values of actual evapotranspiration derived from the Grindley approach show the CRAE's values of actual evapotranspiration from the catchment as a whole during the months of July and August are markedly different from those estimated by the Grindley procedure. Interestingly, the annual values of actual evapotranspiration differ by only 14%.

A 1.5°C temperature rise appears to have little effect on either the monthly values (max. 6.2%, mean 3.0%) or on the annual values (max. 3.1%, mean 2.7%). The biggest increases are in the autumn (September/October) where % increases were 6.2 and 5.5% respectively. The CRAE model thus predicts an annual increase in actual evapotranspiration of around 3% (16.8mm) with monthly variations of between 1 and 6%. This is somewhat lower than the 9% increase (47.1 mm) predicted by the Grindley approach. However, if the CRAE estimate of increases in actual evapotranspiration is applied to the rainfall-runoff data provided by Kallel (1986), it can be seen that a 3% increase in actual evapotranspiration will result in a fall in riverflow of between 7 and 44% for individual river basins and of 11.7% for the total runoff to the Ichkeul–Bizerte region (Table 15.13).

Table 15.13 Rainfall-runoff relationships for the Ichkeul–Bizerte region and the likely impact of a 16.8 mm rise in actual evapotranspiration after the projected 1.5°C rise in mean annual temperature (after Kallel, 1986).

Catchment	Present conditions			+1.5° (+16.8 mm AEt)	
	Rainfall mm	Runoff mm	% runoff	Runoff mm	% change
Sedjenane	825	230	28	213	−7.3
Djoumine	575	143	25	126	−11.7
Tine	520	38	8	21	−44.2
Melah	720	233	32	216	−7.2
Rhezala	680	190	28	173	−8.8
Douimis	635	174	27	157	−9.7
Ben Hassine	505	50	10	33	−33.6
Ichkeul/Bizerte	640	144	22	127	−11.7

On the basis of these calculations it is safe to conclude that actual evapotranspiration in the Ichkeul–Bizerte region will increase by up to 10% when mean air temperature rises by 1.5°C. Similarly, an estimate of 10% for the resulting fall in riverflow can be seen to be conservative.

Potential evapotranspiration was calculated to rise by 14% whilst open water evaporation was shown to be likely to rise by 12%. It is therefore both safe and conservative to project a 12% increase in evapotranspiration will take place after the rise in temperature from those parts of the catchment which are open water or saturated marshland or irrigated fields.

3.2 The Level and Salinity of Garaet El Ichkeul

The existing UCL Hydrological Simulation Model for Ichkeul (Hollis, 1986) was modified to explore the impacts of the projected temperature rise and sea-level rise on Ichkeul. The elements changed were as follows:

Sea-Level Rise

The level of equilibrium between Ichkeul and Lac de Bizerte was raised to 32.5 cm NGT and the Oued Tindja rating curve was modified as shown in Fig. 15.11. The increase in discharge for a given head was introduced

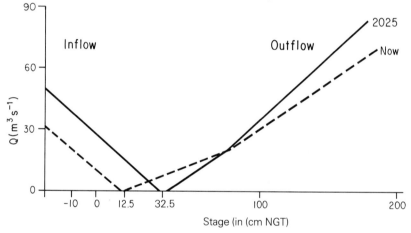

Fig. 15.11 The stage discharge relationship for the Oued Tindja at present and in 2025.

because of the reduction in bed friction produced by the deepening of water in the channel. It is, however, possible that the Oued Tindja channel will simply aggrade to leave the cross-section unchanged! The simulation model automatically increased the area of the lake with the rise in water level because the hypsometric curve of the marshland and surrounding lands is already built into the model. As shown above, this increase in area will necessarily increase evaporation losses from the lake (Fig. 15.12).

Infilling of Reservoirs with Sediment

The conclusion derived in section 3.3 that the reservoirs are likely to fill at the rate of up to 1% per annum was incorporated into the simulations because a reduction in the storage capacity of the reservoirs is likely

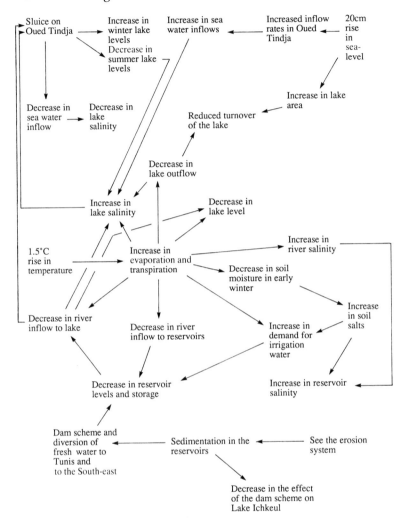

Fig. 15.12 The impact of a 20 cm rise in sea-level, a 1.5°C rise in temperature, the implementation of the dam and water resource scheme, and the sluice on the outflow upon the level and salinity of Lake Ichkeul.

to lead to a greater frequency of overflow and therefore a reduction of their impact on Ichkeul. The definitive simulations incorporate the conservative assumptions that the reservoir capacities will have fallen by 25% during the 37 years between 1988 and 2025.

Wetland Evaporation
This was increased by 5, 10 and 20% to provide an illustrative spectrum of the likely impacts of the rise in open-water evaporation and potential evapotranspiration. In the final, and definitive, set of simulations a rise of 12% was simulated in open-water evaporation, potential evapotranspiration on the marshes and reservoir evaporation (in appropriate simulations).

Riverflow
This was decreased by 10 and 20% in line with the conclusions of the foregoing section on evaporation and evapotranspiration.

Irrigation Demand
This was increased by 5, 10 and 20% in parallel with the illustrative changes made for wetland evaporation and in the definitive simulations it was increased by 12%.

The possible range of simulations involving various assumptions about changes, the existence of various numbers of dams, and the possibility of various types of sluices being built is too large to contemplate. Consequently, the simulations were run in four groups. The intention was to demonstrate the scale of impact of each of the changes when acting alone and a sample of likely scenarios involving a mix of factors. The final definitive simulations take the 12% increase in evaporation and potential evapotranspiration, 12% increase in irrigation demand, 25% reduction in reservoir capacity and a 10% fall in riverflow (Fig. 15.12).

The Group 1 simulations (Table 15.14) show that a simple 20 cm sea-level rise would have a minimal impact upon the ecology of Ichkeul save for raising its level and area. The simulated salinity regime after the sea-level rise is almost exactly the same as natural conditions 1952–1982 and the regime of levels is as before plus 20 cm at the lower levels and reduced slightly at the higher levels because of the simulated increase in outflow efficiency of the deepened Oued Tindja. The simulated increases in evapotranspiration and reductions in riverflow have the very predictable effects, but it is important to note that even in the final simulation in this group (evapotranspiration +20% and riverflow −20%), with assumptions which are more severe than predicted, the impacts upon Ichkeul's ecological determinants of level and salinity are not nearly as great as for the predicted impact of the first three dams in the water resource scheme (Table 15.7). Indeed the scale of change with evaporation up 20% and riverflow down 20% is less than is predicted to follow the implementation of the best possible sluice- and reservoir-release scheme after the completion of the third dam.

The simulations in Group 2 (Table 15.15) show that with changes brought about only by climatic shifts, the lower quartile salinity of Ichkeul will rise from 8.2 to 12.6 g/l for the most likely climatic scenario, but the minimum salinity, i.e. during the greatest floods, will only rise from 2.6 to 3.5 g/l. The predicted change in the level regime, after the climatic changes, is minimal save for the general rise in the level of the lake.

The most significant features of the Group 3 simulations, which incorporate the water resource dams and the climatic changes (Table 15.16), are the salinization of the Ichkeul National Park and the severe reduction in water resources available from the reservoirs. Ichkeul becomes a sebkha system with water salinities in excess of sea-water strength save for infrequent large floods which can reduce the minimum salinity (10.2 g/l) to a value akin to the lower quartile salinity (8.2 g/l) in the period 1952–1982. The mean storage in the dams is cut by 7.8 and 26% for the Djoumine and Sedjenane dams respectively. Perhaps more importantly, the proportion of time when the reservoirs have less than 25×10^6 m^3 of water left in them

Table 15.14 Group 1 simulations: the factors acting singly on the natural system

Simulation	*Lake level at Ichkeul (cm NGT)*						*Lake salinity at Ichkeul (g/l)*					
	Mean	*Max.*	*Min.*	*Med.*	*UQ*	*LQ*	*Mean*	*Max.*	*Min.*	*Med.*	*UQ*	*LQ*
Natural conditions 1952–1982	32.5	330	−6	119.1	52.9	2.5	17.2	49.8	2.6	15.0	24.5	8.2
Sea-level + 20 cm	48.0	329	12.4	36.1	64.8	822.1	17.4	46.1	2.7	15.7	24.2	9.1
Wetland evap. +5%	32.0	330	−7.2	18.5	52.6	2.1	18.6	53.1	2.7	16.2	26.3	8.9
Wetland evap. +10%	31.5	330	−8.2	18.1	52.2	1.8	20.0	56.5	2.8	17.4	28.2	9.6
Wetland evap. +20%	30.5	329	−10	17.1	51.5	1.0	22.7	63.4	3.1	19.9	31.9	11.1
Wetland evap. +10% Riverflow −10%	29.2	299	−8.2	17.0	48.5	1.7	21.9	59.3	3.3	19.5	30.2	11.4
Wetland evap. +20% Riverflow −20%	25.9	268	−10	15.0	43.8	0.7	27.9	70.9	4.3	25.4	37.7	15.7

Notes: Med. Median. UQ = Upper quartile. LQ = Lower quartile. NGT = Tunisian datum.

Table 15.15 Group 2 simulations: the factors of change to 2025 acting together

Simulation	Lake level at Ichkeul (cm NGT)						Lake salinity at Ichkeul (g/l)				
	Mean	Max.	Min.	Med.	UQ	LQ	Mean	Max.	Min.	Med.	UQ	LQ
Natural conditions 1952–1982	32.5	330	−6.1	19.1	52.9	2.5	17.2	49.8	2.6	15.0	24.5	8.2
Sea-level + 20 cm Wetland evap. + 12% Riverflow −10%	44.7	301	10.0	34.2	60.7	20.4	22.4	56.1	3.5	20.5	30.3	12.6
Sea level + 20 cm Wetland evap. + 12% Riverflow −20%	42.7	273	10.0	33.4	57.6	20.2	24.7	59.1	4.1	23.0	32.9	14.8

Notes: Med. = Median. UQ = Upper quartile. LQ = Lower quartile. NGT = Tunisian datum.

Table 15.16 Group 3 simulations: the factors of change to 2025 plus the dams on the Djoumine, Rhezala and Sedjenane plus the agricultural improvement scheme

Simulation	Lake level at Ichkeul (cm NGT)						Lake salinity at Ichkeul (g/l)					
	Mean	Max.	Min.	Med.	UQ	LQ	Mean	Max.	Min.	Med.	UQ	LQ
Natural conditions 1952–1982	32.5	330	−6.1	19.1	52.9	2.5	17.2	49.8	2.6	15.0	24.5	8.2
Sea-level +20 cm Evap. and irrigation +12% Riverflow −10% (plus 3 dams and agric. imp.)	32.8	243	10.8	30.1	40.2	20.4	56.2	109	10.2	53.7	70.4	42.7
Sea level +20 cm Evap. and irrigation +12% Riverflow −10% (plus 3 dams with 25% sediment infill and agric. imp.)	33.2	244	10.8	30.2	40.3	20.4	53.8	107	9.3	51.3	68.1	38.9

	Storage in the Djoumine Dam ($10^6 m^3$)				Storage in the Sedjenane Dam ($10^6 m^3$)			
	Mean	UQ	LQ	% time <25	Mean	UQ	LQ	% time <25
Present conditions	97.5	119	83	0.8	79.4	99	60	3.4
Evap. and irrigation +12% Riverflow −10%	90.1	114	72	5.4	58.8	84	32	19.0
Evap. and irrigation +12% Riverflow −10% Sed. infill 25%	61.1	82	46	9.9	43.8	62	23	24.7

increases from tiny percentages to 5.4 and 19.0% for the Djoumine and Sedjenane dams respectively.

The impact of the 25% filling of the reservoirs with sediment is minimal on the Ichkeul National Park but highly significant on the performance of the reservoirs. The mean storage in the reservoirs falls to 63 and 55% of that pertaining at present for the Djoumine and the Sedjenane but the percentage of time when the reservoirs have less than 25×10^6 m^3 of water left in them increases to 10 and 25% of the time.

The Group 4 simulations (Table 15.17) gives an estimate of the actual conditions likely in 2025. In addition to the climate-related changes in evapotranspiration, riverflow, irrigation demand and sea-level; three large dams and the agricultural improvement scheme are in operation, the reservoirs have a 25% sediment infill, a carefully controlled sluice is operational in the Oued Tindja outflow from the National Park, and reservoir releases are made except in the very driest of years. Table 15.17 shows that the conflagration of human and climatically induced changes will bring about a complete change in the ecology of the wetland National Park. The sluice and reservoir releases are unable to halt the salinization of the ecosystem. The lower quartile salinity rises from a natural level of 8.2g/l to 38.7g/l. The mean salinity rises from a natural level of 17.2g/l to 58.1g/l.

The increase in soil-moisture deficits and the increase in potential and actual rates of evapotranspiration has been related to a projected increase in soluble salts in the soils. Once these salts have begun to accumulate, it is reasonable to suppose that both percolation to ground water and runoff to rivers will tend to become slightly richer in dissolved salts. This greater solute loading of water transfers would result from both the greater concentration of salts in the soil and the increased solubility of the salts because of the elevated temperatures. It is not possible to predict the extent of this effect and so no simulation tests were attempted. However, Table 15.18 shows how the average salinity of rivers increases from the moist north to the more arid south of Tunisia. The rise in temperature at Ichkeul will certainly give its rivers the salinity levels of streams that are much further south at present.

3.3 Physical Alteration of Terrestrial and Coastal Systems
The geomorphological effects of the rises in temperature and sea-level are considered in separate sections.

3.3.1 Rise in temperature: slopes
It is probable that northern Tunisia already suffers from some of the highest erosion rates on Earth. A recent survey by Walling and Webb (1983) concluded that there was still great uncertainty about erosion rates, mainly because of the difficulties in measuring them. Their world map nevertheless shows northern Tunisia to have rates in the range of 750 to 1000 tons/km^2. Although they acknowledged that the evidence was poor, they concluded that there did seem to be a direct relation between mean annual rainfall and the rate of erosion. This implies that a full evaluation of the effects of future climatic change will have to pay special attention to rainfall.

Table 15.17 Group 4 simulations: The factors of change to 2025 plus the dams on the Djoumine, Rhezala and Sedjenane plus the agricultural improvement scheme and a sluice in the Oued Tindja

Simulation	Lake level at Ichkeul (cm NGT)					Lake salinity at Ichkeul (g/l)				
	Mean	Max.	Min.	Med.	UQ	LQ	Mean	Max.	Min.	Med.	UQ	LQ
Natural conditions 1952–1982 Sea-level + 20 cm	32.5	330	−6.1	19.1	52.9	2.5	17.2	49.8	2.6	15.0	24.5	8.2
Evap. and irrigation + 12% Riverflow −10% (plus 3 dams with 25% sediment infill and agric. imp. and the 1988 Sluice + 20 cm)	48.7	284	6.1	37.1	60.7	20.6	58.1	89.4	9.3	53.3	72.1	38.7

Note: The 1988 sluice + 20 cm follows a regime with the controlled top water level (cm NGT) at − Sep. 70, Oct. 70, Nov. 80, Dec. 100, Jan. 120, Feb. 100, Mar. 90, Apr. 85, May 80, Jun. 75, July 70, Aug. 70. During the period from May onwards the actual water level in the lake is often lower than this controlled level because of evaporation. Indeed, when the lake level falls below +10 cm the sluice gates are opened to allow sea water to flow into Ichkeul. In addition there are reservoir releases of 0.4×10^6 m³ per day when the lake is below +7.5 cm, but these releases cease during drought years with a frequency (based on 2025 data) of at least one year in five. Med. = Median. UQ = Upper quartile. LQ = Lower quartile. NGT = Tunisian datum.

Table 15.18 Mean salinity (g/l) of rivers in the hydrological regions of Tunisia (after Kallel, 1984)

	Baseflow	Surface runoff
Extreme north (Zouara etc)	0.68	0.5
Ichkeul/Bizerte region	1.6	0.48
Medjerdah catchment	2.25	1.0
Miliane catchment	5.5	2.85
Cap-Bon region	2.2	0.83
Sebkhet Kelbia system	4.2	0.7
Sahel region	4.6	1.5
Centre/Sud region	3.3	1.04
Sud region	4.8	0.5

Walling and Webb's investigation of erosion rates for the range of 600 to 800 mm mean annual rainfall, which is the range in northern Tunisia, shows that erosion should rise from about 100 to about 200 tons/km²/year as the runoff rises from 200 to 400 mm/year. Their graphs (Fig. 15.13) indicate a minimum erosion rate at about 650 mm/yr^{-1} mean annual rainfall and at about 200 mm/year mean annual runoff; below these figures for rainfall and runoff, erosion rates increase again. Ichkeul–Bizerte basin rainfall is presently 650 mm and runoff is 144 mm (Kallel, 1986).

However, Walling and Webb note that the picture would be complicated by seasonality of rainfall, which is the case in northern Tunisia. Seasonal

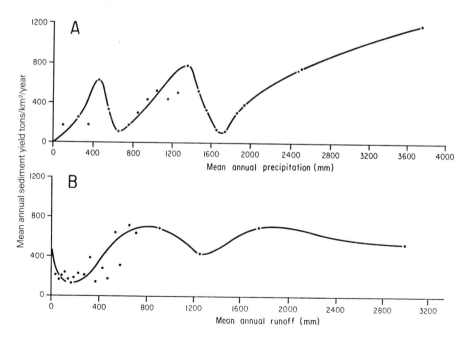

Fig. 15.13 The relationship of sediment yield to mean annual precipitation and mean annual runoff (after Walling and Webb, 1983).

rainfall would be much more erosive than in a less seasonal climate. In Morocco, which was the area closest in character to northern Tunisia among the cases they examined in detail (after Heusch and Millies-Lacroix, 1971), there was a continuous increase in erosion with rainfall from 400 mm/year to 1200 mm/year (Fig. 15.14). Moroccan rates in the 600 to 800 mm mean annual precipitation range were again from about 400 to 1000 tons/km²/year; these high figures appear to confirm the high erosivity of seasonal rainfall. Tayaa and Brooks (1985) found even higher rates of erosion in the Rif Mountains of northern Morocco: up to 20,000 tons/km²/year.

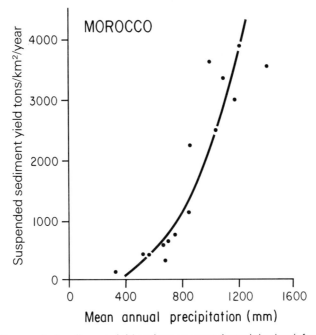

Fig. 15.14 Suspended sediment yield and mean annual precipitation (after Heusch and Millies-Lacroix, 1971).

In view of these findings, the way in which erosion would react to a change in temperature is uncertain (Fig. 15.15). Schumm (1965) reasoned that as temperatures rose, so less of the precipitation would be available for runoff. This view is supported in section 3.2 of this chapter when considering annual runoff totals. This is the left-hand side of Fig. 15.13 and if this happened, erosion rates should fall. There could, however, be two other effects which might increase erosion rates. First, a rise in temperature might inhibit plant growth (section 3.6), so that erosion might increase locally on steep, dry slopes. Secondly, greater soil induration at the surface might encourage runoff, and this might encourage rill and gully erosion. All in all, erosion rates on slopes might increase slightly, but even if this happened, a reduction in runoff would prevent the sediment from reaching channels, or, if reaching them, from moving far down the channel net. The result would be that while small basins might experience

Garaet el Ichkeul/Lac de Bizerte and climatic change 641

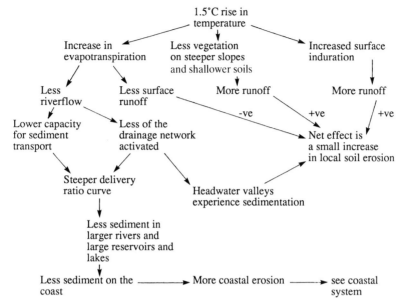

Fig. 15.15 Possible changes in the water-sediment system at Ichkeul.

more sedimentation, the rate of sediment entering medium- and large-sized basins would decline.

Despite Walling and Webb's re-emphasis of how difficult it is to measure erosion rates and the absence of good evidence; it still is widely believed that erosion rates have been accelerated in Tunisia in historical times, as forests have been removed. The growth of the Medjerda Delta, and the consequent silting of Roman ports is usually cited as evidence (Paskoff, 1985). It is also widely believed that erosion is still an important problem in Tunisia although there is again very little good evidence (e.g. United States National Parks Service, 1980). Changes to a more conservation-minded agriculture, in response to alarm of this kind, and the growth of tree plantations might reduce erosion rates significantly by 2025, and thus further reduce the siltation consequent on a rise of temperature that has been predicted above.

3.3.2 Rise in temperature: channels

With the temperature rise of 1.5°C there will be less runoff in total (section 3.2). If that happened channels could become narrower, and the channel net should contract. On the other hand, if there was less vegetation, storm runoff could be faster, and this, together with the greater supply of sediment to streams, might produce more seasonal, wider, shallower, sediment-clogged channels.

How much of the new sediment leaving the slopes would reach channels, and how far down the channel system it would reach is very hard to determine. Walling and Webb's review showed that what studies there have been of this system show that most sediment is trapped high in the drainage net. The sediment leaving a third- or fourth-order stream can be as little as 10% of that leaving slopes in the basin (Fig. 15.16).

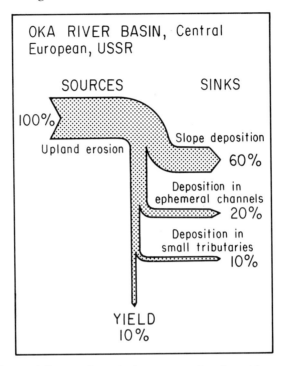

Fig. 15.16 Sediment delivery ratios – the low output of sediment from eroding river basins (after Walling and Webb, 1983).

3.3.3 Rise in temperature: reservoirs, lakes, estuaries

Sediment delivery ratios (tons of sediment per square kilometre of basin area) almost always decline as the basin size increases. Thus, reservoirs or lakes in small catchments trap proportionately more sediment than those in large basins. The relationship holds very well in the Medjerda valley, near to the Ichkeul basin, and in general for barrages throughout Tunisia (Table 15.19). The calculated sediment inflow to Ichkeul is 255.96 tons/km²/year (Hollis, 1986) which also supports the idea of steep

Table 15.19 Sedimentation in reservoirs in Tunisia (after Ghorbel, 1980)

Reservoir	Basin area km²	Reservoir volume 10⁶m³	Years of operation	% filled each year	Sediment delivery tons/km²/year	Sediment conc. kg/m³
Mellégue	10300	268	21.3	0.8	695	38
Nebaana	855	86.4	10	1.5	2300	47.5
Bézirk	84	6.5	14.8	1.76	2430	45.6
Chiba	64	7.9	12	2.81	4220	72.5
Masri	40	6.8	7.5	2.58	6050	85
Kasseb	101	81.9	7.5	0.5	2865	53
Lakhmess	131	8.0	9.3	2.69	5070	86

sediment-delivery ratio curves because, of course, Ichkeul is at the bottom of the catchment system.

It is probable that the delivery ratio against basin area becomes even steeper in more arid climates where fewer flows reach right through a drainage system. Thus, when temperature rises and runoff decreases, the delivery ratio curve will steepen, and even less sediment will find its way down through the system.

If there were to be an increase in erosion on slopes, it seems likely that most of it would be trapped in small reservoirs high up the valleys. Tayaa and Brooks (1985) found that reservoirs in the Rif were accumulating between 20 and 70 m^3/ha/year. Farther down the valleys, in the Lac de Bizerte, for instance, sedimentation rates will probably decrease. The sediment inflow into Ichkeul will certainly dwindle significantly partly because of the processes so far discussed but mainly through the trapping of virtually all of the rivers' sediments in the reservoirs.

Ghorbel's (1980) figures allow us to estimate that the major barrages in the Ichkeul basin will already be about 25% filled with sediment by 2025. This reduced storage capacity will mean that the amount of flood water reaching through to the lower channels will gradually increase. This, in itself, might compensate for the decreased runoff because the relatively clear water of the overflows will rapidly erode the alluvial banks of the downstream channels!

3.3.4 Rise in temperature: coasts

If less sediment is delivered to the coast by streams, then coastal erosion may be stimulated as longshore drift systems, starved of sediment, begin to erode unlithified coastlines (Fig. 15.15). This process would only be important on beaches fed by sediment coming through the canal at Bizerte, and it is very doubtful if this has been a major source of sediment. However, since the riverflow from all other Tunisian rivers will be declining and reservoir construction will both exacerbate the declining flows and further limit the sediment flows, it is likely that the total terrestrial input of sediment to the coastal system will fall substantially. Since "almost the entire Tunisian coastline is currently eroded . . . (and) . . . beach erosion is a serious problem in Tunisia" (Paskoff, 1985), it seems likely that an intensification of beach erosion is likely by 2025.

3.3.5 Changes in sea-level: hard coasts and sand beaches

Coastal erosion is already causing the loss of archaeological sites and tourist facilities (Paskoff, 1985) and this has been partly ascribed to a small historical rise in sea-level. Many authorities divide coasts into "Prograding" and "Retrograding". In general, coastal erosion, driven by longshore drift, removes sediment from retrograding coasts and builds up prograding ones. On most coasts there are sedimentary cells in which there is an input, transport, and export of sediment. It is now believed that the post-glacial rise in sea-level (Bloom, 1977) produced large quantities of sediment on coasts worldwide (Pye, 1984). This is because the form of beaches is in dynamic equilibrium with the coastal erosional processes. A rise in level brings the sea into contact with different profile forms, the adjustment to the new equilibrium liberates new supplies of sediment.

On retrograding, mostly hard-rock coasts as on the north coast of Tunisia where there are some soft Oligocene sandstones (Paskoff, 1985), the rise of 20 cm will produce new supplies of sediment.

The beaches near Bizerte are probably net receivers of this sediment, and might prograde with the new supply from eroding cliffs, their beach profiles might become less steep. Paskoff (1985) shows the sandy coast near Rmel to be retrograding, but the dunes behind the coast there suggest that it has been a net collector of sediment in the recent past. The same must apply to the sandy coves on the north coast. On prograding coasts which are across the prevailing north-westerly winds of winter, as at Rmel, increased progradation and gentler beach slopes could contribute to the renewed growth of sand dunes. This is what happened to the eastern Australian coastal dune systems during the post-glacial rise in sea-level (Pye and Bowman, 1984). Even without a rise in temperature, the plant cover of the dunes might not be able to withstand a great increase in incoming sand, and the new dunes might grow and migrate south-eastward. With a rise in temperature, and the consequent rise in both potential and actual evapotranspiration, it seems even more unlikely that they would be able to cope. This renewed dune movement might happen at Rmel and begin to impact the oil refinery and the extensive forestry plantation.

Erosion by waves on coasts subjected to a rise in sea-level within the Lac de Bizerte might liberate small amounts of sediment. This sediment might find its way onto beaches elsewhere in the lake or be taken through the canal to the beaches on the main coast.

In Lake Ichkeul, the increased salinity (section 3.2) will probably inhibit the growth of *Phragmites* and *Potamogeton*. This will expose the eastern coasts of the lake to more erosion, and this might create even wider shallower beaches than at present, and therefore more mud-flats at low water.

3.4 Coastal and Wetland Ecosystems, Soils and Vegetation

3.4.1 Lake Ichkeul
At present the lake is dominated by two macrophytes – *Potamogeton pectinatus* and *Ruppia cirrhosa* – with a limited amount of the complex alga *Lamprothamnium papulosum* and the marine grass *Zostera noltii*. The presently known ecological limits for Ichkeul are shown in Table 15.20. The response of the wetland vegetation, within these limits, is outlined according to four hydrological scenarios derived from section 3.2:
1) sea-level +20 cm (see Table 15.14);
2) sea-level +20 cm, 12% increase in evapotranspiration and 10% decrease in riverflow (see Table 15.15);
3) Djoumine, Rhezala and Sedjenane Dams plus sea-level + 20 cm, 12% increase in evapotranspiration and 10% decrease in riverflow (see Table 15.16);
4) Djoumine, Rhezala and Sedjenane dams plus sea-level +20 cm, 12% increase in evapotranspiration, 10% decrease in riverflow, and the 1988 sluice + 20 cm scheme with its associated management measures (see Table 15.17).

Table 15.20 The Ecological limits for the natural ecosystem at Ichkeul (1988 sea-level conditions used)

a) The marshes up to 80 cm NGT must be flooded almost every year to facilitate the growth of *Scirpus maritimus* in this zone.
b) The marshes up to 120 cm NGT should be flooded occasionally to compensate for the loss of river water inundation from the Djoumine and Rhezala rivers.
c) The lake should not fall below −20 cm because this would destroy large areas of *Potamogeton pectinatus* and allow the cattle to graze the *Phragmites* to destruction.
d) The winter lake salinity should occasionally be below 5 g/l to allow a strong growth of *Potamogeton pectinatus*.
e) The winter lake salinity should normally be below 10 g/l to allow *Potamogeton pectinatus* to grow.
f) Prolonged periods with salinities over 10 g/l should be avoided to safeguard the regeneration possibilities of *Potamogeton pectinatus*.
g) High salinities when the lake is low should be minimized to prevent *Scirpus litoralis* outcompeting *maritimus* on the Melah marsh.
h) The lake level should not rise above 2.0 m NGT because of the threat of flooding to surrounding farmland.
i) There should be an outflow of fresh water between February and May to attract the inward migration of fish.
j) There should be an inflow of sea water between August and September to allow the outward migration of fish.

The hydrological model suggests that a sea-level rise of 20 cm on present conditions will only affect lake level without any significant change in the salinity regime at all (Table 15.14). Therefore the area of *Potamogeton pectinatus* would increase as a result of a 20 cm height equivalent landward migration. The lakeward limit is unlikely to change since it appears to be controlled by wind fetch rather than by water depth (Hollis, 1986).

Under scenario (2) the salinity regime does change with increases in the lower quartile salinities from 8.2 to 12.6 g/l. This will involve an increase in lake salinity at a critical period when *P. pectinatus* should be at peak production. Many studies have shown that any increase in mean winter salinity or salinity in the growing season substantially diminishes biomass, turion and seed production which are all essential food sources for the waterfowl (Teeter, 1965; Hollis, 1986). Therefore the effects of simply the climatically induced changes on Ichkeul will be a reduction, and probably a significant one, in the biomass of *Potamogeton*, and therefore in the food available to the wintering waterfowl and the fish.

The effect of the 20 cm sea-level rise is trivial when compared with the effects of the first three dams and the changes induced in evapotranspiration and riverflow by the atmospheric warming. Even when the assumption of a 25% filling of the reservoirs with sediment is taken (Table 15.16), scenario (3) has a lower quartile salinity of 38.9 g/l, a minimum salinity of 9.3 g/l and an upper quartile level of only 40.3 cm. The latter will be less than 10 cm above the level of the lake when it is in equilibrium with the Lake de Bizerte. Therefore under scenario (3), Ichkeul will have lost its brackish/fresh nature and it will have become a saline sebkha with a fairly static water level save when large rare floods temporarily raise its level and sweeten the water a little.

Section 2.1 (Table 15.7) showed that it will be possible to minimize the ecological changes to the present system under the present circumstances by very careful manipulation of the sluice and the release of water from the reservoirs. It was shown in Table 15.7 that this so called "1988 Sluice" regime could limit the rise in the lower quartile salinity to 15.3 g/l and could maintain a very good variation in level in the lake. As such, the management scheme is on the very edge, indeed some would argue it is beyond the limit, of maintaining the present ecological character of Ichkeul. However under scenario (4) with the climatic changes likely by 2025, the lower quartile salinity rises to 38.7 g/l and once again it is clear that Ichkeul is highly likely to develop a saline sebkha flora and fauna even if the sluice is operated in the best possible manner and if reservoir releases, except during one-in-five-year droughts, prove to be politically feasible.

Some compensatory changes could occur but this would be dependent on the resultant salinity range (Fig. 15.17). *Ruppia cirrhosa* may increase as would *Zostera noltii* and *Zannichellia*. However, *Zostera* is unlikely to become a dominant food plant since it would be restricted to sandy substrates which are infrequent at Ichkeul. When the lake is highly saline *Posidonia* could invade but this again requires sandy/rocky substrates. Overall, any increase in mean lake salinities is likely to reduce the available food resources for the overwintering waterfowl.

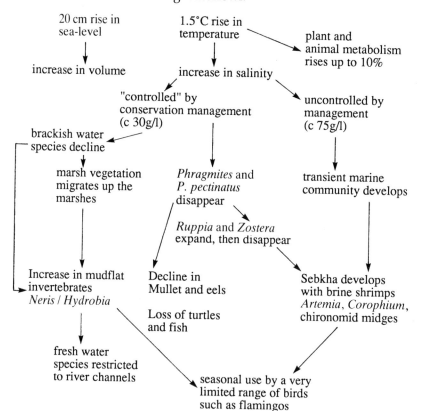

Fig. 15.17 Impacts of the sea-level and temperature rises on the aquatic ecosystem.

3.4.2 The Ichkeul Marshes

The marsh vegetation will be affected by changes in marsh inundation patterns and in the salinity of the inundating water. These changes will primarily affect the lake-water inundated marshes (i.e. Melah, Sedjenane and Douimis). Here, changes in inundation patterns, specifically average water depth and length of inundation, will move the *Scirpus* zones up the marsh (Fig. 15.17). The extent to which either an increase or decrease in available sedge habitat occurs will depend on *Scirpus* distribution/hypsometric curve relationships for the various marshes. Given the inevitable rise in lower quartile water levels because of the sea-level rise and the retention of fresh water in the spring by the sluice, those areas of the Melah marsh less than –20 cm mean sea-level will be at risk (approx 10%). This could be compensated by a movement up the marsh. The Sedjenane marsh is unlikely to be affected since none of the *Scirpus* communities occupy marsh heights much below –20cm mean sea-level. The response of *Phragmites* to these changes is problematical but a migration up the marsh by 20 cm height equivalent could be expected.

When average water and soil salinities begin to increase within scenarios (3) and (4), then major changes will occur in the composition of the *Scirpus* communities. The present trend of the Melah marsh to become dominated by *S. maritimus* at the expense of *S. litoralis* will cease and a reverse trend will occur since experimental and vegetational studies have shown that *S. maritimus* is not as saline tolerant as *S. litoralis*. However, if

Table 15.21 Hypsometric data for the Ichkeul marshes. Percentage of the marshes between classes in cm NGT

	<–40	<–20	0	20	40	60	80	100	120	140	>140
Djoumine	0.0	0.0	0	8.8	10.7	10.1	7.6	10.3	11.6	11.8	29.8
Southern		0.0	0.0	0	0.0	0.0	4.8	25.6	47.9	10.8	10.9
S. Melah	0.0	0.0	2.4	7.4	8.5	35.1	17.1	25.8	1.0	1.0	2.7
Melah	3.9	14.7	42.3	10.3	8.0	7.9	6.3	6.7	0.0	0.0	0.0
Sedjenane	5.5	39.5	15.4	14.9	17.1	3.8	3.8	0.0	0.0	0.0	0.0

Height distribution of *Scirpus maritimus* at Ichkeul

Height range (cm)		Djoumine	Area in hectares Southern	Melah	Sedjenane
–40 to	–20	—	6.43	—	—
–20 to	0	—	4.24	206.45	20.96
0 to	20	18.41	9.01	18.15	1.28
20 to	40	23.94	2.96	—	—
40 to	60	36.42	66.80	—	—
60 to	80	37.84	36.42	—	—
80 to	100	58.82	6.31	—	—
100 to	120	47.11	5.53	—	—
120 to	140	33.98	—	—	—
140 to	160	12.99	—	—	—

soil salinities increase beyond the tolerance of this plant then the floristic changes will be even more dramatic. They will be dependent on the interaction between mean inundation, length of inundation and salinity of the inundating water. With lower quartile salinities in excess of 30 g/l within scenario (4), then large areas of the medium inundated *S. maritimus* marsh will probably turn to bare ground and *Crypsis aculeata*, while lower inundation areas will become dominated by *Arthrocnemum* communities. Therefore the projected increase in inundating salinity and resultant soil salinities will degrade Ichkeul further as a food resource for greylag geese. If the currently planned conservation measures are not fully implemented and lower quartile salinities begin to exceed sea-water salinities (scenario 3) then the present tolerance limits of most of the flora and fauna will be exceeded and a Sebkha will occur.

3.4.3 *Invertebrates*
If the only change being considered were an increase in water temperature at Ichkeul then the effects would be to increase production. Using the standard mean values for Q10 (Bullock, 1955), by which all metabolic rates double for every 10°C rise in temperature, an increase of 1.5°C could be expected to raise the metabolic rate by about 10% at most. The basic assumption here is that the relationship between metabolic rate and growth will remain unchanged.

Such a 10% increased growth rate would give:
1) higher turbidity earlier in the season;
2) higher productivity of algae;
3) higher production of *Potamogeton* and *Ruppia*;
4) increased yield of planktonic and benthic crustacea and fish.

An increase in the level and volume of the lake would, if there were no other factors involved, increase the production and potential yield of mullet, which use Ichkeul for growth of immature fish. The higher metabolic rates and higher primary production discussed above would, in the short term, add substantially to this enhancement of mullet populations. For eels the same would apply with the increase as much due to the increased volume of the lake as rises in primary production (Fig. 15.17).

When Lake Ichkeul changes to a median salinity of around 50 g/l and a lower quartile salinity of 38 g/l (Table 15.16) after the accumulated impacts of the dams and the climatic change then it will have Sebkha communities with salt-tolerant algae and crustacea and salt-tolerant insect larvae only. Figure 15.18 shows that all of the molluscs in Ichkeul are found in more saline lagoons. However, Figure 15.18 does suggest that *Sphaeroma* may become less common and *Balanus, Corophium*, and *Tapes* may become more common as the salinities rise. The mollusc populations will, therefore, remain valuable to wading birds.

As the salinity rises, the recruitment of marine fauna from Lake de Bizerte will be quite rapid. The turtles in Ichkeul would be lost almost immediately as the salinity began to rise and fish stocks would both diversify and decline in yield. Mullet which uses the brackish water plant species and crustacea for the growth of the immature fish would be especially hard hit. There should be an increase in *Cerastoderma edule* and an

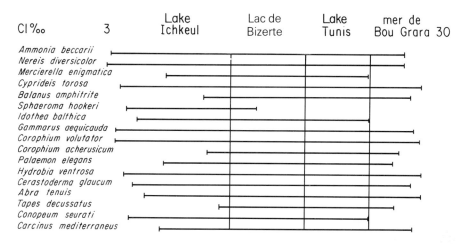

Fig. 15.18 Distribution of molluscs in Tunisian lagoons (after Zaouali, 1975).

invasion of shore crabs *Carcinus mediterraneus* whilst the benthic crustacea should decline only slowly.

Flooding of the marshes by fresh water will be reduced in frequency and duration and so the flush of fresh-water species that is presently found on the marshes each spring will disappear. It will be replaced by sand-eels for a time which will then disappear leaving a completely depleted fauna typical of Sebkha conditions.

3.4.4 Birds

For the birds of the area, the projected changes in salinity seem likely to be much more important than changes in water levels. When the inflow of fresh water decreases, and the salinity rises, the lake will tend towards a salt lake (or Sebkha), so common elsewhere in Tunisia. This will have effects on bird populations at different times of the year:

In winter, the international importance of Ichkeul has been expressed mainly in terms of birds that winter there, in particular greylag geese and several duck species. When the salinity of the water and surrounding marshes increases, food for these birds will be less readily available, and the birds (especially the geese) will adapt to feeding on agricultural crops, especially winter wheat as they have done in northern Europe. This could cause major conflicts with agriculture.

Spring migrants from sub-Saharan Africa, notably garganey, blacked tailed godwit and ruff, but also passerines such as swallows, stop on their way north in spring. A decrease in fresh water will degrade the vegetation on which these migrants depend.

Summertime breeding populations of vulnerable species such as whiteheaded duck, marbled teal and purple gallinule would be affected if vegetation cover from fresh-water aquatic vegetation disappeared.

Returning migrants in autumn, especially waders, use the muddy edges of the lake to feed. The simulations of future conditions at Ichkeul (Table 15.17) show that there are likely to be increased areas of mud flats exposed during the late summer and early autumn. These mud flats are likely to be periodically inundated by lake water as seiches, driven by the wind, pile water up at one end of the lake and then allow an oscillation of level when the wind effect is removed. It has already been argued that some species of invertebrates will benefit from the rise in salinity, it is likely that some wading birds and other species feeding on salt-loving or salt-tolerant organisms will also benefit from these changes in lake level and salinity. Flamingoes, shelducks, stilts, avocets and slender-billed gulls are all very common on existing wet Sebkhas. With the high salinities suggested in Table 15.14, the wading birds who will find Ichkeul less attractive will be sandpipers and godwits.

In concluding this section on birds, three things are important. Firstly, observation of the bird species, their number and feeding ecology will be a particularly effective method of monitoring the extent, direction and rapidity of change at Ichkeul. Secondly, most bird species are very adaptable and so it is likely that large numbers of waterfowl, for instance, will continue to visit Ichkeul each winter. However, given the projected changes they are likely to utilize the large open water area but it is unlikely to provide them with an adequate food supply. Consequently, there is likely to be a large increase in agricultural damage by the water birds. Since this will coincide with a stressful period in agriculture as the temperature effects begin to bite, it is possible that such an increase in bird damage could be a further serious threat to the integrity of agriculture on the Plain of Mateur. Hunting could reduce the predation to some extent and at some cost. However, the birds may then adopt night-feeding as they have done in parts of Europe where hunting is widespread.

Thirdly, it is not known to what extent the general rise in temperature of 1.5°C will affect the whole pattern of bird migration. Whilst trans-Saharan migrations may continue, it is possible that many of the birds which presently winter at Ichkeul, will be able to stay in Europe where the winter weather will presumably be much less frosty and where winter will be much shorter. This is simply an area of speculation at present but it does have a significant bearing on the evaluation of the importance of the likely loss of food resources for winter migrants at Ichkeul.

3.5 Precipitation Patterns
The reliable prediction of the changes in the precipitation patterns on the regional base of the Mediterranean is not yet possible with existing General Circulation Models (Wigley, 1988).

The elements of the precipitation regime that would be relevant to an update of this case study would be:
1) annual rainfall amount;
2) number of rain days;
3) seasonality of rainfall;
4) intensity-frequency-duration relationships;
5) inter-annual variability and reliability.

3.6 Terrestrial Ecosystems

3.6.1 Natural Vegetation

In the low-lying coastal areas, especially around Rmel, soil salinization and renewed dune activity could adversely affect the pine plantations. In the case of the pristine *Quercus coccifera/Juniperus* communities that occur along the coast, then renewed dune activity could destroy large areas of coastal forest and woodland (Fig. 15.19).

In the river catchments above the influence of saline sea water and away from the effects of renewed dune activity, the natural vegetation consists of an *Olea/Ceratonia/Pistacia* community dominating the lower slopes with a gradation into *Quercus suber* on the higher land. The highest mountains in the region have *Quercus faginea*. Assuming no change in the rainfall regime, predictions can be made on the effect that a 1.5°C rise in mean temperature would have on the terrestrial vegetation by recalculating the pluviothermic coefficient (Emberger et al., 1963). This moves from 75 to 70 and together with an increase in the mean winter temperature by 1.5°C, this would move the Ichkeul bioclimatic region right onto the shoulders of the sub-humid/semi-arid bioclimatic zone. However, since the effects of the temperature rise on rainfall are unknown, and since

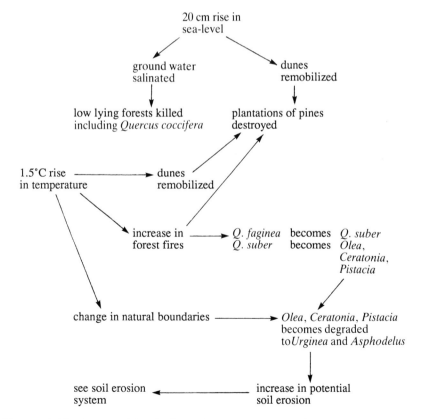

Fig. 15.19 The impact of the rise in sea-level and increase in temperature on forests.

the index and the vegetation is more sensitive to changes in rainfall than in temperature, the resultant bioclimatic zone that will be established at Ichkeul can only be estimated. The results of the soil-moisture model in Table 15.10 are, however, important here. They show that in the lowland areas with relatively low rainfall and high evapotranspiration, and even with the increase in mean annual temperature, there will continue to be zero soil-moisture deficit in average years. Therefore, it is likely that the extent of the change in the natural vegetation around Ichkeul will be quite modest because there will continue to be sufficient soil moisture available during the late winter and spring when the plants are growing. Since the vegetation is already adapted to a strongly seasonal regime with severe desiccation in the summer, the intensification of this effect is unlikely to lead to a change to any other vegetation type.

Small increases in xericity would cause the present vegetation of *Olea/Ceratonia* and *Pistacia* to expand in the middle reaches of the catchment replacing *Q. suber*. In the mountain regions the lowermost communities of *Q. faginea* may be converted to *Q. suber*, a move enhanced by the increased likelihood of fires. The northern Djebel Ichkeul vegetation may be further degraded with increased amounts of *Asphodelus* and *Urginea* occurring. However, the lake via its vapour blanket presumably offers a buffering of these xeric effects.

3.6.2 Soils
The rise in sea-level will bring the salinization of soils near to the coast almost instantaneously. Since so much of the coastal area is low-lying or sandy or both, then these salinization effects could be quite extensive. The rise in temperature, Fig. 15.20, is likely to have a series of interlinked effects which will, overall, tend to foster more soil erosion and greater concentrations of salts in the soils. It is clear, therefore, that the rises in sea-level and in temperature will add significantly to the existing environmental problems in Tunisia as catalogued by the United States National Parks Service (1980).

3.6.3 Agriculture
Farming systems will respond within the 37 years to the projected changes but greater agricultural change might be stimulated by changing directions of Tunisian export markets.

The agricultural impacts of the rise in temperature are likely to be felt mainly through the increase in soil salts and the warmer climate. Efforts to increase irrigation rates will tend to contribute to the soluble salts and the development of crusts or hardpans (Fig. 15.20).

With the rise in temperature and the associated fall in runoff and the reduction in soil-moisture status, the Bizerte–Ichkeul–Mateur area is going to become rather similar to the coastal Tunisian Sahel of today. The characteristics of the agricultural systems in that area are:
- olives dominant;
- semi-arid cereals with low yields compared to northern Tunisia;
- vines and tree fruits;
- extensive sheep and goat grazing;
- selection of salinity-tolerant crops (Table 15.22).

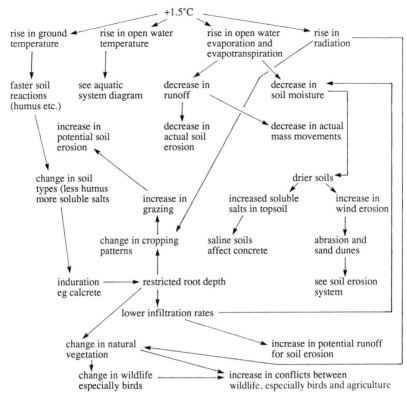

Fig. 15.20 The impact of a 1.5°C rise in temperature on the soil and related systems. (It is assumed that there is no change in precipitation and that only changes manifest within 37 years are to be depicted.)

A great increase in olive cultivation could be anticipated in the plain of Mateur except that the expected EEC olive products import quotas will limit this change. The existing agricultural development scheme has already brought an increase in the area of irrigated fruit trees and this would appear likely to continue. Similarly, there may be a modest increase in vineyard area. Vegetable production can probably survive the projected changes in soil moisture. The projected development of sugar beet in the Plain of Mateur to service the new sugar beet factory is likely to be able to go ahead because that crop is especially resistant to the effects of salinity in the soil (Table 15.22). The relative resistance of sugar beet to salinity effects and its adverse effects on cereals should lead to a decrease in cereal cultivation and a more substantial decline in yield.

The most likely agricultural change, given the elevated temperatures and increased aridity, is a huge increase in grazing which could greatly increase the likelihood of soil erosion (Figs 15.15 and 15.20). This change will be accentuated by the decreasing top-soil quality (more soluble salts, induration of $CaCO_3$, and lower humus contents) and less favourable soil-moisture conditions (lower water holding capacity, greater actual evapotranspiration, a shorter period of moist soil conditions – Table 15.10).

Table 15.22 US Department of Agriculture ratings of relative plant tolerance to salt

	Plant group	High salt tolerance	Medium salt tolerance	Low salt tolerance
Highest tolerance in group – – Lowest tolerance	Fruit crops	Date palm	Pomegranate Fig Olive Grape	Apple Orange Grapefruit Plum Almond Apricot Peach
Highest tolerance in group – Lowest tolerance	Vegetables	$EC_e = 12$ Kale Asparagus	$EC_e = 10$ Tomato Pepper Sweet corn Onion Squash $EC_e = 4$	$EC_e = 4$ Radish Celery Green beans
Highest tolerance in group – – Lowest tolerance	Forage crops	$EC_e = 18$ Salt grass Fescue grass Canada wild rye Barley (hay) Bird's-foot trefoil $EC_e = 12$	$EC_e = 12$ Peren. rye grass Alfalfa Tall fescue Rye (hay) Oats (hay) Reed canary $EC_e = 4$	$EC_e = 4$ White clover Red clover Burnet $EC_e = 2$
Highest tolerance in group – Lowest tolerance	Field crops	$EC_e = 16$ Barley (grain) Sugar beet Rape $EC_e = 10$	$EC_e = 10$ Rye (grain) Wheat (grain) Rice Sunflower Castor beans $EC_e\ 4$	$EC_e = 4$ Field beans Flax $EC_e = 3$

Note: EC_e values (mS/cm) correspond to 50% decrease in yield.

3.7 Impacts on Coastal Activities

Figure 15.4 shows the location of the major urban, industrial and commercial complexes in the area. Several major industrial installations are located in low-lying areas, often reclaimed marshland near the coast or Lac de Bizerte.

By developed world standards the intensity of industrialization and urbanization is modest but "the industrial North" of Bizerte and environs is one of Tunisia's major industrial centres. Most of the commercial centre of Bizerte is only about 1 m above sea-level, the oil refinery to the south of Bizerte is built on the back of a sand dune at about sea-level and the chain of coastal hotels between Bizerte and Cap Blanc are also within a couple of metres of sea-level.

The renewed dune movement at Rmel may impact the oil refinery and will devastate the extensive forestry plantation.

Direct flooding by the sea is likely to be of minor importance because relatively little of the urban areas are actually within 20 cm of sea-level and in any case flood defence walls could be strengthened and raised.

The areas of agricultural land that will be flooded are presently rather poor because of their existing close association with the saline sea water. Most of the shore of Lac de Bizerte consists of a small cliff of between 20 and 35 cm which has been eroded since the introduction of tides in the lake. It is very unlikely therefore that a significant amount of the agricultural land around this lake will suffer any direct inundation by the sea.

The rise in level of Lake Ichkeul will certainly reduce the area of marshland but since the limit of the National Park was set at 95 cm NGT, a rise in water level of 20 cm will not seriously affect the liability to inundation of the agricultural zone. Indeed large parts of the improved agriculture in the Plain of Mateur are already protected from existing flood risks from the lake by an embankment, a drainage ditch and five pumping stations that will lift drainage water over the embankment into the National Park. The effectiveness of the main drainage system that has been upgraded with the agricultural improvement of the Plain of Mateur will obviously be reduced as the lake-level rises but the hydraulic gradients in the main drains will be affected far more by the winter operational regime of the sluice being constructed in the Oued Tindja than by the residual water level during the summer months.

One impact of the sea-level rise will be a tendency for the backing up of water and sewage in the sewers in built-up areas. This will result in increases in the frequency and severity of sewage flooding. Menzel Bourguiba already suffers from this problem and the rise in sea-level will exacerbate the problem.

A second, and potentially destructive consequence of the rise in sea-level, will be the saturation and salinization of the foundations of buildings (Fig. 15.21). This change will be caused by coastal ground-water levels rising by a least 20 cm in parallel with the rise in sea-level and then the heightened potential evapotranspiration increasing the extent and severity of capillary rise of water towards the surface. Many buildings in Bizerte, the oil refinery and the coastal hotels could be affected. There are now well-documented studies of the effects of high salt content/aggressive ground-water conditions in built environments, e.g. Benghazi Plain and Suez City (Cooke et. al., 1982). The effects are:
– internal attack on concrete because of poor aggregates, sediment or poor quality mixing water; and
– external attack on concrete by aggressive ground water.

The main chemical processes are solution; sulphate and chloride reactions; and salt weathering. In addition there may be ground instability in sebkhas due to clay heaving in response to tidal fluctuations. The old hydrographic chart clearly marks a sebkha on the southern side of the Bizerte canal in an area which is now covered with industry and housing (Fig. 15.3). These type of effects have already been seen in the coastal zone around Sousse and southwards. There are practices adopted by

656 *Climatic Change in the Mediterranean*

Fig. 15.21 The impact of a 20 cm rise in sea-level on human activities. (It is assumed that these effects will only affect land up to 2 m above sea-level.)

government contractors to minimize effects, most notably the use of special low-permeability cement imported from Spain and the coating of foundations in bitumen. The widespread use of either technique would increase costs. If the special cement is imported, foreign exchange will be needed and it is also important that salt-free construction aggregates should be found and good quality water used for mixing.

The salinization of clay soils can lead to swelling with associated structural instability. Where this is a current or potential problem the clays must either be pre-loaded for 6 to 12 months or the swelling clays must be excavated and a non-swelling fill used instead. In both cases costs are increased considerably. This problem is probably already present in some of the low-lying areas of Bizerte and Menzel Bourguiba and the rise in sea-level is certain to make it more apparent.

3.7.1 Planning

A Tunisian Ministry of the Environment has recently been created. This should give greater weight to environmental considerations in the planning process. The Ministry or Division will need a long-term planning horizon beyond those presently employed. However, by 2025, the current tendency towards regional devolution may have developed further. This could lead to a greater local input in environmental planning.

3.8 Distribution and Socio-Economic Dynamics of Population in the Littoral Zone

This is the most difficult aspect because of the large number of linkages between the proposed environmental changes and changes in population and settlements. In addition to higher temperatures, population and settlements will be subject to a wide range of environmental, social, economic, epidemiological and political pressures (see sections 2.0, 2.1 and 2.2).

The changes in temperature and sea-level will initially have a direct impact upon the natural environment in Tunisia and only thence an impact upon the economy and the people. The environmental changes forecast to occur by 2025 are simply added to a wide range of environmental problems suffered by Tunisia. The extent of the socio-economic impact of the changes in temperature and sea-level will depend to a large degree on the extent to which Tunisia tackles its existing problems. The problems, as described by the United States National Parks Service (1980), can be characterized as a non-sustainable "mining" of the environment. Tunisia has not yet fully responded to the calls of the World Conservation Strategy (WWF/IUCN/UNEP, 1980) and embarked on the development of a fully sustainable form of development. Therefore the environmental problems enumerated in sections 3.1 to 3.7 will tend, in general, to exacerbate Tunisia's difficulties.

The environmental changes will occur slowly. Without crises that can be tied to a single cause, there are unlikely to be any major policy changes at a national or regional level. Neither is there likely to be any major investment to counteract the adverse effects of the changes because of the almost imperceptible speed of change and the limited finances available. Most of the changes in socio-economic conditions are therefore likely to result from decisions by individuals rather than by groups or governments.

The population of Tunisia is growing at a rate of 2.3% per annum and the urban fraction, 52% in 1984 (Drysdale and Blake, 1985), is enlarging annually at 4.7% (United States National Parks Service, 1980). Almost all of the scenarios in the Mediterranean Blue Plan envisage a doubling of the population by 2025. It seems unlikely that the rise in temperature and sea-level will contribute directly to slowing these rates of rise although the generally adverse environmental effects of the changes may have an effect through the impoverishment of sections of the population who had a per capita GNP of only $1390 in 1982 (Drysdale and Blake, 1985). Since the most immediate impacts are likely to be felt by farmers and inland fishermen, it is likely that they will join the already rapid flow of people into the coastal cities whose growth will therefore be accelerated. There may also be an intensification of the flow of migrants to work temporarily or to live permanently outside of the country.

No major settlements in this region of northern Tunisia are threatened with inundation nor are any transport arteries likely to be impacted by the changes. The marginal rise in sea-level will deepen the shipping channels in and around Bizerte but the saving in dredging costs is probably insignificant.

The heavy industry of the region will not suffer from inundation, because it can probably defend itself against the rise in sea-level. The insidious problem of salt attack upon foundations is likely to be a more serious threat but most of the factories and installations will be due for rebuilding before the end of this study's 37-year perspective. The significant reductions in water resources that will follow the rise in temperature are not likely to have a major impact on industrial users since present priorities rate industry above domestic and particularly above agricultural water users. Consequently, industrial jobs in Bizerte may prove to be even more attractive and so here again there may be a further pull factor drawing population into the major urban centres. The reduction in fresh-water flows will reduce the dilution effects upon industrial effluents from towns such as Mateur but the factories of Bizerte will be little affected by this change since their effluents are discharged into salt water.

It is certain that agriculture will feel the greatest impact of the projected changes. The rise in evapotranspiration, the associated salinization of soils, increased induration and possibly increased erosion, the reduced availability of irrigation water and the climatic pressures to switch towards grazing on marginal agricultural land will all make farming less attractive and less profitable. This will hasten the rural-to-urban migration even further.

Forestry too will probably see its economic value and its stabilizing effect on the environment reduced. The large pine plantation south of Bizerte will be seriously threatened by elevated evapotranspiration, remobilized dunes and rising levels of saline ground water. The forests on the hills and mountains around the headwaters of the Ichkeul catchment are likely to survive but there will be a tendency to move along a gradient from *Quercus faginea* to *Q. suber* to *Olea, Pistacia* and *Ceratonia* to *Urginea* and *Asphodelous*. This will further depress the economic returns on forestry and perhaps more significantly reduce the amount of charcoal produced and used in the countryside.

Inland and lagoon fishing should in theory benefit from the elevated temperatures but in the case of Ichkeul the changes associated with the dams and the sluice will may well massively offset this theoretical gain. The large reduction in the inflow of fresh water and associated nutrients to the Lake de Bizerte, following the completion of the dam scheme, will probably depress fishery productivity to some extent. The coastal fisheries should benefit from the rise in sea temperature so long as the loss of fresh water, nutrient and detritus inputs from the land do not deplete the coastal ecosystem.

Tourism is relatively little developed in the Bizerte region. The hotels between Bizerte and Cap Blanc are likely to retain their beaches but these will be narrower as the sea rises across the lower parts of the sands. Eco-tourism to the National Park at Ichkeul has not yet taken off and the projected impacts of the dams, the agricultural development scheme and the rise in temperature may mean that there is very little of special ecological interest remaining at Ichkeul in 2025. The fact that Europe will be 1.5°C warmer too may mean that fewer Europeans will wish to cross the Mediterranean in search of sun, sand and sea.

4 Conclusions

1) Garaet El Ichkeul and the Lac de Bizerte in northern Tunisia are surrounded by "the industrial north" with its oil refinery, port, cement works and steel plant, a major agricultural zone, a series of dams of national significance, military installations, and some coastal hotels. Ichkeul is one of the Mediterranean's most important wetlands.

2) By 2025 when sea-level has risen by 20 cm, Tunisia's population will have doubled; the economic growth rate will be lower than, and maybe only half that of, recent times; a massive increase in the demand for water by a burgeoning tourist trade and industry will have taken place; and a modest proportion of GDP will derive from agriculture with a significant element under irrigation.

3) The region has undergone enormous changes in the post-Pleistocene era. Superimposed upon these longer term changes are the human-induced changes deriving from forest clearance, the opening of Lake de Bizerte to the sea, channelization of the rivers, and the prospective effects of the diversion of a large part of the region's fresh water. The effects of the global rise in temperature and sea-level will be wrought on an environment already under stress and still adjusting to the cumulative impacts of human activity.

4) The 1.5°C rise in temperature will increase actual evapotranspiration by up to 10%. A corresponding 10% fall in riverflow is a very conservative estimate. Potential evapotranspiration will rise by 14% whilst open water evaporation is likely to rise by 12%. Therefore a 12% increase in evapotranspiration will take place after the rise in temperature from those parts of the catchment which are open water, saturated marshland or irrigated fields. Thus the region will have fewer surface-water resources, less recharge of ground water and a greater propensity for the accumulation of salts by evapotranspiration and inflows of sea water.

5) Table 15.23, for the Ichkeul National Park, summarizes the scale of changes in the ecologically important variables of lake salinity and upper quartile lake level. The former determines the type of ecosystem present in the lake and the latter characterizes the extent to which the marshlands around the lake will maintain their character through inundation. A big change in salinity took place when the Bizerte canal was cut between Lac de Bizerte and the sea in 1895. The envisaged climatic changes will have only a very modest impact upon the salinity regime of the lake and will be equivalent to the effects of just the first two dams built around Ichkeul

Table 15.23 The salinity and water level regime of the Ichkeul National Park in the past, the present and for various future conditions. ( ) indicate a lake level not affected by the projected 20 cm rise in sea-level

	Lower quartile salinity (g/l)	Minimum salinity (g/l)	Upper quartile lake level (cm NGT)
Pre-1895	3.4	1.4	(52.9)
1952–82	8.2	2.6	(52.9)
+20 cm sea-level rise	9.1	2.7	64.8
Wetland evapotranspiration up 10%, riverflow down 10%	11.4	3.3	(48.5)
+ 20 cm sea-level rise Wetland evapotranspiration up 12%, riverflow down 10%	12.6	3.5	60.7
Djoumine and Rhezala dams and the agricultural improvement scheme	12.7	3.2	(40.6)
Djoumine, Rhezala and Sedjenane Dams, the optimal sluice operating regime and reservoir rels. except in one-in-five-year droughts	15.3	5.4	(82.2)
Djoumine, Rhezala and Sedjenane dams and the agric. improvement scheme	25.0	5.3	(25.1)
Djoumine, Rhezala and Sedjenane dams with 25% sediment infill, the agric. improvement scheme, + 20 cm sea-level rise, wetland evapo-transpiration up 12%, riverflow down 10%, the optimal sluice operating regime and reservoir rels. except in one-in-five-year droughts	38.7	9.3	60.7
Djoumine, Rhezala and Sedjenane Dams with 25% sediment infill, the agric. improvement scheme, + 20 cm sea-level rise, wetland evapo-transpiration up 12% and riverflow down 10%	38.9	9.3	40.3

between 1983 and 1984. The most effective management of the sluice and the feasible reservoir releases may just maintain the present ecosystem in the face of three dams.

6) The lower part of Table 15.23 shows that, through its lower-quartile salinities in excess of 25.0, and even 40.0 g/l, the cumulative impact of the dam scheme and the projected climatic shifts are so great that, even

with conservation management, Ichkeul will become a salty sebkha with a water level that fluctuates only minimally. Infrequent floods will occasionally raise the water level but even they will not reduce the minimum salinity to less than the 10 g/l thought necessary for the present aquatic plants to grow.

7) The reduction in water resources, as a result of the rise in temperature and the related hydrometeorological changes, is reflected in the mean storage in the Djoumine and Sedjenane dams being cut by 7.8 and 26%. More importantly, the proportion of time when the reservoirs have less than 25×10^6 m^3 of water, increases from tiny percentages to 5.4 and 19.0% for the Djoumine and Sedjenane dams.

8) Existing trends show that the major barrages in the Ichkeul basin will already be about 25% filled with sediment by 2025. This sedimentation of the reservoirs has a minimal effect on the Ichkeul National Park but the mean storage in the reservoirs falls to 63 and 55% of that pertaining at present for the Djoumine and the Sedjenane. The percentage of time when the reservoirs have less than 25×10^6 m^3 of water is increased to 10% and 25%.

9) Erosion rates on slopes might increase slightly, but even if this were to happen, the reduction in runoff might prevent the sediment produced in this way from reaching channels, or, if reaching them, from moving far down the channel net. The result will be that while small basins might experience more sedimentation, the flow of sediment entering large basins would decline.

10) With less sediment delivered to the coast by streams, coastal erosion will be stimulated, as longshore drift systems, starved of sediment, begin to erode soft coasts. Therefore, an intensification of beach erosion is likely by 2025.

11) As the salinity rises, the recruitment of marine fauna to Ichkeul from Lake de Bizerte will be quite rapid. The turtles in Ichkeul will be lost almost immediately as the salinity begins to rise and fish stocks will both diversify and decline in yield. Mullet uses the brackish water plant species and crustacea and so the commercial fishery will be especially hard hit.

12) As salinity rises, the lake will tend towards becoming a sebkha, so common in Tunisia. In winter the wintering waterfowl will not be able to find both their roosting and feeding requirements inside the National Park. They are likely to fly out to feed on agricultural land, so stimulating further conflicts and hunting. Spring migrants are unlikely to find a ready food source and Ichkeul's rare breeding birds will not survive the disappearance of their favoured vegetation cover. Autumn migrant waders may find adequate food on the more extensive mud banks and salt-loving species, e.g. flamingoes, shelducks and slender billed gulls should increase their use of Ichkeul.

13) It is likely that the change in the natural vegetation around Ichkeul will be modest because there will continue to be sufficient soil moisture available during the late winter and spring when the plants are growing. Since the vegetation is already adapted to a strongly seasonal regime with severe desiccation in the summer, the intensification of this effect is unlikely to lead to a change to any other vegetation type.

Extent of change :	*** very significant; ** significant; * minor - negligible; ( ) change results from other factors over-riding the effects of sea-level and temperature changes	
AMELIORATION		DETERIORATION
	Existing environmental problems	
	loss of agricultural land	**
	degradation of range and forest land and erosion	**
	overcrowding of the urban centres	***
	industrial pollution	*
	the potential of increasing ground-water salinity	***
	lack of adequate protection for native flora and fauna	(**)*
	Economic activity	
	rain-fed agriculture	***
	irrigated agriculture	**
	inland and lagoon fisheries	(***)
*	sea fisheries	
-	heavy industry	-
-	light industry	-
	international tourism	*
	Water Resources	
	reservoir storage (quantity)	***
	reservoir storage (quality)	**
	coastal ground-water resources	***
	upland ground-water resources	**
	river flows	**
	soil moisture (early growing season)	*
	soil moisture (growing season)	-
-	soil moisture (summer)	-
	irrigation demand	**
	Transport	
-	roads	-
-	railways	-
*	shipping	
	Settlements	
	sea flooding	-
	sewage flooding	**
	foundations of coastal buildings	***
	population density	*(**)
	domestic water supply	*
	quality of urban life	*
	Ecosystems	
	coastal dunes	**
-	beaches	-
-	hard rock coasts	-
	erosion in small headward river basins	*
-	sediment yield from large river basins	-
	landforms	-
	coastal lagoons (temperature rise)	***(***)
	coastal lagoons (sea-level rise)	-(***)
	terrestrial ecosystems	*
*	marine ecosystems	
	soils	**

Fig. 15.22 Summary of likely impacts of a 20 cm sea-level rise and a 1.5°C temperature rise on Garaet El Ichkeul and Lake de Bizerte, Tunisia from 1988 to 2025.

14) The main effects of the elevation of temperature on agricultural soils will be more soil erosion and a greater concentrations of salts in the soils.

15) The most likely agricultural change, given the elevated temperatures and increased aridity, is a large rise in grazing which could greatly increase the likelihood of soil erosion.

16) No major settlements are threatened with inundation nor are any transport arteries likely to be impacted by the changes. The marginal rise in sea-level will deepen the shipping channels in and around Bizerte but the saving in dredging costs is probably insignificant.

17) Urban areas may suffer from problems of sewage flooding as the free discharge of sewers into the sea is inhibited. A second potentially destructive consequence of the rise in sea-level will be the saturation and salinization of the foundations of buildings with a consequent rotting of concrete structures.

18) Since the most immediate impacts are likely to be felt by farmers and inland fishermen, it is likely that they will join the already rapid flow of people into the coastal cities whose growth will therefore be accelerated. It seems certain that this trend coupled with the doubling of national population by 2025 will create serious problems in Bizerte, Menzel Bourguiba and other towns. There will also be pressure for further international migration.

19) The heavy industry of the region will not suffer from inundation, because it can defend itself against the rise in sea-level.

20) Inland and lagoon fishing should in theory benefit from the elevated temperatures. But in the case of Ichkeul the changes associated with the dams and the sluice will offset this theoretical gain. In the case of Lake de Bizerte, the large reduction in the inflow of fresh water and its associated nutrients that will follow the completion of the dam scheme will probably depress fishery productivity to some extent. Coastal fisheries should benefit from the rise in sea temperature if the loss of fresh water, nutrient and detritus inputs from the land do not deplete the coastal ecosystem.

21) The environmental changes will occur slowly and particular crises will not be tied simply to a single cause. There are therefore unlikely to be any major policy changes at a national or regional level. Neither is there likely to be any major investment to counteract the adverse effects of the changes because of the almost imperceptible speed of change and the limited finances available. Most of the changes in socio-economic conditions are therefore likely to result from decisions by individuals rather than by groups or governments.

In general the findings have been a suggested direction of change and an indication of the likely extent of that change. The conclusions can therefore be expressed in a tabulation which indicates improvements and declines (Fig. 15.22).

5 REFERENCES

Barić, A. and Gašparović, E., 1988. Implications of climatic changes on the socio-economic activities in the Mediterranean coastal zone. *UNEP Joint Meeting of the Task Team on Implications of Climatic Changes in the Mediterranean and the Co-ordinators of Task Teams for the Caribbean, South-East Pacific, East Asian Seas and South Asian Seas Regions,* Split, October 1988, p. 88.

Bloom, A.C., (compiler), 1977. *Atlas of Sea Level Curves,* International Geological Correlation Programme Project 61.

Bonniard, F., 1934. Les Lacs de Bizerte: Etude de géographie physique. *Revue Tunisienne,* **17**, 93–143.

Bouchet, R.J., 1963. Evapotranspiration réele et potentielle, signification climatique. *Int. Assoc. Sci. Hydr. Proc.* Berkley, Calif., **62**, 134–142.

Bullock, T.H., 1955. Compensation for temperature in the metabolism and activity of poikilotherms. *Biological Review,* **30**, 311–342.

Cooke, R.U., Brunsden, D, Doornkamp, J.C., and Jones, D.K.C., 1982. *Urban Geomorphology in Drylands.* Oxford University Press, p. 324.

Drysdale, A. and Blake, G.H., 1985. *The Middle East and North Africa: A political geography,* Oxford, p. 367.

Emberger, L., Gaussen, H., Ksassas, M. and De Phillipis, A., 1963. Carte bioclimatique de la région Méditeranéene. FAO/UNESCO, Rome/Paris, 2 maps, 1:5,000,000).

Ghorbel, A., 1980. Les transports solides en Tunisie. Division des Ressources en Eau, Tunis, Mimeo, p. 7.

Gimazane, J-P. 1981 La reproduction de la moule *Mytilus galloprovincialis Lamark,* (Mollusque Lamellibranche) dans le lac de Bizerte, Tunisie septentrionale. *Bull. Off. Natn. Pêche Tunisie,* **5**(2), 115–128.

Griffiths, J., 1980. *Climates of the World: Volume 10 Africa.* Elsevier.

Heusch, B., and Millies-Lacroix, A., 1971. Une methode pour estimer l'ecoulement et l'erosion dans une bassin. Application au Maghreb, *Mines et Geologie* (Rabat), 33.

Hollis, G.E. (ed.), 1986. The modelling and management of the internationally important wetland at Garaet El Ichkeul, Tunisia. *International Waterfowl Research Bureau Special Publication No. 4,* p. 121.

Kallel, M.R., 1984. La salinité des eaux de surface en Tunisie. *Direction des Ressources en Eau, Tunis,* Mimeo, p. 55.

Kallel, M.R., 1986. Bilan Global des Eaux de Surface des Regions de L'Extreme Nord Tunisien (Tabarka-Nefza et lac Ichkeul). *Direction des Ressources en Eau, Tunis,* Mimeo, p. 35.

Kassab, A, and Sethom, H., 1980. Géographie de la Tunisie: Le pays et les hommes. *Pub. Univ. Tunis. Deuxième Série,* Volume XII, p. 278.

Mawdsley, J.A. and Ali, M.F., 1985. Modelling non-potential and potential evapotranspiration by means of the equilibrium concept. *Water Res. Research,* **21**, 383–391.

Morton, F.I., 1983a. Operational estimates of evapotranspiration and their significance to the science and practice of hydrology. *J. of Hydrol.,* **66**, 1–76.

Morton, F.I., 1983b. Comment on 'Comparison of techniques for estimating annual lake evaporation using climatological data'. *Water Res. Research,* **19**, 1347–1354.

Morton, F.I., 1986. Comment on 'Modelling non-potential and potential evapotranspiration by means of the equilibrium concept'. *Water Res. Res.,* **22**, 2115–2118.

Paskoff, R. 1985 Tunisia. In: Bird, E.C.F. and Schwartz, M.L. (eds). *The World's Coastline,* Van Nostrand Reinhold, New York, 523–527.

Pye, K., 1984. Models of transgressive coastal dune building episodes and the relationship to Quaternary sea-level changes: a discussion with reference to

evidence from eastern Australia. In: Clark, M., (ed.), *Coastal Research: UK Perspectives*, Geobooks, Norwich, 81–104.
Pye, K., and Bowman, G.M., 1984. The Holocene transgression as a forcing function in episodic dune activity on the eastern Australian coast, In: Thom, B.G. (ed.). *Coastal Geomorphology in Australia*, Academic Press, Sydney, 179–196.
Ramsar Bureau, 1987. Review of National Reports submitted by Contracting Parties. *Document C 3.6, Third Meeting of the Conference of the Contracting Parties to the Convention on Wetlands of International Importance especially as Waterfowl Habitat*, Regina, Saskatchewan, May 1987.
Schumm, S.A., 1965. Quaternary palaeohydrology. In: Wright, H.E. Jr., and Frey, D.G. (eds), *The Quaternary of the United States*, Princeton University Press, Princeton, 738–794.
Service de la Météorologie Nationale, 1967. Climatologie de la Tunisie, Secretariat d'état aux travaux publics et à l'habitat, Mimeo, p. 15.
Stevenson, A.C., Phethean, S.J. and Robinson, J.E., 1986. Palaeoecological investigations of a 6m Holocene sediment core from Garaet El Ichkeul, N.W. Tunisia. *Final Report on Contract ENV–838-UK(AD) to the Commission of the European Communities*, p. 17.
Tayaa, M.H., and Brooks, K.N., 1985. Erosion and sedimentation in the Rif Mountains of northern Morocco, In: O'Loughlin, C.l., and Pearce, A.J. (eds.), *Symposium on the Effects of Forest Land Use on Erosion and Slope Stability*, 1984, Honolulu, East–West Center, Environment and Policy Institute, Honolulu, 23–29.
Teeter, J.W., 1965. Effects of NaCl on the sago pondweed. *J. Wildlife Manage.*, **29**, 838–845.
Thorsell, J., 1987. Endangered spaces. *IUCN Bulletin*, **18**(7–9), 7–8.
Unesco, 1986. Coastal lagoons along the Southern Mediterranean coast (Algeria, Egypt, Libya, Morocco, Tunisia) Description and Bibliography. Unesco Reports in Marine Science 34, p. 223.
United States National Parks Service, 1980. Draft Environmental Profile on Tunisia. Mimeograph report on National Park Service Contract CX–0001–0–0003 by Arid Lands Information Centre of the University of Arizona, Tucson, p. 73.
Walling, D.E., and Webb, B.W., 1983. Patterns of sediment yield. In: Gregory, K.J. (ed.). *Background to Palaeohydrology; A Perspective*, John Wiley & Sons, pp. 69–100.
Wigley, T.M.L., 1988. Future Climate of the Mediterranean Basin with particular emphasis on changes in precipitation. *UNEP Joint Meeting of the Task Team on Implications of Climatic Changes in the Mediterranean and the Co-ordinators of Task Teams for the Caribbean, South-East Pacific, East Asian Seas and South Asian Seas Regions*, Split, October 1988, p. 26.
WWF/IUCN/UNEP, 1980. World Conservation Strategy. p. 64.
Zaouali, J., 1975. Influence des facteurs thermiques et halins sur la faune malacologique de quelques lagunes tunisiennes (lac Ichkeul, lac de Bizerte, lac de Tunis, mer de Bou Grara). *Rapp. Comm. int. Mer Médit.*, **23**(3), 99–101.

Index

Abuqir   538ff, 549, 554, 566, 568, 572, 586, 591
Adriatic Sea   73, 159, 237ff, 270, 290, 291, 299, 428ff
Aegean Sea   236, 238, 252, 269, 271, 273, 274, 275, 291
Agriculture/Activitiés Agricoles   4, 10, 14, 90, 100, 133, 134, 137, 140, 148, 149, 166, 205, 206ff, 290, 295, 296, 301, 304, 305, 307, 323, 324, 332, 337, 344ff, 350ff, 428, 429, 439, 442ff, 476, 487, 488, 495, 501, 520ff, 527ff, 532, 535, 536, 548, 551ff, 584, 588, 603, 606, 608, 609, 620, 652, 653, 655, 658, 663
Albania   1ff, 140, 141, 144ff, 149, 152, 160, 163, 164, 167, 176, 182, 191, 195, 204, 209ff, 223, 224, 275, 291, 298
Albedo   21, 115
Alexandria   55, 56, 295, 535ff, 548ff, 553, 556, 572, 590
Algal blooms   458
Algeria   1ff, 72, 136ff, 144ff, 149, 152ff, 156, 160, 163, 164, 167, 168, 171, 176, 182, 193, 194, 204, 208, 210, 213ff, 221, 222ff, 267, 269, 270, 275, 287
Aliakmon River   495ff
Anaerobic (conditions)   323, 325, 336, 476, 526, 527
Aquaculture   2, 10, 161ff, 168, 308, 323, 324, 350, 435, 476, 488, 520ff, 554, 594
Aquifers (Aquifére)   14, 61ff, 165, 166, 314, 316ff, 389ff, 470, 472, 495, 516, 527, 582, 592
Archaeological data   247ff
Archaeological sites   247ff, 643
Aridity   69, 101, 106ff, 111, 125, 165, 166, 175, 202, 213, 527
Aswan Dam   295, 535, 536, 551, 572, 576ff, 582, 584, 588
Axios Delta   495ff
Axios River   291, 495ff

Baltim   543ff, 549
Bardawil   535, 548, 581, 582, 587, 591
Baroclinic response   239
Barotropic response   233, 235, 237, 239
Beach erosion (*see also* erosion, coastal/shoreline)
Beach(es)   93, 158, 282, 285, 287, 291, 305, 383, 398ff, 429, 435, 439, 446, 448, 467ff, 477, 482ff, 489, 518, 542, 644
Bioclimatic (zones)   182, 183ff
Biostasis   175, 218
Birds   289, 292, 295, 296, 201, 304, 305, 306, 319, 336, 405ff, 477, 485, 519, 530, 586, 587, 603, 609, 612, 645, 646, 649ff, 661
Blue Plan   132, 134, 139, 140, 142, 146ff, 155, 162, 442
Bora (wind)   235, 236, 240, 457, 481
Burullus   535ff, 542ff, 548, 553, 554, 565, 576, 581, 582, 586, 591, 592, 593

Caliche   107
Carbon, organic (*see also* Organic matter)   109
Carbon dioxide (CO_2)   17, 26, 30ff, 47, 64, 67, 68, 93, 97, 99, 102, 203, 206, 218, 220, 224, 226, 227, 284, 328, 428, 429, 527, 535
Cereals   195, 198, 202, 203, 209ff, 225, 226, 439, 518
Channels   641
Chlorofluorocarbons (CFC's)   17
Circulation (oceanic)   368ff, 454, 456, 481, 502ff, 526, 590
Climate/climat   1, 10, 13, 15, 65, 71, 151, 175ff, 308, 330, 333, 354ff, 480ff, 446ff, 596
Climate change   10, 13, 14, 47ff, 97ff, 323, 338, 370, 477ff
Cloud cover   25, 26, 625
Coastal defense   282, 301, 300, 324, 334, 414, 433, 434, 439, 482, 532
Coastal dynamics (processes)   322ff, 330, 333, 334, 460ff, 467ff, 482, 568, 572ff, 590ff, 596
Coastal lowlands   59, 248, 282ff, 428, 474ff, 524, 643
Coastal zone   15, 132, 134, 135, 140, 142ff, 146ff, 154, 157, 161, 162, 165, 170, 226, 377ff, 380ff, 384ff, 401ff, 432ff, 474ff, 495, 518, 523ff, 548, 561, 592, 614, 616, 644ff, 651, 654, 655
Coastline   1, 2, 132, 134, 145, 146, 166, 460, 461
Comacchio (Lagoon)   437ff, 445
Communal infrastructure   172
Communications   557
Conservation   476, 619
Crete   193, 264, 267, 268, 270, 275
Crops   100, 105, 175, 195, 206, 209, 215, 487
Currents   14, 309, 453ff, 561ff
Cyclone   28ff, 43, 239, 558
Cyprus   1ff, 140, 141, 144ff, 149, 152ff, 156, 160, 163, 164, 166, 167, 169, 171, 176, 182, 193, 194, 202, 204, 208, 213ff, 222ff, 226, 252, 269, 270

Damietta   295, 535ff, 545ff, 548, 549, 551, 553, 554, 568, 569, 575, 585, 591, 592
Dams/reservoirs   151, 154, 218, 282, 289, 291, 298, 314, 467, 484, 495, 501, 521, 602, 612, 619, 620, 645, 660, 661
Deforestation   402
Degradation (of land)   608
Deltas/Deltaic areas   10, 53, 56, 145, 172, 248, 282, 284, 286, 291, 296, 328ff, 428ff, 495ff
Dense (deep) water formation production   240ff
Desertification/desertization   113ff, 169, 175, 202, 207
Desiccation   625
Dikes   15, 334, 432ff, 469, 526, 531, 535, 545, 595
Drainage basins   91, 301
Drought   72, 100, 193
Dune(s)   145, 282, 284, 292, 296, 298, 305, 329, 332, 334, 398ff, 401, 439, 460, 538, 541, 542, 545, 548, 596, 616, 655

Eau profonde   369
Ebro Delta   162, 289, 296, 304ff
Ebro River   146, 218, 296, 304ff
Ecological implications   60, 103, 319, 477ff, 489
Economy/economic impact   175, 331, 343ff, 440, 486, 489, 531, 593, 594, 396, 662
Ecosystem(s)   4, 14, 93, 107, 108, 151, 159, 183ff, 195, 320, 323ff, 329, 395ff, 401,

Index 669

476ff, 484ff, 495, 516, 527ff, 586ff, 596, 609, 645, 646, 651, 661, 662
Egypt 1ff, 28, 42, 43, 136ff, 144ff, 148ff, 152, 156ff, 160, 163, 164, 166, 167, 169, 170, 176, 182, 193, 194, 198, 202, 204, 208, 210, 213, 221, 222, 223, 226, 252, 295, 299
El Burg 542ff, 591
Energy 154ff
Environment (*see also* ecosystems) 134, 140ff, 331, 332, 431, 448, 484ff, 595, 657, 662, 663
Erosion 97ff, 106, 110, 166, 168, 504, 637, 639, 663
Erosion, coastal/shoreline 14, 94, 171, 285, 290, 292, 295, 296ff, 298, 301, 313, 322, 333, 334, 384ff, 409, 428, 431, 467ff, 482, 504, 531, 532, 535, 536, 551, 568ff, 572ff, 588, 590, 595, 643, 661
Erosion, soil 14, 113, 115, 127, 148, 169, 173, 215, 216, 322, 483, 489, 639, 640
Estuaries/estuaire 145, 387ff, 499ff, 519ff
Etesian wind 235, 236, 504
Eutrophication 173, 532
Evaporation 25, 48, 64ff, 126, 132, 147, 165, 169, 173, 576, 535, 561, 584, 592, 595, 603, 622ff
Evapotranspiration 10, 11, 66ff, 78ff, 90, 99, 175, 206, 309, 322, 470, 480, 527, 536, 577, 589, 603, 622ff, 653, 660
Evolution du rivage 384ff
Exchangable sodium percentage (ESP) 103ff

Fauna 333, 371ff, 395ff, 401ff, 617
Fires (forest) 43, 97ff, 109ff, 153, 166, 169, 202, 216, 305
Fish 10, 159ff, 395, 485, 521, 536, 566, 587, 593, 616, 645, 648
Fisheries/fishing 14, 92, 159ff, 285, 289, 295, 296, 304, 306, 308, 324, 235, 421, 428, 440, 441, 442ff, 548, 553ff, 588, 594, 603, 609, 658, 663
Floods/flooding 14, 68, 71, 227, 285, 301, 306, 322, 325, 428, 429, 457, 472, 473, 482ff, 488, 500, 524, 535, 633, 655
Flora (*see also* vegetaion) 395ff, 401ff
Forests/forestry 10, 102, 107ff, 133, 140, 148ff, 151, 152, 166, 175, 195, 198, 202, 207, 208, 209, 213ff, 218, 225ff, 477, 519, 603, 658
France 1ff, 73, 115, 136ff, 144ff, 148ff, 152ff, 156ff, 160, 163, 164, 167, 171, 176, 182, 193, 194, 198, 204, 209ff, 223, 224, 226, 274, 290, 328ff

Garert el Ichkeul 602, 609ff
General Circulation model (GCM) 10, 17ff, 64, 69ff, 93, 99ff, 203, 224, 321, 480, 589, 590
Geology 373ff, 459ff, 566ff
Ghyben-Herzberg formula 61, 62
Global positioning staellite (GPS) 56, 58
Grasslands 627
Greece 1ff, 72, 73, 128, 114, 115, 136ff, 144ff, 149, 152ff, 156ff, 160, 163, 164, 166, 167, 168, 170, 171, 176, 182, 193, 194, 198, 204, 209ff, 223, 226, 237, 238, 252, 264, 270, 271, 274, 276, 277, 291ff, 495ff
Greenhouse effect 10, 17ff, 93, 133, 175
Groins/breakwaters/jetties 248, 290, 298, 469ff, 482
Gross Domestic Product (GDP) 9, 135, 136ff, 223, 275, 548, 606, 608
Gross National Product (GNP) 4, 135, 136, 137, 139, 142, 594
Ground water 15, 61ff, 92, 132, 165, 173, 311, 316, 317ff, 322, 463, 472ff, 483, 516, 517, 527, 535, 577, 582ff, 592, 596, 609, 626, 637
Gulf du Lion 290, 328ff

Harbours   14, 54, 145, 154, 156, 158, 262, 282, 290, 292, 296, 301, 324, 428, 430, 432, 440, 445ff, 486, 595, 616
Health (risks)   60
Historic settlements   7, 173ff
Holocene   312ff, 321, 342ff, 373ff, 459ff, 499, 567
Hydrologic cycle   284
Hydrology   4, 60ff, 120, 368ff, 375ff, 389ff, 483ff, 576ff, 612, 617, 645

Idku, Lake   539, 541, 553, 581, 586, 587
Industry/Industrialisation   4, 14, 137, 140, 154, 290, 307, 332, 351, 421, 428, 430, 440, 445ff, 486, 506ff, 521, 556, 595, 606, 617, 655, 658, 663
Intrusion/infiltration (of sea water) 523, 527, 532
Invertebrates   648, 649
Ionian Sea   238, 241
Irrigation   4, 11, 150, 151, 162, 166, 168, 170, 215, 221, 306, 311, 314, 316, 325, 332, 464, 470ff, 483, 487, 518, 520, 526, 527, 528, 532, 551, 553, 576ff, 584, 603, 620, 633
Isonzo delta   431, 432, 464
Israel   1ff, 42, 53, 90, 92, 117, 136ff, 144ff, 149, 152ff, 156ff, 160, 162, 163, 166, 167, 171, 172, 176, 182, 193, 194, 202, 204, 208, 213ff, 222, 223, 226, 252, 263, 264, 268, 277, 286, 294ff, 298
Italy   1ff, 93, 114, 123, 136ff, 144ff, 148ff, 149ff, 152ff, 156ff, 160, 163, 164, 166, 167, 171, 176, 182, 193, 194, 204, 209ff, 223, 224, 226, 248, 252, 263, 264, 267, 268, 275, 287, 290, 428ff

Lac de Bizerte   296, 602, 621ff
Lagoon(s)   93, 282, 286, 287, 291, 304, 305, 311, 329, 335, 387ff, 424ff, 428ff, 467, 474ff, 482, 484ff, 519, 525, 531, 535ff
Lake(s)   292, 500
Land use   204, 208ff, 213ff, 225, 282, 287ff, 486, 588
Lebanon   1ff, 176, 182, 153, 194, 202, 204, 208, 213ff, 222ff, 292ff
Levantine Intermediate Water   240ff
Levantine Sea   240, 241
Libya   1ff, 60, 90, 92, 136ff, 144ff, 152ff, 156ff, 160, 162, 163, 164, 167, 171, 176, 182, 194, 204, 208, 213ff, 221, 222ff, 287, 295
Lido   436, 440, 451
Ligurian current   333
Ligurian Sea   73, 235, 240, 368ff
Lion, Gulf of/golfe du   328ff
Littoralization   146ff
Lowlands   4

Malta   1ff, 140, 141, 144ff, 156ff, 160, 163, 164, 182
Manzala, Lake   535, 545ff, 548, 554, 568, 581, 585, 586, 593
Marano-Grado (Lagoon)
Mariut, Lake   538ff, 549, 553, 581, 582
Marseille   53, 55, 154, 271
Marsh(es)   282, 285, 286, 291, 292, 300, 329, 335ff, 428, 467, 477, 500, 530, 548, 609, 612, 620, 632, 645, 647, 649
Mediterranean Action Plan (MAP)   135, 143, 234
Medjerda delta   296
Methane (CH_4)   17
Mistral   235, 236, 309, 378
Monaco   1ff, 72, 123, 125, 140, 141, 144ff, 163

Morocco   1ff, 116, 136ff, 144ff, 149, 152ff, 156ff, 160, 163, 164, 167, 168, 171, 176, 182, 193, 194, 204, 208, 213ff, 221, 222, 223, 226, 227, 252, 263, 297

Navigation   92
Nile Delta   1, 4, 47, 53, 56, 148, 172, 286, 287, 295, 298, 299, 535ff
Nile River   1, 43, 148, 166, 170, 295, 535ff
Nitrous oxide (N_2O)   17
Nutrient(s)   102, 112, 309, 312, 324, 481

Occupation(s)   344, 345
Organic matter (soil)   107ff, 115, 117, 118, 120, 121, 162, 394, 476, 528
Oved Djoumine   606, 619, 620, 633, 636, 638, 644
Oved Sedjenane   606, 620, 633, 636, 638, 644, 647
Oved Tindja   631, 632, 633
Oxygen depletion (*see* anaerobic)
Ozone (O_3)   17

Park (national)   320, 606, 619, 633, 641, 655
Plant(s)   107, 110, 112, 151
Plate boundaries   267ff
Po Delta   47, 166, 172, 271, 287, 290, 296, 299, 428ff
Po River   218, 241, 290, 297, 428ff
Pollen   618
Pollution   10, 102, 151, 159, 161, 162, 165, 166, 172, 306, 458, 474ff, 489, 522, 536, 553, 554, 563ff, 582, 589, 592, 594, 595, 596, 608, 616
Population   4, 8, 10, 13ff, 90, 92, 132, 145, 146, 147, 163, 173, 179, 199ff, 202ff, 215, 221, 224, 227, 304, 348ff, 431, 436, 440ff, 479, 495, 496, 497, 522ff, 538, 545, 548, 549, 554, 589, 594, 595, 596, 608, 657, 662, 663
Port Said   55, 295, 299, 535ff, 545, 547, 548, 549, 551, 554, 570, 575, 591
Potential evapotranspiration (PEP)   177, 183, 195, 206ff, 212, 213, 218, 220
Precipitation   1, 13, 15, 17, 22, 25, 27, 28, 35ff, 48, 64ff, 90, 99ff, 147, 151, 161, 169, 177, 183, 195, 206ff, 218, 220, 322, 359ff, 428, 449ff, 480ff, 497, 505, 506, 528, 590, 640, 650
Priority Actions Plan (PAP)   143, 144
Protected areas   170ff

Quaternary   48ff, 312, 459, 503

Rainfall (rainy)   1, 3, 14, 30, 43, 97, 98, 164, 165, 177, 181, 183, 189, 195, 202, 206, 212, 215, 218, 226, 308, 321, 330, 469, 480, 527, 528, 532, 558, 561, 590, 605, 607, 627, 630, 639
Range land   198ff, 202, 207, 226
Ravena   429, 440, 441, 446, 453, 462, 464, 474, 482, 489
Reclaimed areas   282, 301, 430, 432, 439, 482, 489, 501, 538, 545, 551ff, 585, 588
Recreation   92, 287, 301, 305
Reservoir(s)   90, 576, 603, 620, 631, 637, 642, 646
Resources   4, 441
Rhexistasis   175, 218
Rhone Delta   55, 166, 172, 264, 271, 290, 328ff
Rhone River   146, 218, 290, 328ff
River(s)   5, 97, 99, 101, 115, 125, 126, 165, 301, 431, 470ff, 639
River runoff (discharge)   1, 65, 67ff, 72, 82ff, 90ff, 102, 113, 119, 121, 169, 301, 304, 309, 315, 322, 470ff, 484, 501ff, 527, 631, 633, 637, 639, 641
Romagna   439, 440, 444, 445, 453, 464, 467, 469, 472, 474, 484, 489

Rosetta   295, 536ff, 541, 542, 548, 549, 554, 568, 569, 572, 577, 591, 592

Sahel   22
Salinity (saline)   14, 63, 106ff, 161, 162, 165, 335, 336, 391, 453, 455, 481, 561ff, 581, 609, 613, 618, 621, 631, 634, 635, 638, 639, 645, 660
Salinization   132, 166, 168, 170, 322, 484, 516, 527, 532, 536, 548, 553, 589, 592, 595, 655ff
Salt balance/accumulation   97, 101, 103ff, 637, 654
Salt water (sea water) intrusion   13, 60ff, 93ff, 165, 286, 301, 305, 387, 483, 484, 527, 535, 584, 585
Saltworks   155, 308, 324
Sea level, mean   47ff, 62, 155, 243, 247ff, 288, 299, 331
Sea level, rise/variation   4, 12, 13ff, 47ff, 60, 63, 71, 72, 92, 93, 132ff, 145, 147ff, 151, 155, 156, 165, 171, 173, 175, 234, 238, 247ff, 282ff, 299, 300, 305, 322ff, 331, 333ff, 338, 380, 428, 429, 452, 457ff, 479ff, 497, 523ff, 588ff, 603, 622ff, 628, 631, 633, 634, 635, 638, 467, 651, 655, 657
Sebkha(s)   295, 633, 648ff
Sediment   99, 115, 145, 218, 266, 290, 376ff, 516, 566, 631
Sediment budget   13, 125, 291
Sediment input   216, 304, 314ff, 321, 503
Sediment load/yield   120, 121, 126, 291, 294, 295, 298, 306, 314ff, 467, 469, 501, 576, 638ff
Sedimentation   106, 301, 374, 376ff, 392ff, 459, 499, 502, 525, 526, 568, 570, 619, 642, 661
Sewage (*see also* pollution)   92, 172, 173, 506ff
Shelf, continental   558ff
Shrubland   102, 170, 180, 195, 198, 207, 208, 209, 214, 216, 218, 225ff
Sirrocco (winds)   235, 236, 457
Siami (North)   547, 551, 553, 586
Socio-economic (activities, consequences)   132ff, 143, 151, 173, 307, 331, 337, 341ff, 439ff, 536, 538, 548ff, 657
Soil(s)   15, 97ff, 124, 133, 140, 151, 164, 167, 169, 175, 464ff, 516, 528, 570ff, 595, 596, 644
Soil degradation   97ff, 165, 170, 207
Soil moisture   43, 64, 67, 109ff, 114, 148, 169, 595, 532, 626, 637, 653
Soil types   166, 464ff, 652, 653
Solution notches   253, 266
Spain   1ff, 42, 72, 73, 102, 112, 113, 114, 118, 120, 123, 124, 136ff, 144ff, 149, 152ff, 156ff, 160, 163, 164, 167, 168, 171, 176, 182, 193, 194, 198, 204, 209ff, 223, 224, 226, 252, 263, 275, 287ff, 304ff
Storm(s)   14, 27, 71, 72, 93ff, 116, 123, 169, 261, 286, 301ff, 428, 432, 453, 467, 469, 484, 532, 542, 558, 572, 589
Storm surges   286, 325ff, 334, 339
Strait of Gibraltar   237, 247, 263
Strait of Otranto   238
Straits of Sicily   238
Subsidence   13, 47, 55ff, 168, 263ff, 268, 270, 299, 304, 381, 383, 384ff, 462ff, 465, 490, 536, 538, 568ff, 590, 596
Suez Canal   295
Sunshine   624
Surface water   470, 474ff, 576ff, 592
Syria   1ff, 136ff, 144ff, 149ff, 152ff, 156ff, 160, 163, 164, 167, 171, 176, 182, 193, 194, 204, 208, 213ff, 221, 222, 226, 252, 263, 264, 292ff
Swamp (*see* marsh(es))

Tagliamento Delta   431, 432
Tectonic movements   248ff, 299
Temperature, air   1, 10, 14, 17, 97ff, 113, 132ff, 177, 189, 193, 195ff, 211, 308, 321, 330, 335ff, 448ff, 558, 561, 589, 605, 607
Temperature, rise in   1, 10, 14, 17ff, 47ff, 120, 127, 148, 151, 156, 162, 165, 169, 171, 172, 175, 186, 187, 193, 195, 203, 220, 227, 234, 398, 428, 477ff, 497, 526, 535, 588ff, 604, 622ff, 651, 657
Temperature, soil   115
Temperature, water   14, 64, 161, 162, 310, 405, 452, 455, 561ff, 581, 595, 613
Thermaikos Gulf   495ff
Thessaloniki   53, 55, 291, 495ff
Tide gauge   50, 247ff, 271, 282, 299, 300
Tides   253ff, 331, 457, 458, 502, 563ff, 616
Tourism/tourisme   10, 12, 14, 54, 140, 144, 145, 155ff, 169, 226, 284, 287, 296, 304, 308, 324, 332, 337, 346ff, 413, 422, 428, 429, 430, 440, 446, 482, 486, 522, 523, 536, 551, 594, 605, 659
Traffic   134, 158, 422
Transport   140, 154, 556, 662
Tunisia   1ff, 42, 60, 73, 90, 92, 136ff, 144ff, 148ff, 152ff, 156ff, 160, 163, 166, 167, 168, 169, 171, 176, 182, 193, 194, 202, 204, 208, 213ff, 221ff, 226, 264, 270, 297ff, 298, 602ff
Turkey   1ff, 136ff, 144ff, 148ff, 152ff, 156ff, 160, 163, 164, 166, 167, 168, 171, 176, 182, 193, 194, 202, 204, 208, 213ff, 221, 222, 226, 227, 252, 263, 264, 268, 269, 274, 291ff, 298

Uplift   50
Urbanization   133, 135, 145, 147, 169, 536, 549ff, 553, 554, 588, 608

Vegetation   43, 64, 94, 107, 120, 124, 133, 165, 169, 170, 175, 175, 177, 179, 188ff, 206, 320, 401ff, 478, 485, 518, 609, 616, 644, 646, 651ff, 661
Vegetation zones   100, 617
Venice   15, 53, 55, 154, 174, 284, 299, 428ff

Wadi(s)   294
Wastewater   147, 156, 172, 173, 490, 549, 596
Water   10, 13, 15, 113, 133ff, 140, 149, 151, 504ff, 572
Water availability   220
Water balance   65ff, 97, 102ff, 304, 311, 375ff, 391ff, 625, 626, 627, 630, 641, 645
Water resources   14, 60ff, 64, 65ff, 73, 88ff, 147, 155, 156, 162ff, 172, 221, 470ff, 483ff, 516, 526ff, 536, 548, 553, 576ff, 584, 588, 608, 612, 617, 620, 662
Water supply   206, 207, 285, 305, 452, 589, 596, 606, 617
Water quality   92, 484
Waves   253, 292, 298, 321, 323, 453, 457, 502, 561ff, 572ff, 581
Wells   62ff
Wetlands   6, 15, 93, 145, 170, 282, 285, 290, 300, 306, 479, 484, 488, 518, 519, 609, 619, 632, 634, 635, 644
Wind erosion   125, 166
Wind forcing   235, 309, 502
Winds/vents   25, 132, 173, 175, 330, 366ff, 378ff, 449, 452, 481, 504ff, 558, 582, 589

Yugoslavia   1ff, 72, 136ff, 144ff, 149, 152ff, 156ff, 160, 163, 164, 167, 168, 171, 176, 182, 194, 204, 209ff, 223, 224, 226, 275, 291